W0112361

Topics in
Current Physics

3

# Topics in Current Physics Founded by Helmut K. V. Lotsch

Volume 1 **Beam-Foil Spectroscopy**
Editor: S. Bashkin

Volume 2 **Modern Three-Hadron Physics**
Editor: A. W. Thomas

Volume 3 **Dynamics of Solids and Liquids by Neutron Scattering**
Editors: S. W. Lovesey and T. Springer

Volume 4 **Electron Spectroscopy for Surface Analysis**
Editor: H. Ibach

# Dynamics of Solids and Liquids by Neutron Scattering

Edited by S. W. Lovesey and T. Springer

With Contributions by
R. Comès B. Dorner J. M. Loveluck
S. W. Lovesey R. D. Mountain H. G. Smith
T. Springer N. Wakabayashi J. W. White

With 156 Figures

Springer-Verlag Berlin Heidelberg New York 1977

Dr. Stephen W. Lovesey

Inst. Laue-Langevin, B. P. 156, Centre de Tri,
F-38042 Grenoble Cedex, France

Professor Dr. Tasso Springer

Institut für Festkörperforschung der Kernforschungsanlage Jülich GmbH,
Postfach 365, D-5170 Jülich 1, Fed. Rep. of Germany

and

Inst. Laue-Langevin, B. P. 156, Centre de Tri,
F-38042 Grenoble, Cedex, France

ISBN-13:978-3-642-81115-9 e-ISBN-13:978-3-642-81113-5
DOI: 10.1007/978-3-642-81113-5

Library of Congress Cataloging in Publication Data. Main entry under title: Dynamics of solids and liquids by neutron scattering. (Topics in current physics ; 3). 1. Lattice dynamics. 2. Liquids. 3. Neutrons-Scattering. I. Lovesey, Stephen, W. II. Springer, Tasso. III. Dorner, B. IV. Series. QC176.8.L3D9 530.4'1. 77-4740.

This work is subject to copyright. All rights are reserved, whether the whole or part of the material is concerned, specifically those of translation, reprinting, re-use of illustrations, broadcasting, reproduction by photocopying machine or similar means, and storage in data banks. Under § 54 of the German Copyright Law where copies are made for other than private use, a fee is payable to the publisher, the amount of the fee to be determined by agreement with the publisher.

© by Springer-Verlag Berlin Heidelberg 1977.
Softcover reprint of the hardcover 1st edition 1977

The use of registered names, trademarks, etc. in this publication does not imply, even in the absence of a specific statement, that such names are exempt from the relevant protective laws and regulations and therefore free general use.

2153/3130-543210

# Preface

Inelastic neutron scattering is a well established and important technique for studying the dynamical properties of condensed matter at the atomic level. Often, as is the case of experiments designed to study motions of hydrogen atoms, or magnetic excitations, it may yield information obtainable in no other way.

Our aim in assembling this book is to produce an overview of some research topics which have come to the fore recently with the development of high neutron fluxes and high performance inelastic scattering spectrometers. The topics discussed here are, by and large, developing rapidly and have not reached the stage at which definitive accounts are always possible. Authors have not therefore attempted to make an extensive review of their topic, and the papers quoted in the text are, in general, those which are seen as having been important in its development (they date, roughly, from the 1971 IAEA conference on neutron scattering held in Grenoble). Basic phenomena are illustrated for the most part by the discussion of one, or two, typical examples.

The authors hope that the book will be useful to researchers who are not yet fully aware of the diverse range of problems to which the technique can be applied, and to students beginning research work. For this reason, the first chapter by *S.W. Lovesey* gives a brief introduction to that part of the theory essentially used for the interpretation of measurements on the various research problems discussed in later chapters. This chapter serves also to introduce the formalism and notation adopted in the book.

Reading through the chapter titles, we come next to a chapter by *H.G. Smith* and *N. Wakabayashi* which deals with progress in understanding the detailed lattice dynamics of complicated structures, such as rutile compounds, and quasi two-dimensional systems. The second set of authors, *B. Dorner* and *R. Comès*, discuss lattice phase transitions, soft-modes and critical scattering. These authors also emphasise the complementary nature of diffuse X-ray scattering and neutron spectroscopy in the study of lattice phase transition. *T. Springer* then treats hydrogen diffusion in metals, and studies of molecular rotations, in particular studies of tunneling transitions which is a relatively new and rapidly developing field of high resolution neutron spectroscopy. *J.W. White* surveys progress with a number of problems in physical chemistry in which neutron scattering experiments are making a significant contribution. Particular attention is paid to recent advances in the understanding of the lattice dynamics of molecular crystals and polymers. The chapter on classical liquids by *R.D. Mountain* discusses collective motions in monatomic liquids for wavelengths of 10 to 500 Å. These are long wavelength phenomena on the scale of wavelengths probed by conventional neutron spectrometers, but they are well within the reach of some advanced spectrometers; it should be possible for neutron scattering measurements to overlap in wavelength with light scattering studies. The final chapter, by *S.W. Lovesey* and *J.M. Loveluck*, contains a survey of some magnetic dynamical problems which have been investigated by inelastic neutron scattering, including collective excitations in low-dimensional magnetic compounds, and singlet ground state systems.

While neutron diffraction measurements have proved very useful in polymer science and biophysics, there are, as yet, very few examples of the use of inelastic neutron scattering in these fields of research. The reader who is interested in the application of neutron scattering techniques to problems in polymer science and biophysics will find the book edited by *Ivin* (1976), and the review article by *Jacrot* (1976) good starting points for further reading.

In putting together this book we have benefitted from discussions and comments from our colleagues and visitors at the "Institut für Festkörperforschung" in Jülich and the Institute "Max v. Laue - Paul Langevin", and to all these people we are very grateful.

February 1977

*S.W. Lovesey*, Grenoble
*T. Springer*, Jülich and Grenoble

---

Jacrot, B. (1976): Rept. Progr. Phys. 39, 911
Ivin, K.J. (1976): *Structural Studies of Macromolecules by Spectroscopic Methods* (John Wiley and Sons, New York)

# Contents

# List of Contributors

COMES, ROBERT

Laboratoire de Physique des Solides, BT. 510, Université de Paris-Sud, Centre d'Orsay, 91405 Orsay, France

DORNER, BRUNO

Inst. Laue-Langevin, B.P. 156, Centre de Tri, 38042 Grenoble Cedex, France

LOVESEY, STEPHEN W.
LOVELUCK, JAMES M.

Inst. Laue-Langevin, B.P. 156, Centre de Tri, 38042 Grenoble Cedex, France

MOUNTAIN, RAYMOND D.

United States Department of Commerce, National Bureau of Standards, Institute for Basic Standards, Heat Division, Washington, D.C. 20234/USA

SPRINGER, TASSO

Institut für Festkörperforschung der Kernforschungsanlage Jülich GmbH, Postfach 365, 5170 Jülich, Fed. Rep. of Germany, and
Inst. Laue-Langevin, B.P. 156, Centre de Tri, 38042 Grenoble Cedex, France

SMITH, HAROLD G.
WAKABAYASHI, NOVO

Solid State Division, Oak Ridge National Laboratory, Oak Ridge, Tenn. 37830, USA

WHITE, JOHN W.

Inst. Laue-Langevin, B.P. 156, Centre de Tri, 38042 Grenoble Cedex France

# 1. Introduction

S. W. Lovesey

With 16 Figures

The aim of the first chapter is to summarize the formalism used in the interpretation of inelastic neutron scattering experiments. The general form of the partial differential neutron cross-section is given together with various examples of inelastic nuclear and magnetic scattering processes, e.g., phonon scattering and spin wave scattering.

## 1.1 Prologue

The usefulness of neutron scattering as a technique to study the dynamic properties of condensed matter stems from the fact that neutrons interact with both nuclei and magnetic, or unpaired electrons, and that slow neutrons have energies and wavelengths that match those of excitations and collective modes, e.g., rotational energy levels in molecules, and acoustic and optic phonons.

The theory of thermal neutron scattering has been reviewed by several authors, and the following texts on nuclear and magnetic scattering are particularly useful: GLAUBER (1962), DE GENNES (1963), TURCHIN (1965), LOMER and LOW (1965), EGELSTAFF (1967), GUREVICH and TASAROV (1968), IZYUMOV and OZEROV (1970), MARSHALL and LOVESEY (1971), SRINGER (1972) and BACON (1975). A bibliography of papers concerned with neutron scattering published between 1934-1974 has been compiled by LAROSE and VANDERWAL (1974).

The partial differential cross-section is a measure of the response of the target to the perturbation created by the incident neutrons, and therefore reflects the dynamic properties of the target as functions of frequency and wave-vector. It is usual to describe the response of a system to a weak frequency and wave-vector dependent perturbation in terms of a dynamic susceptibility, and there is a simple relation between the dissipative part of the dynamic susceptibility and the partial differential cross-section which goes under the name of the fluctuation-dissipation theorem. We emphasise the relation between the cross-section and dynamic susceptibility, and related quantities, because it affords the proper route for interpretation of data in terms of the dynamic properties of the target system. Moreover, these quantities are the natural starting point for many theoretical studies.

Neutron polarization effects have not been exploited in the past largely because of a problem of intensity which has been ameliorated by the advent of high-flux reactors. In view of this advance, we review the various possible uses of polarization effects in inelastic scattering experiments. A review of modern neutron polarization techniques was given by HAYTER (1975).

The peak in the intensity distribution of thermal neutrons occurs at an energy of ~25 meV which corresponds to a wavelength of ~1.8 Å. However, it is often desirable to study processes that involve energies considerably larger and smaller than this, and consequently some reactors are fitted with facilities which enhance the large or small energy region of the spectrum. For example, the intensity distributions from the hot and cold source facilities of the Institute Laue Langevin peak at 170 meV and 2 meV, respectively. The corresponding wavelength can be obtained from the formula

$$\lambda = 0.286\ E^{-1/2}\ \text{Å}$$

with E in eV.

Because of the wide range of energy units used by researchers, the following *approximate* conversion data are useful

$$\begin{aligned} 25\ \text{meV} &\simeq 6\ \text{THz} \\ &\simeq 300\ \text{K} \\ &\simeq 40 \times 10^{12}\ \text{s}^{-1} \\ &\simeq 200\ \text{cm}^{-1} \end{aligned}$$

More precise results are : 1 meV = 11.609 K = 0.241 THz.

Let us turn now to the form of the partial differential cross-section. The geometry of a scattering experiment is illustrated in Fig.1.1. Calculation of the cross-section is tantamount to calculating the transition probability for a scattering process in which the neutron changes from the state defined by the wave-vector $\underline{k}$ to the state $\underline{k}'$, accompanied by a change in the target system. This probability is, in the Born approximation, proportional to the square of the matrix element of the interaction potential $\hat{V}$, taken between the initial and final plane-wave states of the neutron, multiplied by the density of final scattering states per unit energy range. We define the change in the neutron wave-vector by $\underline{Q}$, and the change in the energy by *$\hbar\omega$, i.e.,

$$\underline{Q} = \underline{k} - \underline{k}' \tag{1.1}$$

*$\hbar = h/2\pi$ (normalized Planck's constant)

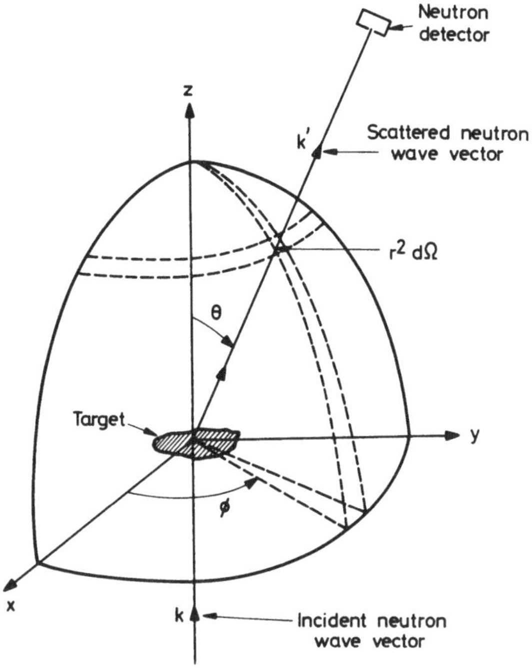


Fig. 1.1 The incident and scattered neutrons are described by plane waves with wave-vectors $\underline{k}$ and $\underline{k}'$, respectively, and the scattering vector $\underline{Q} = \underline{k} - \underline{k}'$

and

$$\hbar\omega = \frac{\hbar^2}{2m}(k^2 - k'^2) = E - E' \tag{1.2}$$

where m is the mass of the neutron. To be quite general we must consider also the spin state of the neutron. We shall denote the initial state by the index $\sigma$, and the final state by $\sigma'$. The initial and final states of the target are described by wave functions $|\lambda\rangle$, and $|\lambda'\rangle$, respectively. If the energy of these states is $E_\lambda$ and $E_{\lambda'}$, then $\hbar\omega = E_{\lambda'} - E_\lambda$ by the conservation of energy. With this notation the partial differential neutron cross-section is, within the Born approximation (we set $\hbar = 1$ in the remainder of this chapter),

$$\frac{d^2\sigma}{d\Omega\, dE'} = \frac{k'}{k}\left(\frac{m}{2\pi}\right)^2 \sum_{\lambda,\sigma} p_\lambda p_\sigma \sum_{\lambda',\sigma'}$$

$$* \left|\langle\sigma',\lambda'|\int d\underline{r}\, \exp(-i\underline{k}'\cdot\underline{r})\, \hat{V}(\underline{r})\, \exp(i\underline{k}\cdot\underline{r})|\sigma,\lambda\rangle\right|^2 \tag{1.3}$$

$$* \delta(\omega + E_\lambda - E_{\lambda'}) \quad .$$

The delta function in (1.3) expresses the conservation of energy. The initial states of the target, and neutron spin states are distributed with probabilities $p_\lambda$ and $p_\sigma$, respectively.

The specific forms of $\hat{V}$ for nuclear and magnetic scattering of neutrons are discussed in Sec. 1.3 and 1.5, respectively. There is, in addition, a weak interaction with the electric field produced by nuclei and atomic electrons but we shall not discuss it here (see, for example, MARSHALL and LOVESEY (1971), Sec.10.3).

Several concepts and results which appear frequently in the interpretation of neutron scattering data are illustrated nicely in the simple example of scattering by a single nucleus bound in a harmonic oscillator potential, and for this reason we discuss this example before discussing more general features of the cross-section. The nucleus is taken to have mass M, and bound in a harmonic oscillator potential of natural frequency $\omega_0$. The interaction between neutrons and nuclei is discussed in Sec.1.3, and for the present example the interaction potential $\hat{V}$ in (1.3) is simply,

$$\hat{V}(\underline{r}) = \frac{2\pi}{m} b\delta(\underline{r} - \underline{u}) \tag{1.4}$$

where b is a scattering length, which is of order $10^{-4}$ Å, and $\underline{u}$ denotes the displacement of the nucleus from its equilibrium position, which is taken as the origin of the coordinates.

Referring to (1.3) for the cross-section, it is evident that we need to calculate the matrix elements of the operator $\exp(i\underline{Q} \cdot \underline{u})$ taken between the eigenstates of the harmonic oscillator. It is well known that the eigenstates can be labelled by a quantum number $n = 0,1,2,\ldots$, and that the eigenvalues are $E_n = \omega_0(n + 1/2)$. In terms of the notation used in (1.3), $|\lambda\rangle \equiv |n\rangle$, $E_\lambda \equiv E_n$, and $p_\lambda \equiv \exp(-\beta E_n)/Z$ where $\Sigma p_\lambda = 1$, and the temperature $T = (k_B\beta)^{-1}$. Because there is no dependence on the neutron spin all dependence of the cross-section on $\sigma$ must disappear, and $\Sigma p_\sigma = 1$ by definition. After a straightforward calculation of the matrix elements the cross-section can be written.

$$\frac{d^2\sigma}{d\Omega\, dE'} = \frac{k'}{k} b^2 S(Q,\omega) \quad . \tag{1.5}$$

We call the function $S(Q,\omega)$ the scattering function and it describes, as we shall see, the dynamic properties of the target system. In the present case $S(Q,\omega)$ is given explicitly by

$$S(Q,\omega) = \exp\left[-2W(Q) + \frac{1}{2}\omega\beta\right] \sum_{n=-\infty}^{\infty} I_n(y)\delta(\omega - n\omega_0) \tag{1.6}$$

where the Debye-Waller factor, $\exp[-W(Q)]$, is determined by the quantity,

$$W(Q) \equiv \frac{1}{2}\langle(\underline{Q} \cdot \underline{u})^2\rangle = Q^2 \frac{1}{4M\omega_0} \coth(\frac{1}{2}\beta\omega_0) \tag{1.7}$$

and

$$y = 2W(Q)/\cosh\left(\tfrac{1}{2}\beta\omega_0\right)$$

and $I_n(y)$ is a modified Bessel function of the first kind. Eq. (1.6) tells us that the nucleus can be excited ($n > 0$) or de-excited ($n < 0$) in the scattering process with a corresponding neutron energy change of $n\omega_0$. The $n^{th}$ order process in the scattering function is weighted by a factor

$$\exp\left(\tfrac{1}{2}n\omega_0\beta\right) I_n(y)$$

and we note that $I_n = I_{-n}$. The mean-square displacement is non-zero at absolute zero, expressing the phenomenon called zero-point motion, and consequently the Debye-Waller factor is non-zero at $T = 0$; in fact

$$2W(Q) \to \frac{Q^2}{2M\omega_0} \quad , \qquad T \to 0$$

and

$$y \to Q^2 \exp\left(-\tfrac{1}{2}\omega_0\beta\right)/M\omega_0 \quad , \qquad T \to 0 \quad .$$

The small-argument expansion of the Bessel function is $I_n(y) \sim (0.5\ y)^n/n\,!$, so that weighting factor for the $n^{th}$ order process is proportional to $(Q^2/M\omega_0)^n$ in the limit of low temperatures. At high temperatures, where $\beta\omega_0 << 1$,

$$2W(Q) \to \frac{Q^2}{M\omega_0^2} \cdot k_B T \quad , \qquad T \to \infty$$

and also

$$y \to \frac{Q^2}{M\omega_0^2} \cdot k_B T \quad , \qquad T \to \infty \quad .$$

We conclude that $W(Q)$ increases linearly with temperature, in the limit of high temperature. The asymptotic form of the Bessel function is $I_n(y) \sim y^{-1/2} \exp(-y)$, so the weighting factor is independent of the quantum number n for sufficiently high temperatures.

Several general features of the cross-section (1.5) and scattering function (1.6) merit explicit mention. First, the details of the interaction mechanism have factored out in the cross-section thereby enabling us to define a scattering function which describes the dynamic properties of the target system. VAN HOVE (1954) was the first author to recognise the importance of the factorization, and to stress the relation between the scattering function and the dynamic susceptibility. We make this latter point more precise by anticipating a result from Sec. 1.2 which relates the scattering function $S(Q,\omega)$ to the dissipative part of the dynamic susceptibility[1] $\tilde{\chi}''(Q,i\omega)$, namely

$$S(Q,\omega) = \frac{\omega}{1 - \exp(-\beta\omega)} \left[\frac{-\tilde{\chi}''(Q,i\omega)}{\pi\omega}\right] \quad . \tag{1.8}$$

This relation is often referrred to as the fluctuation-dissipation theorem, as mentioned earlier (KUBO, 1966, Sec.7).

We see from our example, Eq. (1.6), that the dissipative part of the susceptibility is an odd function of $\omega$, whereas the scattering function satisfies the condition

$$S(Q,\omega) = \exp(\omega\beta)\, S(Q, -\omega) \tag{1.9}$$

which is usually referred to as the condition of detailed balance bacause it relates the process described by $Q$ and $\omega$ to the process described by $Q$ and $-\omega$.

The integral of $S(Q,\omega)$ over $\omega$ for fixed $Q$ is a measure of the scattered intensity. We denote this integral by $S(Q)$, the structure factor, and for the present example we obtain the result

$$S(Q) = \int_{-\infty}^{\infty} d\omega \;\; S(Q,\omega) \equiv -\frac{1}{\pi} \int_{0}^{\infty} d\omega \coth\left(\tfrac{1}{2}\beta\omega\right) \tilde{\chi}''(Q,i\omega) = 1 \quad . \tag{1.10}$$

The structure factor is unity for our example because we are considering only one scattering centre; the same result would be obtained for a system of identical, non-interacting particles. For real systems where there is significant correlation between the particle positions $S(Q)$ displays pronounced structure at wavelengths of the order of the mean separation of particles. The integral of $S(Q,\omega)$ against $\omega$ is a measure of the average energy transfer, and

$$\int_{-\infty}^{\infty} d\omega \cdot \omega\, S(Q,\omega) \equiv -\frac{1}{\pi} \int_{0}^{\infty} d\omega \cdot \omega\tilde{\chi}''(Q,i\omega) = \frac{Q^2}{2M} \quad . \tag{1.11}$$

This result is quite general in that it is found to hold for any system in which the particles interact via velocity-independent forces. Note that the average energy transfer is independent of the interaction potential between particles in the target, and the temperature.

We conclude the discussion by examining the low and high temperature forms of the scattering function. As $T \to 0$, it follows from (1.8) that

---

[1]The dynamic susceptibility was calculated for several simple systems of interest by MARTIN (1968).

$$S(Q,\omega) = -\frac{1}{\pi}\,\tilde{\chi}''(Q,i\omega)\ , \quad \omega \geq 0$$
$$= 0,\ \omega < 0 \quad . \tag{1.12}$$

We conclude that, in the limit of zero temperature, when clearly only neutron energy loss processes ($\omega \geq 0$) can occur, the scattering function is proportional to the susceptibility. For the example of a nucleus bound in a harmonic potential, the weight associated with the $n^{th}$ order scattering process was found to be proportional to $(Q^2/M\omega_0)^n$, as $T \to 0$, and this in turn is proportional to $(Q^2 \langle u^2 \rangle)^n$. Hence if $Q^{-1}$ is large compared to the chracteristic dimension of the system, only low-order processes contribute to the scattering function.

In the limit of high temperatures we find from (1.8)

$$S(Q,\omega) = -\frac{1}{\pi\beta\omega}\,\tilde{\chi}''(Q,i\omega),\ T \to \infty \tag{1.13}$$

so that $S(Q,\omega)$ is an even function of $\omega$ at sufficiently high temperatures. The isothermal susceptibility[2] $\chi(Q)$ measures the response of a system to a wavelength dependent probe, and it can be obtained from $\tilde{\chi}''$ using the relation,

$$\chi(Q) = -\frac{1}{\pi}\int_{-\infty}^{\infty} d\omega \cdot \frac{1}{\omega}\tilde{\chi}''(Q,i\omega) \quad . \tag{1.14}$$

Combining the definition of $S(Q)$, (1.10), and (1.13) and (1.14) we find the relation

$$\chi(Q) = \beta S(Q)\ , \quad T \to \infty \quad , \tag{1.15}$$

which gives added insight to the significance of the structure factor $S(Q)$.

## 1.2 Correlation Functions

The interpretation of the cross-section (1.3) is often made easier by writing it in terms of correlation functions formed from operators associated with the target system. We stress that the use of correlation functions can be a practical advantage, and not only a gain in mathematical elegance. The condensed notation afforded by correlation functions is often a help in itself, but probably the most important advantage stems from the fact that some powerful theoretical techniques are couched naturally in terms

---

[2]The physical significance of $\chi(Q)$ will be discussed further in Sec.1.2 and an explicit formula is given in (1.39).

of correlation functions. For these reasons we devote a section to rewriting the cross-section in terms of correlation functions, and summarising some of its important properties. Furthermore, we elaborate on the relation between the cross-section and dynamic susceptibility. We assume throughout this section that the neutrons are unpolarized; polarization effects are discussed separately in Sec.1.4.

If we label the scatterers by the index j, then the interaction potential in the master formula, (1.3), is

$$\hat{V}(\underline{r}) = \sum_j \hat{V}_j(\underline{r} - \underline{R}_j) \tag{1.16}$$

where $\hat{V}_j$ is the potential associated with the $j^{th}$ scatterer at position $\underline{R}_j$. The cross-section (1.3) then reduces to

$$\frac{d^2\sigma}{d\Omega\, dE'} = \frac{k'}{k}\left(\frac{m}{2\pi}\right)^2 \sum_{\lambda,\lambda'} p_\lambda |\langle\lambda'| \sum_j \hat{V}_j(\underline{Q})|\lambda\rangle|^2 \delta(\omega + E_\lambda - E_{\lambda'}) \tag{1.17}$$

where

$$\hat{V}_j(\underline{Q}) = \int d\underline{r}\; \hat{V}_j(\underline{r}) \exp(i\underline{Q}\cdot\underline{r}) \quad . \tag{1.18}$$

Using the result

$$\langle\lambda'|\sum \hat{V}|\lambda\rangle^* = \langle\lambda|\sum\hat{V}^+|\lambda'\rangle$$

where $\hat{V}^+$ is the Hermitean conjugate of $\hat{V}$, and the integral form of the delta function,

$$\delta(x) = \frac{1}{2\pi}\int_{-\infty}^{\infty} dt\, \exp(ixt)$$

the double summation in (1.17) can be written

$$\frac{1}{2\pi}\int_{-\infty}^{\infty} dt\, \exp(-i\omega t)\sum_{\lambda\lambda'} p_\lambda \langle\lambda|\sum\hat{V}^+|\lambda'\rangle\langle\lambda'|\sum\hat{V}|\lambda\rangle * \exp[it(E_{\lambda'} - E_\lambda)] \quad . \tag{1.19}$$

We define Heisenberg operators by

$$\hat{V}(t) = \exp(it\hat{H})\,\hat{V}\,\exp(-it\hat{H}) \tag{1.20}$$

where $\hat{H}$ is the Hamiltonian of the *target* system. Moreover, we take $p_\lambda$ to be a Boltzmann distribution

$$p_\lambda = \exp(-\beta E_\lambda)/Z \tag{1.21a}$$

with $\Sigma p_\lambda = 1$, and define a correlation function of two arbitrary operators $\hat{A}$ and $\hat{B}$ by

$$\begin{aligned} \langle\hat{A}\hat{B}\rangle &= \sum_\lambda p_\lambda \langle\lambda|\hat{A}\hat{B}|\lambda\rangle \\ &= \sum_{\lambda\lambda'} p_\lambda \langle\lambda|\hat{A}|\lambda'\rangle\langle\lambda'|\hat{B}|\lambda\rangle \quad . \end{aligned} \tag{1.21b}$$

With this notation we can write (1.19) in the form

$$\frac{1}{2\pi} \int_{-\infty}^{\infty} dt \, \exp(-i\omega t) \, \langle \sum \hat{V}^+(o) . \sum \hat{V}(t) \rangle$$

and the cross-section, (1.3), becomes

$$\frac{d^2\sigma}{d\Omega \, dE'} = \frac{k'}{k} \left(\frac{m}{2\pi}\right)^2 \cdot \frac{1}{2\pi} \int_{-\infty}^{\infty} dt \, \exp(-i\omega t) \sum_{ij} \langle \hat{V}_i^+(\underline{Q},o) \, \hat{V}_j \, (\underline{Q},t) \rangle \tag{1.22}$$

which is the desired result.

Looking at the cross-section, it is evident that it is determined by a scattering function of the form

$$S(\omega) = \frac{1}{2\pi} \int_{-\infty}^{\infty} dt \, \exp(-i\omega t) \, \langle \hat{A}^+(o)\hat{A}(t) \rangle \quad . \tag{1.23}$$

The cross-section must be real, clearly, and this can be verified with the aid of the two identities

$$\langle \hat{A}(o)\hat{B}(t) \rangle = \langle \hat{A}(-t)\hat{B}(o) \rangle \tag{1.24}$$

and

$$\langle \hat{A}(o)\hat{B}(t) \rangle^* = \langle \hat{B}^+(t)\hat{A}^+(o) \rangle \tag{1.25}$$

The detailed balance condition (1.9) follows from the identity

$$\langle \hat{A}(t)\hat{B}(o)\rangle = \langle \hat{B}(o)\hat{A}(t + i\beta)\rangle \quad . \tag{1.26}$$

Our second task in this section is to relate the cross-section to the dissipative part of the dynamic susceptibility, a result which we anticipated in the prologue. For ease of notation the $\underline{Q}$ dependence of the quantities is suppressed. In order to develop the concept of a dynamic susceptibility, consider an isolated target system initially in equilibrium to which a weak time-dependent perturbation is applied. At time $t = -\infty$ the system is described by the Hamiltonian $\hat{H}$; at some later time t the total Hamiltonian is taken to be

$$\hat{H} - \hat{A}^{+}f(t) \quad . \tag{1.27}$$

For example, $\hat{A}^{+}$ might be a magnetic moment and f an applied field. The response of the system is to be measured by the change in the variable $\hat{A}$. For a linear response we can write the change in $\hat{A}$ at time t, denoted by $\delta A\ (t)$, as

$$\delta A\ (t) = \int_{-\infty}^{t} d\bar{t}\ \phi(t - \bar{t})f(\bar{t}) \tag{1.28}$$

where $\phi$ is called the response function.[3] Let us choose f(t) to be of the form

$$f(t) = f\ \exp(i\omega t + \varepsilon t)\ , \quad \varepsilon \to {}^{+}o$$

where f and $\varepsilon$ are real constants. We then find that (1.28) reduces to

$$\delta A(t) = f(t)\ \tilde{\chi}\ (i\omega + \varepsilon) \tag{1.29}$$

where the dynamic susceptibility $\tilde{\chi}$ is the Laplace transform of the response function, namely,

$$\tilde{\chi}(s) = \int_{0}^{\infty} dt\ \exp(-st)\ \phi(t) \quad . \tag{1.30}$$

Eq. (1.29) relates the change in the variable $\hat{A}$ caused by the perturbation $-\hat{A}^{+}f(t)$ to the susceptibility $\tilde{\chi}$, and it can be used as a working definition from which to calculate the susceptibility.

---

[3]Physically $\phi(t)$ should be zero for $t < 0$ because of causality. However, it is convenient in the following development to have $\phi(t)$ defined for negative times, and this we shall do by using (1.31).

The response function $\phi(t)$ introduced in (1.28) can be calculated by a straightforward application of time-dependent perturbation theory. The result is (see, e.g., KUBO, 1966)

$$\phi(t) = i\langle\hat{A}(t)\hat{A}^+(0) - \hat{A}^+(o)\hat{A}(t)\rangle \quad . \tag{1.31}$$

The identity (1.25) can be used to show that $\phi(t)$ is a purely real function, as it should be. Using the identities (1.24) and (1.26) we can show that

$$[1 - \exp(-\beta\omega)]S(\omega) = \frac{1}{2\pi i} \int_{-\infty}^{\infty} dt \exp(i\omega t)\ \phi(t) \quad . \tag{1.32}$$

Because $S(\omega)$ is real, it follows from this last result that the response function must satisfy

$$\phi(t) = -\phi^*(-t) = -\phi(-t) \tag{1.33}$$

where the second equality follows because $\phi$ is real. We conclude that $\phi(t)$ is an *odd* function of time, and consequently we obtain from (1.30) and (1.32) the result,

$$[1 - \exp(-\beta\omega)]S(\omega) = \frac{1}{\pi}\tilde{\chi}''(-i\omega) = -\frac{1}{\pi}\tilde{\chi}''(i\omega) \tag{1.34}$$

where $\tilde{\chi}''$ is the imaginary part of the dynamic susceptibility. The relation (1.34) between the imaginary, or dissipative, part of the dynamic susceptibility and the scattering function $S(\omega)$ is usually referred to as the fluctuation-dissipation theorem (KUBO, 1966) because it relates the fluctuation properties of a system in equilibrium, described here by $S(\omega)$, to the response $\tilde{\chi}''(i\omega)$ of the system to an external time-dependent perturbation.

It is often more convenient in an actual calculation to study the relaxation function $R(t)$ instead of the response function $\phi(t)$. The relaxation function was introduced by KUBO (1957), and it is related to $\phi(t)$ through the equation,

$$\frac{\partial R(t)}{\partial t} = -\phi(t) \tag{1.35}$$

from which it follows that

$$-R(o) + s\,\tilde{R}(s) = -\tilde{\chi}(s) \quad . \tag{1.36}$$

Setting $s = i\omega$, and assuming $R(0)$ to be real, we find

$$\tilde{\chi}''(i\omega) = -\omega\tilde{R}'(i\omega) \tag{1.37}$$

so that $S(\omega)$, and hence the cross-section, is proportional to the real part of the Laplace transform of the relaxation function evaluated for $s = i\omega$.

An explicit representation of $R(t)$ is,

$$R(t) = \int_0^\beta d\lambda \langle \hat{A}^+(-i\lambda)\hat{A}(t)\rangle - (\beta/Z) \sum_m \exp(-\beta E_m)\, A^+_{mm} A_{mm} \quad . \tag{1.38}$$

In (1.38) $A_{mm}$ is the matrix element of $\hat{A}$ taken between the exact eigenstates of the Hamiltonian, and $E_m$ is the corresponding eigenvalue. The second term in (1.38) is the constant of integration which is chosen to ensure that $R(\infty) = 0$. In most cases of practical interest $R(0)$ is equal to the isothermal susceptibility

$$\chi = \int_0^\beta d\lambda \langle \hat{A}^+(-i\lambda)\hat{A}\rangle - \beta\langle \hat{A}\rangle\langle \hat{A}^+\rangle \quad . \tag{1.39}$$

The difference between $R(0)$ and $\chi$ lies in the second terms on the right-hand side of (1.38) and (1.39). We might expect that for most systems

$$\langle \hat{A}^+(o)\hat{A}(t)\rangle \rightarrow \langle \hat{A}^+\rangle\langle \hat{A}\rangle \quad , \quad t \rightarrow \infty \quad .$$

However, a direct calculation shows that

$$\lim_{t \rightarrow \infty} \langle \hat{A}^+(o)\hat{A}(t)\rangle = \frac{1}{Z} \sum_m \exp(-\beta E_m)\, A^+_{mm} A_{mm}$$

which is, apart from the factor $\beta$, the second term in (1.38). However, we are safe in most practical problems in assuming the equivalence of $R(0)$ and $\chi$, in which case

$$\chi = R(o) = \frac{1}{\pi} \int_{-\infty}^{\infty} d\omega \cdot \tilde{R}'(i\omega) = -\frac{1}{\pi} \int_{-\infty}^{\infty} d\omega \cdot \frac{1}{\omega}\, \tilde{\chi}''(i\omega) \tag{1.40}$$

where the last equality follows from (1.37).

We illustrate the use of some of the foregoing relations by considering the example of scattering by a single nucleus bound in a damped, harmonic oscillator potential. There is no exact, general solution to this problem. However, a case of special interest in neutron scattering is solved quite simply, and this is the case where we consider just the first excited state of the system, and employ classical mechanics.

In the following section we show that the cross-section for this problem is of the same form as (1.5) with

$$S(Q,\omega) = Q^2 \cdot \frac{1}{2\pi} \int_{-\infty}^{\infty} dt \exp(-i\omega t) \langle \underline{u}(o) \cdot \underline{u}(t)\rangle \tag{1.41}$$

where $\underline{u}$ is the displacement of the nucleus from equilibrium. Strictly speaking we should include the Debye-Waller factor in (1.41); it is omitted for simplicity here but must be inserted in the final result. The scattering function (1.41) is clearly an example of the function $S(\omega)$ that we defined in the preceding discussion of the fluctuation-dissipation theorem.

We calculate $S(Q,\omega)$ by calculating the dynamic susceptibility. Formula (1.29) tells us that we need to calculate the change in the position of the nucleus caused by a perturbation linear in u and of strength h, and we want also to include the effect of damping. The appropriate equation of motion for the displacement of a particle of mass M is

$$\ddot{u} + \omega_0^2 u + \Gamma\dot{u} + \frac{h}{M} = 0$$

where $M\Gamma$ is the strength of the damping. Take h to vary with time as $\exp(i\omega t)$, and then set $u(t) = u\exp(i\omega t)$ to obtain from the equation of motion, and (1.29), the following result for the susceptibility

$$\tilde{\chi}(i\omega) = -\frac{u}{h} = -\left[M(\omega^2 - \omega_0^2 - i\omega\Gamma)\right]^{-1} \tag{1.42a}$$

with

$$\tilde{\chi}''(i\omega) = \frac{-\omega\Gamma}{M[(\omega^2 - \omega_0^2)^2 + (\omega\Gamma)^2]} \quad . \tag{1.42b}$$

From (1.34) and (1.42b) it follows immediately that the scattering function (1.41) is given by

$$\begin{aligned} S(Q,\omega) &= -\frac{1}{\pi}\frac{Q^2}{[1 - \exp(-\beta\omega)]}\tilde{\chi}''(i\omega) \\ &\simeq \frac{-1}{\pi}\cdot\frac{Q^2}{\beta\omega}\cdot\tilde{\chi}''(i\omega) \quad . \end{aligned} \tag{1.43}$$

In the second line of (1.43) we have assumed that $\omega\beta << 1$ which is consistent with our use of classical mechanics. We see that for small damping $S(Q,\omega)$ has peaks at $\pm(\omega_0^2 - \Gamma^2/2)^{1/2}$, with a half-width at half-height of $\Gamma/2$. The structure factor $S(Q)$, defined as the integral of $S(Q,\omega)$ over all $\omega$ for fixed Q, is

$$S(Q) = \exp[-2W(Q)]\cdot\frac{Q^2}{M\beta\omega_0^2} \quad . \tag{1.44}$$

The Debye-Waller factor has been inserted in (1.44) for completeness. We conclude from the result (1.44) that the scattered neutron intensity increases with wave-vector as $Q^2$, and decreases with increasing natural frequency as $\omega_0^{-2}$. Note that the intensity is independent of the damping, apart from a contribution to the Debye-Waller factor which can be neglected in most cases. These features of the intensity are characteristic of nuclear scattering processes involving the excitation of a single quantum state.

It is in fact often useful to consider the classical limit of correlation functions. We defer a detailed discussion of the conditions under which a classical model is a good approximation to the target system until Subsec.1.3.4. However, one clear-cut case is when we wish to interpret computer simulation data on classical systems. In view of this interest in classical systems, we record here the key correlation functions, and their physical significance, for a classical *fluid* with a mean number density n.

To compare computer simulation data, for example, with measurements we need a prescription to relate the measured quantities, namely the scattering functions discussed in the next section, to the computed quantities,

$$F_s(Q,t) = \langle \exp[-i\underline{Q} \cdot \underline{R}_0(o)] \exp[i\underline{Q} \cdot \underline{R}_0(t)] \rangle \tag{1.45}$$

and

$$F(Q,t) = \frac{1}{N} \langle n^*(\underline{Q},o)n(\underline{Q},t) \rangle \tag{1.46}$$

where the correlation functions are computed according to classical statistical mechanics and $n(\underline{Q})$ is the Fourier transform of the microscopic density function. $F_s$ and F are real, and even functions of t, and from this follows that their temporal Fourier transforms are real, and even functions of $\omega$. We denote the Fourier transforms of (1.45) and (1.46) by $\tilde{S}_s(Q,\omega)$ and $\tilde{S}(Q,\omega)$, respectively.

SCHOFIELD (1960) pointed out that the prescriptions

$$S_s(Q,\omega) \simeq \exp(\tfrac{1}{2}\omega\beta)\, \tilde{S}_s(Q,\omega) \tag{1.47a}$$

and

$$S(Q,\omega) \simeq \exp(\tfrac{1}{2}\beta\omega)\, \tilde{S}(Q,\omega) \tag{1.47b}$$

lead to (approximate) scattering functions which fulfill the detailed balance condition, (1.9). The prescriptions (1.47) are almost always used to compare computed and experimentally determined scattering functions.

It is often useful to consider the spatial Fourier transforms of the correlation functions (1.45) and (1.46) because they have a simple interpretation as probability distributions. For example,

$$G_s(r,t) = \left(\frac{1}{2\pi}\right)^3 \int d\underline{Q} \exp(-i\underline{Q} \cdot \underline{r}) \, F_s(Q,t)$$
$$= \langle\delta[\underline{r} - \underline{R}_0(t) + \underline{R}_0(o)]\rangle \tag{1.48}$$

is the probability that, given a particle at the origin at time t = 0, the same particle is at $\underline{r}$ at time t. Similarly,

$$G(r,t) = n + \left(\frac{1}{2\pi}\right)^3 \int d\underline{Q} \exp(-i\underline{Q} \cdot \underline{r}) \, F(Q,t)$$
$$= \sum_j \langle\delta[\underline{r} - \underline{R}_j(t) + \underline{R}_0(o)]\rangle \tag{1.49}$$

is the probability that, given a particle at the origin at time t = 0, any particle (including the original particle) is to be found at the position $\underline{r}$ at time t. Note that for t = 0,

$$G_s(r,o) = \delta(\underline{r}) \quad \text{and} \quad G(r,o) = \delta(\underline{r}) + ng(r) \tag{1.50}$$

where g(r) is the pair distribution function. In the opposite limit $t \to \infty$,

$$G_s(r,\infty) = 0 \ , \quad \text{and} \quad G(r,\infty) = n \quad . \tag{1.51}$$

The significance of g(r) is perhaps appreciated best by noting that the quantity

$$N(R) = 4\pi n \int_0^R dr \, r^2 g(r)$$

is the mean number of particles within a sphere of radius R (HUTCHINSON et al., 1971).

The structure factor S(Q) is

$$S(Q) = \int d\underline{r} \exp(i\underline{Q} \cdot \underline{r}) \, [G(r,o) - n]$$
$$= 1 + n \int d\underline{r} \exp(i\underline{Q} \cdot \underline{r})[g(r) - 1] \tag{1.52}$$

where the last line follows from (1.50). The quantity S(0) is related to the isothermal compressibility $B_T^{-1}$ by the identity (LANDAU and LIFSHITZ, 1959)

$$S(o) = \frac{n}{\beta B_T} = \langle(\Delta N)^2\rangle/N \tag{1.53}$$

which shows also that S(0) is proportional to the mean-square, thermodynamic fluctuation in the particle number N. From (1.53) it follows that S(Q) increases dramatically on approaching the liquid-gas phase transition where fluctuations in the particle density assume almost macroscopic values.

The integral of the scattering functions against $\omega^2$ can be calculated from (1.11) by expanding the detailed balance factor in (1.8) in terms of $\omega\beta$ which is appropriate in the classical limit. The results are

$$\int_{-\infty}^{\infty} d\omega \cdot \omega^2 \tilde{S}_s(Q,\omega) = \int_{-\infty}^{\infty} d\omega \cdot \omega^2 \tilde{S}(Q,\omega) = \frac{Q^2}{M\beta} \quad . \tag{1.54}$$

The integral of the scattering functions against $\omega^4$ has been evaluated by de GENNES (1959) for the case of two-particle interactions. He found that[4]

$$\int_{-\infty}^{\infty} d\omega \cdot \omega^4 \tilde{S}(Q,\omega) = \frac{Q^2}{M\beta} \omega_\ell^2 \tag{1.55}$$

where

$$\omega_\ell^2 = \frac{3Q^2}{M\beta} + \Omega^2(0) - \Omega^2(Q) \tag{1.56}$$

and

$$\Omega^2(Q) = \frac{n}{M} \int d\underline{r}\ g(r) \cos(Qx) \frac{\partial^2}{\partial x^2} u(r) \quad . \tag{1.57}$$

In (1.57), u(r) is the pair interaction potential. Notice that if $T \simeq 0$, and the pair distribution function g(r) is replaced by its form for a perfect crystal, viz.

$$g(r) = \frac{1}{n} \sum_\ell \delta(\underline{r} - \underline{\ell})$$

then (1.56) is of the same form as (1.67) for the eigenfrequencies of a harmonic solid. The combination $\Omega^2(0) - \Omega^2(Q)$ in (1.56) ensures that the quantity $\omega_\ell^2$ tends to zero as $Q \to 0$, as $Q^2$. For a solid, the invariance of the potential energy and its derivates against an infinitesimal rigid body displacement of the crystal requires that $\Sigma D(\ell) = 0$, which ensures that acoustic phonon frequencies vanish in the limit of zero wave-vector.

---

[4] The corresponding result for a quantum liquid has been given by PUFF (1965).

The result for $S_s(Q,\omega)$ which corresponds to (1.55) is

$$\int_{-\infty}^{\infty} d\omega \cdot \omega^4 \tilde{S}_s(Q,\omega) = \frac{Q^2}{M\beta}\left[\frac{3Q^2}{M\beta} + \Omega^2(0)\right] \equiv \frac{Q^2}{M\beta} \cdot \omega_s^2 \tag{1.58}$$

where the last line defines $\omega_s^2$.

## 1.3 Nuclear Scattering

We know from experimental results that a nucleon-nucleon interaction has a range of order $10^{-5}$ Å. Because the range, and the nuclear radius, is much less than the wavelength of slow neutrons, the neutron-nucleus scattering contains only s-wave components, i.e., the scattering is isotropic, and characterized by a single parameter b, called a scattering length. The details of the neutron-nucleus interaction for slow neutron scattering are therefore condensed into this one parameter. There is a potential difficulty in using the master formula for the cross-section, (1.3), because it corresponds to the first term in a perturbation expansion in terms of the interaction potential, and the neutron-nucleus interaction is known to be very strong, as well as short ranged. However, FERMI (1936) pointed out that we can obtain the correct result using (1.3), or (1.22), if we introduce a pseudo-potential defined as

$$\hat{V}(\underline{r}) = \frac{2\pi}{m} \sum_j b_j \, \delta(\underline{r} - \underline{R}_j) \quad . \tag{1.59}$$

In this equation $\underline{R}_j$ denotes the position of the $j^{th}$ nucleus. Substituting (1.59) into (1.22) shows that the cross-section for nuclear scattering is proportional to the temporal Fourier transform of the quantity

$$\sum_{ij} b_i^* b_j Y_{ij}(\underline{Q},t) \quad , \tag{1.60}$$

where the correlation function $Y_{ij}(\underline{Q},t)$ is defined by

$$Y_{ij}(\underline{Q},t) = \langle \exp[-i\underline{Q} \cdot \hat{\underline{R}}_i(0)] \exp[i\underline{Q} \cdot \hat{\underline{R}}_j(t)] \rangle \quad . \tag{1.61}$$

We have assumed in (1.60) that the scattering lengths are not dynamical quantities, and can therefore be taken outside the thermal average; this step would not be valid if the scattering lengths were significantly energy dependent, for example.

The scattering lengths vary from element to element, and isotope to isotope. They depend also on the relative orientations of the neutron and nuclear spins. The cross-section is averaged over all these possible states, and we denote the averaging procedure by a horizontal bar. If there is no correlation between values of $b_i^*$ and $b_j$ associated with different nuclei then,

$$\overline{b_i^* b_j} = \overline{b_i^*} \cdot \overline{b_j}, \quad \text{if } i \neq j \quad .$$

But for i = j,

$$\overline{b_i^* b_i} = \overline{|b_i|^2} \quad .$$

In view of these results it is natural to write the nuclear cross-section as the sum of two parts by writing

$$\overline{b_i^* b_j} = \overline{b_i^*}\ \overline{b_j} + \delta_{i,j}\,(\overline{|b_i|^2} - |\bar{b}_i|^2) \tag{1.62}$$

where the Kronecker delta dunction $\delta_{i,j}$ is zero $i \neq j$, and unity for $i = j$. Using the result (1.62), together with (1.60), the nuclear cross-section is written

$$\frac{d^2\sigma}{d\Omega\, dE'} = \left(\frac{d^2\sigma}{d\Omega\, dE'}\right)_{coh} + \left(\frac{d^2\sigma}{d\Omega\, dE'}\right)_{incoh} \tag{1.63a}$$

where the coherent cross-section is given by

$$\left(\frac{d^2\sigma}{d\Omega\, dE'}\right)_{coh} = \frac{k'}{k} \cdot \frac{1}{2\pi} \int_{-\infty}^{\infty} dt\, \exp(-i\omega t) \sum_{ij} \overline{b_i^*}\ \overline{b_j}\ Y_{ij}(\underline{Q},t) \quad , \tag{1.63b}$$

and the incoherent cross-section is

$$\left(\frac{d^2\sigma}{d\Omega\, dE'}\right)_{incoh} = \frac{k'}{k} \cdot \frac{1}{2\pi} \int_{-\infty}^{\infty} dt\, \exp(-i\omega t) \sum_{j} \overline{|b_j - \bar{b}_j|^2}\ Y_{jj}(\underline{Q},t) \quad . \qquad 1.63c)$$

We now consider the cross-sections for the particular case of a system which contains one element. The temporal evolution of the operators in the correlation functions can be taken to be independent of the isotope and nuclear spin distributions to an excellent approximation. This means that averaged scattering lengths are independent of the site label. If the various isotopes of an element all have zero nuclear spin, then

$$\bar{b} = \sum_{\xi} c_{\xi} b_{\xi} \quad ,$$

and

$$\overline{|b|^2} = \sum_{\xi} c_{\xi} |b_{\xi}|^2 \quad ,$$

where the $\xi$th isotope occurs with a fractional concentration $c_{\xi}$. To be completely general we must allow each isotope to have a nuclear spin $I_{\xi}$. The nuclear spin dependence of the scattering arises because the neutron, with spin 1/2, interacts with a nucleus in states of total spin $I \pm 1/2$, and with scattering lengths $b^{(\pm)}$. The more general form for the quantities in (1.62) and (1.63) for a particular atomic type are then

$$\bar{b} = \sum_{\xi} c_{\xi} \left[ (I_{\xi} + 1)\, b_{\xi}^{(+)} + I_{\xi}\, b_{\xi}^{(-)} \right] / (2I_{\xi} + 1) \tag{1.64a}$$

and,

$$\overline{|b|^2} = \sum_{\xi} c_{\xi} \left[ (I_{\xi} + 1)\, |b_{\xi}^{+}|^2 + I_{\xi}\, |b_{\xi}^{(-)}|^2 \right] \Big/ (2I_{\xi} + 1) \quad . \tag{1.64b}$$

It is conventional to define single atom coherent and incoherent cross-sections, $\sigma_c$ and $\sigma_i$, respectively, by

$$\sigma_c = 4\pi\, |\bar{b}|^2 \tag{1.65a}$$

and

$$\sigma_i = 4\pi\, \overline{|b - \bar{b}|^2} \quad . \tag{1.65b}$$

The coherent and incoherent partial differential cross-sections, (1.63b) and (1.63c), can be expressed, with great advantage, in the form used in (1.5)

$$\left( \frac{d^2\sigma}{d\Omega\, dE'} \right)_{coh} = \frac{k'}{k} \cdot \frac{\sigma_c}{4\pi} \cdot NS(\underset{\sim}{Q},\omega) \tag{1.66a}$$

with the coherent scattering function

$$S(\underset{\sim}{Q},\omega) = \frac{1}{2\pi N} \int_{-\infty}^{\infty} dt\, \exp(-i\omega t) \sum_{ij} Y_{ij}(\underset{\sim}{Q},t) \tag{1.66b}$$

where N is the total number of nuclei, and

$$\left(\frac{d^2\sigma}{d\Omega\, dE'}\right)_{incoh} = \frac{k'}{k} \cdot \frac{\sigma_i}{4\pi} \cdot S_s(\underline{Q},\omega) \tag{1.67a}$$

with the incoherent scattering function defined by

$$S_s(\underline{Q},\omega) = \frac{1}{2\pi} \int_{-\infty}^{\infty} dt\, \exp(-i\omega t) \sum_j Y_{jj}(\underline{Q},t) \quad . \tag{1.67b}$$

We can also rewrite the coherent scattering function (1.66b) in terms of the microscopic particle density operator,

$$\hat{n}(\underline{r},t) = \sum_j \delta\left[\underline{r} - \hat{\underline{R}}_j(t)\right] \tag{1.68}$$

namely,

$$S(\underline{Q},\omega) \frac{1}{2\pi N} \int_{-\infty}^{\infty} dt\, \exp(-i\omega t) \int d\underline{r}\, \exp(i\underline{Q} \cdot \underline{r}) \int d\underline{r}' * \langle \hat{n}(\underline{r}' - \underline{r},o)\, \hat{n}(\underline{r}',t)\rangle \quad . \tag{1.69}$$

Moreover it is often useful in discussing the scattering from homogeneous, isotropic systems to introduce the Fourier components of the fluctuation in the particle density about its mean value $n = N/V$. We define

$$\hat{n}(\underline{r},t) = n + \frac{1}{V} \sum_{\underline{q}} \exp(i\underline{q} \cdot \underline{r})\hat{n}(\underline{q},t) \tag{1.70a}$$

or

$$\hat{n}(\underline{q},t) = \sum_j \left\{ \exp[-i\underline{q} \cdot \hat{\underline{R}}_j(t)] - \delta_{\underline{q},o} \right\} \quad . \tag{1.70b}$$

From (1.70a) it follows that the coherent scattering function (1.69) can be written

$$S(Q,\omega) = N\delta_{\underline{Q},0}\delta(\omega) + \frac{1}{2\pi N} \int_{-\infty}^{\infty} dt\, \exp(-i\omega t) * \langle \hat{n}^+(\underline{Q},0)\hat{n}(\underline{Q},t)\rangle \quad . \tag{1.71}$$

The first term corresponds to strictly elastic scattering with the scattering vector $\underline{Q} = 0$. Because this term does not represent a real scattering process, since $\underline{k}$ is fixed equal to $\underline{k}'$, it is usual to drop it in discussions of the dynamic properties of homogeneous, isotropic systems such as simple liquids.

Formulae (1.64) to (1.71) apply to the particular case of one type of atom. When there is more than one type of atom it is usually necessary to allow for possible correlation between the position operators $\hat{\underline{R}}_j(t)$, and the distribution of the different types of atom. We consider the simplest case of two types of atom, denoted by 1 and 2.

We define the function $p_j$ to be unity if the site j is occupied by an atom of type 1 and zero otherwise. The quantity $\bar{b}_j$ in the coherent cross-section can be written

$$\bar{b}_j = \bar{b}_2 + p_j(\bar{b}_1 - \bar{b}_2) \quad ,$$

where $\bar{b}_1$ and $\bar{b}_2$ are calculated according to (1.64a), and the coherent cross-section (1.63b) is proportional to the temporal Fourier transform of the sum of weighted correlation functions, namely,

$$\begin{aligned} &|\bar{b}_2|^2 \sum_{ij} Y_{ij}(\underline{Q},t) + \bar{b}_2^*(\bar{b}_1 - \bar{b}_2) \sum_{ij} p_j Y_{ij}(\underline{Q},t) \\ &\quad + (\bar{b}_1^* - \bar{b}_2^*)\bar{b}_2 \sum_{ij} p_i Y_{ij}(\underline{Q},t) \\ &\quad + |\bar{b}_1 - \bar{b}_2|^2 \sum_{ij} p_i p_j Y_{ij}(\underline{Q},t) \quad . \end{aligned} \tag{1.72}$$

For incoherent scattering we can write

$$\sigma_i(j) = \sigma_i(2) + p_j\,[\sigma_i(1) - \sigma_i(2)]$$

and obtain from (1.49c)

$$\sigma_i(2) \sum_j Y_{jj}(\underline{Q},t) + [\sigma_i(1) - \sigma_i(2)] \sum_j p_j Y_{jj}(\underline{Q},t) \quad . \tag{1.73}$$

Each term in (1.72) and (1.73) must be averaged over the various atomic configurations.

The averaging is particularly simple if the correlation between the position operators and distribution of atoms can be neglected, and the atoms are randomly arranged, for then we have only to replace $p_j$ and $p_i p_j$ in (1.72) and (1.73) by their average values,

$$\bar{p}_j = c$$

and

$$\overline{p_i p_j} = c^2 + \delta_{i,j}\, c(1 - c)$$

where c is the fractional concentration of type 1 atoms. With these results we find that the *coherent* cross-section (1.63b) is proportional to,

$$N|c\bar{b}_1 + (1 - c)\bar{b}_2|^2\, S(\underset{\sim}{Q},\omega) + c(1 - c)|\,\bar{b}_1 - \bar{b}_2|^2 S_s(\underset{\sim}{Q},\omega)$$

with $S(\underset{\sim}{Q},\omega)$ and $S_s(\underset{\sim}{Q},\omega)$ defined by (1.66b) and (1.67b), respectively. Note that the quantity $c\bar{b}_1 + (1-c)\bar{b}_2$ is simply the average scattering length for the mixture of atoms. The second term describes disorder scattering; it is usual to call this contribution the diffuse cross-section to distinguish it from the incoherent scattering which occurs even for a single component system. The corresponding result for the incoherent cross-section (1.63c) is simply

$$[c\sigma_i(1) + (1 - c)\sigma_i(2)]\; S_s(\underset{\sim}{Q},\omega)$$

where the factor multiplying the incoherent scattering function is the average, single atom incoherent cross-section.

In the prologue we discussed the cross-section for scattering by a single nucleus bound in a harmonic potenial. The results given there can be obtained from (1.67) for the scattering function for nuclear scattering. The correlation function can be calculated with the use of Bose annihilation and creation operators. The algebra is simplified by use of the following identity for any linear function of Bose operators, $\hat{B}$, namely,

$$\langle\exp(\hat{B})\rangle = \exp(\tfrac{1}{2}\langle\hat{B}^2\rangle) \quad . \tag{1.74}$$

A proof of this result is given, for example, by MESSIAH (1962).

### 1.3.1 Scattering by Phonons

We consider now the cross-sections for scattering processes which involve the annihilation or creation of phonons. Because the harmonic theory of lattice vibrations and the quantization of the classical harmonic oscillator equations of motion are discussed in almost every text on solid-state physics we do not give an introduction to the subject. In fact we shall merely record those aspects of the theory of phonons that are necessary to define a notation. Detailed accounts of the theory of phonons were given by PEIERLS (1955), COCHRAN and COWLEY (1967) and HORTON and MARADUDIN (1974). Experimental problems were reviewed by WILLIS and PRYOR (1975).

Neutron scattering affords a unique technique for the measurement of phonon dispersion curves in crystals. For crystals with a simple, symmetric structure (one or two atoms per unit cell) it is often possible to measure polarization vectors, and changes in the one-phonon line shape due to variations in temperature, or pressure, for example. Crystals with a low symmetry and several atoms per unit cell are much more difficult to study thoroughly. It is usually necessary, in such cases, to perform a paper feasibility study of the phonon cross-section, based on a realistic model of the force constants, which is then refined during the course of the measurements until a consistent picture of the data is established. Structural phase transitions are usually heralded by an anomalous temperature dependence of certain phonon groups. Neutron scattering experiments have provided key information on the microscopic origin of various first-order and continuous structural phase transitions, e.g., the phase transition in $BaTiO_3$ which is accompanied by the onset of ferroelectricity, and the phase transition in the A-15 structure compound $Nb_3Sn$. A general review of structural phase transitions was given by SCOTT (1974).

Our discussion in this section is restricted to one-phonon scattering processes in perfect crystals.[5] In terms of the harmonic oscillator example discussed in the prologue the restriction to one-phonon processes is equivalent to considering the terms with $n = \pm 1$, in which the neutrons change energy by a single quantum of energy $\omega_0$. The fundamental difference between the harmonic oscillator example and harmonic phonon theory lies in the fact that the nuclei in a solid interact, and undergo collective oscillations characterized by a wave-vector $\underline{q}$ and angular frequency $\omega_j(\underline{q})$. The periodicity of the crystal requires that phonons propagate with a wave-vector $\underline{q}$. We find that for phonon annihilation or creation processes to contribute to the cross-section there must be conservation of wave-vector analogous to the conservation of energy encountered in the example of the harmonic oscillator.

Crystal lattices and reciprocal lattices are defined in an annex to this chapter. The position vector of a nucleus is

$$\underline{R}_{\ell d} = \underline{\ell} + \underline{d} + \underline{u}(\ell d)$$

where $\underline{u}(\ell d)$ is the displacement of the $d^{th}$ nucleus in the $\ell^{th}$ unit cell, there being $r$ nuclei in each cell. In the harmonic theory of lattice vibrations it is assumed that the Hamiltonian describing the vibrational motion of the nuclei can be expanded about the equilibrium configuration in terms of the displacements, and that it is sufficient to retain just the lowest-order terms which are quadratic in the displacements.

---

[5] Phonons in disordered materials were discussed by ELLIOTT et al. (1974).

For a crystal described by a single Bravais lattice, and therefore with one atom per unit cell, the classical motion of the nuclei in the harmonic approximation are described by a Hamiltonian,

$$H = \sum_{\alpha,\ell} \frac{1}{2M} P_\alpha^2(\ell) + \sum_{\alpha,\ell} \sum_{\beta,\ell'} u_\alpha(\ell) D_{\alpha\beta}(\ell,\ell') u_\beta(\ell') \quad . \tag{1.75}$$

Here $\alpha$ and $\beta$ denote Cartesian components of a vector, $P_\alpha$ is the momentum conjugate to $u_\alpha$, and $D_{\alpha\beta}(\ell,\ell')$ are the second derivatives of the potential energy. The $D_{\alpha\beta}(\ell,\ell')$ satisfy several important relations. By definition

$$D_{\alpha\beta}(\ell,\ell') = D_{\beta\alpha}(\ell',\ell) \quad , \tag{1.76}$$

and clearly they can depend only on the relative positions of the relevant cells, i.e.,

$$D_{\alpha\beta}(\ell,\ell') = D_{\alpha\beta}(\ell - \ell') \quad . \tag{1.77}$$

Also, since every atom in a Bravais lattice is a centre of inversion symmetry

$$D_{\alpha\beta}(\ell,\ell') = D_{\alpha\beta}(\ell',\ell) = D_{\beta\alpha}(\ell,\ell') \quad . \tag{1.78}$$

The equation of motion for the $\ell^{th}$ nucleus is

$$\frac{\partial}{\partial t} P_\alpha \equiv \dot{P}_\alpha(\ell) = -\frac{\partial H}{\partial u_\alpha(\ell)} = M\ddot{u}_\alpha(\ell)$$

$$M\ddot{u}_\alpha(\ell) = -\sum_{\beta,\ell'} D_{\alpha\beta}(\ell - \ell')\, u_\beta(\ell') \tag{1.79}$$

on using (1.75) and (1.76). We now make the substitution

$$u_\alpha(\ell) = \sigma_\alpha M^{-1/2} \exp(i\underline{q} \cdot \underline{\ell} - i\omega_0 t) \quad , \tag{1.80}$$

and (1.79) becomes

$$\omega_0^2 \sigma_\alpha = \frac{1}{M} \sum_{\ell,\beta} D_{\alpha\beta}(\ell) \exp(-i\underline{q} \cdot \underline{\ell})\sigma_\beta \quad . \tag{1.81}$$

For each value of $\underline{q}$ there are three solutions for $\omega_0^2$ corresponding to the three values of $\alpha$. We shall denote this fact by labelling the frequencies with an index j having three values. To each $\omega_j(\underline{q})$ there corresponds an eigenvector, or polarization vector, $\sigma_\alpha$, and we shall henceforth write as $\sigma_\alpha^j(\underline{q})$. Thus the eigenvalue equation (1.81) now reads

$$\omega_j^2(\underline{q})\sigma_\alpha^j(\underline{q}) = \sum_\beta \mathcal{D}_{\alpha\beta}(\underline{q})\sigma_\beta^j(\underline{q}) \tag{1.82}$$

where

$$\mathcal{D}_{\alpha\beta}(\underline{q}) = \frac{1}{M}\sum_\ell \mathcal{D}_{\alpha\beta}(\ell)\ \exp(-i\underline{q}\cdot\underline{\ell}) \tag{1.83}$$

is usually called the dynamical matrix. In general,

$$\mathcal{D}^*_{\alpha\beta}(\underline{q}) = \mathcal{D}_{\alpha\beta}(-\underline{q}) \tag{1.84}$$

but if each atom is a centre of inversion symmetry

$$\mathcal{D}_{\alpha\beta}(\underline{q}) = \mathcal{D}_{\alpha\beta}(-\underline{q}) \tag{1.85}$$

and the dynamical matrix is purely real. The eigenvectors $\sigma_\alpha^j(\underline{q})$ and $\sigma_\alpha^{j*}(-\underline{q})$ satisfy the same equation, and are therefore identical except for an arbitrary phase factor. If we adopt the convention

$$\sigma_\alpha^{j*}(\underline{q}) = \sigma_\alpha^j(-\underline{q}) \tag{1.86}$$

then the eigenvectors are real and even functions of $\underline{q}$ for a crystal in which each atom is a centre of inversion symmetry. The eigenvectors can be constructed so as to satisfy

$$\sum_j \sigma_\alpha^{j*}(\underline{q})\ \ \sigma_\beta^j(\underline{q}) = \delta_{\alpha,\beta} \tag{1.87}$$

and

$$\sum_\alpha \sigma_\alpha^{j*}(\underline{q})\ \sigma_\alpha^{j'}(\underline{q}) = \delta_{j,j'} \quad . \tag{1.88}$$

In order to calculate the cross-section we need an expression for a displacement in terms of the normal coordinates. Moreover, we must convert from classical to quantum mechanics. Because this task is truly a text book problem we give only the final result, namely,

$$\hat{u}_\alpha(\ell,t) = \sum_{j\underline{q}} \left[2NM\omega_j(\underline{q})\right]^{-1/2}\ \sigma_\alpha^j(\underline{q})\ \exp(i\underline{q}\cdot\underline{\ell}) \quad *$$

$$* \left\{ \exp[-i\omega_j(\underline{q})t]\hat{a}_j(\underline{q}) + \exp[i\omega_j(\underline{q})t]\hat{a}_j^+(-\underline{q}) \right\} \tag{1.89}$$

The operators $\hat{a}$, $\hat{a}^+$ satisfy the Bose commutation relation

$$[\hat{a}_j(\underline{q}), \hat{a}^+_{j'}(\underline{q}')] = \delta_{j,j'}\, \delta_{\underline{q},\underline{q}'} \tag{1.90}$$

and,

$$\langle \hat{a}^+_j(\underline{q})\hat{a}_{j'}(\underline{q}')\rangle = \delta_{j,j'}\delta_{\underline{q},\underline{q}'}n_j(\underline{q}) \tag{1.91}$$

where $n_j(\underline{q})$ is the Bose occupation function

$$n_j(\underline{q}) = \left\{ \exp[\beta\omega_j(\underline{q})] -1 \right\}^{-1} \quad . \tag{1.92}$$

Thus far we have concentrated on the special example of a crystal described by a single Bravais lattice, but the extension to the general case of a lattice of several (r) interpenetrating Bravais lattices of identical structure, each containing nuclei of different masses, is straightforward, though cumbersome in notation. The total number of degrees of freedom is now 3rN and hence the quantum number j takes 3r values. The substitution (1.80) can still be made, with the mass and polarization vector carrying the label d to distinguish the r atoms in a unit cell. It is worth mentioning that some authors introduce an additional phase factor $\exp(i\underline{q}\cdot\underline{d})$; both conventions were discussed, for example, by MARADUDIN in the book edited by HORTON and MARADUDIN (1974). We denote the displacement operator for the $d^{th}$ atom in the $\ell^{th}$ unit cell by $\hat{\underline{u}}(\ell d;t)$, and (1.89) remains correct if a label d is attached to the mass and polarization factors. In the remaining part of our discussion we shall give formulae for the general case of r atoms per unit cell, unless it is stated otherwise.

With this brief survey of the harmonic theory of lattice vibrations, we are in a position to calculate the one-phonon inelastic cross-sections. The first step is to expand the correlation functions in the cross-sections (1.66) and (1.67) in terms of displacements. One-phonon terms arise from terms quadratic in the displacements, since the expression (1.89) for $\hat{u}_\alpha(\ell d;t)$ shows that the correlation function $\langle \hat{u}_\alpha(\ell d;0)\hat{u}_\beta(\ell' d';t)\rangle$ contains terms of the form

$$\exp(-i\omega_j t)n_j, \quad \text{and} \quad \exp(i\omega_j t)(1 + n_j)$$

which, when Fourier transformed with respect to time, contribute to the cross-section terms proportional to $\delta[\omega+\omega_j(\underline{q})]$ and $\delta[\omega-\omega_j(\underline{q})]$, respectively. The expansion is most conveniently achieved after rearranging the correlation functions into the form of a single exponential operator. The result,

$$\begin{aligned} &\langle \exp[-i\underline{Q}\cdot\hat{\underline{u}}(\ell d;0)]\ \exp[i\underline{Q}\cdot\hat{\underline{u}}(\ell' d';t)]\rangle \\ &= \exp[-W_d(\underline{Q})-W_{d'}(\underline{Q})]\ \exp\langle\underline{Q}\cdot\hat{\underline{u}}(\ell d;0)\underline{Q}\cdot\hat{\underline{u}}(\ell' d';t)\rangle \end{aligned} \tag{1.93}$$

can be obtained with the aid of (1.74) and

$$\exp(\hat{A})\cdot\exp(\hat{B}) = \exp(\hat{A}+\hat{B})\ \exp\left[\frac{1}{2}[\hat{A},\hat{B}]\right] \tag{1.94}$$

which holds when the commutator $[\hat{A},\hat{B}]$ is a c-number. The quantity $\exp[-W(\underline{Q})]$ in (1.93) is usually called the Debye-Waller factor, and it is defined by

$$\exp[-W_d(\underline{Q})] = \langle\exp[-i\underline{Q}\cdot\hat{\underline{u}}(\ell d,0)]\rangle = \exp\left\{-\frac{1}{2}\langle[\underline{Q}\cdot\hat{\underline{u}}(\ell d;0)]^2\rangle\right\} \tag{1.95}$$

where the second equality follows from (1.74). For a harmonic solid the Debye-Waller factor has the $\underline{Q}$, and temperature dependence obtained for the example of a harmonic oscillator discussed in the prologue.

The result (1.93) is to be substituted into the cross-sections and the exponential operator expanded in a power series. The first term is independent of time, and gives rise therefore to a term proportional to $\delta(\omega)$, which corresponds to purely elastic scattering. The second term in the expansion gives rise to one-phonon scattering processes described by cross-sections

$$\begin{aligned} \left(\frac{d^2\sigma}{d\Omega\, dE'}\right)^{(\pm)}_{coh} &= \frac{k'}{k}\ \frac{(2\pi)^3}{2v_0}\ \sum_{\underline{\tau}}\sum_{j\underline{q}}\ |G_j(\underline{q},\underline{Q})|^2\ \frac{1}{\omega_j(\underline{q})} \\ &\quad * \left[n_j(\underline{q}) + \frac{1}{2} \pm \frac{1}{2}\right]\delta\left[\omega \mp \omega_j(\underline{q})\right]\delta(\underline{Q} \mp \underline{q} - \underline{\tau}) \end{aligned} \tag{1.96a}$$

where the structure factor,

$$G_j(\underline{q},\underline{Q}) = \sum_d \bar{b}_d \exp\left[-W_d(\underline{Q}) + i\underline{Q}\cdot\underline{d}\right]\left[\underline{Q}\cdot\underline{\sigma}^j_d(\underline{q})\right]M_d^{-1/2} \quad . \tag{1.96b}$$

In obtaining the result (1.96a) we have used the result (A.9) for the lattice sum, $|\Sigma \exp(i\underline{\ell}\cdot\underline{k})|^2$, and $v_0$ is the volume of the crystal unit cell.

The upper sign in (1.96a) denotes a scattering process in which a phonon is created, and this can occur even at zero temperature. The lower sign denotes the phonon annihilation process, and the cross-section for this process is zero at zero temperature. Calculation of the structure factors $G_j$ requires a knowledge of the polarization vectors, which in general need to be calculated from a model of the force constants. In fact, for all but the simplest crystals, it is usual to calculate the structure factors before making a neutron experiment in order to determine positions in reciprocal space for optimum intensity. However, if only limited information is required, it is often sufficient to calculate the polarization vectors at positions of high symmetry where use of group theory can replace a numerical computation. Some examples are discussed in Chs. 2 and 3.

The two delta functions in (1.96a) can be considered as representing conservation of energy and wave-vector. It is the restrictions imposed on the scattering process by these two delta functions which enable the phonon dispersion $\omega_j(\underline{q})$ to be measured. Suppose we use a monoenergetic beam of neutrons and measure the energy of the scattered beam. From the conservation of energy we know that the changes in the energy of the neutrons are equal to the 3r phonon energies. Choose one of the neutron groups; knowing E' we have $\underline{k}'$ and from the conservation of wave-vector we have $\underline{q} - \underline{\tau}$. But $\underline{q}$ must lie in the first Brillouin zone so we get both $\underline{q}$ and $\underline{\tau}$. Hence the experiment provides both $\underline{q}$ and $\omega_j(\underline{q})$ for a phonon, i.e., point on the phonon dispersion curves.

Neutron spectrometers, and their various modes of operation, are discussed in Chs. 2 and 3. A detailed account, together with copious references, was given by DOLLING in the book edited by HORTON and MARADUDIN (1974).

We consider now the form of the one-phonon incoherent scattering cross-sections, which are in general less useful because there is no conservation of wave-vector associated with the scattering process. The one-phonon incoherent cross-section is readily calculated from (1.93) by merely setting $\underline{\ell} = \underline{\ell}'$ and $\underline{d} = \underline{d}'$. We find the cross-section is the sum of two terms, which correspond to phonon annihilation and creation, namely,

$$\left(\frac{d^2\sigma}{d\Omega\, dE'}\right)^{(\pm)}_{\text{incoh}} = \frac{1}{2}\frac{k'}{k}\sum_d \exp[-2W_d(\underline{Q})]\ \overline{|b_d - \bar{b}_d|^2}\ \frac{1}{M_d}$$
$$* \sum_{j\underline{q}} |\underline{Q}\cdot\underline{\sigma}_d^j(\underline{q})|^2 \frac{1}{\omega_j(\underline{q})}\left[n_j(\underline{q}) + \frac{1}{2} \pm \frac{1}{2}\right]\delta[\omega \mp \omega_j(\underline{q})] \quad . \tag{1.97}$$

The incoherent one-phonon cross-sections can be used to measure the density of phonon states in simple crystals. We discuss only the case of a single Bravais lattice with one atom per unit cell. The phonon denstiy of states $Z(\omega)$ for $\omega > 0$ is defined by

$$Z(\omega) = \frac{1}{3N}\sum_{j\underline{q}} \delta[\omega - \omega_j(\underline{q})]$$
$$\equiv \frac{2\omega}{3N}\sum_{j\underline{q}} \delta[\omega^2 - \omega_j^2(\underline{q})] \quad , \tag{1.98}$$

and for negative arguments we define $Z(\omega) = Z(-\omega)$. Using (1.98) in (1.97) we have, for both energy loss and energy gain,

$$\left(\frac{d^2\sigma}{d\Omega\, dE'}\right)^{(\pm)}_{\text{incoh}} = \frac{N\sigma_i}{8\pi M} \cdot \frac{k'}{k} \cdot Q^2 \cdot \exp[-2W(Q)] * \frac{Z(\omega)}{\omega}\,[n(\omega) + 1] \quad . \tag{1.99}$$

In deriving this result we have replaced the factor $|\underline{Q} \cdot \underline{\sigma}^j(\underline{q})|^2$ averaged over all modes, by $Q^2/3$, which is correct for cubic lattices. In general the density of states is weighted in the one-phonon incoherent cross-sections by factors $|\underline{Q} \cdot \underline{\sigma}^j_d(\underline{q})/M_d^{1/2}|^2$ which must be determined separately.

### 1.3.2 Phonon Interactions

In this and the following subsection we consider some of the effects of phonon-phonon and electron-phonon interactions on the coherent one-phonon cross-section. Macroscopically, the phonon-phonon interactions give rise to thermal expansion and a finite thermal conductivity, and they arise from higher-order terms in the expansion of the Hamiltonian in terms of the displacements. The presence of the higher-order terms prevents the diagonalization of the Hamiltonian in terms of phonon annihilation and creation operators, but when the additional terms are small compared to the harmonic part of the Hamiltonian their effects can be calculated by perturbation theory. In this instance, it is still meaningful to use the concept of a phonon of wave-vector $\underline{q}$, say, but the phonons can no longer be regarded as forming a non-interacting Bose gas. The effect of the interactions on a phonon is to change its frequency, and also to give it a finite lifetime. For temperatures much less than the maximum phonon frequency, both these effects are small because the magnitude of the displacements is small, and hence the harmonic part of the Hamiltonian is dominant. Furthermore, we should expect the effects on low frequency phonons to be small because the phonon density of states is small for frequencies much less than maximum phonon frequency and thus the probability of the scattering of two phonons is itself also small.

Higher-order terms modify the one-phonon cross-section in several ways. The most significant effect is usually described by the replacement of the energy delta function in (1.96) by a Lorentzian function. The change in the line shape can be described in terms of a peak width and shift away from harmonic phonon frequency. The form of the Debye-Waller factor is also modified by the appearance of higher-order correlation functions of the operator $\underline{Q} \cdot \hat{\underline{u}}$. These higher-order terms are usually a very small effect, however. A close examination of the one-phonon cross-section reveals the possibility of asymmetry in the line-shape due to phonon interference effects, but it is usually unimportant. However, the effect must clearly be taken into account in analysing the change in line shape with temperature or pressure, for example. Finally, the higher-order terms in the Hamiltonian can trigger a variety of structural phase transitions. The study of ferroelectric and non-ferroelectric phase transitions is an active area of research, and a survey of progress in our understanding of the various types of transition using neutron, and diffuse X-ray, scattering techniques is given in Ch.3.

It is convenient to describe the effects of phonon-phonon interactions in terms of the modifications to a dynamic susceptibility $\tilde{\chi}(\underline{q},i\omega)$ of the form given by (1.42). The appropriate form of the susceptibility, for a crystal described by a single Bravais lattice, is

$$M\tilde{\chi}_j(\underline{q},i\omega) = \left\{-\omega^2 + \omega_j^2(\underline{q}) + \omega_j(\underline{q})[\Delta_j(\underline{q},\omega) + i\Gamma_j(\underline{q},\omega)]\right\}^{-1} \tag{1.100}$$

where $\Delta$ and $\Gamma$ describe the effects of phonon-phonon interactions. The coherent, one-phonon cross-section is then

$$\left(\frac{d^2\sigma}{d\Omega\, dE'}\right)_{coh} = \frac{k'}{k}\frac{(2\pi)^3}{2v_0}\sum_{\underline{\tau}}\sum_{j\underline{q}} |G_j(\underline{q},\underline{Q})|^2\delta(\underline{Q} - \underline{q} - \underline{\tau}) * \left(\frac{2M}{\pi}\right)n(-\omega)\tilde{\chi}_j''(\underline{q},i\omega) \tag{1.101}$$

where $n(\omega)$ is defined by (1.92).

In calculating the leading order contributions to $\Delta$ and $\Gamma$ from the higher-order terms in the Hamiltonian it is necessary to retain terms that are cubic and quartic in the displacements. The cubic term does not, in fact, contribute in lowest order, when $\Delta$ is proportional to the strength of the quartic terms in the Hamiltonian, and $\Gamma$ is zero. It follows from this that, to be consistent in the calculation of $\Delta$ and $\Gamma$, we must treat the cubic terms to second order and the quartic terms to first order. The calculation reveals that $\Delta$, an even function of $\omega$, is proportional at $\omega = 0$ to the strength of the quartic terms in the Hamiltonian. $\Gamma$ is an odd function of $\omega$, and proportional to the square of the strength of the cubic terms. At high temperatures both $\Delta$ and $\Gamma$ are proportional to temperature. Explicit expressions for $\Delta$ and $\Gamma$ were given, for example, by THOMPSON (1963) and BALCAR (1970). Fig.1.2 shows $\tilde{\chi}''$, $\Delta$ and $\Gamma$ as functions of frequency for a model of face-centered argon. Note the significant structure in both $\Delta$ and $\Gamma$, and that they are small compared to the harmonic phonon frequency for this example.

Let us assume that, in the range of $\omega$ of interest, we can make the replacements $\Delta \simeq$ constant, and $\Gamma \simeq \omega\Gamma_0/\omega_0$ to a good approximation, where $\omega_0$ denotes the harmonic phonon frequency. When $\Delta$ and $\Gamma_0$ are small compared to $\omega_0$, the imaginary part of the susceptibility (1.100) reduces to

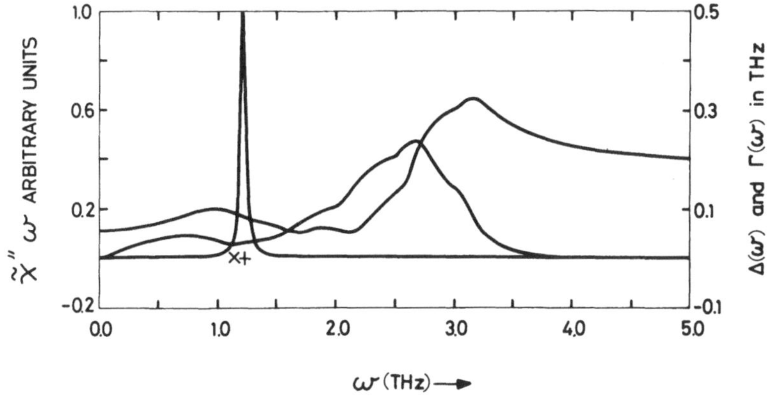


Fig. 1.2 The real and imaginary parts of the self-energy, $\Delta(\omega)$ and $\Gamma(\omega)$, and the dissipative part of the susceptibility equation (1.86) are shown as a function of $\omega$ for a model of face-centered argon. From BARRON and KLEIN (1974)

$$M\tilde{\chi}''(i\omega) = \frac{1}{2\omega_0}\left[\frac{\Gamma_0/2}{(\omega+\omega_0+\Delta/2)^2+(\omega\Gamma_0/2\omega_0)^2} - \frac{\Gamma_0/2}{(\omega-\omega_0-\Delta/2)^2+(\omega\Gamma_0/2\omega_0)^2}\right] \tag{1.102}$$

The wave-vector dependence of the various quantities in (1.102) has been suppressed for ease of notation. We see from the result (1.102) that, when $\Delta$ and $\Gamma_0$ are small compared to the harmonic phonon frequency, the collective mode frequency is shifted to $\omega_0 + \Delta/2$, and the peak has a half-width at half-height of $\Gamma_0/2$. In the limit $\Delta$, $\Gamma_0 \to 0$ the right-hand side of (1.102) reduces to a pair of delta functions at $\omega = \pm\omega_0$, and the coherent cross-section (1.101) reverts to the earlier result.

The higher-order terms in the Hamiltonian modify the expression for the Debye-Waller factor. $W(\underline{Q})$ in (1.95) is replaced by

$$W(\underline{Q}) = \frac{1}{2}\langle\hat{\theta}^2\rangle - \frac{i}{6}\langle\hat{\theta}^3\rangle - \frac{1}{24}\left[\langle\hat{\theta}^4\rangle - 3\langle\hat{\theta}^2\rangle^2\right] + \ldots \tag{1.103}$$

with

$$\hat{\theta} = \underline{Q}\cdot\underline{\hat{u}} \quad .$$

The first term is the usual harmonic form for W. The second term vanishes if every atom is a centre of inversion symmetry. The third term vanishes if the displacement amplitudes have the Gaussian distribution due to harmonic forces. But in the presence of higher-order terms in the Hamiltonian this term is non-zero. However, calculations for simple solids suggest that it represents only a minor correction in most cases. WILLIS and PRYOR (1975) gave a detailed discussion of the Debye-Waller factor for complex crystals.

Higher-order terms in the Hamiltonian describing lattice displacements produce a distortion of the harmonic, one-phonon line shape. The effect arises from interference between one-phonon and multi-phonon processes. For example, a single phonon initially excited in the scattering process can decay, via the higher-order terms, into two phonons. The contribution of this process to the cross-section contains the one-phonon susceptibility, (1.103), and the appropriate conservation of wave-vector, and it is therefore to be included in the one-phonon cross-section. There is also the possibility of two phonons combining into a single phonon, and equivalent processes involving three or more phonons and the decay or creation of a single phonon. In general, the interference terms lead to both an asymmetry and change in the intensity of the harmonic one-phonon line shape. However, the contribution (evaluated in lowest order) to the cross-section vanishes for Q equal to a reciprocal lattice vector, or a vector midway between two Bragg positions (GLYDE, 1974). The interference contribution to the cross-section changes sign as $\underline{Q}$ goes from being just less than a reciprocal lattice vector to being just greater than a reciprocal lattice vector.

All these features are illustrated nicely by the computer simulation of $S(\underline{Q},\omega)$ for potassium at 162.5 K, by KLEIN and collaborators, shown in Fig.1.3. All the

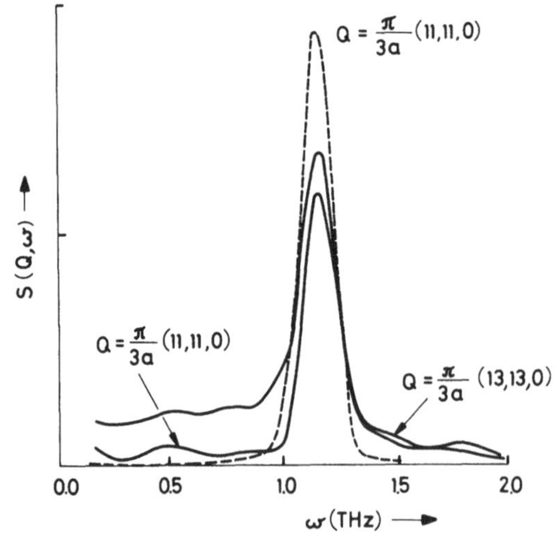


Fig. 1.3 The curves (———) are the full $S(Q,\omega)$ for potassium at 162.5 K, for $Q = 2\pi(11/6,11/6,0)/a$, and $2\pi(13/6,13/6,0)/a$, obtained by computer simulation, (using the potential calculated by DAGENS et al., 1975) by KLEIN and collaborators. The curve (---) is the one-phonon approximation to $S(Q,\omega)$, for $Q = 2\ (11/6,11/6,0)/a$. All curves have been convoluted with a Gaussian of full width at half height of 0.131 THz, which allows a direct comparison with the neutron scattering measurement on potassium by MEYER et al. (1976)

curves have been convoluted with a Gaussian of full-width at half-height of 0.131 THz. The asymmetry in the curves, and the differences in intensities either side of the (2,2,0) Bragg position, are revealed quite clearly, and are consistent with neutron scattering measurements by MEYER et al. (1976). Also shown is the one-phonon approximation for $\underset{\sim}{Q} = 2\pi(11/6,11/6,0)/a$. We conclude that, although the interference contribution to the cross-section is small in most cases, and can be safely neglected, the effect must be considered in a study of the line shape as a function of temperature or pressure, say.

Phonon-phonon interactions are believed to be responsible for a large number of observed structural phase transitions. The basic concept is that a particular phonon becomes unstable due to phonon-phonon interactions and its frequency decreases with temperature. If the frequency reaches zero at some critical temperature $T_0$ the phonon is regarded as condensing into the lattice to cause a structural phase transition. Clearly, this is equivalent to saying that, the force constants associated with the phonon in question decrease, or soften, with decreasing temperature and tend eventually to zero at $T_0$. There is a corresponding increase in the size of the lattice displacement, of course, as can be seen from the $\omega_0^{-1/2}$ dependence of the displacement operator (1.89). Below $T_0$ the unstable phonon is stabilized by higher-order terms in the Hamiltonian of the lattice displacements. The soft mode theory of structural phase transitions was introduced by COCHRAN (1959).

The preceding discussion can be made slightly more quantitative with the aid of results given earlier for the effect of phonon-phonon interactions on the one-phonon cross-section. From (1.100) for the dynamic susceptibility it follows that the energy of the observed collective mode is given approximately by

$$\omega = (\omega_0^2 + \omega_0\Delta)^{1/2} \quad .$$

If the phonon in question is unstable then $\omega_0^2$ is a negative quantity. We remarked earlier that $\Delta$ is proportional to temperature T for large T. The collective mode energy can therefore be written

$$\omega \simeq \alpha(T - T_0)^{1/2} \quad , \quad T > T_0 \tag{1.104}$$

where $\alpha$ is a constant. This picture is satisfactory for a wide variety of ferroelectric crystals which undergo a first-order phase transitions at $T_0$. In most of these crystals, the soft mode is overdamped even at temperatures well above $T_0$. An example is the cubic perovskite $BaTiO_3$ in which an optic phonon, which is coupled strongly to an acoustic phonon, becomes soft at the phase transition (HARADA et al., 1971). A ferroelectric must involve condensation of a phonon that is both polar and long wavelength because the ferroelectric state must have a macroscopic polarization. The connection between the onset of ferroelectricity and the existence of a soft phonon is seen in the relation between the zero frequency dielectric function $\varepsilon(0)$ and the energy of transverse and longitudinal optic modes, $\omega_0$ and $\omega_L$, respectively. In its simplest form, the relation is

$$\varepsilon(0)\ \alpha (\omega_L/\omega_0)^2 \quad .$$

Hence, the onset of ferroelectricity, which is characterized by $\varepsilon(0) \to \infty$ as $T \to T_0$, can arise from an anomalous decrease in the energy of a transverse optic phonon.

In marked contrast to $BaTiO_3$, the perovskite $SrTiO_3$ undergoes a non-ferroelectric, *continuous* phase transition at about 105 K which involves the softening of a short wavelength optic mode. The crystal changes at this transition from a high temperature cubic phase to a lower symmetry tetragonal phase. The primitive cell is doubled in a continuous way by means of a soft optical phonon with wave-vector equal to the position of the Brillouin zone boundary in the (111) direction. The transition involves an out-of-phase rotation of adjacent oxygen octahedra in the (100) planes. Because the rotations of adjacent octahedra are out of phase, the phonon mode associated with the transformation must have a wave-vector on the zone boundary. The neutron scattering data of SHAPIRO et al. (1972) show that the soft phonon is quite distinct near the phase transition, i.e., under-damped, which contrasts with the situation in $BaTiO_3$, and most other ferroelectric transitions.

The data also show a strong central component which is known to grow in intensity as the transition temperature is approached from above. Moreover, the phonon energy $\omega_0$ does not go to zero as $T \to T_0$. SHAPIRO et al. interpreted their data in terms of a dynamic susceptibility proposed by COWLEY (1970), in which the self-energy in (1.100) has a resonant structure of the form,

$$\omega_0[\Delta(\omega) + i\Gamma(\omega)] \simeq \omega_0\Delta + i\omega\Gamma_0 - \frac{\Omega^2}{1 + (i\omega\tau)} \tag{1.105}$$

where $\Delta$, $\Gamma_0$, $\Omega^2$ and $\tau$ are constants. It is to be understood that the wave-vector, which is not shown explicitly in (1.105), corresponds to the (111) Brillouin zone boundary position. The imaginary part of the self-energy is proportional to $\omega$ for both $\omega\tau >> 1$ and $\omega\tau << 1$.

On substituting (1.105) in (1.100) we obtain the results[6]

$$S(Q,0) = \frac{Q^2}{\pi\beta M} \cdot \frac{\Gamma_0 + \tau\Omega^2}{(\omega_\infty^2 - \Omega^2)^2} \tag{1.106}$$

and

$$S(Q) = \int_{-\infty}^{\infty} d\omega\, S(Q,\omega) = \frac{Q^2}{\beta M}\left(\frac{1}{\omega_\infty^2 - \Omega^2}\right) \tag{1.107}$$

with $\omega_\infty^2 = \omega_0^2 + \omega_0\Delta$. Hence, if the quantity $\omega_\infty^2 - \Omega^2$ tends to zero at some temperature $T_0$, say, a strong central peak develops in $S(Q,\omega)$, and the total intensity diverges. The half-width of the peak, clearly, tends to zero as $(\omega_\infty^2 - \Omega^2)$ tends to zero. An analysis of the energy dependence of $S(\underline{Q},\omega)$ shows that for high energies, where $\omega\tau << 1$, the collective mode peak occurs at $\omega_\infty$, whereas the resonance occurs at a lower energy, $\omega_\infty^2 - \Omega^2/(2\omega_0)x$ for $\omega\tau << 1$.

SHAPIRO et al. (1972) analysed their data in terms of the preceding model and found that

$$\omega_\infty^2 - \Omega^2 = C(T - T_0)^\gamma \; , \quad T > T_0$$

where C is a constant and $\gamma = 2.0 \pm 0.5$. The value of $\Omega^2$ at $T_0$ is in good agreement with a calculation by SILBERGLITT (1972) which treats the effect of quartic components in the Hamiltonian to second order. The width of the central component was found to be less than the available energy resolution, and so critical narrowing could not be observed directly. MÜLLER and BERLINGER (1971) have measured the angle of rotation of the octahedra below $T_0$ by EPR and find that the temperature variation follows a simple power law

$$\phi \propto (T_0 - T)^\beta \; , \quad T < T_0$$

with an exponent $\beta = 0.33 \pm 0.02$. The departure of the indices $\gamma$ and $\beta$ from mean-field values, $\gamma = 1$ and $\beta = 1/2$, is evidence for the transition being continuous and not first order, as found in $BaTiO_3$, for example. The direct observation of the divergence of $S(\underline{Q},0)$, accompanied by a vanishing line-width, would confirm the incipient instability of the crystal as the temperature approaches $T_0$.

---

[6] We note also that the integral of $S(\underline{Q},\omega)$ against $\omega^2$ gives the result $Q^2/M\beta$, cf. (1.54).

### 1.3.3 Electron-Phonon Interactions

The electron-phonon interaction, which is responsible for the electrical conductivity at high temperatures, contributes to the width of phonon peaks. It leads also to anomalies in the dispersion curves. This was pointed out first by KOHN (1959), and the phenomena he described have become known as Kohn anomalies. The physics behind the anomalies in the dispersion curves is as follows. In a real metal the motions of the conduction electrons and ions are clearly not independent of each other; when ions move from their equilibrium positions they scatter the electrons. Because typical phonon frequencies are much less then electronic frequencies of importance, the motion of the ions appears to the electrons as a static perturbation varying in space with a wave-vector $\underline{q}$. A consequence of this is that the Coulomb potential between the ions is screened by the electrons, the "bare" potential being essentially multiplied by the susceptibility of the conduction electrons of the static perturbation. For a free electron gas the susceptibility possesses a singularity at $|\underline{q}|$ equal to twice the Fermi wave-vector. It follows that the ion-ion potential, and therefore the phonon frequency, reflects this singularity; in fact, as the screening decreases abruptly so the phonon frequencies increase. In real metals the effect on the phonon frequency depends on both the strength of the electron-phonon interaction, and the sharpness and shape of the Fermi surface. For example, for pieces of the Fermi surfaces possessing cylindrical form the singularity in the derivative of the dispersion curve is of the inverse square root kind, and for planar sections of the Fermi surface the most singular part becomes logarithmic.

Electron-phonon interactions are particularly important in understanding the onset of superconductitivy. A number of interesting features of phonon dispersions in superconducting materials are discussed in Ch.2. For example, $Nb_3Sn$ is one of several A-15 compounds that are high temperature superconductors, and it exhibits a structural phase transition at 45 K which is approximately 26 K above the critical temperature for the onset superconductivity (in zero field). Neutron scattering studies of the structural phase transition by SHIRANE and AXE (1971) revealed a number of interesting features which are, at first sight, akin to those observed in $SrTiO_3$. In particular, a central peak is observed which grows in intensity as the transition is approached from above. However, the instability in $Nb_3Sn$ is associated with an acoustic phonon of zero wave-vector. Moreover, only acoustic modes are involved in the central peak in $Nb_3Sn$, in contrast to the situation in $SrTiO_3$ where the peak arises from the decay of an optic phonon into acoustic phonons. SHIRANE and AXE interpreted their data in terms of the model discussed for $SrTiO_3$ but a microscopic picture for the mechanism driving the transition is lacking. A good discussion of the the various theoretical models proposed for A-15 compounds was given by RANNINGER (1975).

## 1.3.4 Monatomic Liquids

We restrict our discussion to monatomic liquids. The limitation of scope should not be taken to imply that there is minimal interest in multi-component liquids, because this is far from the truth. However, there are a number of subtle points which need careful examination and, in the space available, we cannot achieve this for both monatomic and multi-component liquids. Moreover, MARCH and TOSI (1977) devoted much of their book to the discussion of multi-component liquids. Three recent articles in the series Reports on Progress in Physics survey progress in experiment and theory for molecular liquids (ALLEN and HIGGINS, 1973) liquid $^4$He (WOODS and COWLEY, 1973) and classical monatomic liquids (COPLEY and LOVESEY, 1975). Two additional reviews of the theory of dynamic properties of classical monatomic liquids are those by SCHOFIELD (1975) and MOUNTAIN (1976).

In the first part of this section, we discuss some of the key features of the scattering functions for liquid $^4$He and $^3$He at low temperatures. While there has been a substantial effort over the past decade or so aimed at interpreting neutron data on $^4$He, and some features are well understood, a completely satisfactory theory of the scattering function, for the range of Q and $\omega$ of interest, is lacking. Neutron experiments on $^3$He are extremely difficult because of the large absorption cross-section. Some intriguing effects have been observed in the first experiments aimed at measuring the coherent scattering function, and these are summarised in the next subsection. Neutron experiments on classical monatomic liquids really began with the study of lead by BROCKHOUSE and POPE in 1959. Since then there have been many neutron experiments. However, the proper interpretation of coherent scattering data was confused until quite recently. The incoherent scattering functions are discussed briefly in the second subsection, together with some computer simulation and neutron scattering data. We shall also attempt to summarise the salient properties of the coherent scattering function in terms of the underlying physical processes.

*Quantum Liquids*

Liquid $^4$He has been studied extensively with neutron scattering techniques, and the paper by WOODS and COWLEY (1973) provides a good survey of both theory and experiment. Liquid $^3$He is equally interesting but it has a very large neutron capture cross-section ($^3$He is often used to make neutron detectors for this very reason) and only recently have sufficiently intense neutron beams become available to make its study really practicable.

A measure of the quantum nature of a liquid is given by the thermal wavelength

$$\lambda_T = 2\pi\ (3k_BTM)^{-1/2}$$

This should be compared with the position $\sigma$ of the minimum in the interaction potential between the atoms. If $\lambda_T \gtrsim \sigma$, quantum interference effects are important, and this is manifest in a large zero-point motion of the atoms. Consider, for example, the three liquids $^4$He, Ne and Ar. The value of $\sigma$ is of order 3 Å for each of these liquids while $\lambda_T(^4\mathrm{He}, 2.1\ \mathrm{K}) = 4.6$ Å, $\lambda_T(\mathrm{Ne}, 27\ \mathrm{K}) = 0.6$ Å, and $\lambda_T(\mathrm{Ar}, 85\ \mathrm{K}) = 0.2$ Å. From these numbers, we conclude that for $^4$He at 2.1 K quantum interference effects are important, but are negligible for Ar at 85 K. Liquid Ne at 27 K is something of an intermediate case where quantum effects should be included to at least first order. The quantum nature of a liquid is seen also in the exchange effects induced by symmetrization of the many-particle wave function. This effect can be neglected if the temperature, number density n and mass are such that the mean occupation numbers of all states are small compared to unity. Explicitly, this requires (LANDAU and LIFSHITZ, 1959)

$$n(k_BTM)^{-3/2} \ll 1$$

and this condition is indeed satisfied for the three liquids considered in the discussion.

The neutron scattering intensity is proportional to the structure factor

$$S(\underline{Q}) = -\frac{1}{\pi} \int_0^\infty d\omega \cdot \coth(\tfrac{1}{2}\beta\omega)\tilde{\chi}''(\underline{Q},i\omega) \quad . \tag{1.108}$$

The structure factor for $^4$He at 0.79 K, measured with X-rays, is shown in Fig.1.4. S(Q) decreases rapidly with decreasing Q for Q's less than the position of the main

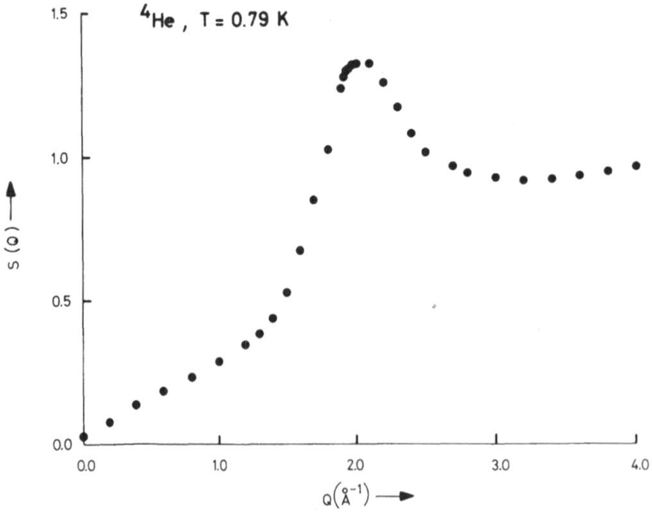


Fig. 1.4 The structure factor for liquid $^4$He at 0.79 K measured by X-ray scattering, after ACHTER and MEYER (1969)

peak, and this makes it difficult to perform accurate neutron measurements at small Q. The isothermal susceptibility $\chi(Q)$, Eq. (1.14), tends in the limit of small Q to a constant value. The result is (PINES and NOZIERES, 1966; WILKS, 1967)

$$\chi(Q) = \frac{1}{Mc^2} \ , \ Q \to 0 \tag{1.109}$$

where M is the nuclear mass and c the isothermal velocity of sound. At zero temperature, (1.14) and (1.108) can be combined with the Schwartz inequality to obtain the condition

$$\chi(Q) \geq \left(\frac{4M}{Q^2}\right) S^2(Q) \quad . \tag{1.110}$$

Using (1.109) we obtain from (1.110) the inequality $S(Q) \leq Q/2Mc$ for small Q, from which deduce $S(0) = 0$ at zero temperature.

We consider now the coherent cross-sections for $^4$He and $^3$He, and begin with some general remarks on the form of $S(Q,\omega)$ at low temperatures. Our discussion is couched in terms of the dynamic susceptibility $\tilde{\chi}(Q,i\omega)$ introduced in Sec.1.2. From (1.34) it follows that

$$S(Q,\omega) = [1 - \exp(-\beta\omega)]^{-1} \frac{-1}{\pi} \tilde{\chi}''(Q,i\omega) \quad .$$

In the limit $T \to 0$ ($\beta \to \infty$), $S(Q,\omega)$ is zero for $\omega < 0$, and proportional to $\tilde{\chi}''(Q,i\omega)$ for $\omega \geq 0$. Because $\tilde{\chi}''(Q,i\omega)$ is an odd function of $\omega$ it follows that the scattering function tends to zero as $\omega \to 0$, in the limit of low temperatures. Moreover, $\tilde{\chi}''$ tends to zero as $\omega \to \infty$.

For large Q, i.e., when S(Q) is close to unity, each atom in a liquid appears as a free, independent scattering centre. In this limit, the scattering function for $^4$He will tend to the result for an ideal Boltzmann liquid (exchange effects being assumed negligible), namely

$$S(Q,\omega) = \left(\frac{M\beta}{2\pi Q^2}\right)^{1/2} \exp\left[-\frac{M\beta}{2Q^2}\left(\omega - \frac{Q^2}{2M}\right)^2\right] \quad . \tag{1.111}$$

This function has a maximum when the energy tranfer $\omega$ equals the recoil energy of an atom $Q^2/2M$. Some neutron data for $^4$He at 1.1 K are shown in Fig.1.5 for a wide range of wave-vectors. While the data at the largest Q are of the form predicted by (1.111), a close examination of the data reveals significant departures. The departures from the free atom result are illustrated in Fig.1.6 which shows the peak width, and deviation of the peak position from $Q^2/2M$, for Q in the range $2.0 \leq Q \leq 9.0$ Å$^{-1}$.

The data in Fig.1.4 for $^4$He below the $\lambda$-transition show that $S(Q,\omega)$ is dominated by a sharp peak for small Q. In view of this observation $\tilde{\chi}''(Q,i\omega)$ can, to a first

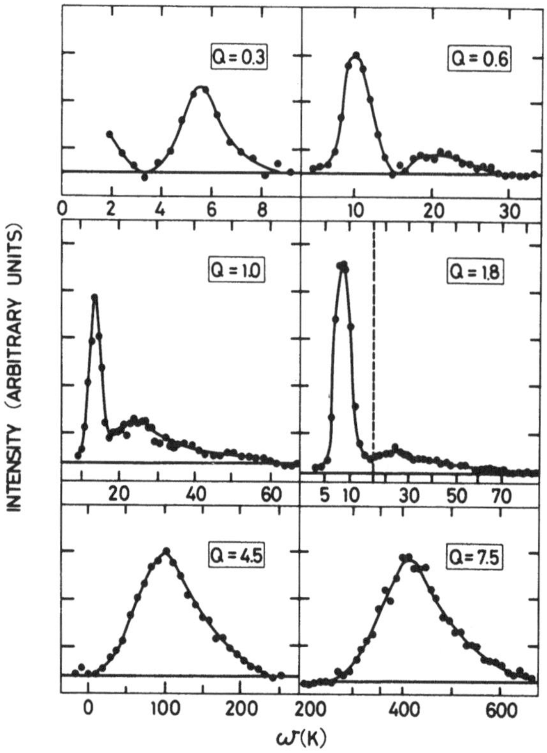


Fig. 1.5 Neutron scattering data for $^4$He at 1.1 K for various values of the scattering vector. Note the abscissa change of scale for Q = 1.8 Å$^{-1}$; after COWLEY and WOODS (1971)

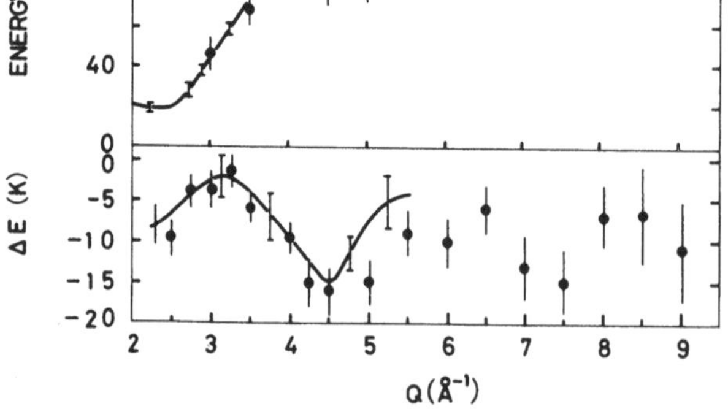

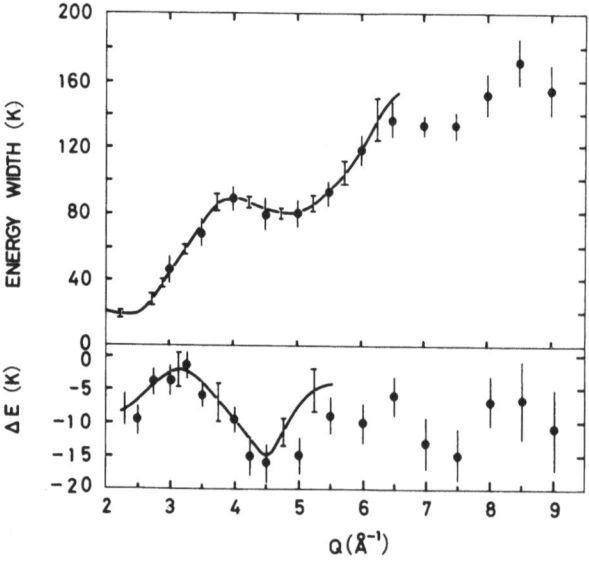

Fig. 1.6 The position of the high frequency peak in Fig.1.5 is shown relative to the free atom result $Q^2/2M$, together with its width; after WOODS and COWLEY (1973)

approximation, be represented by a pair of delta functions at $\pm\omega_0$,

$$\tilde{\chi}''(Q,i\omega) \simeq -C\left[\delta(\omega - \omega_0) - \delta(\omega + \omega_0)\right] \quad . \tag{1.112}$$

The weight C and $\omega_0$ can be chosen by demanding that approximation (1.112) satisfies the relations (1.10) and (1.13). Proceeding in this way, we find for $\omega_0$ the result

$$\omega_0^2 = Q^2/M\chi(Q) \quad , \tag{1.113}$$

and in the limit of small Q it follows from (1.109) that

$$\omega_0 \simeq cQ, \quad Q \to 0 \quad .$$

The approximation (1.113) leads to values of the peak position which are significantly larger than those observed, Fig.1.7. For example, the observed value at the minimum, at $Q \sim 1.9$ Å$^{-1}$, is 0.75 meV, whereas approximation (1.113) gives a value ~1.2 meV (COWLEY and WOODS, 1971) but the approximation (1.113) is reasonable for $Q \leq 0.8$ Å$^{-1}$. Looking back to Fig.1.5 for the neutron distributions we see that there is weight in $S(Q,\omega)$ at high frequencies which is not included in approximation (1.112). For $Q \sim 2.5$ Å$^{-1}$ the two contributions to $S(Q,\omega)$ are, in fact, more or less equal, and increasing Q further results in the total disappearance of the low frequency, collective mode peak. Although the interpretation of these measurements has attracted much attention over the past decade there is no really satisfactory theoretical picture yet. Some of the various methods of approach to the problem were discussed by WOODS and COWLEY (1973).

We turn now to consider briefly some of the conclusions drawn by STIRLING et al. (1976) from their study of $^3$He at 0.63 K. Measurements on $^4$He below the λ-transition show a well-defined collective mode peak for $Q \lesssim 2.5$ Å$^{-1}$, and a mode somewhat

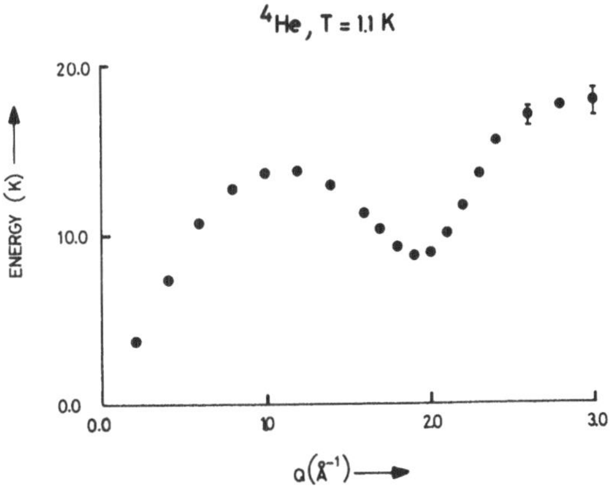


Fig. 1.7 The position of the intense, low frequency peak in Fig.1.9 is shown as a function of scattering vector; after COWLEY and WOODS (1971)

akin to this was predicted for $^3$He by LANDAU (1956). The velocity of propagation of the mode in $^3$He depends on the frequency at which it is driven, a situation similar to what we encountered in the discussion of the structural phase transition in $SrTiO_3$. A low frequency probe excites a mode with a frequency cQ, c being the velocity of ordinary sound, while a high frequency probe excites a mode with a slightly larger velocity, which Laudau called zero sound. The damping of zero sound is small at small Q. AKHIEZER et al. (1962) calculated $S(Q,\omega)$ for $Q \lesssim 0.8\ \text{Å}^{-1}$, and found a large contribution from zero sound, and ALDRICH (1974) predicted a zero sound mode for Q as large as 1.5 $\text{Å}^{-1}$. In the light of this theoretical work, STIRLING et al. made a careful search for a well-defined peak in $S(Q,\omega)$ but found no evidence for it in $^3$He at 0.63 K in the range $0.7 \lesssim Q \lesssim 2.9\ \text{Å}^{-1}$ and $0.4 \lesssim \omega \lesssim 3.0$ meV. Some of the data is shown in Fig.1.8. Substantial inelastic scattering is observed at large Q, with a peak at an energy ~0.8 meV. The inelastic scattering becomes steadily weaker with decreasing Q, and at the lowest wave-vectors no inelastic scattering is observed in the energy range covered by the experiment.

For sufficiently large Q's the scattering function must approach the result for an ideal non-interacting Fermi liquid. At zero temperature, and $Q \geq 2k_f$, where $k_f$ is the Fermi wave-vector, the scattering function for a non-interacting Fermi liquid is non-zero only for frequencies which satisfy ($\tilde{q} = Q/2k_f$ and $\varepsilon_f = k_f^2/2M = Mv_f^2/2$)

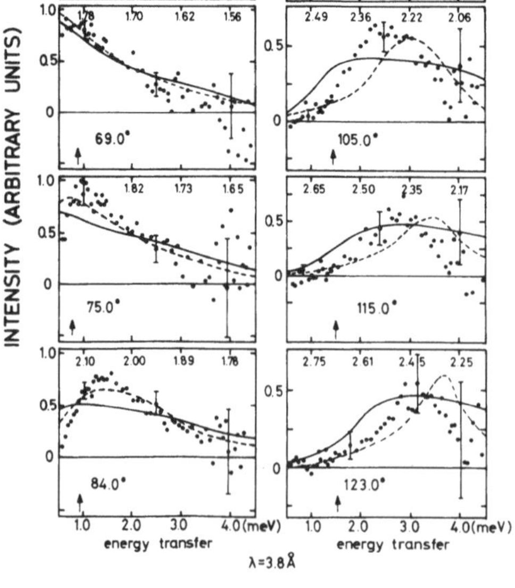

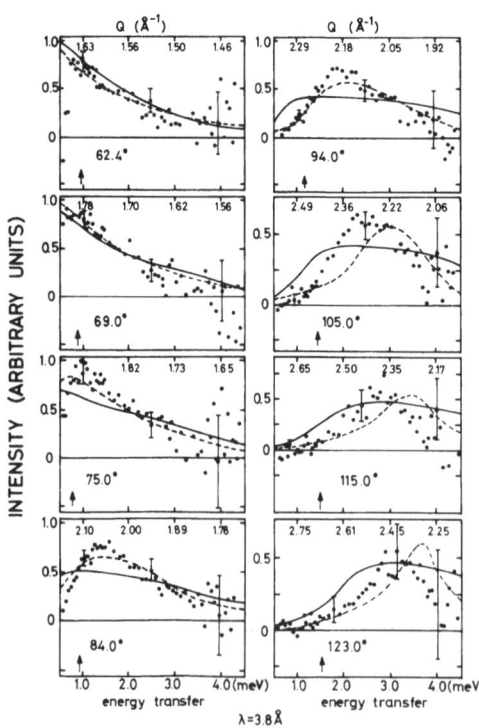

Fig. 1.8 Energy spectra obtained with incident neutrons of wavelength 3.8 Å from liquid $^3$He at 0.63 K. The arrows give the positions of the well-defined excitations in liquid $^4$He and the error bars indicate the typical statistical errors. The scattering vector in $\text{Å}^{-1}$ is marked along the top of each curve. The broken curves denote results calculated at constant scattering angle from a theory due to LOVESEY (1975), and the solid curves are calculated from a modified RPA theory (after STIRLING et al., 1976)

$$4\tilde{q}(\tilde{q} - 1) \le (\omega/\varepsilon_f) \le 4\tilde{q}(\tilde{q} + 1) \quad . \tag{1.114a}$$

Within this range of frequencies, and $\tilde{q} > 1$, PINES and NOZIERES (1966) show that

$$\tilde{\chi}''(Q,i\omega) = -\frac{3m\pi}{4Qk_f}\left[1 - \left(\frac{\omega}{Qv_f} - \frac{Q}{2k_f}\right)^2\right| \quad . \tag{1.114b}$$

The peak in the scattering function is seen to occur at $Q^2/2M$. For a particle density $n = 0.0164$ Å$^{-3}$,

$$k_f = \pi\left(\frac{3n}{\pi}\right)^{1/3} = 0.786 \text{ Å}^{-1}$$

and $\varepsilon_f = 0.428$ meV $\equiv 4.967$ K. With these results we find $Q^2/2M = 2.50$ meV for $Q = 1.9$ Å$^{-1}$. Comparing this result with the data shown in Fig.1.8 we conclude that the effect of particle interactions is significant at the largest wave-vector. The solid curves in Fig.1.8 are results of a calculation by LOVESEY (1975), assuming a two-particle interaction which is approximated by a Lennard-Jones potential. The overall agreement between theory and experiment is satisfactory, and it is noteworthy that the theory does not predict a well-defined collective mode even for Q as small as $k_f$.

If the level activity in the study of liquid $^4$He over the past decade is a guide to future activity in the study of $^3$He by neutron scattering, then we can anticipate a substantial spate of work. The parameters in Landau's quasi-particle theory of $^3$He are sensitive functions of pressure, and this indicates that some interesting features might be observed in pressure experiments. Moreover, there are theoretical predictions for the structure of $^3$He at very low temperatures which have not been investigated thoroughly for largely technical reasons (LEGGETT, 1975; WHEATLEY, 1975). In addition to pressure and temperature effects in pure $^3$He there is a wealth of potentially interesting effects in mixtures of $^3$He and $^4$He (see, e.g., WILKS, 1967).

*Classical Liquids*

The study of the dynamic properties of classical monatomic liquids has received impetus from computer simulation studies of Newtonian particles (typically less than 1000) interacting through pairwise, additive potentials. It can be argued that computer simulation studies have, in fact, set a standard for neutron experiments. Recognising the significance of computer simulation studies in advancing our understanding of classical monatomic liquids leads one to speculate that they will play a vital rôle in developing our understanding of the dynamic properties of multi-component systems.

We consider two approximations to the incoherent scattering function which provide good descriptions of the data and have the added attraction of being easily evaluated. The first result is based on a Gaussian approximation for the wave-vector dependence of $F_s(Q,t)$. We write

$$F_s(Q,t) = \exp\left[-\frac{1}{2}Q^2 w(t)\right] \quad . \tag{1.115}$$

For an ideal, classical gas the form (1.115) is exact and

$$w(t) = t^2/M\beta \quad \text{(ideal gas)}. \tag{1.116}$$

From the definition of $G_s(r,t)$ it follows that the mean square displacement is given by

$$\langle[R(0) - R(t)]^2\rangle = \int dr\, r^2 G_s(r,t) \quad , \tag{1.117}$$

and, using the result (1.115), the right-hand side is simply $3w(t)$. Differentiating the left-hand side of (1.117) twice with respect to time yields the relation

$$2\langle v(0)v(t)\rangle = \frac{\partial^2}{\partial t^2} w(t) \tag{1.118}$$

where $\langle v(0)v(t)\rangle$ is the velocity autocorrelation function. An alternative form for (1.118) is

$$w(t) = 2\int_0^t d\bar{t}\,(t - \bar{t})\,\langle v(0)v(\bar{t})\rangle \quad . \tag{1.119}$$

Since the integral of the velocity correlation function is equal to the diffusion constant D, it follows from (1.119) that in the limit of large times,

$$w(t) \to 2D|t| + \text{constant}, \quad t \to \infty \quad . \tag{1.120}$$

EGELSTAFF and SCHOFILED (1962) noted that it is possible to fabricate $w(t)$ in such a way as to satisfy the short-time and long-time behaviour given by (1.116) and (1.120), respectively, and produce the temporal Fourier transform in terms of a tabulated function. The form suggested was

$$w(t) = 2D\left[(t^2 + c^2)^{1/2} - c\right] \tag{1.121}$$

with $c = M\beta D$. The scattering function can be expressed in terms of a modified Bessel function of the first kind $K_1$,

$$\tilde{S}_s(Q,\omega) = \left(\frac{cQ^2D}{\theta\pi}\right) \exp(cQ^2D)\ K_1(c\theta) \tag{1.122}$$

with

$$\theta^2 = \omega^2 + (Q^2D)^2 \quad .$$

In the limit $c\theta \ll 1$, (1.122) reduces to the result

$$\tilde{S}_s(Q,\omega) = \frac{1}{\pi}\ \frac{Q^2D}{\omega^2 + (Q^2D)^2} \qquad \text{(diffusion)} \tag{1.123}$$

which corresponds to simple diffusion, while in the opposite limit of $c\theta \gg 1$,

$$\tilde{S}_s(Q,\omega) = \left(\frac{c}{2\pi Q^2D}\right)^{1/2} \exp\left(-\frac{c\omega^2}{2Q^2D}\right) \qquad \text{(non-interacting liquid)} \tag{1.124}$$

which coincides for $c = M\beta D$ with the result for a non-interacting classical liquid.

The Egelstaff and Schofield model is compared in Fig.1.9 to computer simulation data for a system of particles interacting through a Lennard-Jones potential

$$u(r) = 4\varepsilon[(\sigma/r)^{12} - (\sigma/r)^6] \quad .$$

The values of the particle density and temperature were chosen so as to simulate liquid argon near its triple point. The agreement between $F_s(Q,t)$ calculated from (1.115), with the choice (1.121) for $w(t)$, and the computer data is satisfactory.

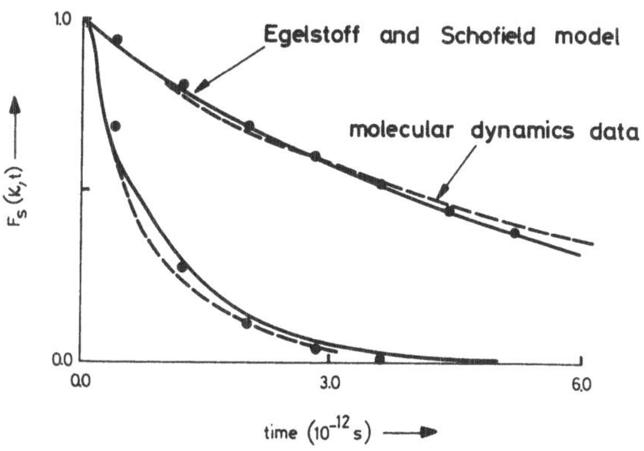


Fig. 1.9 The broken curve is the intermediate scattering function $F_s(Q,t)$, Eq. (1.101), obtained from a computer simulation of a Lennard-Jones liquid representative of liquid argon near its triple point. The points are obtained from the Egelstaff and Schofield model, (1.115) and (1.121) with $c = M\beta D = 0.11 \times 10^{-12}$ s, and the full curve is the temporal Fourier transform of (1.125) evaluated with $\xi = M\beta D\Omega(0) = 0.81$ and $\Omega^2(0) = 0.58 \times 10^{26}\ s^{-2}$. The computer data are from LEVESQUE et al. (1973)

A similar conclusion has been reached by COCKING (1969) from an analysis of incoherent scattering measurements on liquid sodium.

The approximation (1.122) due to Egelstaff and Schofield yields the result (1.58) for the integral against $\omega^4$ with $\Omega^2(0)$ replaced by $3/(M\beta D)^2$. A result due to LOVESEY (1973) represents an improvement in as much that it provides a good description of the experimental data, satisfies (1.58) exactly and retains the advantage of a closed expression for the scattering law. The result is

$$\pi\tilde{S}_s(Q,\omega) = \frac{\tau\omega_0^2(\omega_s^2 - \omega_0^2)}{[\omega\tau(\omega^2 - \omega_s^2)]^2 + (\omega^2 - \omega_0^2)^2} \tag{1.125}$$

where $\omega_s^2$ is given by (1.58), $\omega_0^2 = Q^2/M\beta$ and the relaxation time $\tau$ is given by

$$\tau^{-1} = M\beta D\Omega(0)(\omega_s^2 - \omega_0^2)^{1/2} \quad . \tag{1.126}$$

$F_s(Q,t)$ calculated from (1.125) is shown in Fig.1.9 together with computer simulation data for a system representative of liquid argon near its triple point, and the corresponding quantity evaluated in the approximation suggested by Egelstaff and Schofield. The result (1.125) is compared with neutron data for liquid argon in Fig. 1.10. The comparison between experiment and theory is seen to be good in both cases.

Looking at the measurements of $\tilde{S}_s(Q,\omega)$ for liquid argon shown in Fig.1.10 one is struck by the lack of structure in the scattering function as a function of $\omega$. Varying the wave-vector Q changes $\tilde{S}_s(Q,\omega)$ in a continuous manner; increasing Q decreases the intensity, and increases the width of the distributions. For large Q the difference between the incoherent and coherent scattering functions is mimimal, as is to be expected from physical arguments. Coherent scattering measurements on monatomic liquids show that $\tilde{S}(Q,\omega)$, as a function of $\omega$ for fixed Q, remains structureless down to quite small wave-vectors. There is a pronounced narrowing in the energy spread of $\tilde{S}(Q,\omega)$ when Q coincides with the position of a maximum in S(Q),

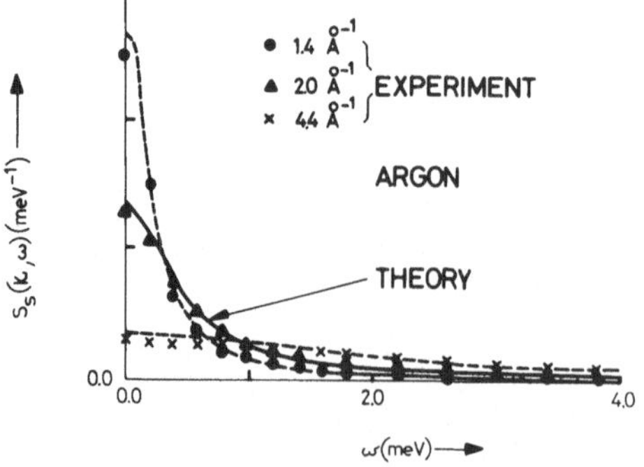


Fig. 1.10 The points represent the unfolded measurements by SKÖLD et al. (1972) of the incoherent scattering function for liquid argon at three different scattering vectors. The continuous curves are obtained from approximation (1.125) evaluated with $\Omega(0) = 1.0 \times 10^{13}$ $s^{-1}$ and $D = 1.94 \times 10^{-5}$ $cm^2\ s^{-1}$

which arises because the mean-square energy transfer is proportional to $Q^2/S(Q)$, for coherent scattering.

Accurate measurements of $\tilde{S}(Q,\omega)$ for small Q demand painstaking care on the part of the experimentalist because of the weak intensity and the accompanying difficulty in making the necessary corrections to the observed data (COPLEY and LOVESEY, 1975). Nonetheless the effort involved can be very rewarding because of the structure in $\tilde{S}(Q,\omega)$ due to the excitation of a collective mode. Measurements by BELL et al. (1973 and 1975) on liquid neon show a side-peak in $\tilde{S}(Q,\omega)$ for wave-vectors in the range $0.06 \leq Q \leq 0.14$ Å$^{-1}$, while COPLEY and ROWE (1974) found that a side-peak persists out to $Q \approx 1.1$ Å$^{-1}$ in liquid rubidium. Some of the neutron data for liquid rubidium is shown in Fig.1.11. The position of the side-peak as a function of Q is shown in Fig.1.12, together with corresponding data from a computer simulation of liquid rubidium by RAHMANN (1974). Both sets of data indicate that the dispersion of the collective mode has a maximum at roughly $Q_0/2$, where $Q_0$(= 1.53 Å$^{-1}$) is the position of the main peak in the structure factor. It is natural to remark on the similarity between the dispersion curve shown in Fig.1.12 and the dispersion of acoustic phonons. While an analogy between the collective mode seen in a liquid with neutron scattering and phonons in a solid might be appealing at first sight it obscures the correct physical interpretation, as we shall now attempt to shown.

A phonon, being associated with sound propagation, is a collective mode which propagates under conditions of local thermodynamic equilibrium, i.e., it is the collective mode which will be excited by a low frequency, long wavelength probe. The response of a liquid to a low frequency, long wavelength probe is described by linearized hydrodynamic equations. We know, from general thermodynamic arguments, that a density fluctuation can be described in terms of fluctuations in two statistically independent variables, such as the pressure and entropy. An analysis of the linearized hydrodynamic equations shows that entropy fluctuations decay with a rate determined by the diffusivity $D_T = \lambda/nC_p$, where $\lambda$ and $C_p$ are the thermal conductivity

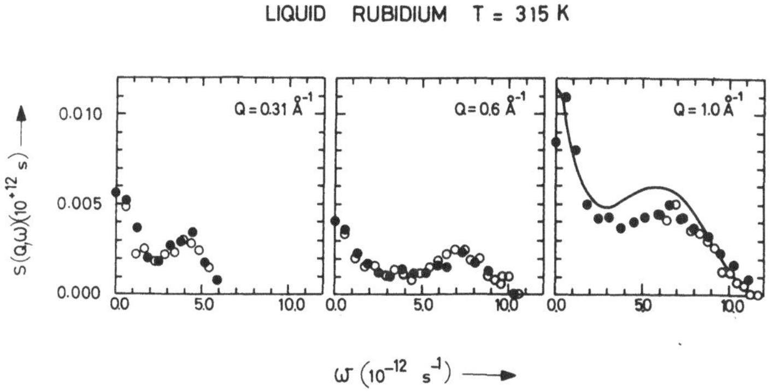


Fig. 1.11 The coherent scattering function of liquid rubidium at 320 K measured by COPLEY and ROWE (1974)

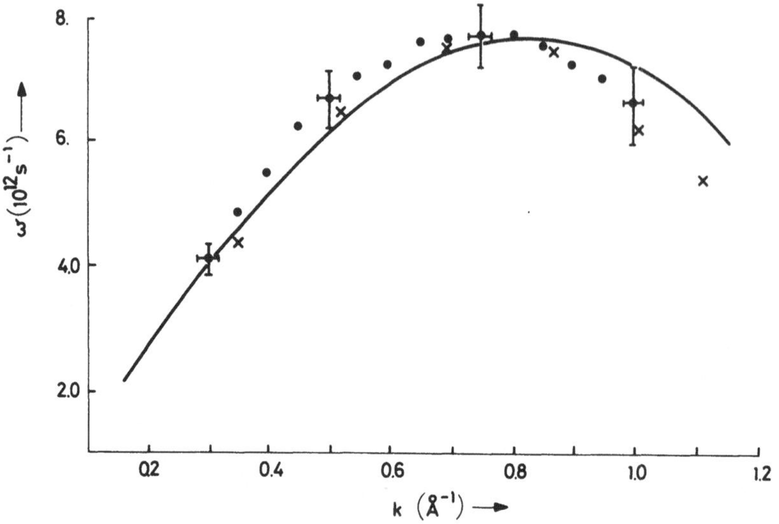


Fig. 1.12 The frequency of the collective excitation in liquid rubidium is shown as a function of wave-vector. The experimental points are due to COPLEY and ROWE (1974). The crosses are from a computer simulation by RAHMAN (1974) and the full curve is calculated from the viscoelastic theory discussed by COPLEY and LOVESEY (1975)

and specific heat constant pressure, respectively. Fluctuations in the pressure decay at a rate determined by the stress tensor and this has a strength measured by the volume and shear viscosities, denoted by $\eta_v$ and $\eta_s$, respectively; the diffusivity for pressure fluctuations is found to be $D_\ell = [(4\eta_s/3) + \eta_v]/Mn$. We note that near the liquid-gas phase transition $D_T \to 0$, because of the divergence of $C_p$. On the other hand, $D_\ell$ is not strongly anomalous at the phase transition.

The actual calculation of the dynamic susceptibility from the linearized hydrodynamic equations was described by MOUNTAIN (1966), i.e.,

$$\tilde{\chi}(Q,i\omega) = -\frac{Q^2}{M}\left\{\omega^2 - (Qv_T)^2 - i\omega Q^2\left[D_\ell + \frac{v_T^2(\gamma - 1)}{\gamma Q^2 D_T + i\omega}\right]\right\}^{-1} . \tag{1.127}$$

Here $v_s$ is the velocity of sound, and $v_T^2 = v_s^2/\gamma$ where $\gamma$ is the ratio of $C_p$ to the specific heat at constant volume. The three poles of the susceptibility (1.127), to order $Q^2$, are

$$\omega = iQ^2 D_T \tag{1.128}$$

and

$$\omega = \pm Qv_s + \frac{iQ^2}{2}\left[D_\ell + D_T(\gamma - 1)\right] \quad .$$

It follows from (1.128) that the scattering function can display a three-peaked structure for sufficiently small values of $D_\ell$, $D_T$ and $\gamma$. The central peak arises from entropy fluctuations and the two side-peaks arise from the decay of a density fluctuation via propagation of sound, or phonons.

Let us turn now to consider the response of a liquid to a high frequency probe. When a mechanical perturbation is applied suddenly to a liquid the liquid must respond elastically at first, just as if it were a solid body. This initial response is determined by the high frequency, or instantaneous, rigidity modulus G which enters the elastic stress tensor. At high frequencies the velocity of the collective oscillation of the atoms will be $(G/Mn)^{1/2}$, as is appropriate for an elastic solid, whereas at low frequencies we have sound propagation with a velocity $v_s$. A more careful argument shows that the velocity of the high frequency mode is larger than, or equal to, $v_s$.

The cross-over from high low frequency regimes is determined essentially by the time $\tau_M = \eta_s/G$, which for simple liquids is typically of order $10^{-12}$ - $10^{-13}$ s. For frequencies $\omega$ such that $\omega\tau_M \ll 1$ there is viscous flow, and the small Q response is described by the dynamic susceptibility (1.127). The limit $\omega\tau_M \gg 1$ corresponds to purely elastic vibrations.

Neutron scattering measurements are usually performed in the viscoelastic régime where $\omega\tau_M \gtrsim 1$. We see that the interpretation of the experiments on neon and rubidium requires a theory of density fluctuations that is valid for what would be termed conventionally as small wave-vectors and intermediate frequencies. The generalization of the hydrodynamic result (1.127) is an approach which suggests itself, and the value of this was discussed by COPLEY and LOVESEY (1975), and in Ch.6 by Mountain.

### 1.3.5 Molecular Gases

The separation of vibrational and rotational energy levels in molecules can be of the same order of magnitude as the energy of slow neutrons. However, in many cases the vibrational states are too high in energy to be excited by thermal neutrons. For example, the 3N - 6 = 9 normal modes of vibration of the spherical top molecule $CH_4$ are energies of 168 meV (threefold degenerate mode), 172 meV (twofold degenerate) 370 meV (singlet) and 391 meV (threefold degenerate), and so the molecule can be assumed to remain in its vibrational ground state. The energy required to excite a molecule out of its normal electron configuration is very large compared with thermal energies. While the electrons do not therefore contribute directly to the neutron cross-section, they are involved insofar as the dynamical states of the nuclei depend on the total symmetry of the molecule, and therefore, on the symmetry of the normal electron configuration. It is found that in the great majority of cases the wave function of the electrons in this state is completely symmetrical, i.e., the wave function is invariant with respect to all the elements of the symmetry group of the molecule. A general discussion of scattering from molecular gases was given by BOUTIN and YIP (1968).

We shall find it convenient here, and in the next section, to introduce the concept of a scattering amplitude operator. For nuclear scattering the operator is

$$\hat{b} = \hat{A} + \frac{1}{2} B \,\hat{\underline{\sigma}} \cdot \hat{\underline{I}} \tag{1.129}$$

where

$$A = [(I+1)\, b^{(+)} + I\, b^{(-)}] \,/\, (2I+1)$$

and

$$B = 2[b^{(+)} - b^{(-)}] \,/ (2I+1) \quad . \tag{1.130}$$

In (1.129), $\hat{\underline{\sigma}}/2$ is the spin of the neutron, and $\hat{\underline{I}}$ that of the nucleus. If the nuclei are randomly oriented and of one kind, averaging $\hat{b}$ is trivial and we have from (1.65a)

$$\sigma_c = 4\pi\, |\bar{\hat{b}}|^2 = 4\pi A^2 \quad .$$

Calculation of the average value of $\hat{b}^2$ is straightforward, and the result is that the incoherent cross-section (1.65b)

$$\sigma_i = \pi B^2 I(I+1) \quad .$$

The calculation of the cross-section for molecular gases entails, of course, the evaluation of the product $\hat{b}_i\hat{b}_j$ averaged over the neutron and nuclear spin orientations. For unpolarized neutrons we find that the average over the neutron spin orientations is equivalent to the replacement

$$\hat{b}_i^+\hat{b}_j \rightarrow A_i^* A_j + \frac{1}{4} B_i^* B_j\, \hat{\underline{I}}_i^+ \cdot \hat{\underline{I}}_j \quad . \tag{1.131}$$

The partial differential cross-section is averaged over the orientations of the molecules and nuclei, and this is denoted by a bar in the following equations. With the result (1.131), the cross-section for unpolarized neutrons is

$$\frac{d^2\sigma}{d\Omega\, dE'} = \frac{k'}{k} \cdot \frac{1}{2\pi} \int_{-\infty}^{\infty} dt \exp(-i\omega t) \sum_{ij} \Big\{ A_i^* A_j\, \bar{Y}_{ij}(\underline{Q},t) + \frac{1}{4} B_i^* B_j\, \overline{\exp \langle[-i\underline{Q} \cdot \hat{\underline{R}}_j(0)]\, \hat{\underline{I}}_i^+ \cdot \hat{\underline{I}}_j \exp[i\underline{Q} \cdot \hat{\underline{R}}_j(t)]\rangle} \Big\} \tag{1.132}$$

with $Y_{ij}$ defined by (1.61).

If we neglect the correlation within a molecule between the allowed nuclear spin states, and vibrational and rotational states of the nuclei which is imposed by symmetry, the nuclear spin operators in the second term of (1.132) can be factored out. If the nuclei are randomly oriented then, clearly,

$$\hat{\underline{I}}_i^+ \cdot \hat{\underline{I}}_j \to \delta_{i,j}\, I(I+1) \tag{1.133}$$

and the calculation of the cross-section (1.132) reduces to the evaluation of the correlation function $Y_{ij}$, with unsymmetrized wave functions, and averaged over the orientations of the molecule.

In calculating the contributions to $Y_{ij}$ from the translational, vibrational and rotational motions of the nuclei it is customary to assume that the motions are uncorrelated. We write the position vector for the $\mu^{th}$ molecule, $\underline{R}_{\ell\mu}$, as

$$\underline{R}_{\ell\mu} = \underline{\ell} + \underline{d}_\mu + \underline{u}_\mu \quad . \tag{1.134}$$

Here $\underline{d}_\mu$ denotes the equilibrium positions of the nuclei relative to the centre of mass of the molecule, and $\underline{u}_\mu$ are their displacements from the equilibrium configuration. The total Hamiltonian for N non-interacting molecules reduces then to the sum N single-molecule Hamiltonians, each of which we take to be

$$\hat{H}_t + \hat{H}_r + \hat{H}_v \quad . \tag{1.135}$$

In (1.135) $\hat{H}_t$ is the Hamiltonian of the translational motion of the centre of mass, and $\hat{H}_r$ and $\hat{H}_v$ are the Hamiltonians that describe the rotational and vibrational motions of the nuclei. Obviously $\hat{\underline{\ell}}$, $\hat{\underline{d}}_\mu$ and $\hat{\underline{u}}_\mu$ do not commute with $\hat{H}_t$, $\hat{H}_r$ and $\hat{H}_v$, respectively. In addition $\hat{\underline{u}}_\mu$ does not commute with $\hat{H}_r$ because the vector $\hat{\underline{u}}_\mu$ rotates with the molecule. With the assumption that translational, vibrational and rotational motions can be separated in the calculation of the correlation function $Y_{ij}$ the latter reduces to the product of correlation functions involving exponential operators of $\hat{\underline{\ell}}$, $\hat{\underline{d}}_\mu$ and $\hat{\underline{u}}_\mu$, and we now consider each of these three correlation functions.

The correlation function describing the translational motion

$$\langle \exp[-i\underline{Q} \cdot \hat{\underline{\ell}}(0)]\, \exp[i\underline{Q} \cdot \hat{\underline{\ell}}(t)] \rangle \tag{1.136}$$

is usually approximated by the result for the free motion of the molecules. In this case (1.136) is replaced by

$$\exp[i\underline{Q} \cdot (\underline{\ell}' - \underline{\ell})] \exp\left[\frac{Q^2}{2\bar{M}} (it\beta - t^2)\right] \tag{1.137}$$

where $\bar{M}$ is the total mass of a molecule.

The vibrational states of a molecule can be calculated within the harmonic approximation by a procedure akin to that sketched in Subsec.1.3.1 for the vibrations of a crystal. If the nuclei are taken to be in their vibrational ground state, the correlation function for the vibrational motion is approximated by (ZEMACH and GLAUBER, 1956)

$$\exp(-Q^2\gamma_{\mu\mu'}) \tag{1.138}$$

where $\gamma_{\mu\mu'}$ is, essentially, a sum of Debye-Waller factors for the $\mu$ and $\mu'$th nucleus.

The calculation of the correlation function describing rotational motion is algebraically cumbersome and we therefore refer the reader to the original papers of GRIFFING (1961), SEARS (1966) and HAMA and NAKAMURA (1971). Various approximation schemes have been proposed, the most widely used of which is the one developed by KRIEGER and NELKIN (1957). Fig.1.13 shows some time-of-flight distributions for methane gas at 20°C for three scattering angles, calculated with the full rotational correlation function and the approximation of Krieger and Nelkin. The approximation is seen to fail badly for incoherent scattering at low angles where is does not reproduce the elastic peak in the distributions, but is satisfactory at the highest scattering angle. The incoherent scattering from methane dominates the coherent scattering because of the presence of the protons.

The calculation of the rotational correlation function used in the calculations discussed in the preceding paragraph was exact within the assumption that the correlation between the nuclear spin and rotational states could be neglected. We should expect the purely quantum mechanical effects of the spin correlation to be negligible at high temperatures and for nuclei with mass >> m. The effect of nuclear spin correlations at low temperatures has been investigated for methane by HAMA and MIYAGI (1973). For this calculation the translational and vibrational contributions to the cross-section were approximated in the manner described above. The rotational correlation function contains the operators $\exp(i\underline{Q} \cdot \hat{\underline{d}})$ and the nuclear spin operators of (1.132). The comparison between the incoherent cross-sections calculated with a complete treatment of the rotational correlation function, and the approximate, uncorrelated calculation shown in Fig.1.14 for a hypothetical gas of methane molecules at 10 K shows that there are significant differences in the rotational peak intensities in the two calculations. However, at room temperature there is minimal difference between the two calculations, even for low scattering angles, as anticipated.

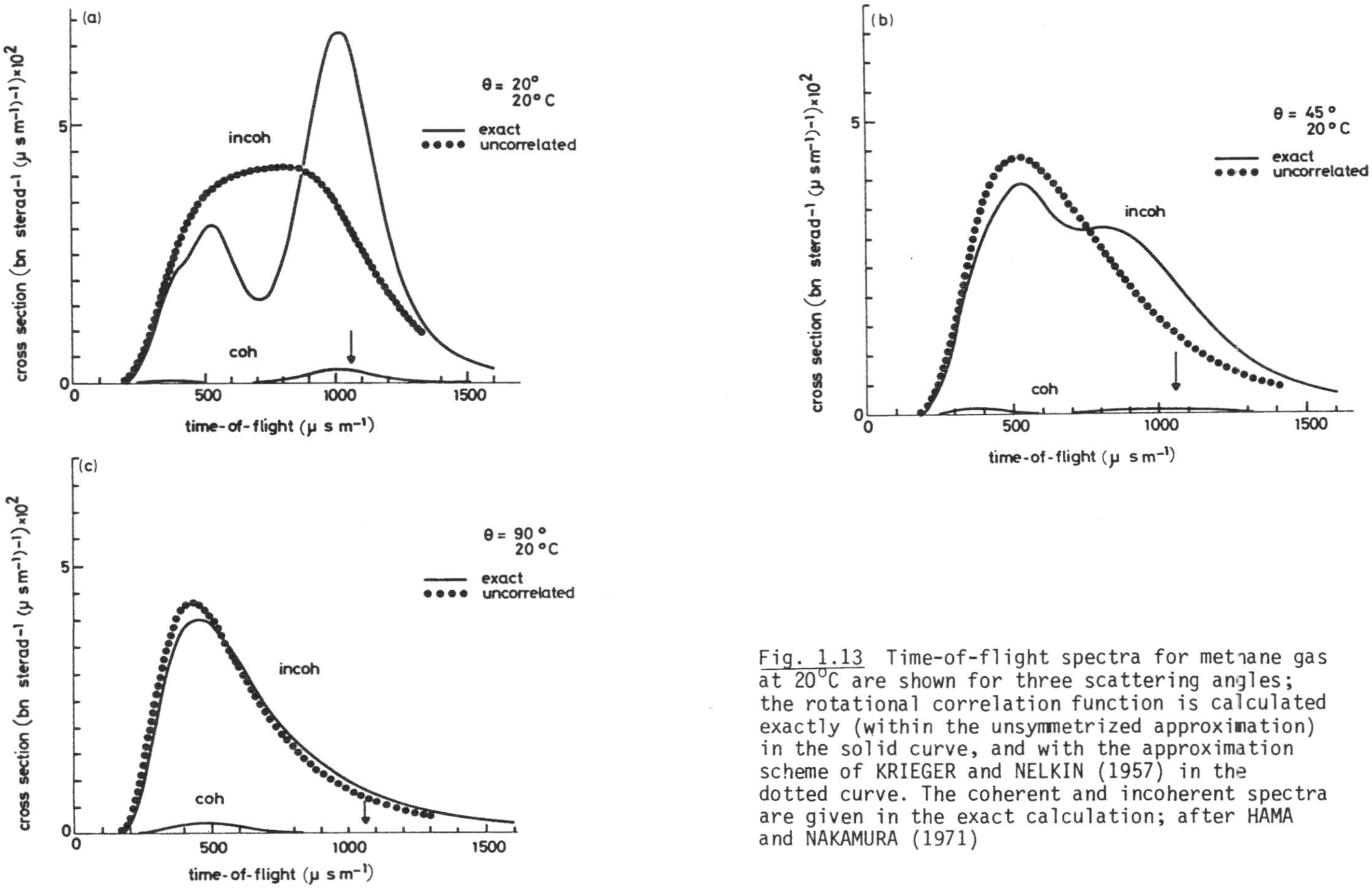


Fig. 1.13 Time-of-flight spectra for methane gas at 20°C are shown for three scattering angles; the rotational correlation function is calculated exactly (within the unsymmetrized approximation) in the solid curve, and with the approximation scheme of KRIEGER and NELKIN (1957) in the dotted curve. The coherent and incoherent spectra are given in the exact calculation; after HAMA and NAKAMURA (1971)

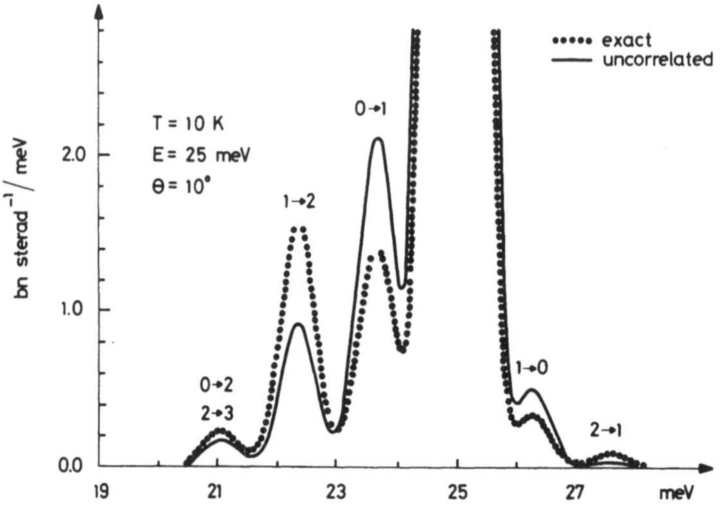


Fig. 1.14 Partial differential cross-section for a hypothetical gas of methane molecules at 10 K; the solid curve is from a calculation using exact, symmetrized rotational correlation functions, and the dotted curve is the unsymmetrized theory used in Fig.1.12; after HAMA and MIYAGI (1973) (see also Fig.5.18)

## 1.4 Polarization Effects

Sometimes the use of polarized neutron beams, and the analysis of the polarization of the scattered beam, can improve the accuracy of measurements, or even provide unique information. Several examples are discussed by MOON et al. (1969). In this section we first generalise the formula for the partial differential neutron cross-section, (1.3), to the case when the incident beam is polarized, and then develop the expression for the polarization of the scattered beam. We assume throughout that the nuclei are randomly oriented. Finite nuclear polarizations can be achieved, for example, in the limit of very low temperatures or by the use of frequency pumping. SCHERMER and BLUME (1968) have evaluated the various contributions to the elastic cross-section, and the polarization of the scattered beam, which arise from finite nuclear polarization.

We define the polarization of a beam of neutrons $\underline{P}$ as twice the average value of the spins in the beam, namely,

$$\underline{P} = 2\langle\hat{\underline{S}}\rangle = \langle\hat{\underline{\sigma}}\rangle \qquad (1.139)$$

where $\hat{\sigma}_\alpha$ are Pauli matrices. For an unpolarized beam $|\underline{P}| = 0$, and for a completely polarized beam $|\underline{P}| = 1$.

A partially polarized beam, i.e., a beam for which $0 < |\underline{P}| < 1$, must be described by a probability distribution. For if the spin state of a neutron is described by a wave function it follows that in some direction in space the neutron has a definite spin value, and is therefore completely polarized. The statistical formalism appropriate for the description of a polarized beam is that of a density matrix. We denote the density matrix for a neutron beam by $\hat{\rho}$. The average value of an operator $\hat{O}$ associated with the beam is

$$\langle\hat{O}\rangle = \mathrm{Tr}\,\{\hat{\rho}\hat{O}\} = \mathrm{Tr}\,\{\hat{O}\hat{\rho}\} \tag{1.140}$$

where Tr{...} denotes the trace operation.

The density matrix for the problem in question is (MARSHALL and LOVESEY, 1971, Ch.10)

$$\hat{\rho} = \frac{1}{2}\,(I + \underline{P}\cdot\hat{\underline{\sigma}}) \tag{1.141}$$

where $I$ is the unit matrix. From the definition (1.141) and the formula (1.140) we find

$$\begin{aligned}\langle\hat{\sigma}_\alpha\rangle &= \mathrm{Tr}\{\hat{\rho}\hat{\sigma}_\alpha\} = \frac{1}{2}\;\mathrm{Tr}\{\hat{\sigma}_\alpha + \sum_\beta\; P_\beta\hat{\sigma}_\alpha\hat{\sigma}_\beta\}\\ &= P_\alpha\end{aligned} \tag{1.142}$$

in agreement with (1.139). The last line of (1.142) follows from the results

$$\begin{aligned}&\mathrm{Tr}\{\hat{\sigma}_\alpha\} = 0,\ \text{for all}\ \alpha\\ \text{and}\quad&\\ &\mathrm{Tr}\{\hat{\sigma}_\alpha\hat{\sigma}_\beta\} = 2\delta_{\alpha,\beta}\quad .\end{aligned} \tag{1.143}$$

The neutron spin dependence of the cross-section (1.3) is all contained in the quantity

$$\begin{aligned}&\sum_{\sigma,\sigma'} p_\sigma\,\langle\sigma|\hat{V}_i^+|\sigma'\rangle\langle\sigma'|\hat{V}_j|\sigma\rangle\\ &\equiv \mathrm{Tr}\{\hat{\rho}\hat{V}_i^+\hat{V}_j\}\end{aligned} \tag{1.144}$$

where the trace operation is taken with respect only to the neutron spin coordinates. Using the result (1.114) in (1.3) for the cross-section, and the rewriting the latter in terms of correlation functions, as we described in Sec.1.2, the key formula for the cross-section generalised to include polarization of the incident beam is

$$\frac{d^2\sigma}{d\Omega\, dE'} = \frac{k'}{k}\left(\frac{m}{2\pi}\right)^2 \frac{1}{2\pi} \int_{-\infty}^{\infty} dt\, \exp(-i\omega t)\, \mathrm{Tr}\Big\{\hat{\rho} \sum_{ij} \langle \hat{V}_i^+(\underline{Q},0)\hat{V}_j(\underline{Q},t)\rangle\Big\} \quad . \tag{1.145}$$

In order to identify new terms in the cross-section (1.145) which arise from the polarization we evaluate the quantity (1.144) with a general form for the interaction operator, namely,

$$\hat{V}_i = \hat{\beta}_i + \hat{\underline{\alpha}}_i \cdot \hat{\underline{\sigma}} \quad . \tag{1.146}$$

For purely nuclear scattering, for example,

$$\hat{\beta} = \exp(i\underline{Q} \cdot \underline{R})A$$

and

$$\hat{\underline{\alpha}} = \frac{1}{2}\exp(i\underline{Q} \cdot \underline{R})B\hat{\underline{I}}$$

where A and B are given by (1.130). The calculation of (1.144) with $\hat{V}$ given by (1.146) is described by MARSHALL and LOVESEY (1971). The result is

$$\begin{aligned}\mathrm{Tr}\{\hat{\rho}\hat{V}_i^+\hat{V}_j\} = {} & \hat{\underline{\alpha}}_i^+ \cdot \hat{\underline{\alpha}}_j + \hat{\beta}_i^+\hat{\beta}_j + \hat{\beta}_i^+(\hat{\underline{\alpha}}_j \cdot \underline{P}) \\ & +(\hat{\underline{\alpha}}_i^+ \cdot \underline{P})\hat{\beta}_j + i\underline{P} \cdot (\hat{\underline{\alpha}}_i^+ \times \hat{\underline{\alpha}}_j) \quad .\end{aligned} \tag{1.147}$$

We see from (1.147) that it is possible to have interference between the two contributions to the scattering operator. The interference terms are zero for purely nuclear scattering, however, because they are linear in the nuclear spin operator $\hat{\underline{I}}$ which has zero average value for randomly oriented nuclei. The last term in (1.147) is also zero for nuclear scattering because if $i \neq j$ each factor is zero, while if $i = j$ we have $\hat{\underline{I}} \times \hat{\underline{I}} = i\hat{\underline{I}}$ which also averages to zero. We conclude therefore, that the cross-section for purely nuclear scattering is independent of the incident polarization. This result is a particular example of a rule which says that the cross-section is independent of $\underline{P}$ if the target systems do not have a preferred direction. Thus, for example, the cross-section for inelastic magnetic scattering from a simple two-sublattice antiferromagnet is independent of $\underline{P}$ for zero applied magnetic field.

Consider now the polarization of the scattered beam $\underline{P}'$. The formula for $\underline{P}'$ must represent the transformation of the spin state of the incident neutron beam, defined by $\underline{P}$, due to the interaction with target system, i.e., we must average the initial spin state of the beam over all possible scattering processes and sum over all possible final states. The result is (MARSHALL and LOVESEY, 1971, Ch. 10)

$$\underline{P}' = \frac{k'}{k}\left(\frac{m}{2\pi}\right)^2 \frac{1}{2\pi}\int_{-\infty}^{\infty} dt\, \exp(-i\omega t)$$

$$* \,\mathrm{Tr}\left\{\hat{\rho}\sum_{ij} \langle \hat{V}_i^+(\underline{Q},0)\hat{\underline{\sigma}}\hat{V}_j(\underline{Q},t)\rangle\right\}\left(\frac{d^2\sigma}{d\Omega\, dE'}\right)^{-1} \tag{1.148}$$

where the partial differential cross-section on the left-hand side is given by (1.145).

The various contributions on the right-hand side of (1.148) are found by evaluating $\mathrm{Tr}\{\hat{\rho}\hat{V}^+\hat{\sigma}\hat{V}\}$ with $\hat{V}$ gvien by (1.146), and we find

$$\begin{aligned} \mathrm{Tr}\left\{\hat{\rho}\hat{V}_i^+\hat{\underline{\sigma}}\hat{V}_j\right\} &= \hat{\beta}_i^+\hat{\underline{\alpha}}_j + \hat{\underline{\alpha}}_i^+\hat{\beta}_j + \hat{\beta}_i^+\hat{\beta}_j\underline{P} + \\ &+ \hat{\underline{\alpha}}_i^+(\hat{\underline{\alpha}}_j \cdot \underline{P}) + (\hat{\underline{\alpha}}_i^+ \cdot \underline{P})\hat{\underline{\alpha}}_j - \underline{P}(\hat{\underline{\alpha}}_i^+ \cdot \hat{\underline{\alpha}}_j) \\ &- i\hat{\underline{\alpha}}_i^+ \times \hat{\underline{\alpha}}_j + i\hat{\beta}_i^+(\hat{\underline{\alpha}}_j \times \underline{P}) + i(\underline{P} \times \hat{\underline{\alpha}}_i^+)\hat{\beta}_j \quad . \end{aligned} \tag{1.149}$$

For the example of purely nuclear scattering, terms linear in $\hat{\underline{\alpha}}$ average to zero for randomly oriented nuclei. We saw also that the vector product $\hat{\underline{\alpha}} \times \hat{\underline{\alpha}}$ is zero when averaged. It can be shown that the fourth and fifth terms on the right-hand side of (1.149) are equal when averaged over the random orientations of the nuclei, and have a value equal to one third of that of the sixth term averaged over nuclear spin orientations. Consequently, the expression (1.149), evaluated for purely nuclear scattering reduces after averaging over nuclear spin orientations to the sum of two terms, each proportional to the incident polarization $\underline{P}$,

$$\mathrm{Tr}\left[\overline{\hat{\rho}\hat{V}_i^+\hat{\underline{\sigma}}\hat{V}_j}\right] \Rightarrow \underline{P}\left[\overline{\hat{\beta}_i^+\hat{\beta}_j} - \frac{1}{3}\,\overline{\hat{\underline{\alpha}}_i^+ \cdot \hat{\underline{\alpha}}_j}\right] \qquad \text{(nuclear scattering)}$$

where the first term on the right-hand side is averaged over the isotope distribution and the second both the isotope distribution and the nuclear spin orientations. For a system containing a single type of atom, the two quantities are

$$\begin{aligned} \langle\overline{\hat{\beta}_i^+\hat{\beta}_j}\rangle &\equiv \overline{A_i^*A_j}\; Y_{ij}(\underline{Q},t) \\ &\equiv \left[|\bar{A}|^2 + \delta_{i,j}(\overline{|A|^2} - |\bar{A}|^2)\right] Y_{ij}(\underline{Q},t) \end{aligned}$$

and

$$\langle\overline{\hat{\underline{\alpha}}_i^+ \cdot \hat{\underline{\alpha}}_j}\rangle = \frac{1}{4}\; \overline{|B|^2 I(I+1)}\; \delta_{i,j} Y_{jj}(\underline{Q},t)$$

where $Y_{ij}(\underline{Q},t)$ is defined by (1.61). The single-atom coherent, and total cross-sections are,

$$\sigma_c = 4\pi \, |\bar{A}|^2$$

and

$$\sigma = 4\pi \left[\overline{|A|^2} + \frac{1}{4} \overline{B^2 I(I+1)}\right]$$

and the incoherent cross-section $\sigma_i = \sigma - \sigma_c$. With these definitions in mind, the preceding results lead to the following expression for the polarization of the scattered beam $\underline{P}'$ expressed in terms of the scattering functions (1.66b) and (1.67b),

$$\underline{P}'\left[\sigma_c S(\underline{Q},\omega) + \sigma_i \, S_s(\underline{Q},\omega)\right] = \underline{P}\left[\sigma_c S(\underline{Q},\omega) + \left(\sigma_i - \frac{4}{3}\sigma + \frac{16\pi}{3}\overline{|A|^2}\right) S_s(\underline{Q},\omega)\right] \quad . \tag{1.150}$$

If there is just one type of isotope, the coefficient of the incoherent scattering function on the right-hand side reduces to $-\sigma_i/3$.

A conventional experiment on a monatomic liquid measures the weighted sum of the coherent and incoherent scattering functions which appears on the left-hand side of (1.150). It is possible to obtain the two scattering laws separately in some cases by using isotopically enriched samples, as was done, for example, by SKÖLD et al. (1972) in their study of liquid argon. The result (1.150) shows that the combination of a conventional measurement of the cross-section, together with a measurement of the change in polarization (as a function of Q and $\omega$) affords a second method of separating the coherent and incoherent scattering functions.

## 1.5 Magnetic Scattering

The formalism required to describe inelastic, magnetic scattering is essentially the same as that used in preceding sections for nuclear scattering. Here we record the expression for the magnetic interaction operator, and discuss the form of the cross-section for spin wave and paramagnetic scattering. Some examples of the type of information that can be obtained using polarized neutrons, or analysis of the polarization of the scattered beam, are also included.

The interaction between a neutron and the magnetic, or unpaired, electrons in a target system is described by the operator

$$-\gamma\mu_N \, \hat{\underline{\sigma}} \cdot \underline{H} \tag{1.151}$$

where $\gamma = -1.91$, $\mu_N$ is the nuclear Bohr magneton and $\underline{H}$ is the magnetic field created by the electrons. There are two types of contribution to $\underline{H}$. First, there is the field which arises from the magnetic moment of an electron, and secondly a term proportional to its velocity. On taking the spatial Fourier transform of (1.151) we obtain the result (see, e.g., MARSHALL and LOVESEY, 1971, Ch.5),

$$\hat{V}(\underline{Q}) = \frac{2\pi}{m}\left(\frac{\gamma e^2}{m_e c^2}\right)\hat{\underline{\sigma}} \cdot \hat{\underline{D}}_\perp(\underline{Q}) \quad . \tag{1.152}$$

The constant

$$\frac{\gamma e^2}{m_e c^2} = -0.54\ 10^{-12}\mathrm{cm}$$

and the operator $\hat{D}_\perp$ is defined by,

$$\hat{\underline{D}}_\perp(\underline{Q}) = \sum_j \exp(i\underline{Q} \cdot \underline{R}_j)\left[\tilde{\underline{Q}} \times (\hat{\underline{s}}_j \times \tilde{\underline{Q}}) - \frac{i}{Q^2}\underline{Q} \times \hat{\underline{p}}_j\right] \quad .$$

In (1.153), $\tilde{\underline{Q}} = \underline{Q}/|\underline{Q}|$, and $\hat{s}_j$ and $\hat{p}_j$ are the spin and linear momentum operators, respectively, of the jth magnetic electron. The first term arises from the electrons magnetic moment, and the second term from its translational motion.

We consider first the form of the matrix elements of $\hat{\underline{D}}_\perp$ for a target system in which the magnetic electrons possess wave functions localised on magnetic ions. In many cases of interest, the total orbital angular momentum of the ions is zero (e.g., the half-filled shell configurations $Mn^{2+}$, $Fe^{3+}$, and $Gd^{3+}$) or quenched by the crystal field. Under these circumstances the matrix elements of the second part of $\hat{\underline{D}}_\perp$ are zero and we can therefore take

$$\hat{\underline{D}}_\perp = \tilde{\underline{Q}} \times (\hat{\underline{D}} \times \tilde{\underline{Q}}) \tag{1.154}$$

with

$$\hat{\underline{D}} = \sum_j \exp(i\underline{Q} \cdot \underline{R}_j)\hat{\underline{s}}_j \quad . \tag{1.155}$$

The sum in (1.155) runs over all sites in the target and all the magnetic electrons associated with the ions. The electrons of each ion couple together to give a ground state with a certain total spin $\hat{\underline{S}}$, and the neutrons will not have sufficient energy to break down this coupling $\hat{\underline{S}}$ being the only vector associated with $\hat{\underline{D}}$, the matrix elements of the latter must be proportional to those of $\hat{\underline{S}}$. We therefore write

$$\langle\lambda'|\hat{\underline{D}}|\lambda\rangle = \sum_{\ell,d} \exp(i\underline{Q} \cdot \underline{R}_{\ell d})F_d(\underline{Q})\ \langle\lambda'|\hat{\underline{S}}_{\ell d}|\lambda\rangle$$

where $F_d(\underline{Q})$ is the form factor defined as the Fourier transform of the normalised spin density associated with the ion on the $d^{th}$ site of the unit cell. By definition, $F_d(0) = 1$.

In the general case a magnetic ion will possess both spin and orbital angular momentum and so it is necessary to include both terms in (1.153) for $\hat{\underline{D}}_\perp$. The calculation of the matrix elements is then no longer a simple task because it is tantamount to an exercise in Racah algebra. Detailed discussions of the attendant algebra were given, for example, by LOVESEY and RIMMER (1969), and MARSHALL and LOVESEY (1971) in their Ch.6. Fortunately, it is usually adequate in interpreting inelastic, magnetic scattering experiments to use an approximation for the matrix in which $\hat{\underline{S}}$ is the actual spin, as in the case of the preceding example, or the total angular momentum $\hat{\underline{J}}$, or some effective spin operator in the case of scattering by ions with partially quenched orbital angular momentum. The approximation leads to the result

$$\hat{V}(\underline{Q}) = \frac{2\pi}{m}\left(\frac{\gamma e^2}{m_e c^2}\right) \sum_{\ell,d} \frac{1}{2} g_d F_d(\underline{Q})\hat{\underline{\sigma}} \cdot \hat{\underline{T}}_{\ell d}(\underline{Q}) \tag{1.156}$$

with

$$\hat{\underline{T}}_{\ell d}(\underline{Q}) = \exp(i\underline{Q} \cdot \underline{R}_{\ell d})\hat{\underline{S}}_\perp(\ell d) \tag{1.157}$$

where $g_d$ is the gyromagnetic ratio of the dth ion in the unit cell, and

$$\hat{\underline{S}}_\perp(\ell d) = \tilde{\underline{Q}} \times (\hat{\underline{S}}_{\ell d} \times \tilde{\underline{Q}}) \tag{1.158}$$

is the component of $\underline{S}_{\ell d}$ which is perpendicular to the scattering vector $\underline{Q}$.

We do not discuss itinerant electron models of magnetism in any detail because to do so would take us too far afield from our main task. A survey of some of the main points is given in Ch.7.

The relative importance of the spin and orbital contributions in scattering from itinerant electron systems has been discussed by LOVESEY and WINDSOR (1971) for a tight binding model of paramagnetic nickel. They found that the spin contribution to the cross-section dominates the orbital contribution for energy transfers less than 50 meV, but for energy transfers exceeding 50 meV the scattering for $q \sim 0$ comes solely from the orbital contribution. In the wave-vector and energy range used in thermal neutron scattering experiments the orbital contribution is however negligible.

The partial differential cross-section for magnetic scattering of polarized neutrons, with approximation (1.156) for the interaction operator, can be written down with the aid of results given in preceding sections. The various contributions to the cross-section are displayed in (1.147). In the present case we have $\hat{\beta} \equiv 0$, and $\hat{\alpha}$ is given by (1.156)

Using the correlation function formalism discussed in Sec.1.2 the cross-section can be written

$$\frac{d^2\sigma}{d\Omega\, dE'} = \frac{k'}{k}\left(\frac{\gamma e^2}{m_e c^2}\right)^2 \sum_{\ell d} \sum_{\ell' d'} \frac{1}{4}\, g_d g_{d'} F_d^*(\underline{Q}) F_{d'}(\underline{Q})$$

$$* \frac{1}{2\pi} \int_{-\infty}^{\infty} dt\, \exp(-i\omega t) \langle \hat{\underline{T}}^+_{\ell d}(\underline{Q},0) \cdot \hat{\underline{T}}_{\ell' d'}(\underline{Q},t) \tag{1.159}$$

$$+ i\underline{P} \cdot \hat{\underline{T}}^+_{\ell d}(\underline{Q},0) \times \hat{\underline{T}}_{\ell' d'}(\underline{Q},t)\rangle$$

where the operator $\hat{\underline{T}}$ is defined by (1.157).

By inspecting (1.149) we see that the polarization of the scattered beam $\underline{P}'$ can be written

$$\underline{P}'\left(\frac{d^2\sigma}{d\Omega\, dE'}\right) = \frac{k'}{k}\left(\frac{\gamma e^2}{m_e c^2}\right)^2 \sum_{\ell d} \sum_{\ell' d'} \frac{1}{4}\, J_d J_{d'} F_d^*(\underline{Q}) F_{d'}(\underline{Q})$$

$$* \frac{1}{2\pi} \int_{-\infty}^{\infty} dt\, \exp(-i\omega t) \langle \hat{\underline{T}}^+_{\ell d}(\underline{Q},0)\underline{P} \cdot \hat{\underline{T}}_{\ell' d'}(\underline{Q},t) + \underline{P} \cdot \hat{\underline{T}}^+_{\ell d}(\underline{Q},0)\hat{\underline{T}}_{\ell' d'}(\underline{Q},t)$$

$$- \underline{P}\, \hat{\underline{T}}^+_{\ell d}(\underline{Q},0) \cdot \hat{\underline{T}}_{\ell' d'}(\underline{Q},t) - i\, \hat{\underline{T}}^+_{\ell d}(\underline{Q},0) \times \underline{T}_{\ell' d'}(\underline{Q},t)\rangle \ . \tag{1.160}$$

The cross-section appearing on the left-hand side of (1.160) is given by (1.159). Note the presence of a term on the right-hand side which is independent of the incident polarization.

The preceding formulae for the cross-section and $\underline{P}'$ simplify when the total spin in the z-direction, say, is a constant of motion. Fortunately, this condition holds, to a good approximation, for many examples of interest. Two examples where the condition is not satisfied are when dipolar forces are important, and when there is hybridization of spin waves and phonons (LOVESEY, 1972).

The simplification of (1.159) and (1.160) comes about because if

$$\left[\hat{S}^z_{tot}\, ,\, \hat{H}\right] = 0 \tag{1.161a}$$

then

$$\langle \hat{S}^{\pm}_{\ell d}(0)\hat{S}^{\pm}_{\ell' d'}(t)\rangle = 0 \tag{1.161b}$$

where $\hat{S}^{\pm}$ are spin angular momentum raising and lowering operators, $\hat{S}^{\pm} = \hat{S}^x \pm i\hat{S}^y$, and

$$\langle \hat{S}^z_{\ell d}(0)\hat{S}^{\alpha}_{\ell' d'}(t)\rangle = 0 \tag{1.161c}$$

for $\alpha$ = x or y. Moreover, the term $\langle\hat{S}^z(o)\hat{S}^z(t)\rangle$ does not lead to single spin wave scattering. With the results (1.161b,1.161c) the cross-section (1.159) for the case of one magnetic ion per unit cell reduces to

$$\frac{d^2\sigma}{d\Omega\, dE'} = \frac{k'}{k}\left(\frac{\gamma e^2}{m_e c^2}\right)^2 \left\{\frac{1}{2}\, JF(Q)\, \exp[-W(Q)]\right\}^2$$

$$* \frac{1}{2\pi}\int_{-\infty}^{\infty} dt\, \exp(-i\omega t)\sum_{\ell\ell'} \exp[i\underline{Q}\cdot(\underline{\ell}' - \underline{\ell})]\left[G^{(+)}\langle\hat{S}_{\ell}^{-}(0)\hat{S}_{\ell'}^{+}(t)\rangle\right. \quad (1.162)$$

$$\left. + G^{(-)}\langle\hat{S}_{\ell}^{+}(0)\hat{S}_{\ell'}^{-}(t)\rangle\right] \quad .$$

The structure factors $G^{(\pm)}$ appearing in (1.162) are defined by

$$G^{(\pm)} = \frac{1}{4}\left[1 + (\tilde{\underline{Q}}\cdot\tilde{\underline{n}})^2 \pm 2(\underline{P}\cdot\tilde{\underline{Q}})(\tilde{\underline{Q}}\cdot\tilde{\underline{n}})\right] \quad (1.163)$$

where $\tilde{\underline{n}}$ is a unit vector in the direction along which the total spin is conserved. If $\underline{Q}$ is perpendicular to $\tilde{\underline{n}}$ then $G^{(\pm)} = 1/4$, independent of the incident polarization.

The cross-section (1.162) is seen to be the sum of two terms, one associated with the correlation function $\langle\hat{S}^-(0)\hat{S}^+(t)\rangle$, and the second with $\langle\hat{S}^+(0)\hat{S}^-(t)\rangle$, and these two correlation functions are related through

$$\langle\hat{S}_{\ell}^{+}(0)\hat{S}_{\ell'}^{-}(t)\rangle = \langle\hat{S}_{\ell'}^{-}(0)\hat{S}_{\ell}^{+}(-t + i\beta)\rangle \quad . \quad (1.164)$$

We now discuss the form of the neutron cross-section for scattering from spin waves in ferro- and antiferromagnetic compounds. The theory of spin waves is reviewed by van KRANENDONK and van VLECK (1958) and KEFFER (1967).

## 1.5.1 Ferromagnetic Spin Waves

The correlation functions in (1.162) are easily evaluated for single spin wave scattering from a Heisenberg magnet described by the Hamiltonian

$$\hat{H} = -\sum_{\ell,\ell'} J(\underline{\ell} - \underline{\ell}')\, \hat{\underline{S}}_{\ell}\cdot\hat{\underline{S}}_{\ell'} - g\mu_B H \sum_{\ell} \hat{S}_{\ell}^{z} \quad . \quad (1.165)$$

The exchange coupling $J(\underline{\ell} - \underline{\ell}')$ is zero for $\ell = \ell'$, and positive for ferromagnetic ordering. For this example, linear spin wave theory leads to the result,

$$\langle\hat{S}_{\ell}^{-}(0)\hat{S}_{\ell'}^{+}(t)\rangle = \frac{2S}{N}\sum_{\underline{q}} \exp[i\underline{q}(\underline{\ell}' - \underline{\ell}) - i\omega_{\underline{q}} t]\, n_{\underline{q}} \quad (1.166)$$

where S is the magnitude of the spin, $n_{\underline{q}}$ is the Bose occupation number, (1.78), for a spin wave with energy,

$$\omega_{\underline{q}} = J\mu_B H + 2S[J(0) - J(\underline{q})] \tag{1.167}$$

where

$$J(\underline{q}) = \sum_{\ell} \exp(-i\underline{q} \cdot \underline{\ell}) J(\underline{\ell}) \quad . \tag{1.168}$$

The result (1.166), together with the identity (1.164), shows that the cross-section for single spin wave scattering is the sum of the two cross-sections

$$\left(\frac{d^2\sigma}{d\Omega\, dE'}\right)^{(\pm)} = \frac{k'}{k}\left(\frac{e^2}{m_e c^2}\right)^2 \left\{\frac{1}{2}\, gF(Q)\, \exp[-W(\underline{Q})]\right\}^2 2S$$
$$* \frac{(2\pi)^3}{v_0} \sum_{\underline{q}\underline{\tau}} (n_{\underline{q}} + \tfrac{1}{2} \pm \tfrac{1}{2})\, \delta(\omega \mp \omega_{\underline{q}})\, \delta(\underline{Q} \mp \underline{q} - \underline{\tau}) G^{(\mp)} \tag{1.169}$$

where the upper sign denotes the cross-section for spin wave annihilation.

The two delta functions in (1.169) arise from conservation of wave-vector and energy, and the restrictions imposed on the scattering process enable the spin wave dispersion to be measured. We refer to the discussion following (1.96) for the one phonon cross-section. If the incident polarization $\underline{P}$ and the scattering vector $\underline{Q}$ are parallel to $\tilde{\underline{n}}$ then the structure factor $G^{(\pm)}$, (1.163), is zero for spin wave creation and unity for spin wave annihilation. On reversing the direction of $\underline{P}$ scattering which involes spin wave annihilation is zero.

The polarization of the scattered beam is readily calculated from (1.160), using the results (1.161b,c), (1.164) and (1.166). The result is

$$\underline{P}'_{(\pm)} = \Big\{-2\tilde{\underline{Q}}(\tilde{\underline{Q}} \cdot \tilde{\underline{n}})[(\tilde{\underline{Q}} \cdot \underline{P})(\tilde{\underline{Q}} \cdot \tilde{\underline{n}}) \mp 1] - \underline{P}[1 - [(\tilde{\underline{Q}} \cdot \tilde{\underline{n}})^2]$$
$$+ 2\underline{n}_\perp \times (\underline{P}_\perp \times \underline{n}_\perp)\Big\} \Big/ G^{(\mp)} \tag{1.170}$$

where

$$\underline{n}_\perp = \tilde{\underline{Q}} \times (\tilde{\underline{n}} \times \tilde{\underline{Q}})$$

is the component of $\tilde{\underline{n}}$ which is perpendicular to $\underline{Q}$. If the incident beam is unpolarized (1.170) simplifies to (MOON et al., 1969)

$$\underline{P}'_{(\pm)} = \frac{\pm 2\tilde{\underline{Q}}(\tilde{\underline{Q}} \cdot \tilde{\underline{n}})}{1 + (\tilde{\underline{Q}} \cdot \tilde{\underline{n}})^2} \quad . \tag{1.171}$$

Thus, the polarization created in the two scattering processes is oppositely aligned; for the creation of a single spin wave the created polarization is parallel to the scattering vector and for the annihilation processes the created polarization is antiparallel to the scattering vector. Complete polarization is achieved if the magnetization of the ferromagnet is aligned parallel to the scattering vector.

### 1.5.2 Antiferromagnetic Spin Waves

A second example of spin wave scattering which is of great interest is that of scattering from a simple, two-sublattice antiferromagnet. In such a magnet the ions are arranged on two identical interpenetrating sublattices, and the exchange coupling between ions on different sublattices is negative. In zero field, the ions on the two sublattices align in opposite directions to minimise the exchange energy. Close inspection of the state in which ions on different sublattices are aligned oppositely reveals that it is not even an eigenstate of the Hamiltonian, and that the quantity $\langle\hat{S}^z\rangle$ - S is never quite zero, even at absolute zero. However, the deviation of the exact ground state from the classical ground state is believed to be small for three-dimensional systems.

Most antiferromagnets possess a single site anisotropy energy of the form $D(\hat{S}^z)^2$ which produces a gap in spin wave dispersion at $\underline{q} = 0$. A magnetic field applied along the z axis creates an imbalance in the energy of the spins on the two sublattices, and at a critical magnetic field, the ions undergo a spin-flop transition to a state in which they lie approximately in a plane perpendicular to the field. However, there is an interesting class of antiferromagnets, called metamagnets, for which the spin-flop phase is absent. On increasing the applied field a metamagnet undergoes a transition, at a critical value of the field, from the antiferro- to a ferromagnet state. One of the best known metamagnets is $FeCl_2$, and the spin wave dispersion in zero field has been measured by BIRGENEAU et al. (1972).

The two branches of the spin wave dispersion in a simple antiferromagnet are

$$\omega_{\underline{q},a} = (-1)^a g\mu_B H + \Sigma_{\underline{q}} \ , \quad \text{with } a = 0, \text{ or } 1 \quad . \tag{1.172}$$

In (1.172)

$$\Sigma_{\underline{q}}^2 = [2SJ(0) + (2S - 1)D]^2 - [2SJ(\underline{q})]^2 \tag{1.173}$$

where $J(\underline{q})$ is the spatial Fourier transform of the intersublattice exchange coupling. For simplicity, we have not included the possibility of intrasublattice exchange coupling. The annihilation and creation cross-sections are the sum of two terms, one for each branch of the spin wave dispersion. We find,

$$\left(\frac{d^2\sigma}{d\Omega\, dE'}\right)^{(\pm)} = \sum_{a=0,1} \left(\frac{d^2\sigma}{d\Omega\, dE'}\right)_a^{(\pm)} \tag{1.174}$$

with

$$\left(\frac{d^2\sigma}{d\Omega\, dE'}\right)_a^{(\pm)} = \frac{k'}{k}\left(\frac{\gamma e^2}{m_e c^2}\right)^2 \left\{\frac{1}{2}\, gF(\underline{Q})\, \exp[-W(\underline{Q})]\right\}^2 2S$$

$$* \; \frac{(2\pi)^3}{v_0} \sum_{\underline{q}\underline{\tau}} (n_{\underline{q},a} + \frac{1}{2} \pm \frac{1}{2})\,\delta(\omega \mp \omega_{\underline{q},a})\,\delta(\underline{Q} \mp \underline{q} - \underline{\tau}) \tag{1.175}$$

$$* \left[u_{\underline{q}}^2 + v_{\underline{q}}^2 + 2u_{\underline{q}}v_{\underline{q}} \cos(\underline{\tau} \cdot \underline{\rho})\right] G_a^{(\mp)} \quad .$$

Here

$$G_a^{(\pm)} = \frac{1}{4}\left[1 + (\tilde{\underline{Q}} \cdot \tilde{\underline{n}})^2 \pm (-1)^a 2(\underline{P} \cdot \tilde{\underline{Q}})(\tilde{\underline{Q}} \cdot \tilde{\underline{n}})\right] \tag{1.176}$$

and the functions $u_{\underline{q}}$ and $v_{\underline{q}}$ are defined by

$$u_{\underline{q}}^2 = [\Sigma_{\underline{q}} + 2SJ(0) + (2S - 1)D]/2\Sigma_{\underline{q}} \tag{1.177a}$$

$$u_{\underline{q}}v_{\underline{q}} = -SJ(\underline{q})/\Sigma_{\underline{q}} \tag{1.177b}$$

with

$$u_{\underline{q}}^2 - v_{\underline{q}}^2 = 1 \quad . \tag{1.177c}$$

The reciprocal lattice vectors $\underline{\tau}$ in (1.175) are those of a sublattice, and $\underline{\rho}$ is the vector joining nearest neigbour magnetic ions on different sublattices.

If $2\underline{\rho} = \underline{m}$, where $\underline{m}$ is some lattice vector of a sublattice, it follows by definition of the reciprocal lattice vectors $\underline{\tau}$ that

$$\underline{\rho} \cdot \underline{\tau} = \frac{1}{2}\underline{m} \cdot \underline{\tau} = \pi \times \text{integer} \quad .$$

Thus

$$\cos(\underline{\rho} \cdot \underline{\tau}) = \pm 1$$

according to whether the integer is even (+) or odd (-).

For instance, consider a b.c.c. lattice. In this case the sublattices are simple cubic lattices, $\underline{\rho} = 0.5(1,1,1)$ and the reciprocal lattice vectors $\underline{\tau}$ are given by

$$\underline{\tau} = \frac{2\pi}{a}(\tau_1,\tau_2,\tau_3)$$

where $\tau_1,\tau_2,\tau_3$ are arbitrary integers. The reciprocal lattice vectors for a b.c.c. structure, however, are given by

$$\underline{\tau}_n = \frac{2\pi}{a}(\tau_1,\tau_2,\tau_3)$$

where $\tau_1 + \tau_2 + \tau_3$ is an even integer. Hence

$$\underline{\rho} \cdot \underline{\tau}_n = \pi \times \text{even integer}$$

and

$$\underline{\rho} \cdot (\underline{\tau}_n + \underline{w}) = \pi \times \text{odd integer}$$

if

$$\underline{w} = \frac{2\pi}{a}(w_1,w_2,w_3)$$

where $w_1 + w_2 + w_3$ is an odd integer. Thus we may write

$$\sum_{\underline{\tau}} \delta(\underline{Q} - \underline{q} - \underline{\tau})\left[u_{\underline{q}}^2 + v_{\underline{q}}^2 + 2u_{\underline{q}}v_{\underline{q}}\cos(\underline{\tau}\cdot\underline{\rho})\right]$$

$$\equiv \sum_{\underline{\tau}_n} (u_{\underline{q}} \pm v_{\underline{q}})^2 \begin{cases} \delta(\underline{Q} - \underline{q} - \underline{\tau}_n) \\ \delta(\underline{Q} - \underline{q} - \underline{\tau}_n - \underline{w}) \end{cases}$$

that is to say, the scattering occurs at all the reciprocal lattice factors; but near the peaks that coincide with the nuclear reflections the intensity of scattering is small $[\sim(u_g + v_g)^2]$ whereas near the *superlattice* reflections the intensity of scattering is large $[\sim(u_q - v_q)^2]$.

Also note that for a b.c.c. lattice the vectors $\underline{R}$ are given by

$$\underline{R} = \frac{1}{2}a\,(R_1,R_2,R_3)$$

where the integers $R_1,R_2,R_3$ are either all even or all odd. Now $\exp(i\underline{w}\cdot\underline{R}) = 1$ when all $R_i$ are even, and equals $-1$ when all $R_i$ are odd, i.e., $\exp(i\underline{w}\cdot\underline{R})$ alternates in sign as R goes from one sublattice to another.

Similarly, a s.c. lattice can be constructed from two sublattices corresponding to the two cases $\underline{m} = a(m_1, m_2, m_3)$ with $m_1 + m_2 + m_3$ an even integer and $\underline{n} = a(n_1, n_2, n_3)$ with $n_1 + n_2 + n_3$ an odd integer. For this lattice

$$\underline{\tau}_n = \frac{2\pi}{a} (\tau_1, \tau_2, \tau_3)$$

and we can choose $\underline{w} = (\pi/a)\ (1,1,1)$.

If no external field is applied to the antiferromagnet we see from (1.172) that the two spin waves branches become degenerate, each having an energy $\Sigma_q$. In this instance, the cross-sections $(d^2\sigma/d\Omega dE')^{(\pm)}$ become independent of the incident polarization. This result could have been anticipated because it is merely an example of the general result that, the cross-section is independent of $\underline{P}$ if the target has no preferred axis. In the present case, the two sublattices are identical in the absence of a magnetic field and as a consequence the two spin wave branches are degenerate.

The polarization of the scattered beam can be shown to be[7] $(H \neq 0)$

$$\underline{P}'_{a(\pm)} = \Big\{-2\tilde{\underline{Q}}(\tilde{\underline{Q}} \cdot \tilde{\underline{n}})[(\tilde{\underline{Q}} \cdot \underline{P})(\tilde{\underline{Q}} \cdot \tilde{\underline{n}}) \mp (-1)^a] - \underline{P}[1 - (\tilde{\underline{Q}} \cdot \tilde{\underline{n}})^2] + 2\underline{n}_\perp \times (\underline{P}_\perp \times \underline{n}_\perp)\Big\}/G_a^{(\mp)} \quad . \tag{1.178}$$

Thus the created polarization for a given scattering process is seen to lie in in opposite directions for the two spin wave branches.

If there is no external field applied to the antiferromagnet, then in place of (1.178) we have the result

$$\underline{P}' = -\underline{P} + \frac{2[\underline{P}_\perp - (\tilde{\underline{n}} \cdot \underline{P}_\perp)\underline{n}_\perp]}{1 + (\tilde{\underline{Q}} \cdot \tilde{\underline{n}})^2} \quad . \tag{1.179}$$

From this we see that there is no created polarization nor any dependence of the final polarization on the scattering process.

### 1.5.3 Paramagnetic Scattering

The paramagnetic state of a magnet in zero field is characterised by the condition $\langle \hat{S}^\alpha_\ell \rangle = 0$, for all $\ell$ and $\alpha = x$, $y$ or $z$, i.e., there is no long-range magnetic order. For a perfect paramagnet, in which there are no interactions and no correlation

---

[7] There is a typographical error in Eq. (10.155) of MARSHALL and LOVESEY, (1971). Comparison of (1.178) with Eq. (10.155) of Marshall and Lovesey shows that a factor 2 multiplies the terms in $G_a(\pm)$ which contains $(-1)^a$.

between magnetic ions, the magnetic cross-section is strictly elastic, spatially isotropic and temperature independent. However, magnetic exchange coupling produces some inelasticity in the scattering process even in the limit of high temperatures when the isothermal susceptibility is a constant, independent of Q and the exchange coupling. This effect is seen in the mean square energy transfer which is proportional to, cf. (1.186),

$$J(0)[J(0) - J(0)]$$

in the limit of infinite temperature. From this result we see that the mean square energy transfer is proportional to the square of the exchange coupling, i.e., it is the same for ferro-, and antiferromagnetically coupled paramagnets. We note also that the mean square energy transfer is a function of the neutron wave-vector.

At temperatures $T \gtrsim T_c$, where $T_c$ is the critical temperature, there can be substantial short range magnetic order which will manifest itself, for example, in a pronounced Q dependence of the isothermal susceptibility. For a ferromagnetically coupled system the susceptibility $\chi_\alpha(Q)$ peaks strongly at $Q = 0$, whereas for an antiferromagnetically coupled system the susceptibility peaks at $\underline{Q} = \underline{w}$, a superlattice vector. In the vicinity of the phase transition at $T_c$ the susceptibility for small wave-vectors is of the form

$$\chi(q) \propto (K^2 + q^2)^{-(1-\frac{1}{2}\eta)} \tag{1.180a}$$

and the inverse correlation length $K$ varies with temperature as

$$K \propto (T - T_c)^\nu \quad . \tag{1.180b}$$

In the case of an antiferromagnet q is measured with respect to a superlattice vector. A molecular field calculation of the susceptibility leads to the result (1.180) with the values $\eta = 0$, and $\nu = 1/2$ for the exponents, whereas experiment and more sophisticated theories indicate that $\eta \sim 0.03 - 0.07$, and $\nu \sim 0.6 - 0.7$ (see, e.g., STANLEY 1971).

The divergence of $\chi(q)$ at magnetic Bragg positions, as T approaches $T_c$, results in a pronounced decrease in the inelasticity, or thermodynamic slowing down of long wavelength spin fluctuations. To see this effect most simply we describe the magnetic scattering in terms of a relaxation function $R^\alpha(\underline{q},t)$ defined in terms of the dynamical variables (cf. Sec. 1.2 and MARSHALL and LOWDE, 1968)

$$\hat{S}^\alpha_{\underline{q}} = \sum_{\ell} \exp(-i\underline{q} \cdot \underline{\ell})\, \hat{S}^\alpha_{\underline{\ell}} \quad . \tag{1.181}$$

With the definition

$$R^{\alpha}(q,t) = \int_0^{\beta} d\lambda \ \langle \hat{S}^{\alpha}_{\underline{q}}(-i\lambda)\hat{S}^{\alpha}_{-\underline{q}}(t)\rangle \tag{1.182}$$

$R^{\alpha}(\underline{q},t = 0)$ coincides with the isothermal susceptibility $\chi_{\alpha}(q)$.

The cross-section for scattering from a paramagnet in zero applied magnetic field is independent of the polarization of the incident neutrons.

From (1.159) we find

$$\frac{d^2\sigma}{d\Omega\, dE'} = \frac{k'}{k}\left(\frac{\gamma e^2}{m_e c^2}\right)^2 \left\{\frac{1}{2}\, gF(Q)\, \exp[-W(\underline{Q})]\right\}^2$$

$$* \sum_{\alpha} (1 - \tilde{Q}^2_{\alpha}) S^{\alpha}(\underline{Q},\omega) \tag{1.183}$$

where the scattering function

$$S^{\alpha}(\underline{Q},\omega) = \frac{1}{2\pi} \int_{-\infty}^{\infty} dt\, \exp(-i\omega t) \sum_{\ell,\ell'} \exp[i\underline{Q}\cdot(\underline{\ell}' - \underline{\ell})]$$

$$* \langle \hat{S}^{\alpha}_{\underline{\ell}}(0)\hat{S}^{\alpha}_{\underline{\ell}'}(t)\rangle$$

$$= \chi_{\alpha}(\underline{Q}) \left[\frac{\omega}{1 - \exp(-\beta\omega)}\right] F_{\alpha}(\underline{Q},\omega) \tag{1.184}$$

and

$$F_{\alpha}(\underline{Q},\omega) = \frac{1}{2\pi} \int_{-\infty}^{\infty} dt\, \exp(-i\omega t)[R^{\alpha}(\underline{Q},t)/\chi_{\alpha}(\underline{Q})] \quad . \tag{1.185}$$

We recall that $F_{\alpha}(\underline{Q},\omega)$ is an even function of $\omega$.

Now, for the Heisenberg magnet described by the Hamiltonian (1.165) with H = 0, it can be shown that

$$\int_{-\infty}^{\infty} d\,\omega \cdot \omega^2\, F_{\alpha}(\underline{Q},\omega) = \frac{4N}{\chi_{\alpha}(\underline{Q})} \sum_{m} J(\underline{m})(1 - \cos \underline{Q}\cdot\underline{m})\langle \hat{S}^z_0 \hat{S}^z_{\underline{m}}\rangle \quad . \tag{1.186}$$

When the exchange coupling is short ranged the numerator on the right-hand side of (1.186) involves near neighbour correlation functions only, and it does not display an anomalous behaviour near $T_c$. The mean square energy transfer for $T \gtrsim T_c$ is

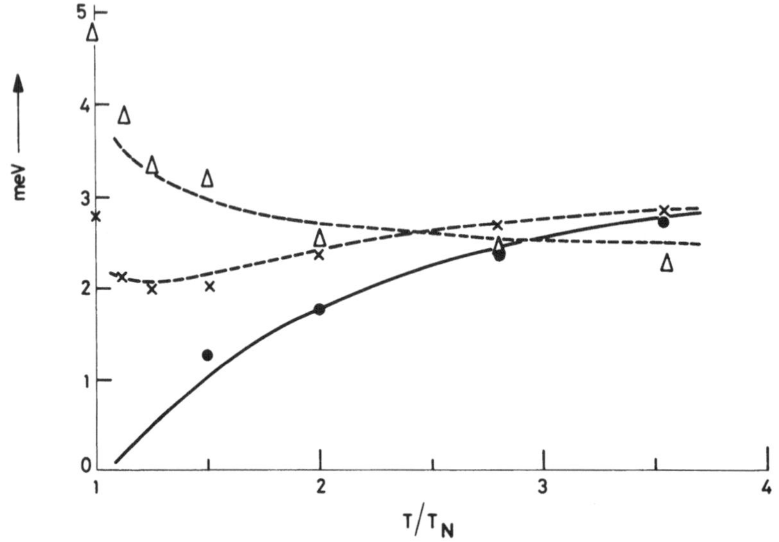


Fig. 1.15 The half-widths at half area of the relaxation shape function $F(Q,\omega)$ Eq. (1.185), measured by TUCCIARONE et al. (1971) for the simple cubic antiferromagnet $RbMnF_3$; Δ (1/4,1/4,1/4), X (3/8,3/8,3/8) and o (1/2,1/2,1/2). The curves are from a calculation by LOVESEY and MESERVE (1973)

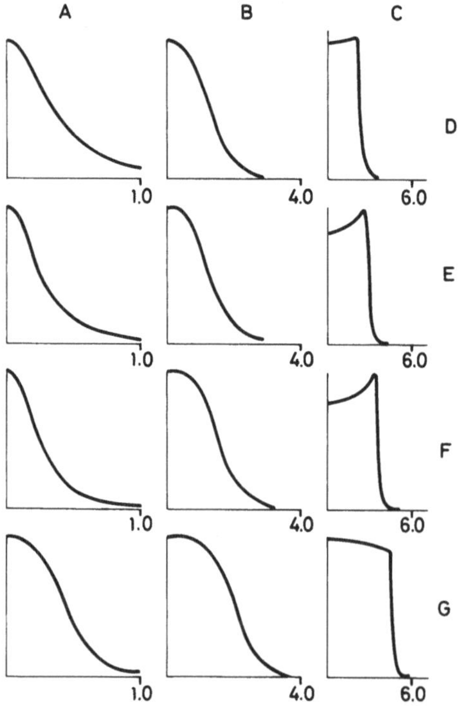


Fig. 1.16 The relaxation shape function $F(Q,\omega)$, Eq. (1.185), calculated by HUBBARD (1971) for a simple cubic ferromagnet. Note that the scales are different for the different graphs: A, q = (1/8,1/8,1/8); B, q = (1/4,1/4,1/4); C, q = (1/2,1/2,1/2); D, $T = \infty$; E, $T = 2.65\ T_c$; F, $T = 1.27\ T_c$; and G, $T = T_c$

therefore determined essentially by the behaviour of $\chi_\alpha(\underline{Q})^{-1}$, and this tends to zero, as $T \to T_c$, for small Q.

The wave-vector dependence of the shape of $F_\alpha(\underline{Q},\omega)$ is characterised by its half-width. In Fig.1.15 we show the half-width at half-area of $F_\alpha(\underline{Q},\omega)$ for the simple cubic antiferromagnet $RbMnF_3$ as a function of temperature for three wave-vectors. The width for $\underline{Q}$ equal to the Bragg vector $\underline{w}$ decreases with temperature, and a careful study shows that it approaches zero as $T \to T_c$. For large values of $\underline{q} = \underline{Q} - \underline{w}$ the widths tend to a finite value at $T_c$. Moreover, the line shape deviates markedly from the Lorentzian shape found for $\underline{Q} \simeq \underline{w}$. The departure of $F_\alpha(\underline{Q},\omega)$ from a Lorentzian shape at large $\underline{q}$ can be understood physically from the following argument. The wave-length of the spin fluctuation is comparable with the distance between neighbouring magnetic ions for large $\underline{q}$, and the temporal evolution of the variable $\hat{S}^\alpha_{\underline{q}}$ is then determined by the correlation between near neighbour ions, which becomes strong as $T \to T_c$. Indeed, the correlation between near neighbour ions might be sufficient to support a collective motion of the spins. This conjecture is supported by calculations of $F_\alpha(\underline{Q},\omega)$ by HUBBARD (1971), for a simple cubic Heisenberg ferromagnet. Fig.1.16 shows some results of Hubbard's calculation for various wave-vectors and temperatures. For small $\underline{q}$ the shape is essentially Lorentzian, but for larger $\underline{q}$ the shape tends to be squarer than Lorentzian, and for the largest $\underline{q}$ the squaring is very pronounced and accompanied by the formation of a weak peak. NATHANS et al. (1968) were the first to observe distinct collective modes above the critical temperature, the sample in their case being $RbMnF_3$.

*Acknowledgements*
Drs. J.R.D. Copley, J. Hayter and J.M. Loveluck kindly read, and criticised, various parts of this chapter. I am grateful to Dr. M. Klein and his colleagues for supplying the data for Fig.1.3. My long time colleague, Dr. Ewald Balcar, perused the final draft, and drew attention to several errors and points of potential confusion; but even with his talented assistance some errors probably remain and these are, like my foibles, my responsiblity only.

## Appendix A

### Crystal Lattices and Reciprocal Lattices

A perfect crystal lattice may be constructed by repeating a unit cell periodically in space. The unit cell is defined in terms of three non-coplanar *basic vectors* $\underline{a}_1$, $\underline{a}_2$ and $\underline{a}_3$ and has a volume $v_0 = \underline{a}_1 \cdot (\underline{a}_2 \times \underline{a}_3)$. In general, the unit cell will contain more than one lattice site. The vector leading from a point in one unit cell to the corresponding point in another cell is called a *lattice vector*. We denote the lattice vectors by $\underline{\ell}$. By definition

$$\underline{\ell} = \ell_1\underline{a}_1 + \ell_2\underline{a}_2 + \ell_3\underline{a}_3 \qquad (A.1)$$

where the integers $\ell_1$, $\ell_2$, $\ell_3$, known as cell indices and frequently denoted for brevity by the single letter $\ell$, take all values.

If the unit cell contains just one lattice site, so that the lattice vectors $\underline{\ell}$ give every lattice site in the crystal ($\underline{R}_\ell = \underline{\ell}$), then we have a Bravais lattice. We note that in a Bravais lattice each lattice point is a centre of symmetry, as is evident from its definition in terms of the lattice vectors $\underline{\ell}$. There are in total fourteen Bravais lattices.[8]

In general, crystal lattices have more than one atom per unit cell and are constructed from several (r) interpenetrating *identical* Bravais lattices (i.e., Bravais lattices with identical basic vectors).

The positions of the r atoms within a unit cell are denoted by the vectors $\underline{d}$, and these can be expressed in terms of the basic vectors $\underline{a}_1$, $\underline{a}_2$ and $\underline{a}_3$ that define the unit cell. The r - 1 non-null vectors $\underline{d}$ are

$$\underline{d} = d_1\underline{a}_1 + d_2\underline{a}_2 + d_3\underline{a}_3 \qquad (A.2)$$

with $0 \le d_i \le 1$, $i = 1,2,3$, the site $\underline{d} = 0$ coinciding with the corner of the unit cell. The position vector $\underline{R}_\ell$ of an atom in the crystal is now given by

$$\underline{R}_{\ell d} = \underline{\ell} + \underline{d} \quad . \qquad (A.3)$$

For every crystal lattice defined by vectors $\underline{\ell}$ as in (A.1), we can also define a reciprocal lattice with vectors $\underline{\tau}$ such that

$$\exp(i\underline{\tau} \cdot \underline{\ell}) = 1 \quad \text{for all } \underline{\ell} \quad . \qquad (A.4)$$

If we write, by analogy with (A.1)

$$\underline{\tau} = t_1\underline{\tau}_1 + t_2\underline{\tau}_2 + t_3\underline{\tau}_3 \qquad (A.5)$$

where the integers $t_1$, $t_2$ and $t_3$ take all possible values, the basic vectors of the reciprocal lattice are given by

---

[8]All the Bravais lattices and their fundamental symmetry properties are given by L.D. LANDAU and E.M. LIFSHITZ: Statistical physics, (Pergamon Press, London 1959) §129

$$\underline{\tau}_1 = \frac{2\pi}{v_0}\,\underline{a}_2 \times \underline{a}_3 \quad , \quad \underline{\tau}_2 = \frac{2\pi}{v_0}\,\underline{a}_3 \times \underline{a}_1 \tag{A.6}$$

and

$$\underline{\tau}_3 = \frac{2\pi}{v_0}\,\underline{a}_1 \times \underline{a}_2$$

as is easily verified.

From (A.6) we notice that the volume of the unit cell of the reciprocal lattice

$$\underline{\tau}_1 \cdot (\underline{\tau}_2 \times \underline{\tau}_3) = (2\pi)^3/v_0 \quad . \tag{A.7}$$

The sum

$$|\sum_{\ell} \exp(i\underline{Q} \cdot \underline{\ell})|^2 \tag{A.8}$$

occurs in the evaluation of the cross-sections for phonon and spin wave scattering. Obviously when the scattering vector $\underline{Q}$ is equal to zero, or indeed any reciprocal lattice vector, the sum is large because all terms add up in phase. However, as $\underline{Q}$ moves away from a reciprocal lattice vector the terms in the sum come out of phase rapidly and the sum has a negligible value. For a large crystal one can therefore write

$$|\sum_{\ell} \exp(i\underline{Q} \cdot \underline{\ell})|^2 \propto \sum_{\underline{\tau}} \delta(\underline{Q} - \underline{\tau}) \quad .$$

The constant of proportionality is determined by integrating both sides over an arbitrary unit cell of the reciprocal lattice. The result is that

$$|\sum_{\ell} \exp(i\underline{Q} \cdot \underline{\ell})|^2 = N\,\frac{(2\pi)^3}{v_0} \sum_{\underline{\tau}} \delta\,(\underline{Q} - \underline{\tau}) \quad . \tag{A.9}$$

## References


Achter,F.K., Meyer,L. (1969): Phys. Rev. 188, 291

Akhiezer,A.I., Akhiezer,I.Ya., Pomeranchuk,I.Ya. (1962) : Soviet Phys. JETP 14, 343

Aldrich,C.H. (1974) : Thesis, University of Illinois, Urbana

Allen,G., Higgins,J.S. (1973) : Rept. Progr. Phys. 36 , 1073

Bacon,G.E. (1975) : *Neutron Diffraction* (Oxford University Press, Oxford)

Balcar,E. (1970) : Acta Phys. Austriaca 31, 300

Barron,T.H.K., Klein,M.L. (1974) : *Dynamical Properties of Solids*, Vol. 1, Chap. 5, ed. by G.K.Horton, A.A.Maradudin (North-Holland, New York)

Bell, H.G., Kollmar,A., Alefeld,B., Springer,T. (1973) : Phys. Lett. 45A, 479

Bell,H.G., Moeller-Wenghoffer,H., Kollmar,A., Stockmeyer,R., Springer,T., Stiller, H. (1975) : Phys. Rev. A11, 316

Birgenau,R.J., Yelon,W.B., Cohen,E., Makovsky,J. (1972) : Phys. Rev. B5, 2607
Boutin,H., Yip,S. (1968) : *Molecular Spectroscopy with Neutrons* (MIT Press, Cambridge)
Brockhouse,B.N., Pope,N.K. (1959) : Phys. Rev. Lett. 3, 259
Cochran,W. (1959) : Phys. Rev. Lett. 3, 412
Cochran,W., Cowley,R.A. (1967) : Phonons in Perfect Crystals. In *Handbuch der Physik*, Vol. XXV/2a, ed. by L.Genzel (Springer, Berlin, Heidelberg, New York)
Cocking,S.J. (1969) : J. Phys. C1, 507
Cooke,J.F., Davis,H.L. (1973) : AIP (Amer. Inst. Phys.) Conf.Proc. 10, 1218
Copley,J.R.D., Lovesey,S.W. (1975) : Rept. Progr. Phys. 38, 461
Copley,J.R.D., Rowe,J.M. (1974a) : Phys. Rev. Lett. 32, 49; (1974b) : Phys. Rev. A9, 1656
Cowley,R.A. (1970) : J. Phys. Soc. Japan Suppl. 28, 239
Cowley,R.A., Woods,A.D.B. (1971) : Can. J. Phys. 49, 177
Dagens,L., Rasott,M., Taylor,R. (1975) : Phys. Rev. B11, 2726
Egelstaff,P.A. (1967) : *An Introduction to the Liquid State* (Pergamon Press, Oxford)
Egelstaff,P.A., Schofield,P. (1962) : Nucl. Sci. Eng. 12, 260
Fermi,E. (1936) : Ric. Sci. 7, 13
de Gennes,P.G. (1959) : Physica 25, 825
de Gennes,P.G. (1963) : *Magnetism III*, ed. by G.T.Rado, H.Suhl (Academic Press, New York)
Glauber,R.J. (1962) : *Lectures in Theoretical Physics*, ed. by Boulder, Vol. IV (Interscience, New York)
Glyde,H.R. (1974) : Can. J. Phys. 52, 2281
Griffing,G.W. (1961) : Phys. Rev. 124,1489
Gurevich,I.I., Tasarov,L.V. (1968) : *Low Energy Neutron Physics* (North-Holland, Amsterdam)
Hama,J., Nakamura,T. (1971) : Progr. Theoret. Phys. 46, 1666
Hama,J., Miyagi,H. (1973) : Progr. Theoret. Phys. 50, 1142
Harada,J., Axe,J.D., Shirane,G. (1971) : Phys. Rev. B4, 155
Hayter,J.B. (1975) : Proceedings of New Methods and Techniques in Neutron Diffraction, Reactor Centrum Nederland (Petten, Holland)
Horton,G.K., Maradudin,A.A. (1974) : *Dynamical Properties of Solids*, Vol.1 (North-Holland, Amsterdam)
Van Hove,L. (1954) : Phys. Rev. 95, 249
Hubbard,J. (1971) : J. Phys. C4, 53
Hutchinson,P., Ross,M., Schofield,P., Conkie,W.R. (1971) : J. Phys. A4, L114
Izyumov,Y.A., Ozerov,R.P. (1970) : *Magnetic Neutron Diffraction* (Plenum Press, New York)
Keffer,F. (1966) : Spin Waves. In *Handbuch der Physik*, Vol. XVIII/2, ed. by H.P.J. Wijn (Springer, Berlin, Heidelberg, New York)
Kohn,W. (1959) : Phys. Rev. Lett. 2, 393
Van Kranendonk,J., Van Vleck,J.H. (1958) : Rev. Mod. Phys. 30, 1
Krieger,T.J., Nelkin,M.S. (1957) : Phys. Rev. 106, 290
Kubo,R. (1957) : J. Phys. Soc. Japan 12, 570
Kubo,R. (1966) : Rept. Progr. Phys. XXIX, 255
Landau,L.D. (1956) : JETP 30, 1058; (1957) : Ibid 32, 59
Landau,L.D., Lifshitz,E.M. (1959) : *Statistical Physics* (Pergamon Press, London)
Leggett,A.J. (1975) : Rev. Mod. Phys. 47, 331
Levesque,D., Verlet,L., Kurkijarvi,J. (1973) : Phys. Rev. A7, 1690
Lomer,W.M. Low,G.G. (1965) : *Thermal Neutron Scattering*, ed. by P.A.Egelstaff (Academic Press, New York)
Lovesey,S.W. (1972) : J. Phys. C5, 2769
Lovesey,S.W. (1973) : J. Phys. C6, 1856
Lovesey,S.W. (1975) : J. Phys. C8, 1649
Lovesey,S.W., Meserve,R.A. (1973) : J. Phys. C6, 79
Lovesey,S.W., Rimmer,D.E. (1969) : Rept. Progr. Phys. 32, 333
Lovesey,S.W., Windsor,C.G. (1971) : Phys. Rev. B4, 3048
Lynn,J.W. (1975) : Phys. Rev. B11, 2624
March,N.H., Tosi,M.P. (1977) : *Atomic Dynamics in Liquids* (Macmillan, London)

Marshall,W., Lovesey,S.W. (1971) : *Theory of Thermal Neutron Scattering* (Clarendon Press, Oxford)
Marshall,W., Lowde,R.D. (1968) : Rept. Progr. Phys. XXXI, 705
Martin,P.C. (1968) : *Many-Body Physics*, ed. by C.De Witt, R.Balian (Ecole d'été de Physique Théorique, Gordon and Breach, New York)
Messiah,A. (1962) : *Quantum Mechanics* (North-Holland, Amsterdam)
Meyer,J., Dolling,G., Scherm,R., Glyde,H.R. (1976) : J. Phys. C
Mook,H.A., Lynn,J.W., Nicklow,R.M. (1973) : Phys. Rev. Lett. 30, 556
Moon,R.M., Riste,T., Koehler,W.C. (1969) : Phys. Rev. 181, 920
Mountain,R.D. (1966) : Rev. Mod. Phys. 38, 205
Mountain, R.D. (1976): Adv. Mol. Relax. Proc. 9, 225
Müller,K.A., Berlinger,M. (1971) : Phys. Rev. Lett. 26, 13
Nikotin,O., Lindgard,P.A., Dietrich,O.W. (1969) : J. Phys. C2, 1168
Peierls,R.E. (1955) : *Quantum Theory of Solids* (Clarendon Press, Oxford)
Pines,D., Nozieres,P. (1966) : *The Theory of Quantum Liquids* (Benjamin, New York)
Rahman,A. (1974) : Phys. Rev. Lett. 32, 52
Ranninger,I. (1975) : J. Phys. (F)5, 1083
Schermer,R.I., Blume,M. (1968) : Phys. Rev. 166, 554
Schofield,P. (1960) : Phys. Rev. Lett. 4, 239
Schofield,P. (1975) : *Specialist Reports - Statistical Mechanics*, Vol. II (The Chemical Society, London)
Scott,J.F. (1974) : Rev. Mod. Phys. 46, 83
Sears,V.F. (1966) : Can. J. Phys. 44, 1299
Shapiro,S.M., Axe,J.D., Shirane,G., Riste,T. (1972) : Phys. Rev. B6, 4332
Shirane,G., Axe,J.D. (1971a) : Phys. Rev.Lett. 27, 1803; (1971b) : Phys. Rev. B4, 2957
Silberglitt,R. (1972) : Solid State Commun. 11, 247
Sköld,K., Rowe,J.M., Ostrowski,G., Randolph,P.D. (1972) : Phys. Rev. A6, 1107
Springer,T. (1972) : Quasielastic neutron scattering for the investigation of diffusive motions in solids and liquids. In *Springer Tracts in Modern Physics*, Vol. 64 (Springer, Berlin, Heidelberg, New York)
Stanley,H.E. (1971) : *Phase Transitions and Critical Phenomena* (OUP, New York)
Stirling,W.G., Scherm,R., Hilton,P.A., Cowley,R.A. (1976) : J. Phys. C9, 1643
Thompson,B.V. (1963) : Phys. Rev. 131, 1420
Tucciarone,A., Corliss,L.M., Hastings,J.M. (1971) : J. Appl. Phys. 42, 1378
Turchin,V.F. (1965) : Slow Neutrons, Israel Program for Sientific Translations
Wheatley,J. (1975) : Rev. Mod. Phys. 47, 415
Wilks,J. (1967) : *Liquid and Solid Helium* (Clarendon Press, Oxford)
Willis,B.T.M., Pryor,A.W. (1975) : *Thermal Vibrations in Crystallography* (Cambridge University Press, Cambridge)
Woods,A.D.B., Cowley,R.A. (1973) : Rept. Progr. Phys. 36, 1135
Zemach,A.C., Glauber,R.J. (1956) : Phys. Rev. 101, 118; 129

# 2. Phonons

H. G. Smith and N. Wakabayashi

With 31 Figures

Phonons play an important role in all phases of solid state physics. Phenomena such as resistivity, thermal conductivity, heat capacity, and superconductivity, for example, are related directly to the phonon density of states. Other properties such as the dielectric function, ultrasonics, the interatomic potentials (through the force constants) are more intimately related to the normal modes of vibration (phonons) of the crystal. These in turn are, of course, ultimately related to the electronic structure of the crystal.

There are many valuable techniques for studying the spectrum of normal modes - the most useful is coherent inelastic scattering of thermal neutrons by the one-phonon process; however, this chapter is not meant to be a review of neutron scattering techniques or a thorough discussion of the lattice dynamics of various types of solids. Space does not permit such a survey here and it is hardly necessary since many excellent review articles are available (HORTON and MARADUDIN, 1974), (VENKATARAMAN, FELDKAMP, and SAHNI, 1975). Rather, the intent is to discuss herein a few recent problems of current interest, which to some extent demonstrate the continuing power and versatility of the triple axis neutron spectrometer in studying the lattice vibrations of complex crystals in greater detail.

The first example is concerned with the phonon spectra and lattice dynamics of ionic crystals of the rutile type, $TiO_2$, $MgF_2$, and $MnF_2$. The crystals are tetragonal with two formula units per primitive cell; hence, there are eighteen branches in the dispersion curves compared to six for the alkali halides. The overall symmetry is lower than the alkali halides which results in many more model parameters to be determined. Group theoretical methods are, therefore, essential to properly identify the branches, and Raman scattering and infrared absorption measurements are extremely valuable in assigning modes at $|\underline{q}| = 0$ in order to arrive at trial models, which in turn are necessary to calculate eigenvectors for mode assignment, particularly where there is much overlapping of the branches.

The second example deals with the phonon spectra of superconductors—metals, alloys, and compounds—and for comparison, related materials which have a low $T_c$ or do not exhibit superconductivity at all. All materials with moderate to high $T_c$ which have been studied by neutron scattering show pronounced anomalies in their dispersion curves. These features, which cannot be explained by any simple models, are believed to be indicative of strong electron-phonon interactions and lattice instabilities.

The third example illustrates the characteristic features of layered compounds which exhibit weak interlayer forces, due to van der Waals interactions, with strong intralayer forces representing covalent bonding. Classic examples are graphite and $MoS_2$, both used extensively as lubricants. Other examples include superconducting $NbSe_2$ and $TaSe_2$.

The fourth and last example illustrates the advantages of using high pressure techniques to study anharmonic effects. Besides observing the inadequacies of the quasi-harmonic model, microscopic mode-Grüneisen parameters and compressibilities are readily determined. It can be particularly informative to study phonon frequency shifts and line shapes at constant volume by varying both the pressure and temperature appropriately. It is then sometimes possible to determine the relative effects of third and fourth order terms in the expansion of the potential energy expression. Two studies are chosen, Ne and Rb, representing two very different types of interatomic potentials, one molecular and the other metallic.

## 2.1 Lattice Dynamics and Neutron Scattering

The field of lattice dynamics has been a very active one in the last twenty years or so and inelastic neutron scattering has played a major role in this development. (A very useful bibliography for the years 1932-1974 has been compiled by LAROSE and VANDERWAL (1974)). The renewed interest in optical spectroscopy, occasioned by the development of laser techniques and the availability of high speed, large capacity computers, has also contributed substantially to the advancements in this field. The significant X-ray scattering experiments of Fe, Cu, and Al by CURIEN, JACOBSEN, and WALKER, respectively, in the mid-fifties confirmed the correctness of the Born-von Kármán theory of lattice dynamics as well as indicating the need for greater-than-nearest-neighbor interactions in even simple monatomic solids. Almost simultaneously BROCKHOUSE was demonstrating the ease and simplicity of determining the phonon dispersion curves of Aℓ, and later Pb, by coherent inelastic neutron scattering, utilizing the triple axis spectrometer and the one-phonon scattering cross section. The technique was readily applied to the alkali halides and the concept of the shell model was successfully tested (and later proved useful in covalent semiconducters as well). The studies on Pb showed that the interatomic interactions were of much longer range and more complex than originally thought, and stimulated thinking in terms of new theoretical models.

2.1.1 General Model

A general introduction to lattice dynamics based on the Born-von Kármán formalism has been given in Ch.1. The solution to the equations of motion expressed in (1.68) and (1.69) can be modified to include more than one atom per unit cell. Written in matrix form, with a slightly different notation, it is

$$\omega^2 \underline{u} = \underline{\underline{D}}\underline{u} \ , \tag{2.1}$$

where

$$D_{\alpha\beta}(dd') = \frac{1}{\sqrt{M_d M_{d'}}} \sum_{\ell'} \phi_{\alpha\beta}(od,\ell'd') \, e^{i\underline{q} \cdot \underline{r}(\ell')} \quad , \tag{2.2}$$

and the $\phi_{\alpha\beta}$ represent the force constans between the atoms. Their exact form will depend on the type of force model used for a particular crystal. The most general is the tensor force model where the force constant tensor between atom (o,d) and ($\ell'$,d') is given by

$$\underline{\underline{\phi}}(od,\ell'd') = \begin{pmatrix} \phi_{xy} & \phi_{xy} & \phi_{xz} \\ \phi_{yx} & \phi_{yy} & \phi_{yz} \\ \phi_{zx} & \phi_{zy} & \phi_{zz} \end{pmatrix} , \tag{2.3}$$

$\phi_{xy}$ = the force constant relating the x component of force on atom (o,d) due to the displacement of atom ($\ell'$,d') in the y direction.

This interaction can be simplified considerably for certain neighbors in simple structures by properly applying the symmetry operations of the crystal, but even then it is necessary to restrict the model to short range interactions in order to limit the number of adjustable parameters. In real crystals forces are often long range and/or complex and, to reduce the number of adjustable parameters, it is usually necessary to make further simplifying assumptions about the nature of the interactions. One assumption commonly made is axially symmetric forces—the interaction involves central forces where the potential is a function only of the distance between the atoms, i.e., where $V(\ell d,\ell'd') = V(r)$

$$r = |\underline{r}(\ell d) - \underline{r}(\ell',d')| \ . \tag{2.4}$$

One obtains

$$\phi_R(\ell d,\ell' d') = \left.\frac{\partial^2 V(r)}{\partial^2 r}\right|_{r=r_0} \quad ,$$

the bond-stretching or radial force constant and

$$\phi_T(\ell k,\ell' k') = \left.\frac{1}{r}\frac{\partial V(r)}{\partial r}\right|_{r=r_0} \quad ,$$

the bond-bending or tangential force constant. Eq.(2.3) is now determined by only two independent parameters. If it is assumed that the atoms are located at the equilibrium positions of V(r) then $\phi_T = 0$ and there is only one parameter per interaction.

### 2.1.2 Rigid-Ion and Shell Models

The dynamical matrix for ionic crystals can be written as a sum of two parts—a Born-von Kármán part, R, restricted to short range interactions and the long range Coulomb interactions, C. Although the Coulomb interactions are of very long range, the form of the potential is explicitly known and the contributions to the dynamical matrix can be readily calculated. This is the Rigid-Ion Model (RIM) and the equations of motion become

$$\omega^2 \underline{u} = (\underline{\underline{R}} + \underline{\underline{Z}}\;\underline{\underline{C}}\;\underline{\underline{Z}})\underline{u} \quad , \tag{2.5}$$

where $\underline{\underline{C}}$ represents the Coulomb coefficients.

Experience has shown that in order to explain some of the dynamical and optical properties of ionic crystals it is necessary to consider the electronic polarizabilities of the ions. The formalism that has been used more extensively than others in neutron scattering investigations is the shell model developed by WOODS, COCHRAN, and BROCKHOUSE (1960) and COWLEY, COCHRAN, BROCKHOUSE, and WOODS (1963).

In the simple shell model each ion is considered to consist of a core and shell connected by a spring and the short range interaction between the ions takes place between the shells only. A more complex shell model is shown in Fig.2.1. The interaction parameters are:

$k_1$ core-to-shell of positive ion
$k_2$ core-to-shell of negative ion
$D$ core of positive ion to core of negative ion
$F_1$ core of positive ion to shell of negative ion

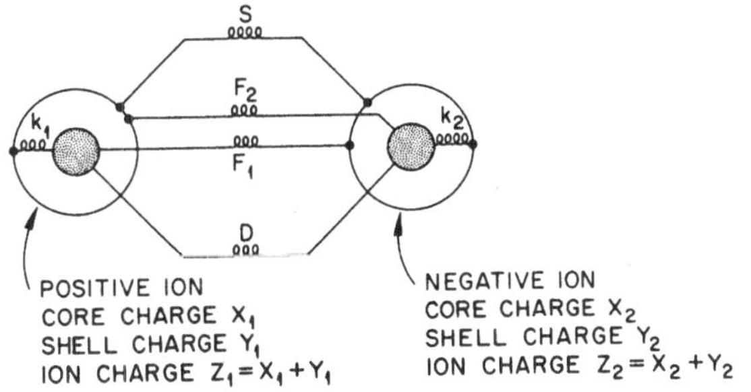


Fig. 2.1 The interactions represented by the parameters of the shell model. Each charge specified in the drawing is in units of e, the electronic charge

$F_2$ core of negative ion to shell of positive ion

S shell of positive ion to shell of negative ion.

To make the dynamical equations symmetrical, which simplifies the analysis, it is assumed that $F_1 = F_2$.

The reader is referred to the original papers for the derivation of the shell model; we merely state here the equations of motion

$$\omega^2 \underline{u} = (\underline{\underline{R}} + \underline{\underline{Z}}\,\underline{\underline{C}}\,\underline{\underline{Z}})\,\underline{u} + (\underline{\underline{T}} + \underline{\underline{Z}}\,\underline{\underline{C}}\,\underline{\underline{Y}})\,\underline{w}, \tag{2.6}$$

$$0 = (\underline{\underline{T}}^{\dagger} + \underline{\underline{Y}}\,\underline{\underline{C}}\,\underline{\underline{Z}})\,\underline{u} + (\underline{\underline{S}} + \underline{\underline{Y}}\,\underline{\underline{C}}\,\underline{\underline{Y}})\,\underline{w}, \tag{2.7}$$

with

$$S_{\alpha\beta}(\underline{q}|dd') = k_d \delta_{dd'} \delta_{\alpha\beta} + S_{\alpha\beta}(\underline{q}|dd'),$$

$$Z_{\alpha\beta}(dd') = Z_d \delta_{dd'} \delta_{\alpha\beta} ,$$

$$Y_{\alpha\beta}(dd') = Y_d \delta_{dd'} \delta_{\alpha\beta} ,$$

where $Y_d$ is the charge of the shell which is bound to the core of the dth atom (ion) by the spring constant $k_d$. $\underline{\underline{R}}$, $\underline{\underline{T}}$, and $\underline{\underline{S}}$ are dynamical matrices corresponding to atom-atom, atom-shell, and shell-shell interactions, respectively. They include only short range forces and can be treated by the models described previously. $\underline{\underline{C}}$ is the matrix of Coulomb coefficients, as before. The electronic dipole moment at an atomic site can be expressed by the relative displacement of the core and shell, $\underline{w}$, as

$$\underline{\rho} = \underline{\underline{Y}}\,\underline{w} .$$

By eliminating $\underline{w}$ from (2.6) and (2.7), the dynamical matrix becomes

$$\underline{\underline{D}} = \underline{\underline{R}} + \underline{\underline{Z}}\,\underline{\underline{C}}\,\underline{\underline{Z}} - (\underline{\underline{T}} + \underline{\underline{Z}}\,\underline{\underline{C}}\,\underline{\underline{Y}})\,(\underline{\underline{S}} + \underline{\underline{Y}}\,\underline{\underline{C}}\,\underline{\underline{Y}})^{-1}\,(\underline{\underline{T}}^{\dagger} + \underline{\underline{Y}}\,\underline{\underline{C}}\,\underline{\underline{Z}}) \; . \tag{2.8}$$

Thus, even when Z = 0, as in the case of Si and Ge, there is a contribution

$$-\underline{\underline{T}}(\underline{\underline{S}} + \underline{\underline{Y}}\,\underline{\underline{C}}\,\underline{\underline{Y}})^{-1}\underline{\underline{T}}^{\dagger} \; ,$$

to the dynamical matrix due to the polarization of the electrons. This term is, of course, long range in behavior. In the simplest form of the shell model, the short range interactions are taken to be due only to shell-shell interactions. In this case, $\underline{\underline{R}} = \underline{\underline{T}} = \underline{\underline{S}}$ and $\underline{\underline{T}}^{\dagger} = \underline{\underline{T}}$.

### 2.1.3 Experimental

The availability of high flux reactors ($10^{14}$-$10^{15}$ n/cm$^2$-s) has permitted neutron inelastic scattering studies of crystals of greater complexity and smaller size (the latter often a determining factor in adequately measuring the phonon and magnon spectra). For determining phonon dispersion curves in high symmetry directions and detailed studies of phonon line shapes, the triple-axis spectrometer generally is preferred. Time-of-flight methods are complementary to the triple axis technique, being particularly useful in studying polycrystalline, amorphous, and liquid substances, and in some cases, phonon and magnons in off-symmetry directions in single crystals. The pros and cons of these and hybrid instruments are discussed in some detail by DOLLING (1974).

The HB-3 triple axis spectrometer at the Oak Ridge High Flux Isotope Reactor (HFIR) is shown schematically in Fig.2.2; it is patterned after the instrument of BROCKHOUSE. The conditions for conservation of crystal momentum and energy for the one-phonon coherent scattering process in single crystals require that

$$\underline{Q} = \underline{k} - \underline{k}' = \underline{\tau} + \underline{q} \; , \tag{2.9}$$

$$^{*}\hbar\omega(\underline{q}) = \hbar^2(k^2 - k'^2)/2m \quad , \tag{2.10}$$

where $\underline{k}$ is the momentum of the incoming neutron and $\underline{k}'$ is the momentum of the scattered neutron, $\hbar\omega(\underline{q})$ is the energy of the phonon with wave vector $\underline{q}$, and $\underline{\tau}$ is a reciprocal lattice vector. The dispersion relation is given by $\omega_j(\underline{q})$ where j denotes

---

*$\hbar = h/2\pi$ (normalized Planck's constant)

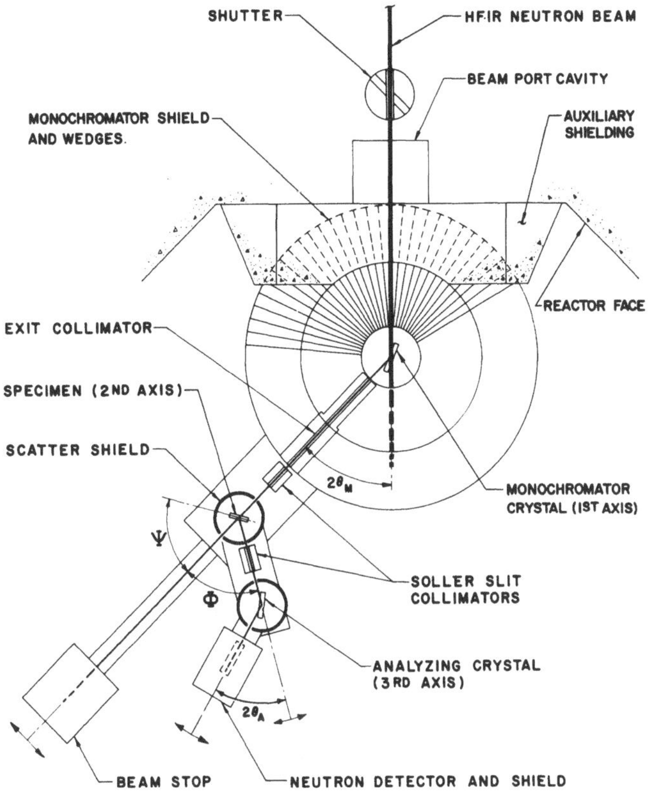


Fig. 2.2 Schematic diagram of HB-3 triple-axis spectrometer

the polarization of the phonons for a given direction in the crystal. These quantities are directly and simply related to the experimentally accessible angle variables in Fig.2.2. Since the momentum vectors of the neutrons are of the same order of magnitude as the phonon wave vectors (in contrast to optical methods) and the energies of the neutrons are the same order of magnitude as the phonon energies (in contrast to X-ray scattering) the changes in these quantities during the scattering process are relatively large and easily determined (1 THz = $10^{12}$ cps = 4.14 meV).

The intensity of coherent one-phonon scattering of thermal neutrons, under the conditions of a "constant-Q" scan over the jth mode, is given in (1.96) and (1.96). The sum is over all the d atoms with coherent scattering amplitude $\bar{b}_d$ in the unit cell with volume $v_0$. The dependence of the neutron scattering intensity on polarization character, wave-vector, and reciprocal lattice vector facilitates both the identification of the different branches of the dispersion relation in a complex crystal structure, and the investigation of individual branches which are in close proximity, without the need for excessively good energy resolution.

Of particular importance, of course, is that, through the structure factor $G_j(\underline{q},\underline{Q})$, the intensities are dependent on the phonon eigenvectors which can be calculated from lattice dynamical models. Thus, the intensity measurement provides a means, at least in principle, of deciding among various lattice dynamical models which are constructed by fitting only to measured phonon frequencies.

There are numerous spurious processes (DOLLING, 1974) that can occur during the scans; however, many of them can be anticipated and avoided or, if unavoidable, they can be identified. Often it is necessary to repeat the scans under different conditions (e.g., change E,$\tau$, or analyzing plane), particularly when the phonon assignments are still uncertain.

## 2.2 Lattice Dynamics of Crystals of the Rutile Type – $TiO_2$, $MgF_2$, and $MnF_2$

Rutile ($TiO_2$) is the structure type for a large number of compounds of composition $MX_2$, where M is a metal and X is an ion, usually F or O. Their physical properties vary widely – from $MgF_2$ and $MnF_2$ being essentially ionic, to $TiO_2$ with mixed ionic-covalent character, to $VO_2$ and $NbO_2$, which exhibit metallic properties in their high temperature rutile structure, but become insulators or semiconductors at lower temperatures, when they transform to a distorted rutile structure. $RuO_2$ retains the rutile structure and metallic behavior down to He temperatures. Some of the compounds have interesting magnetic properties as well, e.g., $MnF_2$, $FeF_2$, and $CoF_2$. $TiO_2$ itself is polymorphic, also having the anatase and brookite structures, with some doubt as to which is the most stable phase thermodynamically.

Rutile and several other compounds of this type have been studied extensively by many physical techniques and, therefore, much is known about their ultrasonic, specific heat, optical (infrared and Raman), and dielectric properties. For example, the static dielectric constants of $TiO_2$ in the $c_0$ and $a_0$ directions are unusually large and anisotropic ($\varepsilon_0$ = 170 and 86, respectively) and are temperature dependent, characteristic of a material with ferroelectric properties.

All of these properties, including chemical bonding, are intimately related to the lattice dynamics of the crystals. However, because of the increased complexity of the phonon dispersion curves in this structure and the uncertainty in identifying which modes the observed phonons belong to, other physical and chemical information can be extremely valuable. The thorough Raman and infrared studies of the three compounds discussed here were particularly valuable in the proper assignment of many of the optic modes.

### 2.2.1 $TiO_2$

Much of this section deals with the symmetry properties of the rutile structure and, therefore, will also be applicable to the following subsections which are concerned with $MnF_2$ and $MgF_2$. The structure is primitive tetragonal (Space Group $D_{4h}^{14}$-$P4_2/mnm$) with six atoms per unit cell, as shown in Fig.2.3. The two titanium ions occupy positions (000) and (1/2,1/2,1/2) and the four oxygen ions occupy positions (uuo),

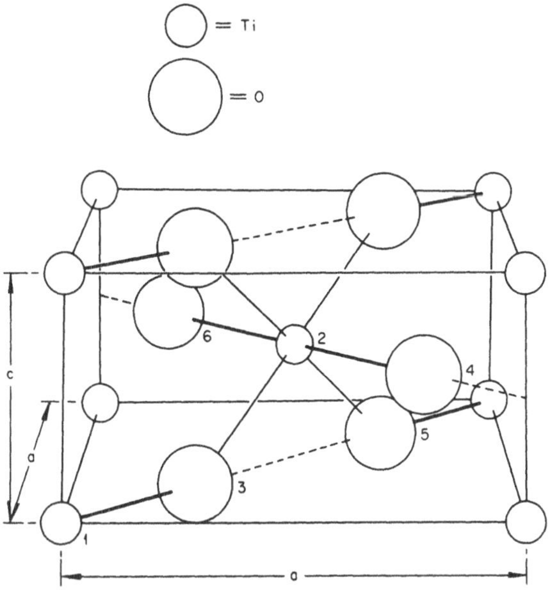


Fig. 2.3 Rutile structure. (TRAYLOR et al., 1971)

(1-u, 1-u, 0), (1/2+u, 1/2-u, 1/2), and (1/2-u, 1/2+u, 1/2), where u = 0.3056. The Ti atoms are located at the centers of distorted oxygen octahedra. There are two near-neighbor Ti-O distances, 1.983 and 1.946 A, and three near-neighbor O-O distances within the octahedron, 2.525, 2.789, and 2.958 A. The shortest Ti-Ti interaction is 2.958 A in the c-direction.

The reader is referred to the paper (TRAYLOR et al., 1971) and thesis (TRAYLOR, 1971) on $TiO_2$ for the detailed description of its lattice dynamics; only some results and conclusions will be mentioned here. Because of the lower symmetry in rutile compared to, say, the alkali halides, the short range Born-von Kármán force constant matrix of (2.3) does not simplify significantly for some of the interactions (e.g., the Ti(2)-O(3) and the O(3)-O(4) interactions) unless some assumptions are made about the nature of the interaction, such as axially symmetric forces. The application of translational and rotational invariance conditions can also impose important constraints on the force constant relations.

In the studies described here the deviations from the simple shell model (T = S = R) did not seem to improve the agreement between calculated and observed frequencies, even though additional parameters were required. Therefore, the dispersion curves presented for $TiO_2$, $MnF_2$, and $MgF_2$ are those based on a simple shell model. The short range forces in $MnF_2$ and $MgF_2$ were limited to axially symmetric forces. The parameters used in the tensor force shell model for $TiO_2$ are shown in Table 2.1. It should be noted that in this model axially symmetric forces were assumed for the O(3)-O(4) type interactions since this reduced the number of parameters for this interaction from nine to two. It was also found that the Ti-Ti interaction

Table 2.1 Independent parameters for the models used to describe the lattice dynamics of rutile

| Parameter | Type | Description | Atom types (dd') |
|---|---|---|---|
| P(1) | $\phi_{xx}$ | Ti-Ti, c-axis | 11, 22 |
| P(2) | $\phi_{xy}$ | Ti-Ti, c-axis | 11, 22 |
| P(3) | $\phi_{zz}$ | Ti-Ti, c-axis | 11, 22 |
| P(4) | $\phi_{xx}$ | Ti-O, same xy plane | 13, 15, 24, 26 |
| P(5) | $\phi_{xy}$ | Ti-O, same xy plane | 13, 15, 24, 26 |
| P(6) | $\phi_{zz}$ | Ti-O, same xy plane | 13, 15, 24, 26 |
| P(7) | $\phi_{xx}$ | Ti-O, out of xy plane of Ti | 14, 16, 23, 25 |
| P(8) | $\phi_{xy}$ | Ti-O, out of xy plane of Ti | 14, 16, 23, 25 |
| P(9) | $\phi_{xz};\phi_{yz}$ | Ti-O, out of xy plane of Ti | 14, 16, 23, 25 |
| P(10) | $\phi_{zx};\phi_{zy}$ | Ti-O, out of xy plane of Ti | 14, 16, 23, 25 |
| P(11) | $\phi_{zz}$ | Ti-O, out of xy plane of Ti | 14, 16, 23, 25 |
| P(12) | $\phi_{r}$ | O-O, out of same xy plane | 34, 36, 45, 56 |
| P(13) | $\phi_{t}$ | O-O, out of same xy plane | 34, 36, 45, 56 |
| P(14) | $\phi_{xx}$ | O-O, same xy plane | 35, 46 |
| P(15) | $\phi_{xy}$ | O-O, same xy plane | 35, 46 |
| P(16) | $\phi_{zz}$ | O-O, same xy plane | 35, 46 |
| P(17) | Z(Ti) | (Charge of Ti ion)/e | |
| P(18)[a] | Z(O) | (Charge of O ion)/e | |
| P(19) | $\alpha$(Ti) | Electonic polarizability of Ti | |
| P(20) | $\alpha$(O) | Electronic polarizability of O | |
| P(21) | d(Ti)/e | (Mechanical polarizability of Ti)/e | |
| P(22) | d(O)/e | (Mechanical polarizability of O)/e | |

[a]P(18) is not independent, but is constrained to be -P(17)/2

had little contribution to the dynamical matrix and was ignored in the latter stages of refinement. Metal-metal interactions were also ignored in the studies on the other two compounds.

The dynamical matrices for rutile are 18 x 18 complex matrices; however, along the principal symmetry directions of the Brillouin zone the 18 x 18 matrices can be factorized into block matrices. The Brillouin zone for rutile is shown in Fig.2.4 and the symmetry decompositions are given in Table 2.2. Various methods of reducing a representation are discussed and demonstrated by numerous authors (the method is

Table 2.2 The irreducible representations for selected points in the Brillouin zone for the rutile structure

| Point in the Brillouin zone | $\underline{q}$ | Irreducible representations |
|---|---|---|
| Γ | (0,0,0) | ${}^1\Gamma_1^+ + {}^1\Gamma_2^+ + {}^1\Gamma_3^+ + {}^1\Gamma_4^+ + 4\,{}^2\Gamma_5^+ + 2\,{}^1\Gamma_1^- + 2\,{}^1\Gamma_4^- + {}^2\Gamma_5^-$ |
| Δ | (ζ,0,0) | $6\,{}^1\Delta_1 + 3\,{}^1\Delta_2 + 6\,{}^1\Delta_3 + 3\,{}^1\Delta_4$ |
| Σ | (ζ,ζ,0) | $6\,{}^1\Sigma_1 + 5\,{}^1\Sigma_2 + 6\,{}^1\Sigma_3 + {}^1\Sigma_4$ |
| Λ | (0,0,ζ) | $3\,{}^1\Lambda_1 + {}^1\Lambda_2 + {}^1\Lambda_3 + 3\,{}^1\Lambda_4 + 5\,{}^2\Lambda_5$ |
| X | (π/a,0,0) | $6\,{}^2X_1 + 3\,{}^2X_2$ |
| M | (π/a,π/a,0) | ${}^2M_{1,2} + {}^2M_{3,4} + 2\,{}^2M_{5,6} + 4\,{}^2M_9^+ + {}^2M_9^-$ |
| Z | (0,0,π/c) | $3\,{}^2Z_1 + 2\,{}^2Z_2 + {}^2Z_3 + 3\,{}^2Z_4$ |

demonstrated for a particular direction in rutile in the thesis of TRAYLOR). The symmetry of the optic modes at zero wave-vector are shown in Fig.2.5. Each mode is labeled with the Mulliken notation and with the Koster notation; a "(2)" before the Mulliken label indicates that the symmetry of the lattice requires there to be two degenerate modes with symmetry similar to that of the drawing. A consideration of the atomic displacements and the geometrical structure factors of the various reciprocal lattice points, along with the infrared and Raman identification of the "active" modes, is extremely valuable during the early stages of a neutron scattering investigation. Even though the rigid ion model is not expected to give

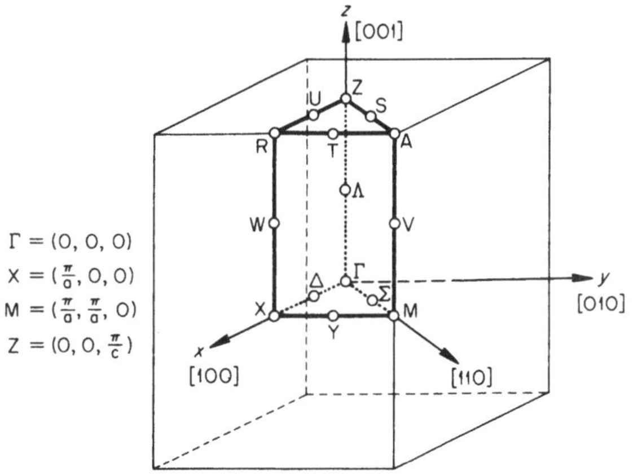


Fig. 2.4 Brillouin zone for the primitive tetragonal lattice. (TRAYLOR et al., 1971)

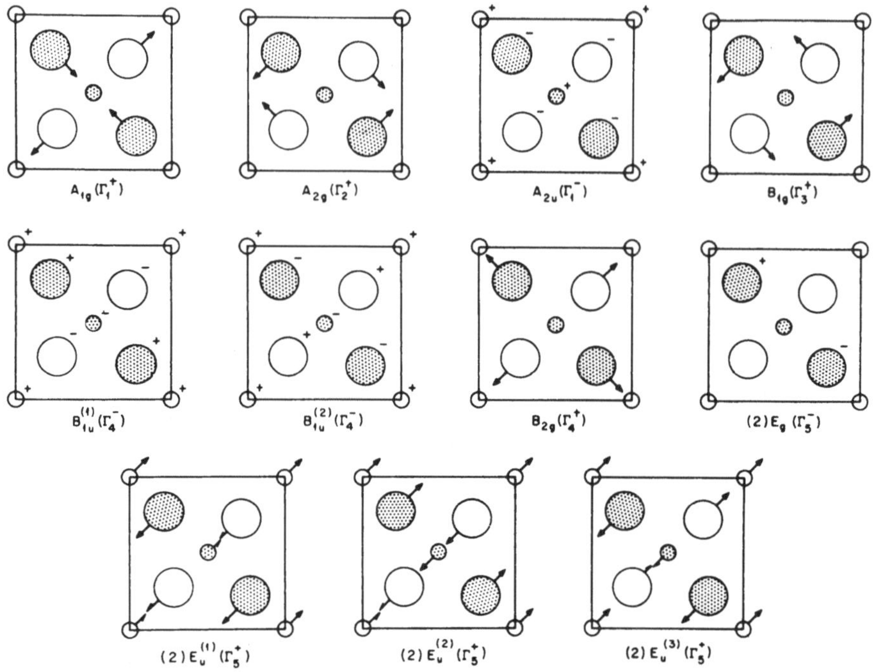


Fig. 2.5 Symmetry of the optic modes of the rutile structure for zero wave-vector. Labeling is in the notation of MULLIKEN and in parentheses, of KOSTER. Doubly degenerate modes are indicated by a "(2)" preceding the MULLIKEN symbol. (TRAYLOR et al., 1971)

a satisfactory fit to the dispersion curves it is usually a good starting point for the calculation of eigenvectors and most of the $|\underline{q}| = 0$ optic frequencies.

For rutile 980 phonon groups were observed of which about 350 were associated with independent wave-vectors. It was necessary to remeasure many of the phonon groups near various lattice points for comparison with the model calculations of the eigenvectors to ensure proper assignment of the branches. Some branches were not found in the experiment because of having small structure factors or being unresolved from branches with very large structure factors. Particular attention was given to the problem of spurious peaks.

A comparison of the experimentally determined frequencies with the calculated dispersion curves based on the shell model with a modified tensor force model is shown in Fig.2.6. Only the lower two thirds of the branches are shown as the upper branches were not completely determined, although most of the modes at $|\underline{q}| = 0$ were observed. These latter frequencies are listed in Table 2.3 and compared with the infrared and Raman determinations.

The rigid ion model and shell model based on axially symmetric forces were rejected because they give incorrect degeneracies at the zone boundaries.

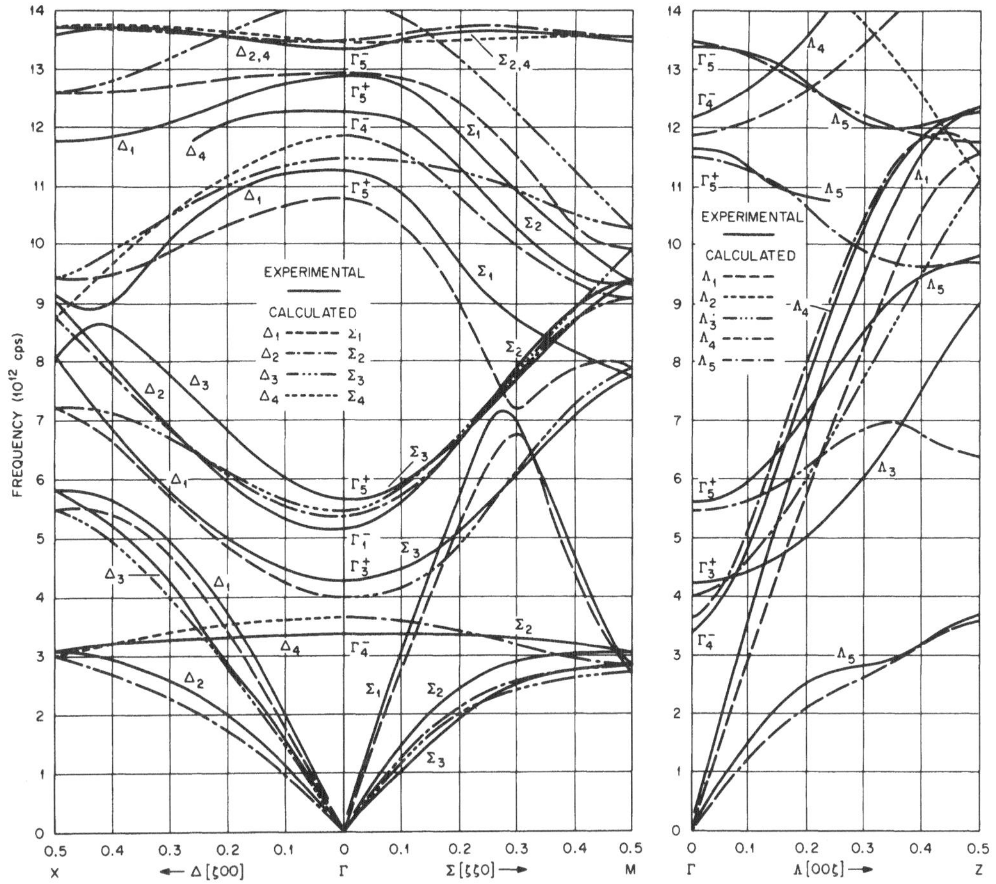


Fig. 2.6 Comparison of the dispersion relation calculated by SM VII and that measured in the neutron experiment (low frequencies). Solid lines are the measured curves and are labeled according to the irreducible representation of the branch. (TRAYLOR et al., 1971)

The chemical bonding in $TiO_2$ is not considered to be representative of an ideal ionic crystal and the Ti-O bond is sometimes described as having covalent character in a Pauling-type description of chemical bonds. Whether the necessity of a tensor force model is a reflection of this bonding, or the neglect of rotational invariance, or elastic constraints can be decided only by further calculations. Additional neutron measurements may be useful but are not essential.

For cubic crystals with two ions per unit cell the Lyddane-Sachs-Teller relation is given by

$$\frac{\varepsilon_0}{\varepsilon_\infty} = \left(\frac{\omega_L^2}{\omega_T^2}\right)_{|\underline{q}|=0} \quad . \tag{2.11}$$

Table 2.3 Optic Phonon Frequencies in Ruile at $\Gamma$

| Koster notation[a] | Mulliken notation | Optical determination[b] ($10^{12}$ cps) | Neutron determination[c] ($10^{12}$ cps) |
|---|---|---|---|
| Optic modes: | | | |
| $\Gamma_4^+$ | $B_{2g}$ | R: 24.78 | 24.72 |
| $\Gamma_5^+$ | $E_u$(LO) | I: 24.18 | 25.24 |
| $\Gamma_1^-$ | $A_{2u}$(LO)[d] | I: 24.33 | NF |
| $\Gamma_2^+$ | $A_{2g}$ | NA | NF |
| $\Gamma_1^+$ | $A_{1g}$ | R: 18.36 | 18.30 |
| $\Gamma_5^+$ | $E_u$(TO) | I: 15.00 | 14.81 |
| $\Gamma_5^-$ | $E_g$ | R: 13.41 | 13.339 |
| $\Gamma_5^+$ | $E_u$(LO) | I: 13.74 | 12.853 |
| $\Gamma_4^-$ | $B_{1u}^2$ | NA | 12.182 |
| $\Gamma_5^+$ | $E_u$(TO) | I: 11.64 | NF |
| $\Gamma_5^+$ | $E_u$(LO) | I: 11.19 | 11.232 |
| $\Gamma_5^+$ | $E_u$(TO) | I: 5.49 | 5.661 |
| $\Gamma_1^-$ | $A_{2u}$(TO)[d] | I: 5.01 | 5.177 |
| $\Gamma_3^+$ | $B_{1g}$ | R: 4.29 | 4.246 |
| $\Gamma_4^-$ | $B_{1u}^1$ | NA | 3.389 |

[a]See J.G. Gay, W.A. Albers, Jr., and F.J. Arlinghaus, J.Phys. Chem. Solids 29, 1449 (1968).
[b]Raman data (R) from S.P.S. Porto, P.A. Fleury, and T.C. Damen, Phys. Rev. 154, 522 (1967), infrared data (I) from D.M. Eagles, J. Phys. Chem. Solids 25, 1243 (1964). (NA = not active.)
[c]NF = not found.
[d]The $\Gamma_1^-$ mode is TO along $\Gamma\Lambda X$ and $\Gamma\Sigma M$, and is LO along $\Gamma\Lambda Z$.

The expression for a uniaxial crystal with more than two ions per unit cell has been generalized by BARKER (1964) and (2.11) becomes

$$\frac{\varepsilon_{c,0}}{\varepsilon_{c,\infty}} = \prod_i \left( \frac{\omega_{c,L_i}^2}{\omega_{c,T_i}^2} \right)_{|\underline{q}|=0} , \qquad (2.12)$$

where c is the unique axis. There is a similar expression for the a-direction. For rutile, there is only one pair of LO-TO polar modes with displacements in the c-direction, namely, the $\Gamma_1^-(A_{2u})$ modes. The temperature dependence of the c-axis static dielectric constant suggested the TO mode with this symmetry should show soft mode behavior. The neutron study of this mode did, indeed, exhibit abnormal behavior; the phonon energies decreased with decreasing temperature with pronounced changes in line shape, both characteristic of anharmonicity. Analysis in terms of a Curie-Weiss law gave a Curie temperature of -540K, reflecting the fact that rutile does not become a ferroelectric. The important area of neutron scattering studies on ferrodistortive and other phase transitions are discussed in detail in Ch.3.

### 2.2.2 $MgF_2$

$MgF_2$ is a much simpler bonding system than $TiO_2$ and it was expected to be a good test of the simple shell model for crystals with lower than cubic symmetry. The optical properties of pure and impure $MgF_2$ have been widely studied. Phonon sidebands of zero-phonon lines have been observed in single crystals containing M centers and related (MOSTOLLER et al., 1971) to the one-phonon density of states based on a rigid-ion lattice dynamics model. ALMAIRAC and BENOIT (1974) have reported lattice dynamics studies using a simple shell model and a partial determination of the phonon dispersion curves by neutron scattering. The study has now been completed, and it is described in the thesis of ALMAIRAC (1975a). The simple shell model with axially symmetric short range interactions was used throughout with excellent results, as shown in Fig.2.7. The neglect of the core-shell displacement of the fluorine ion at equilibrium seems to be justified, or at least not a serious approximation. The two Mg-F interactions were dominant and nearly equal (the interactions are non-equivalent but the two Mg-F distances are nearly equal). They were about 8 to 15 times larger than the F-F interactions.

In a separate paper, ALMAIRAC (1975b) describes the measurement of two eigenvectors for $MgF_2$ and showed that there was good agreement with the lattice dynamical models. It would be a time consuming task to measure the intensities of the majority of the phonon groups, but measurements at special q values, which are strongly force-dependent, should prove useful in deciding between various models that give similar phonon frequencies.

### 2.2.3 $MnF_2$

The phonon spectra of $MnF_2$ have also been extensively investigated by neutron scattering (ROTTER et al., 1975). The bonding is expected to be more similar to $MgF_2$ than $TiO_2$, although the polarizability of the larger Mn may be important. The intensities of the phonons, however, should be, and are, more like those of $TiO_2$ since the coherent neutron scattering amplitudes of Ti and Mn are both negative and similar in value.

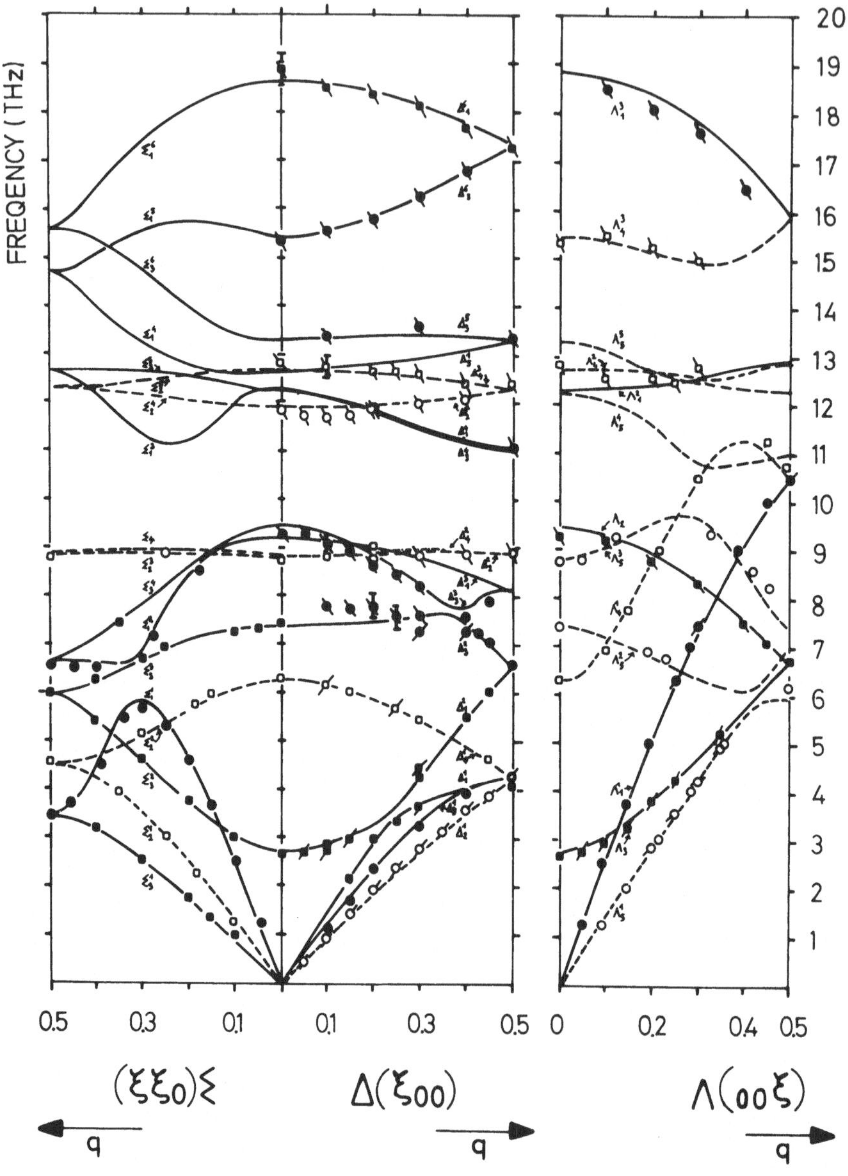


Fig. 2.7 Phonon dispersion relation of $MgF_2$. The solid, dotted, and dashed lines are calculated curves based on a simple shell model. (ALMAIRAC, 1975a)

CRAN and SANGSTER (1974) made lattice dynamical calculations of $MnF_2$ using only $|q| = 0$ optical data and the elastic constants. They included the two Mn-F interactions but only the shortest F-F interactions; however they went beyond the simplest shell model and included a core-shell interaction between the two fluorine ions.

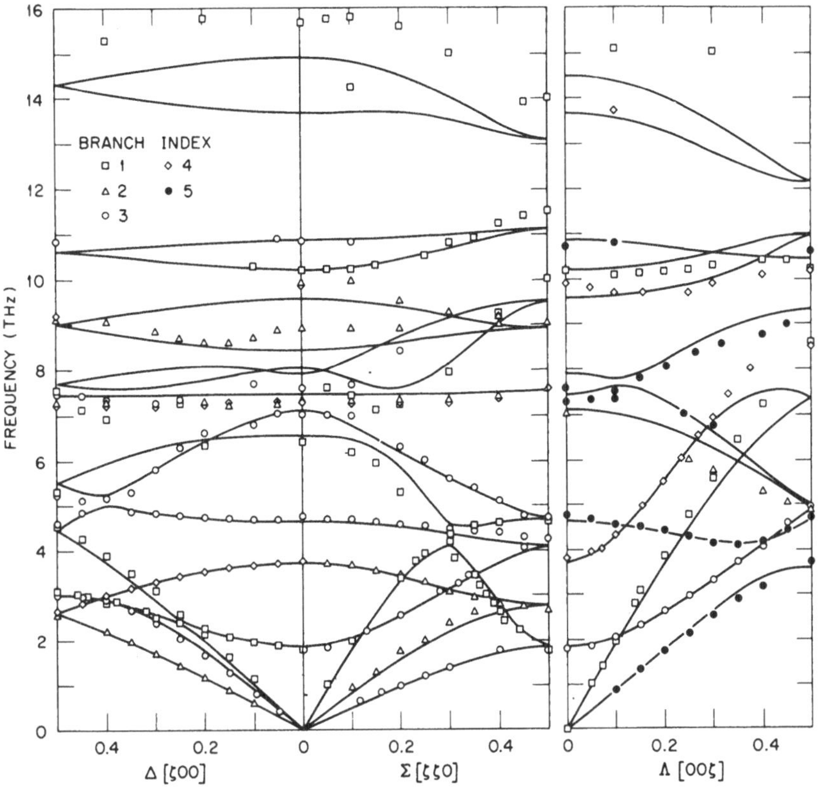


Fig. 2.8 Phonon dispersion curves for $MnF_2$. The solid lines are calculated curves based on a simple shell model. (ROTTER et al., 1975)

They emphasized the inclusion of rotational invariance and constraints on the elastic constants due to the lower site symmetry of the fluorines. Their resultant calculations are in quite good agreement with the observed neutron data.

ROTTER based his calculations on the simple shell model that TRAYLOR used for $TiO_2$ with several modifications: 1) axial symmetry constraints on all short-range interactions; 2) addition of F-F interactions along the c-direction; and 3) application of rotational invariance and elastic constant constraints. The experimental data and the results of his calculations are shown in Fig.2.8. The agreement with his model is also quite satisfactory, for, although the highest frequency modes are not fit as well as with CRAN and SANGSTER's model, the lower modes are fit somewhat better.

$MnF_2$ is a simple antiferromagnet and its magnetic properties are also of interest; the spin waves have been widely studied. Since $Mn^{+2}$ is a "spin only" ion, strong magnon-phonon interactions are not expected, nor have they been observed as they have for $FeF_2$. There is a small degree of covalency in the compound (a small spin density residing on the $F^-$ ions) so perhaps a very careful look at the spin wave spectrum may show deviations from the idealized system.

### 2.2.4 Conclusions

The simple shell model appears entirely adequate for essentially ionic crystals of the rutile structure, such as $MgF_2$, and even $MnF_2$. However, it is questionable for the less ionic $TiO_2$, and presumably even more questionable for the metal-like $VO_2$, $NbO_2$, and $RuO_2$. Some form of electronic screening will be necessary for the latter compounds since there can be no macroscopic electric fields in the crystals.

These results show that crystals more complex than the alkali halides can be studied successfully in great detail and, indeed, investigations of even more complex crystals than rutile are underway at several reactor centers. However, for these more complex structures the number of adjustable parameters becomes prohibitive and it is desirable to reduce the number of parameters if possible. Microscopic models with few parameters have been moderately successful for the alkali halides (BILZ, GLISS, and HANKE, 1974, WAKABAYASHI and SINHA, 1975); perhaps in the near future they can be applied to more complex structures.

## 2.3 Phonon Anomalies in Superconducting Materials

It has become apparent that the complex phonon spectra observed in most superconducting materials are related to their superconducting properties, presumably through the electron-phonon interaction. The anomalous features (pronounced dips in the phonon dispersion curves in various regions of the Brillouin zone) are most noticeable when compared with phonon spectra of related materials that are either non-superconducting or have low values of $T_c$. The first dispersion curves of a superconductor studied by neutron scattering which showed marked departures from their expected shape were those of Pb (BROCKHOUSE et al., 1962), but it was not until some years later (NG and BROCKHOUSE, 1968) that studies of the Pb-Tl-Bi alloys suggested these features were, perhaps, related to superconductivity. Neutron studies of Nb (NAKAGAWA and WOODS, 1963) and Mo (WOODS and CHEN, 1964) also showed pronounced differences when their phonon spectra were compared. In this case strong electron-phonon interactions were more or less expected because of the complex nature of the d-electrons in the conduction band. Sharp "kinks" were observed at several values of the wave-vector in different directions in the crystals. These were interpreted as "Kohn anomalies" (KOHN, 1959) and could be related to images of the Fermi surface of the crystal, cf.1.3.3. This phenomenon had been demonstrated very convincingly by BROCKHOUSE in his studies on Pb, and led to the studies on the Pb-Tl-Bi and Nb-Mo alloy systems (POWELL et al., 1968); however, the reason for the large "dips" could only be speculated upon.

At that time the measured dispersion curves of fcc Pb and bcc Nb could be reproduced theoretically only by assuming appreciable force constants out to many neighbors in a Born-von Kârmân model, requiring a large number of adjustable parameters (perhaps as many as 20). The significant *first principles* calculations of the dispersion curves in Pb by VOSKO, TAYLOR, and KEECH (1965), while not producing a very good fit to the observed data, did exhibit qualitative features which were in agreement with experiment; namely, that the electronic contributions to the dynamical matrix were large and negative in the region of the phonon anomalies. Present day models, phenomenological (WEBER, 1973) and microscopic (SINHA and HARMON, 1975), of the transition metals and their carbides also emphasize large negative contributions to the dynamical matrix in the region of the anomalies.

In the period between the experimental studies on the transition metals and those of the transition metal carbides (SMITH and GLÄSER, 1970, SMITH, 1972), McMILLAN (1967) derived his well-known formula for the transition temperature of strong-coupled superconductors. He showed that the transition temperature depends on the electron-phonon coupling constant, $\lambda$, through the relation

$$T_c = \frac{\theta_D}{1.45} \exp\left[\frac{-1.04(1+\lambda)}{\lambda - \mu^*(1+0.62\lambda)}\right] , \tag{2.13}$$

where $\mu^*$ is the Coulomb pseudopotential of MOREL and ANDERSON and $\theta_D$ is the Debye temperature of the material. $\lambda$ is related to the phonon spectra through the definition

$$\lambda \equiv \int_0^{\omega_0} \alpha^2(\omega) Z(\omega) d\omega/\omega \quad . \tag{2.14}$$

$Z(\omega)$ is the phonon density of states and $\omega_0$ is the maximum phonon frequency. $\alpha^2(\omega)Z(\omega)$ is a complicated function involving phonon frequencies and the atomic pseudopotential, and, although it is difficult to calculate *a priori*, it can be determined from tunneling measurements. It is often compared to $Z(\omega)$ obtained from neutron scattering data. McMILLAN found that $\lambda = N(0)\langle I^2\rangle/M\langle\omega^2\rangle$, where $N(0)$ is the electronic density of states at the Fermi energy and $\langle I^2\rangle$ is the electron-phonon matrix element averaged over the Fermi surface, M is the ionic mass and

$$\langle\omega^2\rangle = \int Z(\omega)\omega d\omega / \int Z(\omega)\omega^{-1} d\omega \ .$$

He proposed that $N(0)\langle I^2\rangle$[1] was approximately a constant within a given class of materials and, therefore, within each class $\lambda$ is related directly to the phonon spectra,

[1] This product is the parameter $\eta$ that HOPFIELD (1969) related to the "chemical" property of a superconducting material.

and furthermore, in the strong-coupling limit the electron-phonon coupling constant has an upper limit of 2. In a recent re-examination of the McMILLAN equation, ALLEN and DYNES (1975) concluded that, while the McMILLAN equation is correct for $\lambda < 1.5$, it must be modified for $\lambda > 1.5$, but that in either case $\lambda$ is not determined solely by the phonon frequencies, for $N\langle I^2\rangle$ is not a constant (DYNES and ROWELL, 1975) and it is probably more responsible for variations in $T_c$ than $M\langle\omega^2\rangle$.

In the subsections below, phonon spectra of several examples from different classes of materials are examined and discussed in terms of superconductivity and lattice instabilities.

### 2.3.1 Transition Metal Carbides

The transition metal carbides are an interesting class of materials. Some of them are moderate-to-high temperature superconductors, as are the nitrides ($T_c$ of $NbC_xN_{1-x} \geq 17K$ for $x \approx 0.30$), and all are very high melting point materials ($T_m$ of TaC = 4256K, one of the highest known). The chemical bonding properties in these compounds are particularly interesting, exhibiting simultaneously features characteristic of covalent, ionic, and metallic bonding. These features have been discussed, both from a chemical point of view (RUNDLE, 1948) and from a band structure point of view (BILZ, 1958).

The phonon dispersion curves of tantalum and hafnium carbide were first reported by SMITH and GLÄSER (1970). They noted that certain phonon branches in TaC, a superconductor, showed anomalous features not apparent in HfC, a non-superconductor. Studies of other transition metal carbides showed similar behavior (SMITH 1972) and it was suggested that the observed anomalous dips, whatever their cause, were indicative of strong electron-phonon interactions and were related to superconductivity. PHILLIPS (1971, 1972) proposed that the lattices of the high $T_c$ carbides and nitrides are unstable (chemically, not thermally) and are perhaps preventing the attainment of higher $T_c$'s in these systems. He suggested the observed anomalous dips or lattice softenings were manifestations of these instabilities due to overscreening of the d-electrons near the Fermi surface.

The phonon dispersion curves of TaC (a superconductor with $T_c \approx$ 10-12K) measured at room temperature and 4.2K for the [001], [110], and [111] directions are shown in Fig.2.9. The very slight differences in frequencies at the two temperatures are believed to be real. They are better illustrated in Fig.2.10 where the measurements of the TA (0.5, 0.5, 0.5) phonon were made under conditions of better resolution, (similar temperature dependences have been observed in Nb (POWELL et al., 1972)). The slight, apparent, increase in width on warming from 4.2 to 28 K, though expected, may be fortuitous. Although one generally observes an increase in phonon frequencies with a decrease in temperature, the slight decrease in frequencies in the neighborhood of the anomalies may be expected if the electronic contributions to the dynamical matrix are large since these terms are negative. Additional measurements under

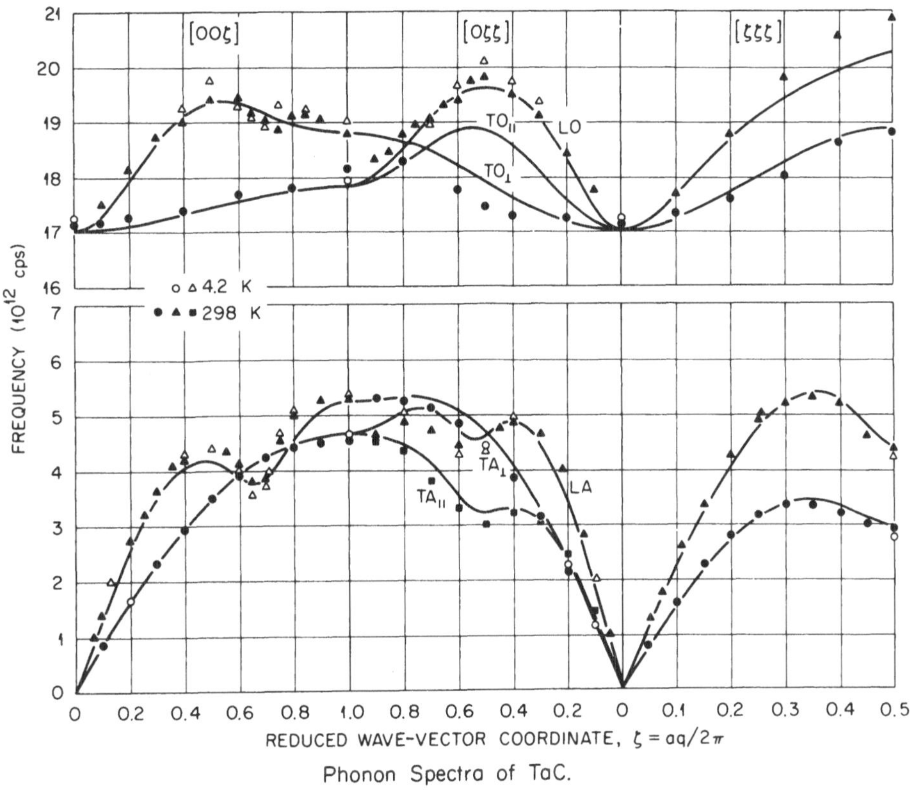


Fig. 2.9 Double-shell model fit to experimental results for the phonon dispersion curves of TaC. The DSM parameters used for the calculation are the second set of values in Tables II and III of WEBER, 1973

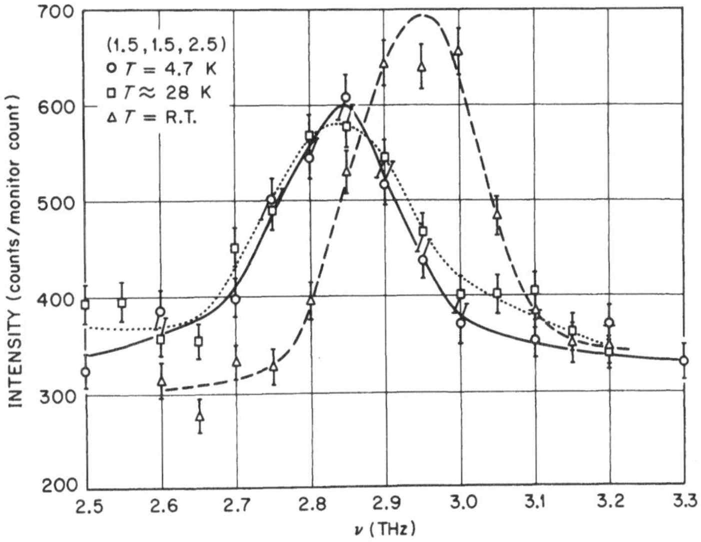


Fig. 2.10 Frequency of TA (0.5, 0.5, 0.5) phonon vs. temperature. (SMITH, 1972)

increased resolution, and better focussing conditions, for frequencies above and below the energy gap would be very desirable. Because of the small crystal sizes, coarse collimation was used in most of the experiments in order to have adequate neutron intensity to measure the phonons in a reasonable length of time. Therefore, sharply defined Kohn anomalies, as have been observed in Pb and Nb, may have been overlooked.

A similar but less complete study on NbC showed effects similar to TaC (Fig.2.11), whereas the phonon spectra of ZrC (a non-superconductor, $T_c < 1.2$) did not exhibit these anomalous features (Fig.2.12) (SMITH et al., 1974).

WEBER (1973) was able to reproduce reasonably well the observed phonon spectra of ZrC and HfC using a simple shell model modified for free electron screening at long wavelengths. The screening was necessary to explain the observed degeneracy of the TO and LO modes at $|\underline{q}| = 0$, characteristic of a metal. However, in order to explain the anomalous features of the dispersion curves in TaC and NbC, he had to invoke an extra shell, or "supershell", of electrons about the metal cores and shells which interact (attractively) with neighboring supershells, including second-neighbors on the metal sublattice. This phenomenological "double shell model" (DSM) is rather complicated with about fifteen adjustable parameters (some were adjusted to be zero), but it does have the advantage of emphasizing the important contributions to the dynamical matrix in the region of the anomalies. WEBER interprets the anomalous regions as a "resonance cube" in the Brillouin zone due to a q-dependent resonance-like behavior of the electronic polarizability of the d-electron charge density,

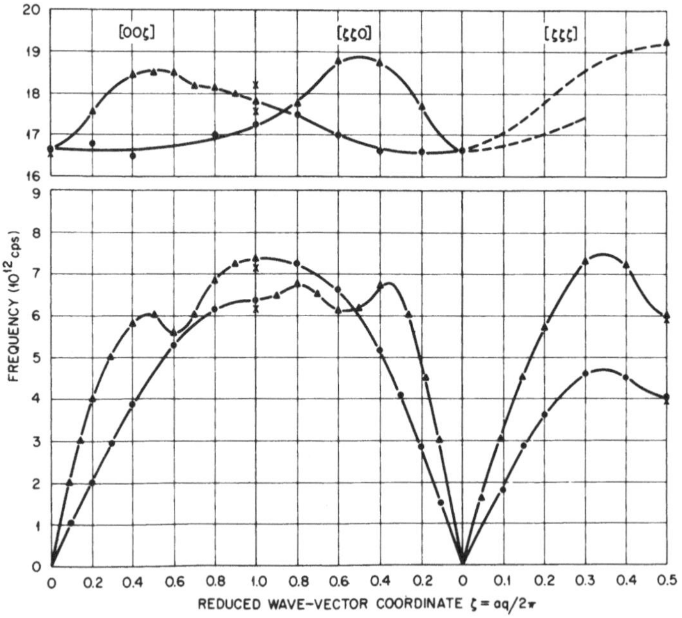


Fig. 2.11 Phonon dispersion curves in NbC. (SMITH et al., 1971)

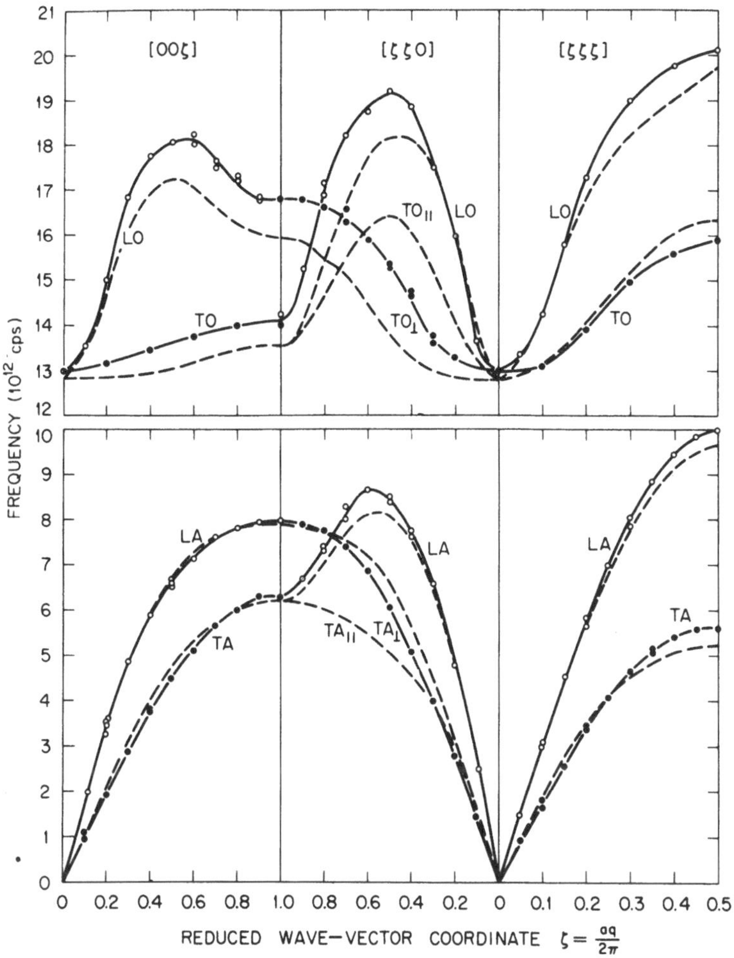


Fig. 2.12 Phonon dispersion curves in ZrC. Dotted lines are calculations based on WEBER's screened shell model. (SMITH et al., 1974)

more specifically, those of $t_{2g}$ symmetry. His calculations predicted phonon softening in the [110] transverse acoustic branch with [1$\bar{1}$0] polarization which was subsequently confirmed experimentally (SMITH, 1972), as were some anomalous regions in off-symmetry directions.

The dispersion curves of TaC and HfC have also been calculated by VERMA and GUPTA (1975) with their three-body-force shell model, obtaining reasonable agreement with experiment. Their model has the advantage of using far fewer adjustable parameters (five), but a disadvantage lies in the fact that the various contributions to the dynamical matrix (and, hence, the phonon frequencies) in the anomalous regions are obscured.

There is encouragement that more first principle calculations with microscopic theories (HANKE 1973, SINHA and HARMON 1975) will be successful in leading to a more detailed understanding of the lattice dynamics of superconductors and the relationship to their electronic properties which, of course, determines ultimately their physical and chemical properties. It will be interesting to see if these models can be applied to the superconducting palladium hydrides when the experimental data become available. It already has been shown that either a simple screened shell model or a Born-von Kårmån model with second-neighbor metal-metal and deuterium-deuterium interactions adequately explains the phonon dispersion curves (ROWE et al., 1974) in non-superconducting $PdD_{0.63}$. This system is a very interesting one, for tunneling measurements on superconducting PdH (EICHLER, WÜHL, and STRITZKER, 1975; DYNES and GARNO, 1975; SILVERMAN and BRISCOE, 1975) indicate the optic modes are playing a more dominant role in the electron-phonon interaction than are the acoustic modes (in contrast to the tunneling studies on TaC). An unexpected feature in $PdD_{0.63}$, revealed in the neutron study, was the large dispersion observed for the optic modes (very similar to that observed in ZrC). It will be of great interest to see if this large dispersion persists in the superconducting hydride. The neutron measurements are imminent.

### 2.3.2 A-15 Compounds

The highest superconducting transition temperatures occur for compounds with the cubic A-15 structure, of which $Nb_3Sn$ is a classic example. Those compounds with high $T_c$ are of great technological importance and have been studied extensively (physically, chemically, and metallurgically). They are known to exhibit elastic softening and undergo structural phase transformations above the superconducting transition (TESTARDI, 1971). These transformations are generally thought to be related to their superconducting properties, e.g., those A-15 compounds with low $T_c$'s do not appear to transform their structure as do the high $T_c$ compounds. (Not all samples of a given high $T_c$ compound transform, but they still show softening of the modes). The single crystals of the interesting A-15's which are currently available are either very small, as for $Nb_3Sn$, or non-existent, as in the case of $Nb_3Ge$. Exceptions are the vanadium compounds, but the small coherent and large incoherent neutron scattering properties of vanadium preclude a complete determination of the dispersion curves.

The elastic and specific heat properties of many A-15 compounds have been thoroughly studied as a function of temperature and attempts have been made to correlate the observed lattice instabilities with their structural and superconducting properties. It has been observed that the elastic shear mode in the [110] direction with [1$\bar{1}$0] polarization, $C' = 1/2\ (C_{11} - C_{12})$, goes soft and is considered an extreme manifestation of a lattice instability driven, perhaps, by an electronic instability. The transverse acoustic modes in $Nb_3Sn$ have been measured by AXE and SHIRANE (1973) as a function of temperature by coherent inelastic neutron scattering techniques. (The crystals were too small to measure the longitudinal acoustic modes or the optic modes). Their studies confirmed the ultrasonic results and, in addition,

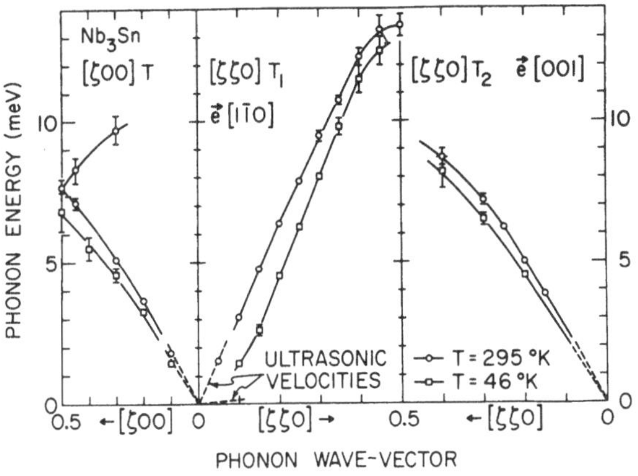


Fig. 2.13 Dispersion relations for transverse acoustic phonons in $Nb_3Sn$ at two temperatures. (AXE et al., 1973)

showed that the softening extended well out into the Brillouin zone. The experimental results are shown in Fig.2.13. It is rather unusual to see all the TA modes decrease with decreasing temperature.

KLEIN and BIRMAN (1970) in their theoretical lattice dynamics study of A-15 compounds proposed that lattice instabilities in the optic modes were responsible for the phase transformation, although their acoustic modes also showed a softening. PYTTE (1970) concluded from his theoretical study that the observed acoustic softening which triggers the phase transformation does not affect the superconducting transition. One would be surprised if some of the optic modes did not show anomalous features as observed in other superconductors, perhaps, with even more pronounced behavior. The A-15 compounds have the cubic β-W structure with eight atoms in the primitive unit cell, so even with sizeable single crystals it may be difficult to measure all twenty-one optic branches, but even a partial determination would be very valuable and of great interest.

In the meantime, phonon densities of states determined by tunneling in superconductors or neutron inelastic *incoherent* scattering experiments on polycrystalline materials are very informative. REICHARDT (1975) and his colleagues at GfK, Karlsruhe, Germany, have demonstrated that, with proper care and techniques, the phonon density of states of polycrystalline *coherent* neutron scatterers can be approximated. Their measurements on $Nb_3Sn$ at room temperature and 4 K are shown in Fig.2.14 where the experimental curves are compared with the tunneling measurements of SHEN (1972). It appears that the phonon density of states above the maximum of the TA modes also shows sizeable downward shifts on lowering the temperature. Presumably these are due to optic modes as well as LA modes. (The reader is reminded that in the carbides, the temperature dependence of the modes in the region of the anomalies was very small even though the dips were quite large).

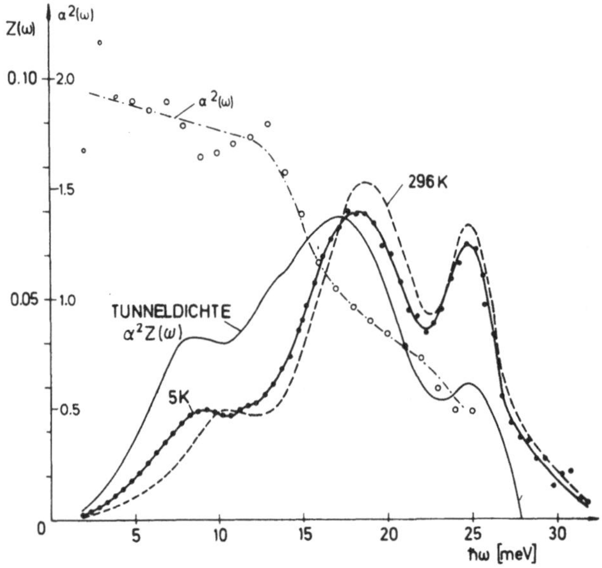


Fig. 2.14 Comparison of the tunneling density $\alpha^2 Z(\omega)$ with the phonon spectra at 296 K and 5 K of $Nb_3Sn$. (REICHARDT, 1975)

ALLEN (1972) proposed that a study of the phonon line widths by neutron scattering can, in principle, provide valuable information concerning the electron-phonon line width contribution, $\gamma(Q)$, to the total line width. He showed that the quantity $\alpha^2(\omega)Z(\omega)$, measured in tunneling experiments, can be expressed in terms of the neutron line widths by the relation

$$\alpha^2(\omega)Z(\omega) = 2/\pi N(0)\omega \sum_Q \gamma(Q)\delta(\omega - \omega_Q), \tag{2.15}$$

where, as before, N(0) is the electronic density of states at the Fermi energy. It would be a formidable experimental task to measure sufficient $\gamma$'s to determine $\alpha^2 Z$ from this expression; however, ALLEN introduced an average phonon width, $\bar{\gamma}$, which is related to the McMILLAN parameter, $\lambda$, by

$$\bar{\gamma}/\bar{\omega} = (\pi/12)(\lambda\bar{\omega}/W) , \tag{2.16}$$

where $\bar{\omega}$ is defined as $\langle\omega^2\rangle^{1/2}$ and W is an effective bandwidth defined by N(0) = N/W; N is the number of atoms in the crystal. He showed that, although this quantity is very small for s-p materials like Pb (and probably experimentally inaccessible), for narrow-band materials such as high-$T_c$ A-15 superconductors the *average* electron-phonon induced width should be comparable with the experimental resolution. This prediction has been verified experimentally (see Fig.2.15) for $Nb_3Sn$ in the neutron scattering work of AXE and SHIRANE (1973), and subsequently for Nb by SHAPIRO et al., (1975). Experiments such as these lead, in addition, to a measurement of the

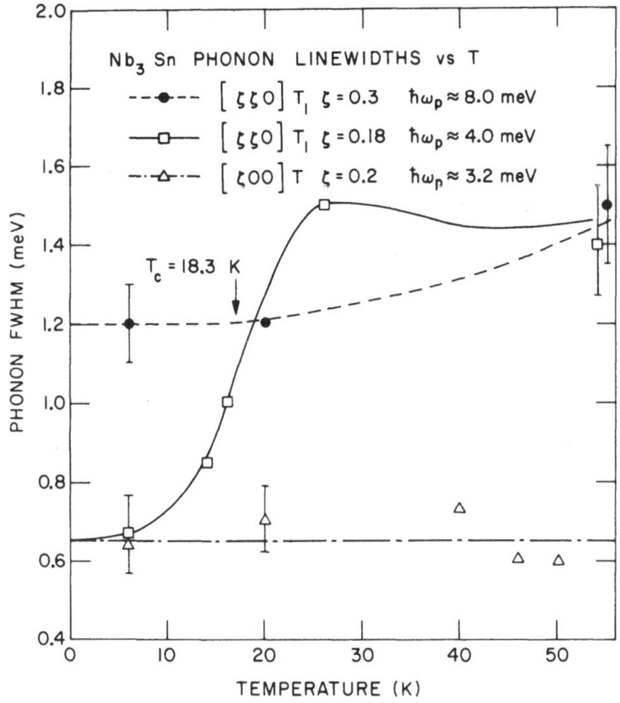


Fig. 2.15 Summary of transverse acoustic phonon line widths in $Nb_3Sn$. The unusually large $[\zeta\zeta 0]T_1$ phonon line width is suppressed if the phonon energy $\hbar\omega_p$ falls below the superconductivity gap energy $2\Delta(T)$. ("Phonon FWHM" is the full width at half maximum of the observed phonon peak.) (AXE et al., 1973)

superconducting energy gap and open up another fruitful area for inelastic neutron scattering investigations of superconducting materials.

### 2.3.3 Niobium - Alloys, Other Compounds, and Impurities

The niobium atom apparently has certain properties that are particularly suitable for superconductivity, for it has the highest $T_c$ of any element and it is contained in, and undoubtably has the dominant role in, the highest superconducting compounds, notably $Nb_3Sn$ and $Nb_3Ge$. A possible understanding of the role it plays in superconductivity has led to extensive studies of various systems (compounds, alloys, impurities) containing it.

In addition to the interesting Nb-Mo system mentioned previously in the introductory subsection, the temperature dependence of the phonon spectra of the metal itself has been studied (POWELL et al., 1972). Not only do some of the modes show unusual temperature behavior, but it is interesting to note that the anomalous features of the dispersion curves that are believed to be related to its superconducting properties near the transition temperature are still apparent over 1000° above $T_c$, and possibly up to the melting point. Further temperature dependent phonon studies are necessary, but it is tempting to suggest (this is not a new suggestion) that when a material condenses, the conditions for superconductivity are already inherent in its basic physical and chemical properties, and probably explains why it sometimes takes other means (application of high pressures, sputtering, etc.) to produce materials that would otherwise be unstable.

Many of the physical properties of metallic Nb are seriously affected by impurities (e.g., O, N, and H); in particular, it is known that a few percent of oxygen in Nb decreases its superconducting transition temperature appreciably. Specific heat measurements by KOCH et al. (1974) demonstrated that the average phonon frequencies (through the $\theta_D$) were increasing with increasing oxygen content. Subsequent inelastic neutron scattering measurements of the dispersion curves in a single crystal of Nb (2.5% O) (SMITH, 1973) confirmed their conclusions in that an overall increase in most phonon frequencies was observed. The greatest increases occurred in the low energy region of the TA modes in the [100] and [110] directions, hence, correlating with the large changes of $\theta_D$ derived from the low temperature specific heat measurements. Similar changes, but to a lesser extent, have been observed in the phonon spectra of $NbD_x$ (ROWE et al., 1975) and $TaD_x$ (WAKABAYASHI, 1976).

It is well known that there is a strong correlation between high $T_c$'s and certain values of the valence electron per atom ratio with a maximum $T_c$ at e/a = 4.7, a minimum at 5.8, and another maximum at 6.5. The curves differ in detail depending on the type of crystal structure. In the $Nb_{1-x}Zr_x$ system a maximum $T_c \approx 12$ K is reached at about $Nb_{0.70}Zr_{0.30}$; however, this is in an unstable two-phase region. The bcc phase boundary occurs near $Nb_{0.85}Zr_{0.15}$, where $T_c$ is about 11 K. Accordingly, a single crystal neutron scattering study was made in an alloy of $Nb_{0.87}Zr_{0.13}$ (TRAYLOR et al., 1975) and the changes in the dispersion curves were readily observable. The phonon frequency shifts are in general opposite to those observed in $Nb_{0.85}Mo_{0.15}$, as expected. The results should prove useful when more precise theories are developed.

Another interesting class of superconductors in which Nb plays a dominant role is the transition metal dichalcogenides which exhibit features characteristic of two-dimensional lattices. For example, $NbSe_2$ is metallic and has a superconducting transition temperature of $\approx 7$ K, compared to $TaS_2$ with $T_c \approx 1$ K, and $MoS_2$, a semiconductor. Its interesting phonon spectra (WAKABAYASHI et al., 1974, MONCTON et al., 1975) are discussed in the next section on layered compounds.

### 2.3.4 Other Superconductors - Tc, Mo-Re, Hg

The high $T_c$ superconductors discussed so far have been concerned with the transition metals, alloys, and compounds centered about the electron/atom ratio of 4.7. As mentioned above, a minimum in $T_c$ occurs at approximately e/a = 5.8 and a secondary maximum exists at about e/a = 6.5. Technetium (e/a of 7) is an interesting example

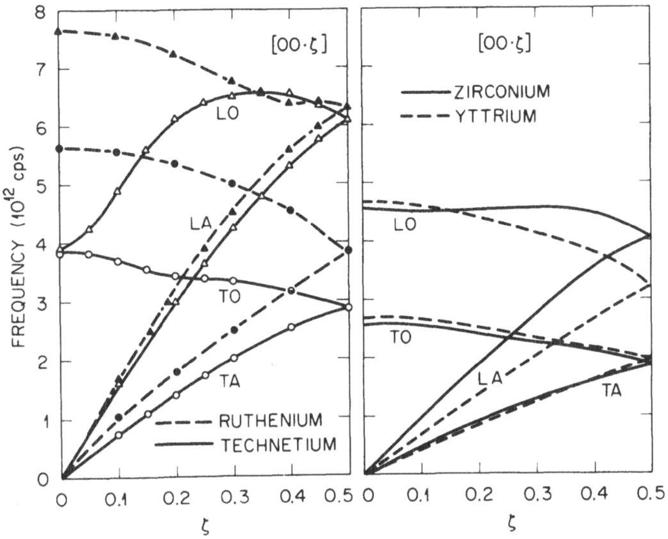


Fig. 2.16 Phonon dispersion curves in c-direction in techneticum, ruthenium, zirconium, and yttrium

in that its e/a falls near this maximum, it has the highest $T_c$ (8 K) of any hcp element, and the $T_c$ of its alloys with Mo and Nb increases rapidly as they are added to the metal. (Technetium does not exist naturally but is a byproduct of the fission process). The phonon spectra (SMITH et al., 1974) of technetium also are rather unusual with large anomalous dips in the dispersion curves but, in different regions of the Brillouin zone, reflecting the different crystal and electronic structure compared to say Nb. The phonon modes in the c-direction are compared (Fig.2.16) with those of other neighboring hcp elements in the same row of the periodic table: Y, Zr, and Ru. It again follows that the anomalous features are most pronounced the higher the superconducting transition temperature.

The superconducting transition temperature of hcp Re (e/a of 7) increases considerably when alloyed with small amounts of Mo and W. However, the most dramatic increase in $T_c$ occurs with alloying Re into bcc Mo. At $Mo_{0.75}Re_{0.25}$, $T_c \doteq 10$ K ($T_c$ continues to increase with increasing Re content, but it is in a two-phase region), and the phonon spectra are drastically different from pure Mo (SMITH and WAKABAYASHI, 1976). It will be interesting to see if the double shell model or some other model can explain these and the technetium spectra.

Last but not least, we discuss the phonon spectrum of mercury. It is very important from a historical as well as scientific point of view. KAMERLINGH ONNES first discovered (1911) superconductivity with Hg, and later (1950) the isotope effect was observed experimentally with the mercury isotopes giving the first experimental evidence of the role of the electron-phonon interaction in superconductivity. Also,

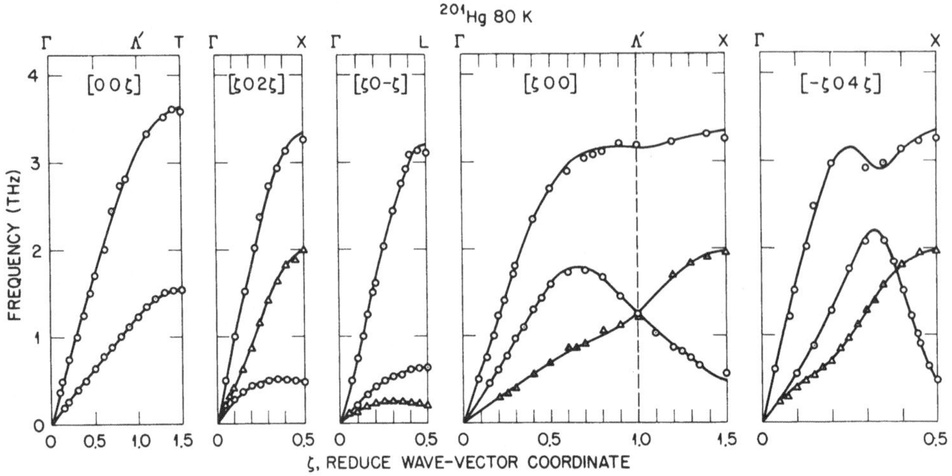


Fig. 2.17 Phonon dispersion curves in $^{201}$Hg at 80 K. (KAMITAKAHARA et al., 1975)

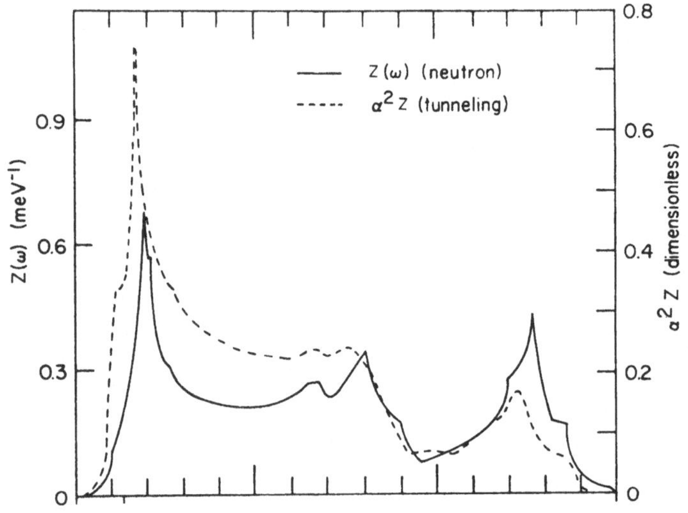


Fig. 2.18 Comparison of $\alpha^2(\omega)Z(\omega)$ from tunneling measurements with $Z(\omega)$ obtained from an eight-neighbor general Born-von Kármán model fitted to the neutron measurements. (KAMITAKAHARA et al., 1975)

Hg has the largest electron-phonon coupling constant, $\lambda = 1.6$, of any element and tunneling measurements indicated pronounced peaks in the phonon density of states at very low frequencies, uncharacteristic of any other superconductors. KAMITAKAHARA et al. (1975) have measured the phonon dispersion curves in a single crystal of the $^{201}$Hg isotope at several temperatures. The crystal structure is trigonal, distorted from an fcc lattice by compression along a cubic [111] direction, with one atom per unit cell. The deviations from cubic symmetry are pronounced, as can be seen from the non-degeneracies observed in the pseudocubic [001] ([$\zeta$,0,2$\zeta$]) and [111] ([$\zeta$,0,-$\zeta$]) directions (Fig.2.17). The data have been accurately fitted with an 8-neighbor Born-von Kármán model, and a reliable phonon density of states has been calculated. Comparison with $\alpha^2(\omega)Z(\omega)$ derived from tunneling measurements reveals a substantial increase in $\alpha^2(\omega)$ at low frequencies (Fig. 2.18).

### 2.3.5 Conclusions

What is the apparent significance of the observed phonon anomalies in relation to superconductivity? There seems to be little doubt that a relationship exists, but are they a cause or an effect of high-$T_c$ superconductivity? There does not seem to be an explicit answer at this time, but indications are that they reflect the detailed electronic properties of the metal, alloy, or compound, particularly the properties relating to chemical bonding. Perhaps not so much with the strong near-neighbor interactions, but with the weaker, more distant neighbor interactions; the stronger near-neighbor bonds are undoubtably responsible for the structure-type of the material and the overall features of the dispersion curves.

Perhaps explicit answers to these questions will be forthcoming since the theoretical physicists are nearly able to include the necessary "chemistry" in their calculations.

## 2.4 Layered Compounds

Graphite may be regarded as a classic example of a layered compound. The temperature dependence of the specific heat, $c_v$, of graphite was known to be proportional to $T^3$ only for $T \leq 3$ K, and nearly proportional to $T^2$ in a rather wide temperature range above this temperature, which is a manifestation of the quasi two-dimensional character of the lattice vibrations in this material. This temperature dependence was originally interpreted on the basis of a model in which the layers were treated as weakly interacting elastic sheets. Subsequent neutron scattering measurements have yielded somewhat more detailed information about the interatomic forces, and $c_v$ can be calculated quite accurately. Molybdenum disulfide has long been known to have similar properties as graphite and it is also widely used as a lubricant. More recently, other transition metal dichalcogenides with layered structures have been studied in connection with lattice instabilities and superconductivity. Interesting features of two-dimensional character also show up in the electronic properties, in addition to the two-dimensional atomic and molecular properties. For example, the critical field, $H_c$, for superconducting $NbSe_2$ is quite anisotropic; $H_c$ has a variation of about two orders of magnitude depending on the direction of the field with respect to the crystallographic axes.

From the point of view of lattice dynamics, interesting types of layered compounds are characterized by extremely anisotropic interatomic interactions. Interactions between atoms in the neighboring layers are also typically two orders of magnitude weaker than those between atoms within a layer. However, some compounds, especially those of atoms with large ionicities, such as $PbI_2$, do not show this anisotropy even though their structures are distinctly layer-like. In this section mainly anisotropic layered compounds are discussed and an example of isotropic layered compounds is presented. In the following, the term "layered compound" is used for "anisotropic layered compound" unless otherwise stated.

The lattice vibrations in layered compounds, then, may be regarded approximately as those of a two-dimensional lattice. This is not unlike the case of molecular crystals that contain tightly bound atoms forming rather rigid molecules which are only weakly bound to neighboring molecules. Most of the vibrational modes can be classified as internal modes corresponding, for example, to stretching or twisting of bonds within the molecule. Only a few low frequency modes are translational or librational, with frequencies determined primarily by weak intermolecular forces. In the case of a layered compound the entire layer may be regarded as a giant molecular unit with a two-dimensional periodic structure, and the normal modes of vibrations in these compounds may be classified in a similar fashion. Since the phonons propagating parallel to the plane of the layer correspond to the "internal" modes, these modes can be approximated by those for two-dimensional lattice vibrations. For wave-vectors perpendicular to the plane of the layer there are vibrational modes whose frequencies are almost independent of the wave-vector and hence given approximately by the values of the two-dimensional vibrations of a single layer at $|q|= 0$. These two classes of modes are commonly called *intralayer* modes. There are also modes in which each layer vibrates almost as a rigid unit. These modes are sometimes called *interlayer* modes, determined mainly by the weak interlayer interactions, and their frequencies are generally very low. In particular, if the layer is a strictly two-dimensional sheet of atoms, such as that of graphite, all the atoms in a layer vibrate as a rigid unit and the interlayer modes in such a case are called *rigid-layer* modes. The distinction between the "internal" or intralayer and "translational" or interlayer modes can be best demonstrated by the phonons propagating in the direction perpendicular to the plane of the layer. If the unit cell extends over n layers, most phonon branches are nearly n-fold degenerate. The separations between the frequencies of the nearly degenerate modes are determined by the weak interlayer interactions. As an example, let us consider a sheet of atoms arranged in a two-dimensional periodic structure. The frequencies $\omega_0(q_x,q_y)$ of an isolated sheet are given by the intralayer interactions as the solutions of

$$\det|D_0(q_x,q_y) - I\omega_0^2| = 0 , \tag{2.17}$$

where $D_0$ is the dynamical matrix for the two-dimensional lattice and I is the unit matrix. When the interlayer interactions are introduced and the unit cell extends over two layers, the phonon frequencies are determined by the secular equation

$$\begin{vmatrix} D_0(q_x,q_y) + D_0' - I\omega^2 & -D'(q_x,q_y,q_z) \\ -D'(q_x,q_y,q_z) & D_0(q_x,q_y) + D_0' - I\omega^2 \end{vmatrix} = 0, \qquad (2.18)$$

where $D_0' = D'(0,0,0)$, and D' represents the effects of the third dimension including, in particular, the interlayer interactions. It can be shown by the perturbation method for degenerate states that, to first order in D', the frequencies can be written as

$$\omega^2 = \omega_0^2(q_x,q_y) + \Delta_0 \pm \Delta(q_x,q_y,q_z) , \qquad (2.19)$$

where $\Delta$ is determined by the interlayer interactions through D', and $\Delta_0 = \Delta(0,0,0)$. There are pairs of modes nearly degenerate as indicated by the double sign. The modes with two types of wave-vectors are of special interest:

1) For $\underline{q} = (q_x,q_y,0)$ , $\quad \omega \cong \omega_0(q_x,q_y)$ , $\qquad$ (2.20)

since $\Delta$ is generally negligible compared with $\omega_0(q_x,q_y)$. Thus the frequencies are approximately given by those for the two-dimensional lattice and can be regarded as the interlayer modes. The exceptions are the acoustic modes with small q's, since $\omega_0(q_x,q_y) \propto q^2$ for such modes.

2) For $\underline{q} = (0,0,q_z)$, $\quad \omega^2 = \omega^2(0) + \Delta_0 \pm \Delta(q_z)$ . $\qquad$ (2.21)

In particular for $\omega_0(0) = 0$, namely the acoustic modes,

$$\omega^2 = \Delta_0 \pm \Delta(q_z) , \qquad (2.22)$$

and is determined by the interlayer interactions to this order of approximation. These modes correspond to the interlayer modes (or the rigid layer modes in this example). On the other hand, for those modes with non-zero $\omega_0(0)$, $\omega \cong \omega_0(0)$, nearly independent of $q_z$. These modes are also the intralayer modes, and the distinction between the inter- and intralayer modes are most apparent in this direction of $\underline{q}$.

The greatest difficulty in measuring phonon dispersion curves for layered compounds has been the lack of large single crystals of reasonably good quality. Together with the very high intralayer mode frequencies, this accounts for the fact that virtually no high frequency modes have been determined with the accuracy commonly found in the measurements for other materials discussed in this chapter. In most of the layered compounds there are several polytypes which are characterized by the difference in the stacking sequence of the layers (as may be expected from the very weak interlayer interactions). The size of the Brillouin zone along the c-direction is different for each polytype, creating new Brillouin zones and new branches as the unit cell size increases. But the relative effect on the phonon frequencies due to the difference in the stacking sequence may be small (of the order of the ratio of the interlayer interaction to the intralayer interaction). The existence of polytypes has a significant effect on the optical methods for measuring phonon frequencies since these methods yield the information about the phonon frequencies at the origin of Brillouin zone. However, a single crystal of any polytype may be used as a sample for the neutron scattering experiment since the phonon frequencies for wave-vectors in the entire Brillouin zone with any symmetry may be studied by this technique.

### 2.4.1 $MoS_2$

$MoS_2$ is a semiconductor and one of the most extensively studied layered compounds. Each layer of this compound consists of a plane of Mo atoms, in the form of a hexagonal network, sandwiched by two planes of S atoms. The Mo atom is located at the center of a trigonal prism formed by the nearest neighbor S atoms. Hexagonal $MoS_2$(2H)

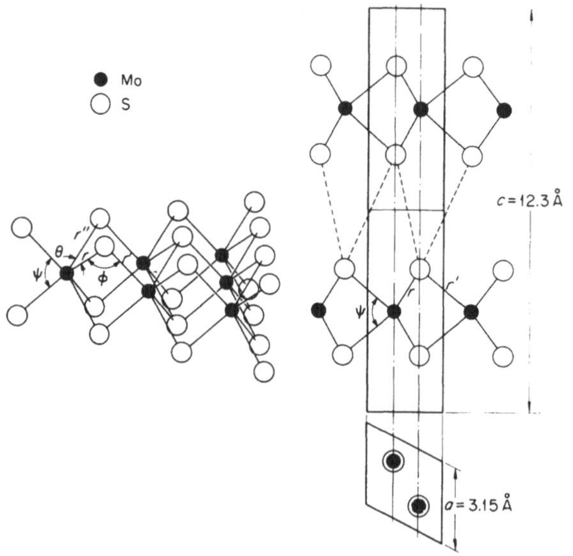


Fig. 2.19 Left drawing is to show the structure of bonding within a layer of hexagonal $MoS_2$. Right drawing shows the relative configuration of neighboring layers. The basal projection of the atomic positions is shown below. The unit cell is outlined by heavy lines. (WAKABAYASHI et al., 1975)

has the space group $D_{6h}^4$ and the unit cell extends over two layers (Fig.2.19). Thus there are two molecular units in the unit cell, and most of the branches should be nearly doubly degenerate according to the discussion in the previous subsection. Large crystals can be found in nature but their quality tends to be poor.

The phonon dispersion curves along two symmetry directions have been measured (WAKABAYASHI et al., 1975) and are shown in Fig.2.20. Since the sample had the shape of a large, very thin, plate, no attempt was made to measure the [100] transverse modes that are polarized in the plane of the layer. As expected, many optic branches are nearly doubly degenerate and separations could not be determined in the measurement. Also, there are very low frequency modes in the [001] direction corresponding to the interlayer modes. For crystals with hexagonal structure the crystal symmetry requires that the velocities for the transverse sound waves propagating in both the [100] and [001] directions be given by the elastic constant $C_{44}$. Since the transverse acoustic mode in the [001] direction is an interlayer mode

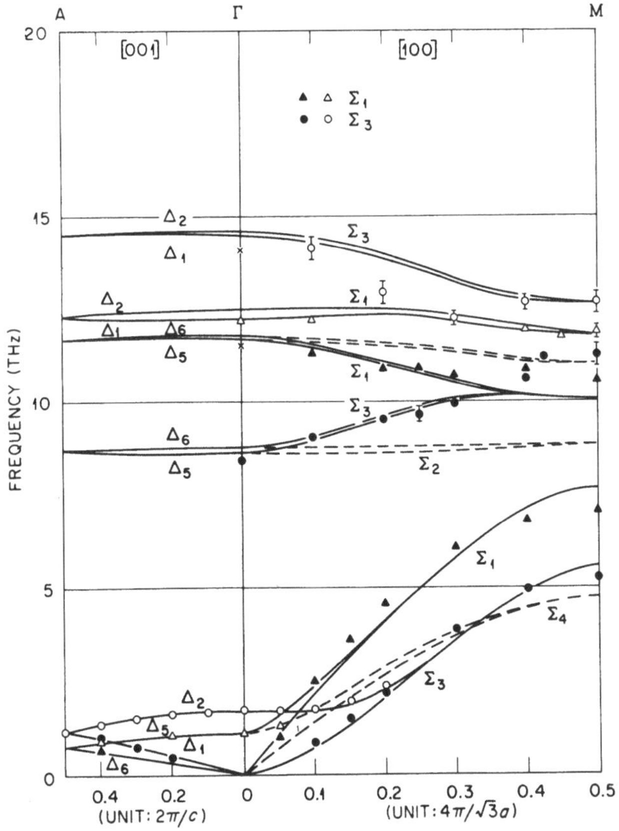


Fig. 2.20 Phonon dispersion curves for hexagonal $MoS_2$ along the [001] and [100] directions. $\Sigma_2$ and $\Sigma_4$ branches cannot be observed due to the geometrical restriction of the shape of the sample crystal. The lines are the calculated curves based on the model described in the text. (WAKABAYASHI et al., 1975)

with very low frequencies, the initial slope, and hence $C_{44}$, is quite small. This requires the [100] transverse branch to remain low near the origin, but this branch rises rapidly to high frequencies since the modes in this direction are, after all, intralayer modes which should have high frequencies determined by the strong covalent bonding between atoms in the layer. These features are direct consequences of the weak interlayer interactions.

The data were initially analysed on the basis of a Born-von Kármán model with axially symmetric forces, but much better agreement between the observed and calculated frequencies was obtained with a valence force model which included the following terms in the potential,

$$\frac{1}{2} K_r (\Delta r)^2 + \frac{1}{2} K_\theta (r_0 \Delta\theta)^2 + \frac{1}{2} K_\phi (r_0 \Delta\phi)^2 + \frac{1}{2} K_\psi (r_0 \Delta\psi)^2 , \qquad (2.23)$$

where $\Delta r$, $\Delta\theta$, $\Delta\phi$, and $\Delta\psi$ are changes in variables defined in Fig.2.19, and in the cross terms such as $K^{\theta}_{rr'}(\Delta r)(\Delta r')$. This model probably reflects the covalent character of the bonding within the layers. The interlayer interactions were assumed to be derived from a central potential, which is valid for a van der Waals interaction. The K's are the model parameters representing the forces associated with the bending and stretching of bonds. These parameters were determined by fitting the model to the experimental data shown in Fig.2.20. The solid lines represent the dispersion curves calculated for the model parameters thus determined. The distinction between the inter- and intralayer modes is clearly demonstrated in the [001] direction. Nearly degenerate optic modes and the upward curvature of the [100] transverse branch are also produced by this model.

Similar features, though somewhat weaker, have been observed in GaSe (JANDL et al., 1976) and $TiSe_2$ (STIRLING et al., 1975); however, the crystals of $TiSe_2$ were too small to measure the *intralayer* optic modes.

### 2.4.2 $NbSe_2$

The structure of 2H-$NbSe_2$ is nearly the same as that of 2H-$MoS_2$ and its space group is also $D^4_{6h}$. The atomic arrangement within the layer is identical to that of $MoS_2$, but the positions of atoms with respect to those of the adjacent layers are slightly different. The structure indicates strong covalent bonding between atoms within the layer, as in the case of $MoS_2$. However, since it is a metal and its superconducting transition temperature is moderately high ($T_c \sim 7$ K), the electron-phonon interaction should be large and the phonon dispersion curves may be expected to show the effects of conduction electrons on the interatomic interactions. The sizes of available single crystals are extremely small ($<0.05$ cm$^3$), and only a limited amount of information exists for the lower frequency branches obtained from measurements on such small samples. In Fig.2.21 are shown the results of the measurements of WAKABAYASHI

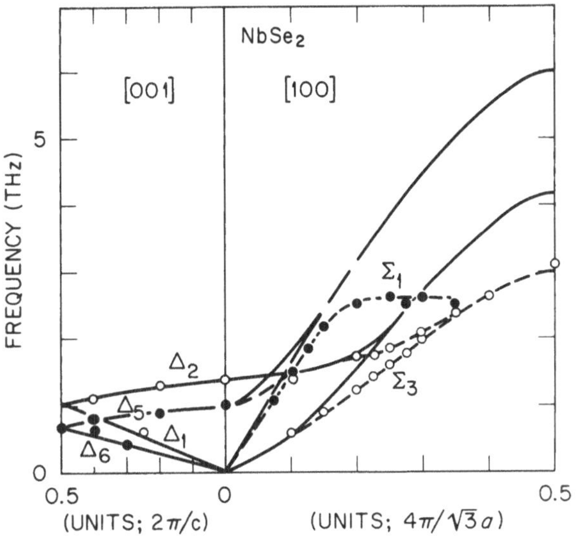


Fig. 2.21 Phonon dispersion curves for the low-frequency branches in $NbSe_2$. Solid lines are the results of the calculation described in the text. (WAKABAYASHI et al., 1974)

et al. (1974). The frequencies were calculated on the basis of the model parameters obtained for $MoS_2$ with proper masses substituted for $NbSe_2$ (solid lines). In the [001] direction the phonon frequencies are reproduced within the experimental uncertainties by these force parameters, indicating that the interlayer interactions are not affected by the conduction electrons within the layer. However, in the [100] direction the comparison shows substantial deviations from the curves expected from such a valence force model. This may be considered to be due to the effects of the Fermi surface whose shape is nearly the same as that of a two-dimensional energy band (MATTHEISS, 1973), i.e., a nearly cylindrical Fermi surface. The screening by the conduction electrons is described in terms of the dielectric function $\varepsilon(Q)$ and the anomalies in $\varepsilon(Q)$ is reflected in the phonon dispersion curves as Kohn anomalies. For a cylindrical Fermi surface, of radius $k_F$, the singular part of $\varepsilon$ is given by

$$\begin{aligned} \varepsilon(Q) &\propto \sqrt{1 - 4k_F^2/Q^2} \quad \text{for} \quad Q > 2k_F \\ &= 0 \qquad\qquad\qquad\quad Q \leqslant 2k_F . \end{aligned} \tag{2.24}$$

The Kohn anomaly is expected to be as shown in Fig.2.22, which is very similar to the observed deviations in Fig.2.21, for the $\Sigma_1$ (nearly longitudinal) branch with $2k_F \sim 0.2$.

The deviation in the $\Sigma_3$ (nearly transverse) branch may possibly be attributed to the same origin. If, furthermore, the Fermi surface is distorted from a cylinder in such a way that some portions are made of sets of two more or less parallel sheets, as found in band calculations, the Kohn anomaly may be strong enough to set up a lattice instability in the form of charge density waves. Such an instability

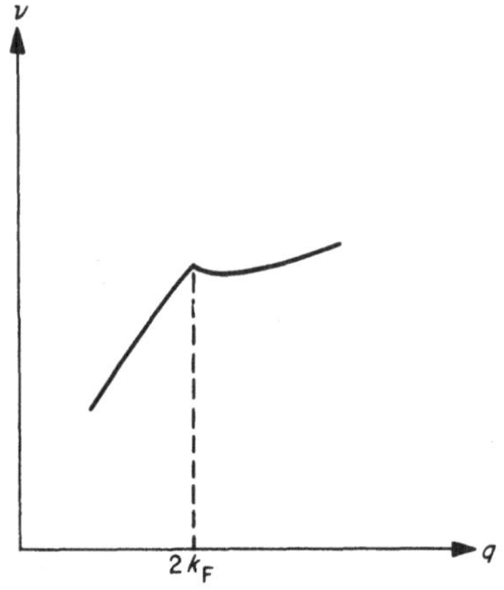


Fig. 2.22 The form of the Kohn anomaly for a cylindrical Fermi surface. (WAKABAYASHI et al., 1974)

was, indeed, observed (MONCTON et al., 1975) in $TaSe_2$ at $\underline{q} \cong (0.3,0,0)$ and to a lesser degree in $NbSe_2$ at low temperatures.

### 2.4.3 Graphite

The graphite structure is shown in Fig.2.23. Each layer of the crystal is a sheet of carbon atoms arranged in a hexagonal network (there are four atoms in the unit cell). A large single crystal is not available at present, and measurements have been performed only on pyrolytic graphite. Pyrolytic graphite is formed by thin graphite crystallites randomly oriented about a common hexagonal axis. The reciprocal

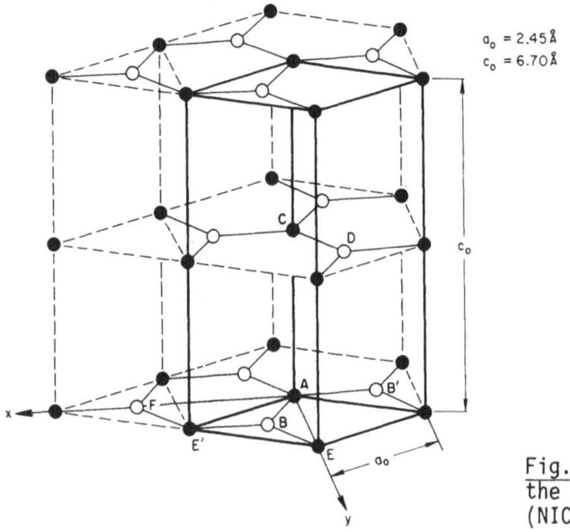


Fig. 2.23 Perspective drawing showing the positions of the atoms in graphite. (NICKLOW et al., 1972)

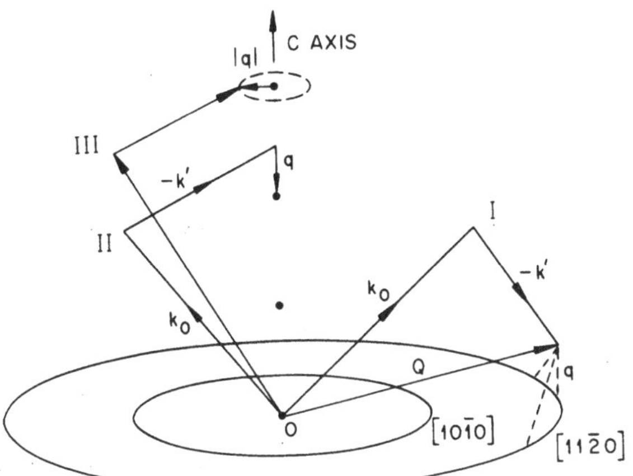


Fig. 2.24 Reciprocal-space diagram for pyrolytic graphite showing three of the neutron-scattering configurations. (NICKLOW et al., 1972)

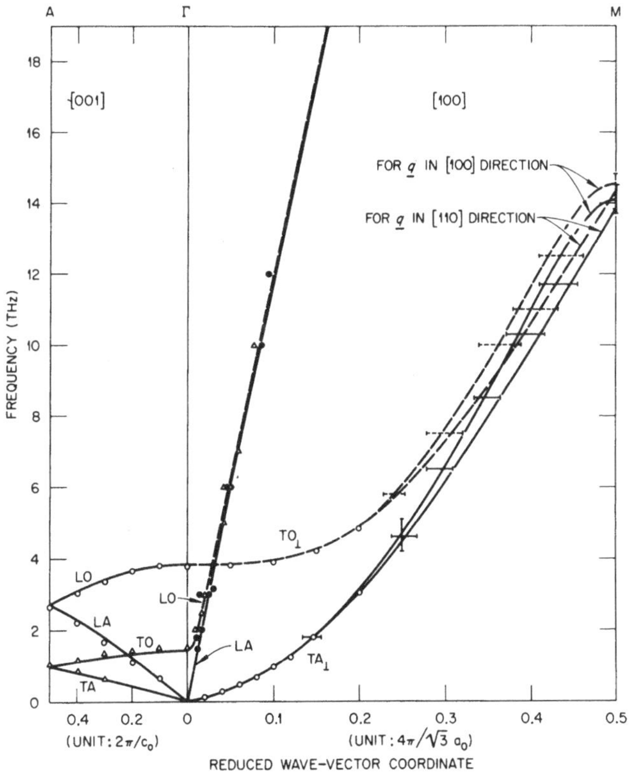


Fig. 2.25 Phonon frequencies measured for pyrolytic graphite. All measurements for q in the basal plane have been plotted with q expressed in units appropriate for the [100] direction. The lines shown represent calculations with the fourth-neighbor AS model. For the $TA_\perp$ and $TO_\perp$ branches calculations for q in both the [100] and [110] directions are shown to illustrate the approximate isotropy of these branches. (NICKLOW et al., 1972)

lattice may be visualized as that of graphite rotated about the c*-axis (Fig.2.24). In the experimental configuration I, the momentum transfer $\underline{Q}$ cannot be reduced to a unique phonon wave-vector $\underline{q}$ by $\underline{Q} = \underline{q} + \underline{\tau}$, since $\underline{\tau}$ is not well defined as it is in the case of a single crystal. Only for $\underline{Q}$ along the c*-direction, given by the configuration II, is $\underline{q}$ unique and the phonon dispersion curves for the longitudinal branches in this direction can be determined unambiguously. The configuration III is intermediate in that the magnitude of $\underline{q}$ is unique in this configuration. However, the phonon dispersion curves of many crystals with hexagonal symmetry have been found to be quite isotropic in the basal plane. Thus measurements for transverse modes in the basal plane, having the eigenvectors along the c-direction, were attempted in this configuration. The experimental results (NICKLOW et al., 1972) confirmed the isotropy of the dispersion surface.

Information about low frequency modes obtained directly, or indirectly, from such measurements are summarized in Fig.2.25. Although the data cover very small portions of the entire dispersion surfaces, the distinction betwen the inter- and intralayer

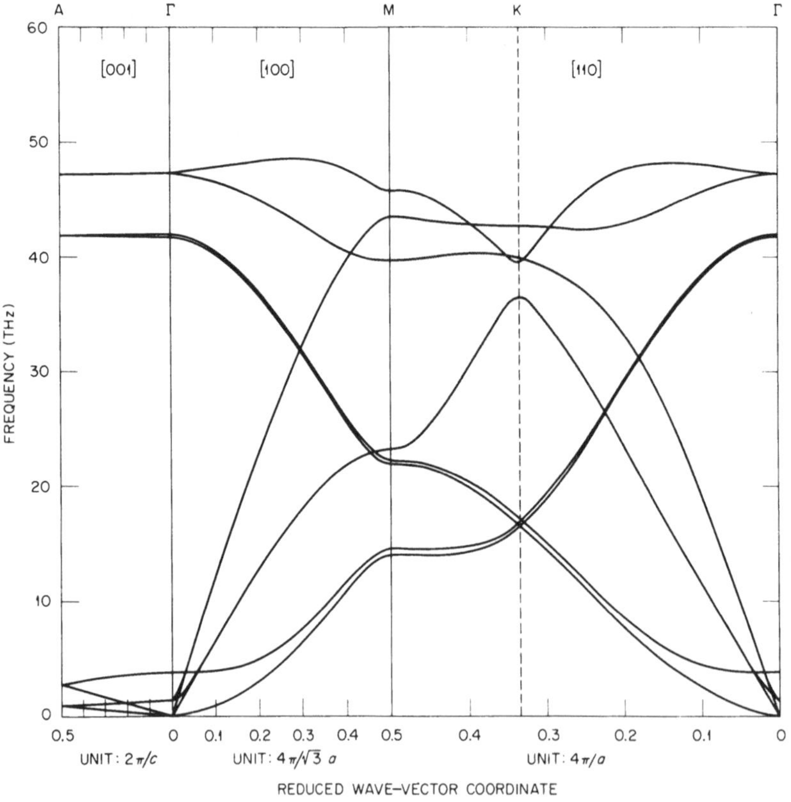


Fig. 2.26 Phonon dispersion relation for graphite in the [001], [100], and [110] directions as calculated with the fourth-nearest-neighbor AS model. (NICKLOW et al., 1972)

modes is clearly demonstrated. Details of the data reduction and various force model analyses are given in the original paper. The solid lines in Fig.2.26 represent the result of a fit of the data with an axially symmetric model. The phonon density of states and the temperature dependence of the specific heat were calculated. Fig.2.27 shows the comparison between the measured and calculated specific heat. The calculation reproduces the experimental results quite well and, in particular, the transition from a $T^3$ to a $T^2$ dependence at a very low temperature, which is another manifestation of the quasi two-dimensional character of the lattice vibrations.

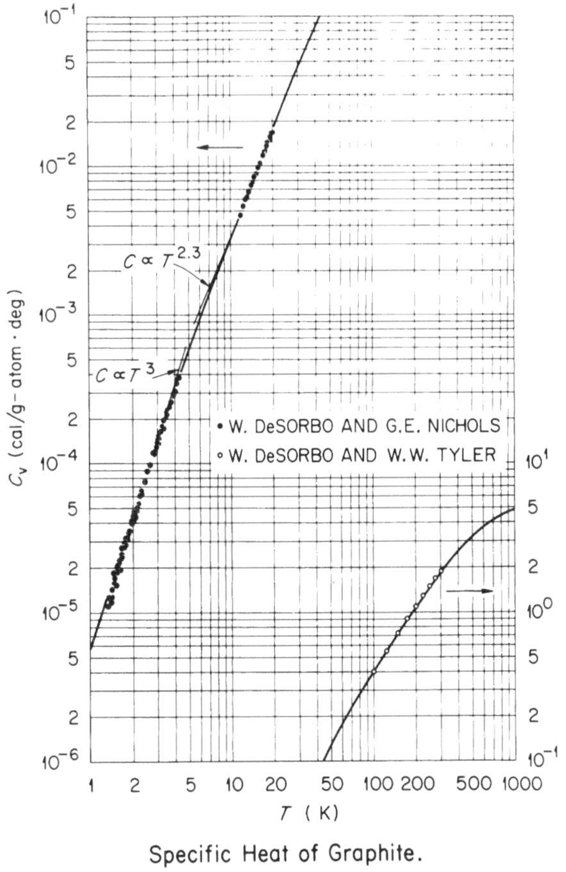


Fig. 2.27 Comparison of the measured specific heat and that calculated from the frequency distribution function. (NICKLOW et al., 1972)

## 2.4.4 $PbI_2$

There is a large group of materials which are often classified as layered compounds but do not show quasi two-dimensional character. $PbI_2$ may be regarded as an example of such materials. Although it is a semiconductor, light scattering experiments indicate that the ionicity of $PbI_2$ is quite large and the electrostatic forces are expected to give large contributions to the interatomic interactions. Because of the large number of electrons on the iodine atoms, the van der Waals interactions between layers are also expected to be much greater than in graphite or $MoS_2$.

It exists in several polytypes (2H, 4H, 12R). The 2H-polytype has the $CdI_2$ crystal structure and contains only one chemical unit in the unit cell. The layer spacing is given by the c-lattice constant (6.979 Å) which is equal to the distance between Pb atoms in two neighboring layers. The distance between Pb atoms within the layer is 4.557 Å and the ratio between those two distances is 1.53, which is to be compared with 1.97 for this ratio in $MoS_2$. The more isotropic atomic arrangement combined with a large ionicity manifests itself clearly in the phonon dispersion curves (Fig.2.28), as measured by HARBEKE et al.(1976), and no two-dimensional character is evident.

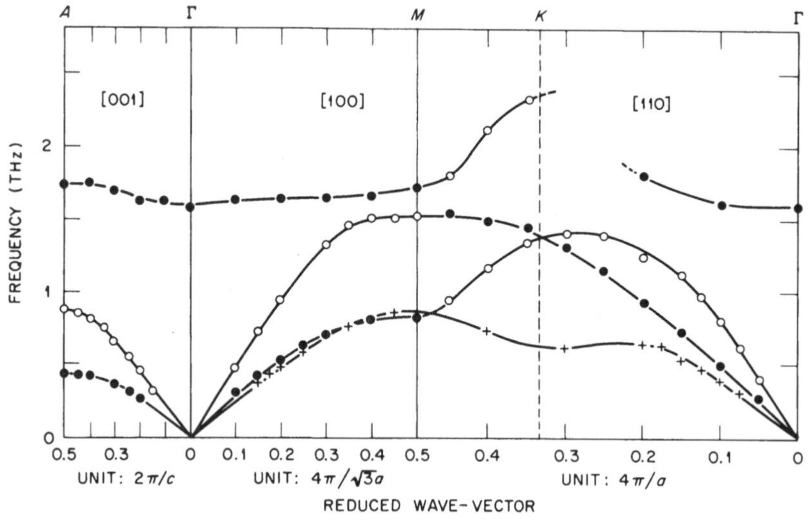


Fig. 2.28 Phonon dispersion curves in $PbI_2$: o measured in longitudinal geometry • in transverse geometry in the hexagonal plane, + in transverse geometry parallel to z. The curves are guidelines for the eye. (HARBEKE et al., 1976)

Iodine is another example of a layered compound that does not show two-dimensional behavior (SMITH et al., 1975) in its phonon spectra. Presumably this is because the van der Waals interactions between the layers are comparable to the weak intermolecular bonding within the layers.

## 2.4.5 Concluding Remarks

Another class of two-dimensional compounds which are highly interesting, but have not been discussed here, is intercalated transition metal dichalcogenides, whereby various organic molecules or alkali atoms are sandwiched between the layers. The optical and electronic properties are affected drastically; for example, $MoS_2$, a semiconductor, becomes a superconductor when intercalated with the alkali metals. The superconducting transition temperature of $TaS_2$ is increased on intercalation with various organic molecules; however, $NbSe_2$ shows a decrease in $T_c$ upon intercalation. Intercalated samples suitable for inelastic neutron scattering have not been available, but hopefully they can be prepared in the near future.

## 2.5. Effects of Pressure on Phonon Spectra

In recent years there has been an upsurge of interest in the Materials Sciences and in the search for new materials which have properties superior to those commonly available and/or for replacement of those that are rare or in short supply. In this regard the synthesis, by high pressure techniques, of new phases and compounds that may be of technological interest is an ever expanding field. However, an understanding of the effects of pressure on their equations of state, and the thermodynamical variables involved, is essential to these developments and, therefore, many basic physical and chemical studies on even relatively simple materials under pressure are necessary to reach these goals.

Neutron scattering methods, both elastic and inelastic, are good probes for studying many of the properties of materials under pressure. Thermal neutrons have great penetrating power and, with a few exceptions, can enter and exit pressure vessels, cryostats, and furnaces without a great loss of intensity (however, excessive scattering by thick vessel walls can be a problem). Elastic scattering of neutrons is valuable not only in studying the structural and magnetic phase changes which take place under applied pressure, but also in readily obtaining the compressibilities of the materials. Nevertheless, it is the inelastic scattering of thermal neutrons which is, perhaps, of greatest value in investigating the microscopic quantities of the equation of state, since it is a sensitive probe of the interatomic potential of the solid (and liquid also). Although the interatomic potential is somewhat of an elusive quantity to determine, except for some very simple substances, the anharmonic properties of crystal lattices, which are measureable, are intimately related to it.

A brief description of anharmonic effects in terms of phonon-phonon interactions is given in Subsec.1.3.2 of the introductory chapter, where the phonon shifts $\Delta$ and widths $\Gamma$ are related to the coherent one-phonon cross-section through the dynamic susceptibility $\tilde{\chi}_j(g,i\omega)$ of (1.100) and (1.102). (Thorough treatments of anharmonic lattices may be found in Vol. 1, HORTON and MARADUDIN (1974) and WALLACE (1972)).

Extensive measurements have been made on the effects of temperature on phonon frequencies, often with the hope of learning something about the anharmonicity of crystal lattices. It is desirable to separate quasi-harmonic effects (due to thermal expansion only) from real anharmonic effects. Varying both pressure and temperature offers a means of accomplishing this. The remainder of this section will be concerned with measuring the effects of pressure and temperature on the phonon frequencies and line widths in neon and rubidium.

### 2.5.1 Experimental Aspects

Both time-of-flight and triple-axis neutron spectrometers have been employed in inelastic neutron scattering investigations of materials under pressure. Generally, the studies on liquids and gases, and quasielastic scattering of solids have utilized the time-of-flight spectrometer and incoherent scattering techniques, whereas the phonon groups and line shapes of single crystals have been determined on triple-axis spectrometers. (There are no inherent reasons for not using time-of-flight instruments for some single crystal studies; in fact, there are some instrumental advantages in doing so).

Because of the large sample volumes usually employed in inelastic neutron scattering studies, the pressure vessels required are rather large and careful choices must be made of the pressure device, vessel material, and pressure transmitting medium. The pressure and temperature range of the experiments will also influence the above choices. For an extensive review, the reader is referred to a report by CARLILE and SALTER (1975) on high pressure techniques for thermal neutron scattering experiments. Most of the experiments described below were carried out at low temperatures and used the techniques described by DANIELS et al. (1967) and KLEB and COPLEY (1975); that is, a high pressure He gas generating system connected to an aluminium pressure vessel, which is mounted in a cryostat, by long, thin high pressure tubing. The advantages of the high pressure He gas or liquid pressure transmitting medium are the ease of obtaining low temperatures and also the ease of changing the sample pressure or temperature at will. The disadvantages are that the pressures attainable are definitely below 10 kb and more than likely less than 6 kb. Also, the fact that the stored energy of such a system is much greater relative to a liquid transmittant under the same pressure requires extra safety precautions.

The principle of the neutron measurements is essentially unchanged, except that, since the changes in phonon line shape are one of the primary goals of the experiment, it is necessary to have higher resolution and proper knowledge of the convolution of the spectrometer resolution function with the experimentally determined dispersion curves.

BLOCH et al. (1976) have recently demonstrated a compact preclamped type device that accepts single crystal specimens, large enough for inelastic scattering studies, at liquid He temperatures with pressures up to 30 kb. The flexibility of pressure variation is somewhat curtailed, but since the changes in the phonon frequencies and widths per unit pressure are small for most materials, such a device should increase substantially the number of substances that can be studied.

## 2.5.2 Phonon Shifts and Widths in Neon

The first experiments to measure anharmonic properties of the phonon dispersion curves at constant volume are those of LIPSCHULTZ et al. (1967), on helium, and LEAKE et al. (1969) on neon. The most extensive measurements are those on neon and these have been analysed in terms of analytic pair potentials. Both conventional perturbation theory and self-consistent phonon theory were employed in the analyses. Fig.2.29 shows the longitudinal and transverse branches in the [001] directions for crystal A ($a_0$ = 4.462 A) and crystal B ($a_0$ = 4.402 A) at 4.7 K. The differences in the phonon frequencies are relatively large considering the modest differences in pressure (0.6 kb) and volume (4%). The microscopic Grüneisen parameters,

$$\gamma_i \equiv -\left(\frac{\partial \ln \omega_i}{\partial \ln V}\right), \tag{2.25}$$

averaged over wave-vector, for the L and T branches are 4.2 ± 0.6 and 2.9 ± 0.5, respectively. Although the changes in phonon frequencies were small, when crystal B was heated to 25 K while keeping the volume constant, it was evident that the L branch increased slightly in frequency, whereas the T branch either decreased slightly or remained constant (the behavior of the latter branch is contrary to theoretical predictions). More extensive calculations led to additional neutron

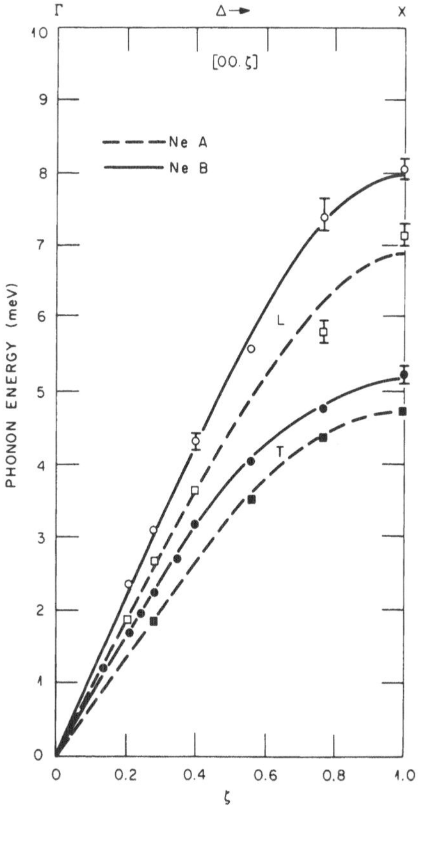


Fig. 2.29 Phonon dispersion curves in the [100] direction in Ne(A) and Ne(B) crystals at 4.7 K. (LEAKE et al., 1969)

scattering studies (SKALYO et al., 1972) of a more precise nature on the anharmonic properties of neon. The later, more precise, measurements on neon confirmed the original findings. In addition, the increased resolution permitted the measurement of the widths of some of the transverse phonons at 25 K. However, the resolution analysis also indicated that the measurements of the high energy zone boundary frequencies (>6 meV) and widths were hampered by multiphonon scattering, which was broad in energy and comparable in intensity to the one-phonon scattering. (Phonon measurements have recently been made on a single crystal of neon with a volume change of 16% by ECKERT et al.(1976) at Brookhaven National Laboratory).

### 2.5.3 Selected Phonons in Rubidium as a Function of Temperature and Volume

The lattice dynamics of rubidium, as a function of temperature, were studied thoroughly by COPLEY and BROCKHOUSE (1973). The observed data, properly corrected for instrumental resolution, were compared with calculated frequencies based on Born-von Kármán and pseudopotential theories, in the harmonic approximation, with good agreement for both models. In a subsequent paper COPLEY (1973) investigated the anharmonic properties of rubidium in which he calculated, to first order, the phonon frequency shifts and widths, Grüneisen parameters, thermal strain, and the lattice heat capacity, using a real space effective interatomic potential derived (PRICE et al., 1970) by fitting an Ashcroft pseudopotential with a self-consistent dielectric screening function to the low temperature (12 K) phonon measurements. Rubidium has a large compressibility and thermal expansivity, and large single crystals can easily be grown; therefore, it was a particularly good candidate for high pressure studies.

Measurements of selected phonons as a function of volume at constant temperature, and as a function of temperature at constant volume, have been reported by COPLEY

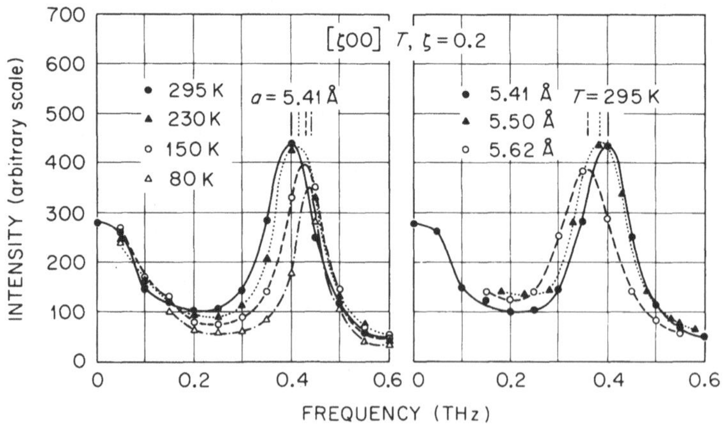


Fig. 2.30 Neutron groups for the [ζ00]T, ζ = 0.2 mode, measured at Q = [2,0.2,0] x 2π/a, normalized to the same number of monitor counts. The lines are drawn to guide the eye. (COPLEY et al., 1974)

et al. (1974). The measurements were made on the HB-4A spectrometer at the HFIR, Oak Ridge National Laboratory. The total volume change during the experiment was about 15% and the temperature range was from 80 to 295 K. Some neutron groups of the best-determined mode, [$\zeta$00]T at $\zeta$ = 0.2, are shown in Fig.2.30. It is clear that, at least for the highest density obtained, the phonon shifts and widths are readily observable and that the width of the mode is also increasing with increasing temperature. Fig.2.31 shows the temperature dependence of four normal modes, for three values of the lattice parameter. Here, a large decrease in frequency of the [$\zeta$00]T mode as the temperature increases is clearly evident for all three volumes. At the highest density, *increases* in the [$\zeta$00]L mode with temperature are indicated. These observations are consistent with those noted in the neon studies. The average values for the zero-pressure mode-Grüneisen parameters were determined to be: [$\zeta$00]L, $\zeta$ = 0.2, 2.15 ± 0.23; [$\zeta\zeta$0]L, $\zeta$ = 0.2, 1.65 ± 0.07; [$\zeta$00]T, $\zeta$ = 0.2, 0.99 ± 0.13; and [$\zeta\zeta$0]T1, $\zeta$ = 0.3, 1.31 ± 0.25. Except for the [$\zeta\zeta$0]T1 mode, they are in excellent agreement with those computed by COPLEY (1973) (2.1, 1.7, 1.1, and 3.4, respectively) from his effective potential. The large discrepancy for the [$\zeta\zeta$0]T1 mode indicates that, perhaps, anharmonic terms of higher order may be important for some branches.

Since the [$\zeta$00]T, $\zeta$ = 0.2 phonon shows a definite dependence on temperature at constant volume, it is clear that the quasi-harmonic approximation is insufficient.

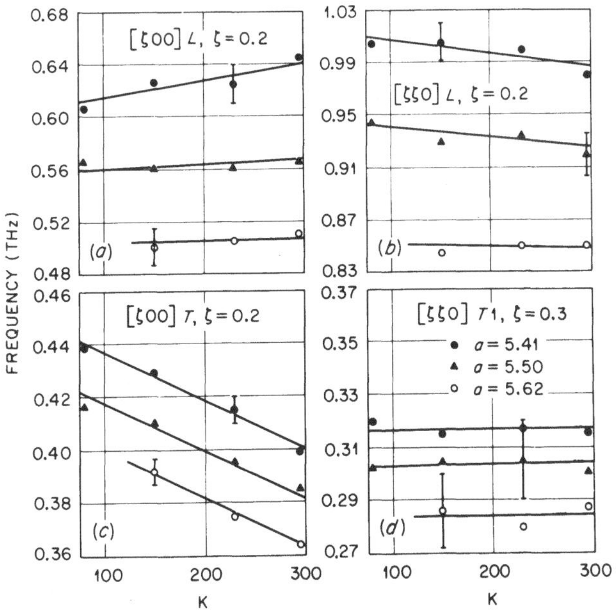


Fig. 2.31 The temperature dependence of four normal modes for three values of the lattice parameter. Typical error bars are shown. Note the different ordinate scales. The modes shown in (a), (b), (c), and (d) were measured at Q's of {[2.2.,0,0], [1.2,1.2,0], and [0.7,1.3,0]} x $2\pi/a$, respectively. The lines are drawn to guide the eye. (COPLEY et al., 1974)

However, it has been pointed out by COPLEY and others that for most branches, to first order, the cubic and quartic contributions to the phonon self-energy are of opposite sign and largely cancel. Thus, the absence of a detectable shift in a mode's frequency with T at constant V, does not imply the *absence* of anharmonicity. His calculations show that for the [ζ00]T branch the two contributions are of the same sign (negative with increasing temperature).

These measurements, although only a first attempt, provide valuable data for comparison with anharmonic calculations. It is highly desirable that a full set of dispersion curves be made at a higher density in order to determine another effective potential for comparison with COPLEY's volume dependent potential calculations. Measurements of the mode-Grüneisen parameters in K (MEYER et al., 1976) and RbI (BLASCHKO et al., 1976) have recently been made at the Institute Laue-Langevin, Grenoble.

### 2.5.4 Conclusions

Hopefully, these two examples illustrate the power and usefulness of phonon measurements under pressure, and will encourage additional theoretical and experimental investigations in this area of research. The rather low pressures attainable with the the He gas systems are adequate for many anharmonic studies on the rare gas systems and on other more complicated molecular crystal systems, as well as the alkali metals potassium and cesium. The availability of a low temperature, higher pressure device capable of permitting single crystal inelastic neutron scattering measurements up to pressures of 30 kb opens the field to other new and exciting problems. The results are eagerly awaited.

*Acknowledgments*

The authors wish to thank M. Mostoller and R. M. Nicklow for many helpful discussions.

### References

Allen,Phillip,B. (1972) : Phys. Rev. B6, 2577
Allen,P.B., Dynes,R.C. (1975) : Phys. Rev. B12, 905
Almairac,R., Benoit,C. (1974) : J. Phys. C7, 2614
Almairac,R. (1975a) : Thesis. Université des Sciences at Techniques de Languedoc, Montpellier, France
Almairac,R. (1975b) : J. de Phys. 36, (C-73)
Axe,J.D., Shirane,G. (1973) : Phys. Rev. B8, 1965
Barker,A.S., Jr. (1964) : Phys. Rev. 136, A1290
Bilz.H. (1958) : Z. Physik 153, 338
Bilz,H., Gliss,B., Hanke,W. (1974) : *Dynamical Properties of Solids*, Vol. 1, Chap.6, ed. by G.K.Horton, A.A.Maradudin (North-Holland, New York) pp. 343-390
Blaschko,O., Ernst,G., Quittner,G., Kress,W., Lechner,R.E. (1975) : Phys. Rev. B11, 3960
Bloch,D., Paureau,Jr., Voiron,J., Parisot,G. (Private communication)
Brockhouse,B.N., Arase,T., Caglioti,G., Rao,K.R., Woods,A.D.B. (1962) : Phys. Rev. 128, 1099
Carlile,C.D., Salter,D.C. (1975) : High Pressure Techniques for Thermal Neutron Shattering Experiments, Rutherford Laboratory Report RL-75-096

Copley,J.R.D., Brockhouse,B.N. (1973) : Can. J. Phys. 51, 657
Copley,J.R.D. (1973) : Can. J. Phys. 51, 2564
Copley,J.R.D., Rotter,C.A., Smith,H.G., Kamitakahara,W.A. (1974) : Phys. Rev. Lett. 33, 365
Cowley,R.A., Cochran,W., Brockhouse,B.N., Woods,A.D.B. (1963) : Phys. Rev. 131, 1030
Cran,G.C., Sangster,M.J.L. (1974) : J. Phys. C, Solid State Phys. 7, 1937
Daniels,W.B., Shirane,B., Frazer,B.C., Umebayashi,H., Leake,J.A. (1967) : Phys. Rev. Lett. 18, 548
Dolling,G. (1974) : *Dynamical Properties of Solids*, Vol.1, Chap.10, ed. by G.K. Horton, A.A.Maradudin (North-Holland, New York) pp. 541-629
Dynes,R.C., Rowell,J.M. (1975) : Phys. Rev. B11, 1884
Dynes,R.C., Garno,J.P. (1975) : Bull. Amer. Soc. 20, 422
Eagles,D.M. (1964) : J. Phys. Chem. Solids 25, 1243
Eckert,J., Daniels,W.B., Axe,J.D. (1976) : Proceedings of Neutron Scattering Conference, Gatlinburg, Tenn
Eichler,A., Wühl,H., Stritzker,B. (1975) : Solid State Commun. 17, 213
Hanke,W. (1973) : Phys. Rev. B8, 4585
Harbeke,G., Dorner,B. (1976): Phys. Stat. Sol. (in the press)
Hopfield,J.J. (1969): Phys. Rev. 186, 443
Horton,G.K., Maradudin,A.A. (ed.) (1974) : *Dynamical Properties of Solids*, Vol. 1 (North-Holland, New York)
Jandl,S., Brebner,J.L., Powell,B.M. (1976) : Phys. Rev. B13, 686
Kleb,R., Copley,J.R.D. (1975) : Rev. Sci. Instr. 46, 1190
Klein,Barry M., Birman,Joseph,L. (1970) : Phys. Rev. Lett. 25, 1014
Koch,C.C., Scarbrough,J.O., Kroeger,D.M. (1974) : Phys. Rev. B9, 15
Kohn,W. (1959) : Phys. Rev. Lett. 2, 393
Larose,A., Vanderwal,J. (1974) : *Scattering of Thermal Neutrons - A Bibliography, Solid State Physics Literature Guides*, Vol. 7, gen. ed. by T.F.Connolly (IFI/Plenum, New York)
Leake,J.A., Daniels, W.B., Skalyo,J., Frazer,B.C., Shirane,G. (1969) : Phys. Rev. 181, 339
Lipschultz,F.P., Minkiewicz,V.J., Kitchens,T.A., Shirane,G., Natans,R. (1967) : Phys. Rev. Lett. 19, 1302
Matheiss,L.F. (1973) : Phys. Rev. B8 , 3719
McMillan,W.L. (1968) : Phys. Rev. 167, 331
Moncton,D.E., Axe,J.D. Disalvo,F.J. (1975) : Phys. Rev. Lett. 34, 734
Mostoller,M., Henderson,B., Sibley,W.A., Wood,R.F. (1971) : Phys. Rev. B4, 2667
Meyer,J., Dolling,G., Klaus,J., Lüscher,E., Vettier,C., Paureau,J. (1976) : *Annual Report Annex* (Institute Laue-Lagevin, Grenoble, France)
Nakagawa,Y., Woods,A.D.B. (1963) : Phys. Rev. Lett. 11, 271
NG,S.C., Brockhouse,B.N. (1968) : In *Neutron Inelastic Scattering*, Vol. 1, p. 253 (IAEA, Vienna)
Nicklow,R.M., Wakabayashi,N., Smith,H.G. (1972) : Phys. Rev. B5, 4951
Phillips,J.C. (1971) : Phys. Rev. Lett. 26, 543
Phillips,J.C. (1972): In *Superconductivity in d- and f-band Metals*. Vol. 4, ed. by D.H. Douglass (AIP, New York)
Porto,S.P.S., Fleury,P.A., Damen,T.C. (1967): Phys. Rev. 154, 522
Powell,B.M., Woods,A.D.B., Martel,P. (1972): *Neutron Inelastic Scattering*. pp. 43-51 (IAEA, Vienna)
Powell,B.M., Martel,P., Woods, A.D.B. (1968) : Phys. Rev. 171, 727
Price,D.L., Singwi,K.S., Tosi,M.P. (1970): Phys. Rev. B2, 2983
Pytte,E. (1970) : Phys. Rev. Lett. 25, 1176
Reichardt,W. (1975) : Progress Report KFK, Kernforschungszentrum, Karlsruhe, Germany
Rotter,C.A., Traylor,J.G., Smith,H.G. (1975) : Bull. Amer. Phys. Soc. 20, 300
Rowe,J.M., Rush,J.J., Smith,H.G., Mostoller,M., Flotow,H.E. (1974) : Phys. Rev. Lett. 33, 1297
Rowe,J.M., Vagelatos,N., Rush,J.J. (1975): Phys. Rev. B12, 2959
Rundle,R.E. (1948) : Acta. Cryst. 1, 180
Shapiro,S.M., Shirane,G., Axe,J.D. (1975) : Phys. Rev. B12, 4899

Shen,L.Y.L. (1972) : Phys. Rev. Lett. 29, 1082
Silvermann,P.J., Briscoe,C.V. (1975) : Phys. Lett. 53A, 221
Sinha,S.K., Harmon,B.W. (1975) : Phys. Rev. Lett. 35, 1515
Skalyo,J., Jr., Minkiewicz,V.J., Shirane,G., Daniels,W.B. (1972) : Phys. Rev. B36, 4766
Smith,H.G., Gläser,W. (1970) : Phys. Rev. Lett. 25, 1611
Smith,H.G. (1972) : In *Superconductivity in d- and f-Band Metals*, Vol. 4, ed. by D.H.Douglas (AIP, New York)
Smith,H.G. (1972) : Phys. Rev. Lett. 29, 353
Smith,H.G. (1974) : Annual Progress Report ORNL-4952
Smith,H.G., Wakabayashi,N., Nicklow,R.M., Mihailovich,S. (1974) : *Proc. Low Temperature Physics*-LT13, Vol. 3, ed. by K.D.Timmerhaus, W.J. Sullivan and E. F.Hammel (Plenum Publishing, New York)
Smith,H.G., Nielsen,M., Clark,C.B. (1975) : Chem. Phys. Lett. 33, 75
Smith,H.G., Wakabayashi,N. (1976) : Bull. Amer. Phys. Soc. 21, 410
Stirling,W.G., Dorner,B., Cheeke,I.D.N. (1975) : (to be published)
Testardi,L.R. (1971) : Phys. Rev. B3, 95
Traylor,J.G., Smith,H.G., Nicklow,R.M., Wilkinson,M.K. (1971) : Phys. Rev. B3, 3457
Traylor,J.G. (1971) : Ph.D. Thesis, University of Tennessee
Traylor,J.G., Wakabayashi,N. Sinha,S.K. (1975) : Bull. Amer. Phys. Soc. 20, 300
Venkataraman,G., Feldkamp,L.A., Sahni,V.C. (1975) : *Dynamics of Perfect Crystals* (MIT Press, Cambridge, Mass.)
Verma,M.P., Gupta,B.R.K. (1975) : Phys. Rev. B12, 1314
Vosko,S.H., Taylor,R., Keekh,G.H. (1965): Can. J. Phys. 43, 1187
Wakabayashi,N., Smith,H.G., Nicklow,R.M. (1975) : Phys. Rev. B12, 659
Wakabayashi,N., Smith,H.G., Shanks,R. (1974) : Phys. Lett. 50A 367
Wakabayashi,N. (1975) : (Private communication)
Wakabayashi,N. (1976) : Annual Progress Report ORNL
Wallace,D.C. (1972) : *Thermodynamics of Crystals* (John Wiley and Sons, New York)
Weber,W. (1973) : Phys. Rev. B8, 5082
Wieting,T.J., Verble,J.L. (1971) : Phys. Rev. B3, 4286
Woods,A.D.B., Cochran,W., Brockhouse,B.N. (1960) : Phys. Rev. 119, 980
Woods, A.D.B., Chen,S.H. (1964) : Solid State Commun. 2, 233

Recent References with Titles:

Sakamoto, M., Chihara, J., Nakahara, Y., Kalotani, H., Sekiya, T., Gotoh, Y. (1976): *Bibliography for Thermal Neutron Scattering*, 5th ed. JAERI-M6857. Division of Technical Information, Japan Atomic Energy Research Institute, Tokai-mura, Naka-gun, Ibaraki-ken, Japan

Smith, H.G., Wakabayashi, N., Mostoller, M. (1976): Phonon Anomalies in Transition Metals, Alloys and Compounds. In *Supersonductivity in d- and f-Band Metals*, ed. by D.H. Douglass (Plenum Publishing Corp., New York)

Sinha, S.K., Harmon, B.N. (1976): Phonon Anomalies in d-Band Metals and Their relationship to Superconductivity. In *Superconductivity in d- and f-Band Metals*, ed. by D.H. Douglass, (Plenum Publishing Corp., New York)

Harmon, B.N., Sinha, S.K. (1976): Calculation of the Electron-Phonon Spectral Function of Niobium. In *Superconductivity in d- and f-band Metals*,ed. by D.H. Douglass (Plenum Publishing Corp., New York)

Hanke, W., Hafner, J., Bilz, H. (1976): Phonon anomalies and superconductivity in transition metal compounds. Phys. Rev. Lett. 37, 1560

Gupta, M., Freeman, A.J. (1976): Role of electronic structure on observed phonon anomalies of transition-metal carbides. Phys. Rev. B. 14, 5205

Chadi, D.J., Cohen, M.L. (1974): Electronic band structures and charge densities of NbC and NbN. Phys. Rev. B 10, 496

Schwarz, K., Rösch, N. (1976): Effects of carbon vacancies in NbC on superconductivity. J. Phys. C: Solid State Phys. 9, L433

Splettstösser, B. (1977): Simple shell model for phonon dispersion of nonstoichiometric niobium carbide. Z. Physik B 26, 151

# 3. Phonons and Structural Phase Transformations

B. Dorner and R. Comès

With 30 Figures

Investigations on structural properties of crystals are best performed with radiations of wavelength comparable to the interatomic distances, that is to say, X-rays or thermal neutrons. The complementary nature of X-ray and neutron scattering techniques is well known for structure determinations. Although less widely used, the techniques are complementary also for investigations dealing with phonons. Neutron scattering allows one to perform a frequency analysis of atomic motions, and the determination of phonon dispersion relations has now become almost routine (at least for simple systems). The earlier, indirect, determinations deduced from X-ray intensities deserve to be mentioned for historical reasons (WALKER, 1956).

In the particular case of structural phase transformations one is often interested in pretransitional effects, which involve not only dynamical aspects, best studied with neutrons, but also long-ranged, spacial, correlations. Because the long-ranged, spacial, correlations give rise to localized scattering in reciprocal space, they can also be investigated by X-ray scattering. For instance, low frequency modes existing in particular, and often unexpected special, directions in reciprocal space are easily detectable with X-rays in photographic patterns, as their intensity is proportional to the mean square displacements of the atoms, or to $\omega^{-2}$ (see (1.44) and (3.13)). Moreover, for both techniques, the variation of intensity from one Brillouin zone to another is governed by a structure factor similar to the static structure factor used for structure determinations, which allows complementary information in order to perform mode determinations. A particular aim of this chapter is to illustrate some studies in which the simultaneous use of X-ray and neutron scattering techniques has proved fruitful.

As an introductory example, we present some results obtained on the ferroelectric material $KNbO_3$ which, like $BaTiO_3$, undergoes three phase transformations with decreasing temperature. As these transformations involve only small distortion (≈1%) of the unit cell we shall, for simplicity, always refer to the cubic axes of the high temperature cubic paraelectric phase (space group Pm3m).

At 435°C a transition is observed to a ferroelectric, tetragonal phase (space group P4mm) with a spontaneous polarization directed along [001]; then at 225°C this phase goes over to a second, ferroelectric phase of orthorhombic symmetry (space group Bmm2) with a polarization directed along [101]; finally, around -50°C a third ferroelectric phase of rhombohedral symmetry (space group R3m) with a spontaneous polarization now directed along [111] is formed.

Structure analysis has shown that, in a first approximation, the spontaneous polarization is produced by opposite displacements of the niobium atoms and the oxygen framework, successively in [001],[101] and [111] directions (see Fig.3.1).

The four X-ray patterns, shown in Fig.3.2 and obtained with a technique described in Sec.3.1, correspond to single domain crystals, in the four phases of $KNbO_3$ (COMES et al., 1970).

Only the pattern of Fig.3.2a, corresponding to the low temperature rhombohedral phase, has the aspect expected from an ordinary, well-behaved crystal: namely broad diffuse spots arising from the small wave-vector acoustic phonons (the intensity is $\propto \omega^{-2}$) which happen to satisfy the reflection conditions for the fixed orientation of the sample, and a *slowly* varying background from the other phonons. From this pattern one can immediately deduce that only the low temperature phase corresponds to a well-ordered stable phase.

A striking feature of the patterns of the three higher symmetry phases (Fig.3.2 b-d) is the appearance of well-defined diffuse streaks going across the whole pattern. These streaks correspond to scattered intensity localized in [001] reciprocal planes. The pattern is modified at each phase transition.

In the orthorhombic phase, only one set of such sheets, namely the [010] reciprocal planes, is observed, as depicted schematically in the reduced Brillouin zone of Fig.3.3b.

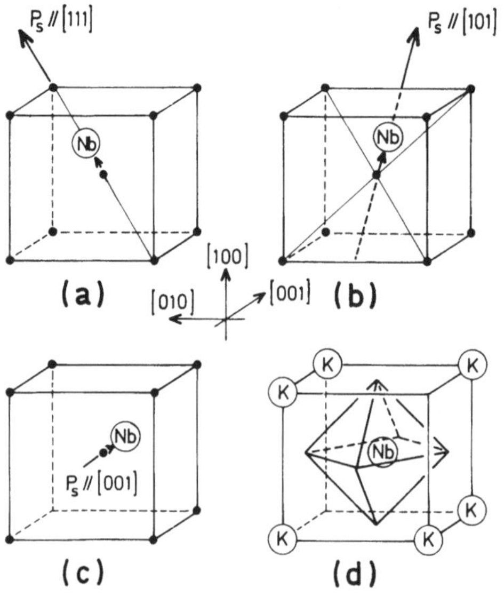


Fig. 3.1a-d Structure of $KNbO_3$ and spontaneous polarization P in the a) rhombohedral, b) orthorhombic, c) tetragonal and d) cubic phase

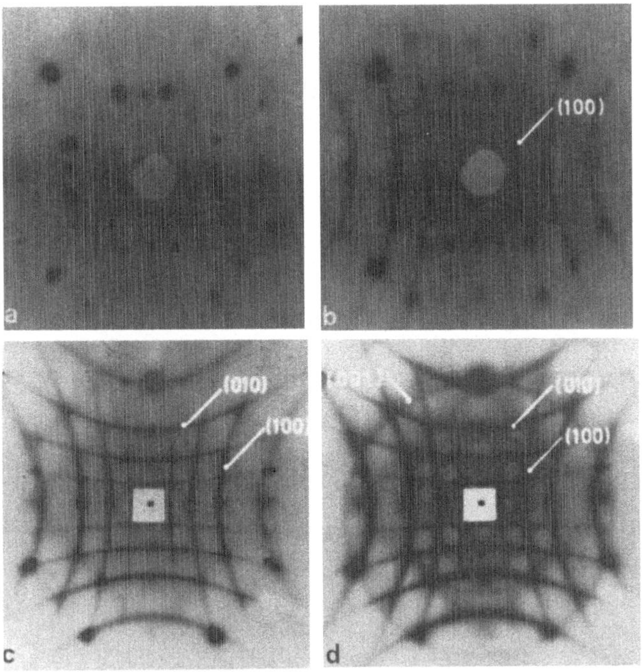


Fig. 3.2a-d Patterns of diffuse X-ray scattering in the a) rhombohedral, b) orthorhombic, c) tetragonal and d) cubic phase (COMES et al., 1970)

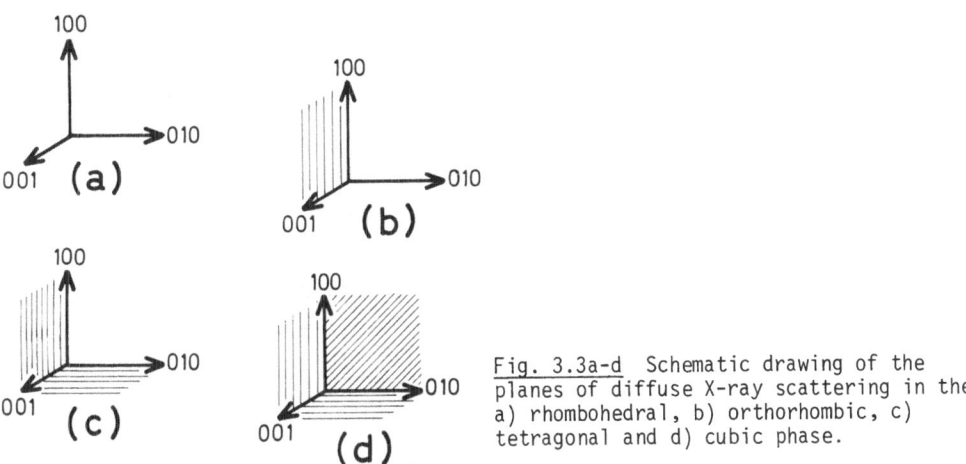

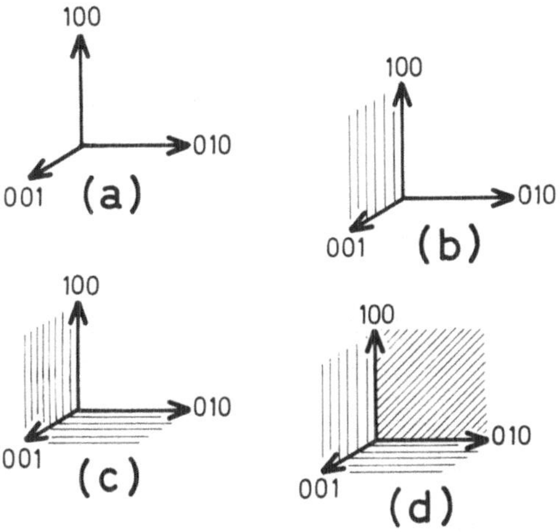


Fig. 3.3a-d Schematic drawing of the planes of diffuse X-ray scattering in the a) rhombohedral, b) orthorhombic, c) tetragonal and d) cubic phase.

In the tetragonal phase, a second set of such sheets becomes visible, so that scattering is now observed in [010] and [100] reciprocal planes (see Fig.3.3c).

A third set of such sheets is observed in the cubic phase. This set appears as rings on the patterns of Fig.3.2d, because it corresponds to intensity localized in the [001] reciprocal planes which are, under the particular experimental conditions, perpendicular to the incident X-ray beam. Extra intensity in this phase is consequently observed in all [100] reciprocal planes, as shown in Fig.3.3d.

As will be briefly shown in Subsec.3.1.3, intensity localized in reciprocal sheets corresponds to one-dimensional correlations in real space. It can also be simply shown that the streaks of the patterns of Fig.3.2. are due to displacive phenomena and not to chemical local order. One can therefore conclude that there are chains of atoms displaced, or moving, in phase independently from each other in the three higher temperature phases of $KNbO_3$.

The low temperature phase of $KNbO_3$ has a spontaneous polarization directed along [111], that is to say, with components along all three pseudo-cubic axes. The analysis of the sets of diffuse sheets, which appear successively when the temperature is increased, shows that additional one-dimensional correlations set in at each phase transition in the direction corresponding to the component of the spontaneous polarization, which disappears. In other words, when the temperature is increased, the long-range three-dimensional *displacive order*, which gives rise to the spontaneous polarization, is replaced by short-range, one-dimensional correlations, successively in the three directions of the cubic axes, until the paraelectric state is reached.

$KNbO_3$ has also been studied by inelastic neutron scattering, and the clearest results have been obtained in the orthorhombic phase (CURRAT et al., 1974). Besides the expected transverse zone center soft optic mode (see Subsec.3.2.1), these investigations have revealed another, unexpected, dynamical feature at the positions of diffuse X-ray intensity.

Figure 3.4 is a spectrum (constant Q scan) measured at a point contained in an orthorhombic reciprocal [010] plane; it clearly shows a well-defined, low frequency

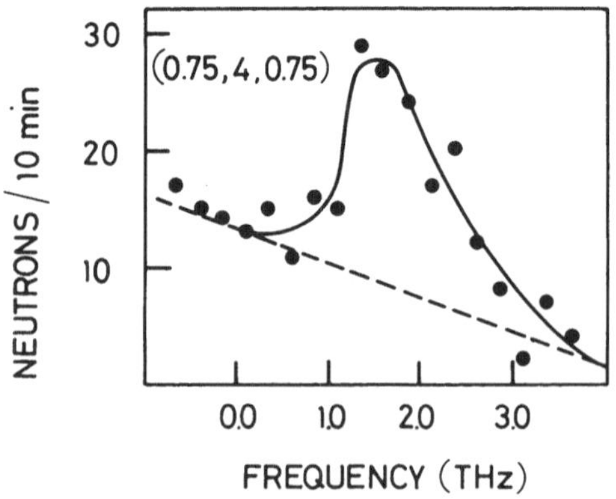


Fig. 3.4 Constant-Q scan through the low frequency mode in $KNbO_3$ in the orthorhombic phase (CURRAT et al., 1974)

excitation at about 1.5 THz. Similar scans measured at positions close to, but slightly away from, the diffuse planes give much higher frequencies. More detailed measurements showed that the dispersion surface of the phonons everywhere within the plane of diffuse X-ray scattering display such low frequency, and show no dispersion in the major part of the Brillouin zone towards the zone boundary (Fig.3.5). In contrast, as soon as measurements are performed away from the plane (Fig.3.6), the

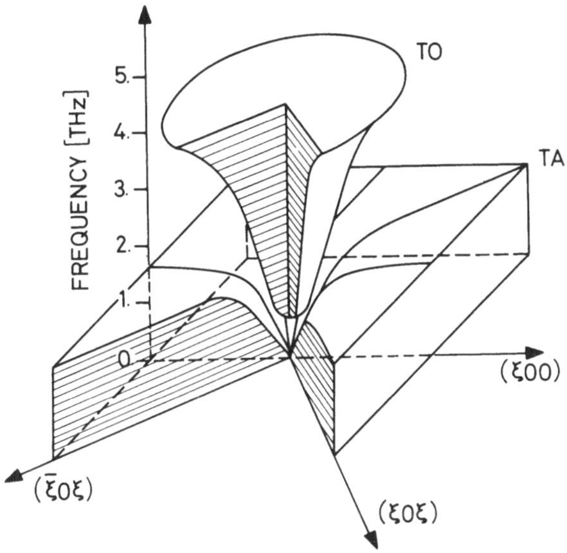


Fig. 3.5 Schematic view of the TA and TO dispersion surfaces of orthorhombic $KNbO_3$ in the (h,0,ℓ) plane ("soft plane") ( CURRAT et al., 1974)

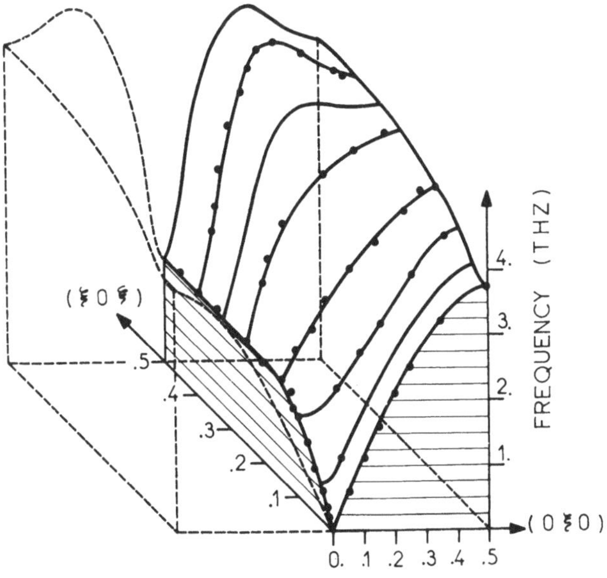


Fig. 3.6 Acoustic dispersion surface of orthorhomic $KNbO_3$ in the (h,k,h) plane. The rigid Nb-chains are oriented parallel to the [0ξ0] direction (CURRAT et al., 1974)

frequencies are found to increase steeply with increasing wave-vector. The intensity of phonons is proportional to $\omega^{-2}$; the low frequency spectra, restricted to the [010] reciprocal planes, thus explain the extra X-ray intensity in such planes.

The polarization of the atomic oscillations for these low frequency modes can be determined as well from the X-ray patterns as from the neutron measurements. From (3.8) and (3.10) we understand that only those atomic motions are visible which have a component of the eigenvector $\underline{\sigma}$ parallel to the momentum transfer $\underline{Q}$. On the X-ray patterns, for example, no intensity was found for $\underline{Q}$ parallel to the diffuse planes (Fig.3.2), which shows unambiguously that the atomic motion is perpendicular to the diffuse plane, in agreement with the neutron measurements which show that the low frequency modes are transverse.

Detailed eigenvector determinations (see Subsec.3.1.1) within the flat dispersion part of the reciprocal planes were performed, both with X-rays and with neutrons, with the conclusion that there is essentially *one atom* moving, with amplitudes about ten times larger than any other atom. This was understood from the fact that the intensity was varying monotonically with increasing Q. In principle, in such a case, it is impossible to determine which atom it is; here comparisons with X-ray patterns performed with other perovskites ($BaTiO_3$, $KTaO_3$) showed that it could only be the atom with the highest scattering factor, namely the niobium atom, which was moving.

The picture which emerges from these investigations is that of [010] chains of Nb atoms. These chains oscillate in the orthorhombic phase independent from each other (Einstein oscillators) with large amplitudes (low frequency) in the [010] direction. This direction is the direction in which one component of the static displacement of the Nb atom condensed out in the rhombohedral phase, establishing that these oscillating chains are part of the pretransitional effects.

The other part of these pretransitional phenomena is the zone center transverse soft optic mode. The investigation of this mode, and more generally of the whole branch labeled TO in Fig.3.5, can be performed exclusively with inelastic neutron scattering; it is foreshadowed in X-ray results by both the Bragg peaks, and the more intense lower frequency branch (labeled TA in Fig.3.5). The corresponding dispersion was also found to be anisotropic in the [010] reciprocal planes, but perhaps more important is the fact that a mode determination performed at finite wave-vector (q = 0.2, 0, 0,2) showed that it was essentially the rest of the lattice (all atoms but niobium) which was moving. The zone center mode of this branch, which is impossible to measure directly and is very likely overdamped, must, however, leave the center of gravity invariant for all atoms in the unit cell. At least qualitatively, it can be easily understood that mixing the character of both transverse branches, as established by the mode determinations performed at higher wave-vectors, can produce the required eigenvectors for the zone center mode, and a pattern which is very close to the static displacements which are added when going over from the orthorhombic phase to the rhombohedral phase.

The conclusion of these combined X-ray and neutron scattering studies of $KNbO_3$ is that both techniques have provided complementary information in the determination of a relatively complete picture of the pretransitional effects in the higher temperature phase. In principle, of course, the whole description could have been obtained from extended neutron measurements alone. In fact, the X-ray observations have motivated a much more general survey of the dispersion, including measurements outside the high symmetry directions. Considering the relative simplicity of such X-ray measurements, one can conclude that they should, generally, be performed prior to a neutron study; even if they do not reveal unexpected features, they can save a considerable amount of time, which is wisely taken into account for expensive neutron investigations.

We have chosen to refer in detail here only to the orthorhombic phase of $KNbO_3$ because it is one of the rare examples, among the ferroelectric perovskites undergoing phase transformations, in which most phonon modes are underdamped. Earlier, extensive investigations on various ferroelectric perovskites in the cubic phases with the exception of $PbTiO_3$ (SHIRANE et al., 1970) have generally revealed sheets of overdamped modes on the location of the X-ray streaks (YAMADA et al., 1969; SHIRANE et al., 1970; NUNES et al., 1971; YELON et al., 1971). Only the incipient ferroelectric $KTaO_3$ showed features in the cubic phase comparable with those described here for orthorhombic $KNbO_3$ (COMES, 1972).

## 3.1 Relation Between Neutron and X-Ray Scattering Techniques

In both techniques we define a wave-vector $\underline{k}$ in the direction of the beam:

$$|\underline{k}| = \frac{2\pi}{\lambda} \quad . \tag{3.1}$$

The wavelength $\lambda$ for thermal neutrons, about 1.8 Å, is comparable to X-ray wavelengths, Cu-$K_\alpha$ = 1.54 Å. The momentum transfer $\underline{Q}$ (exactly *$\hbar\underline{Q}$) is given by (see Fig.1.1 and 3.7)

$$\underline{Q} = \underline{k} - \underline{k}' \tag{3.2}$$

In an inelastic scattering process the energy, $\hbar\omega$, transferred to the sample, and lost by the radiation, is conventionally taken positive, i.e.,

$$\hbar\omega = E - E' \quad . \tag{3.3}$$

*$\hbar = h/2\pi$ (normalized Planck's constant)

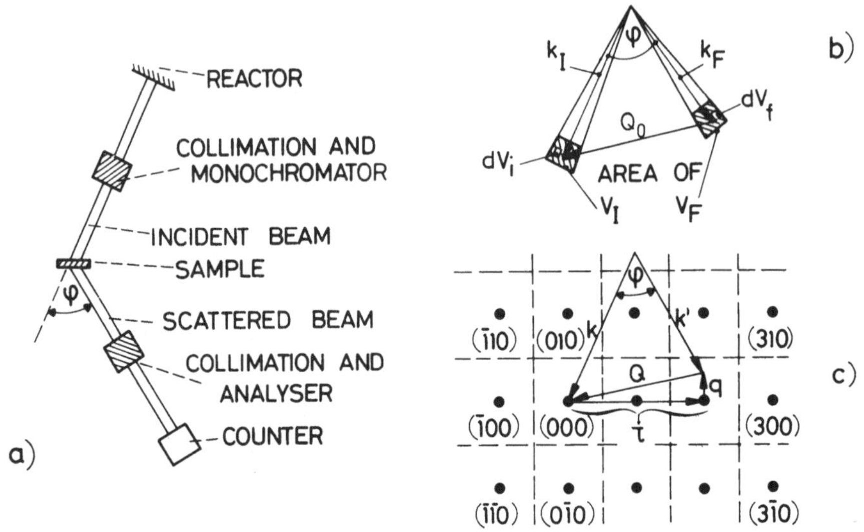


Fig. 3.7a-c Inelastic neutron scattering: a) path of neutrons in real space with "black boxes" for the determination of neutron energy before and after scattering; b) corresponding distribution of neutrons $V_I$ and $V_F$ in reciprocal space around the mean wave-vectors $k_I$ and $K_F$; c) momentum transfer $\underline{Q}$ of the neutron in relation to the reciprocal lattice of the sample (vectors $\underline{\tau}$) and the phonon wave-vectors q

An important difference between neutrons and X-rays is the extremely different energy: for neutrons

$$E = \frac{\hbar^2}{2m} k^2 \tag{3.4a}$$

and for X-rays

$$E = c \hbar k \tag{3.4b}$$

where m is the neutron mass and c the velocity of light. E (thermal neutrons) ≈ 6 THz, whereas E (X-ray Cu-K) ≈ $2.10^6$THz. This means that thermal neutrons have energies comparable to lattice excitations, and inelastic scattering will produce drastic changes in the neutron energy. For X-rays, on the other hand, it will always be found that E ≈ E'. Energy analysis for inelastic X-ray scattering has not yet been performed, except for Mössbauer spectroscopy. It might become possible with the high intensity of synchrotron radiation.

What has been said for energies before, and after, an inelastic process holds similarly for the wave-vectors $\underline{k}$ (see (3.4a) and (3.4b)). It follows for X-rays that elastic, or inelastic, processes always take place with the same momentum transfer $\underline{Q}$, while in the case of neutrons special care has to be taken to keep $\underline{Q}$ constant while varying the energy transfer. This will be discussed in detail in Subsec.3.1.2.

The interaction of a neutron with a nucleus (see (1.4)), described by a scattering length b and a δ-function in space, is different from the interaction of X-rays with matter. X-rays interact with the electrons in the material. Therefore b is to be replaced by the interaction of electromagnetic radiation with one electron, and the δ-function goes over into a density distribution of the electrons. The interaction parameter and the Fourier transform of the electron density are contained in the X-ray form factor f(Q), which decreases with increasing Q. The extension of f(Q) in Q is reciprocal to the diameter of the atom or ion. If we replace b by f(Q), all formulae for nuclear neutron scattering hold as well for X-ray scattering.

A comparison of scattering length for neutrons (b) and scattering factors for X-rays (f) scaled to the same unit is given in Table 3.1. This table already shows advantages, or disadvantages, of the two techniques.

The scattering length for neutrons varies rapidly from element to element (even from isotope to isotope, most often producing unwanted incoherent scattering). The scattering factors for X-rays are proportional to the atomic number, and elements with only slightly different atomic number are indistinguishable with X-rays, but generally have different scattering length for neutrons. Heavy and light elements can have comparable b's for neutrons but very different f's for X-rays. In fact, in X-ray investigations light elements are almost invisible if heavy ions are present. These differences are of interest in structure determinations, and here in eigenvector determinations (see Subsec. 3.1.1).

The scattering lenghts for neutrons, as shown in Table 3.1, are much smaller than the scattering factors for X-rays, and this difference is enhanced by the fact that they enter as squared quantities in the formulas for the scattered intensity. This is one reason why larger samples are required for neutron investigations, which is often a major problem since single crystals are generally needed. A second reason is that X-rays correspond to higher fluxes, X-ray anodes produce strong characteristic lines, narrow in energy and the size of X-ray beams is usually of the order of $1 mm^2$. To reach comparable intensities, neutron beams are often of the order of $5 \times 5$ cm.

A considerable advantage for neutrons is, however, that they are only weakly absorbed by most elements, whereas X-rays, which by definition have energies comparable to the absorption edges of the inner electron shells, can be heavily absorbed and even can produce unwanted X-ray fluorescence. This fact not only permits, in neutron scattering, the use of sufficiently large sample and the investigation of bulk crystals, but also eases considerably measurements with variable temperature, as windows will only weakly absorb the neutron beams; Aluminium, for example, gives an attenuation of 1% per mm thickness.

Table 3.1 Neutron and X-ray scattering data for elements and isotopes

| Atomic number | Element of isotope | Neutrons b $\|10^{-12}$cm$\|$ | $\sigma_{inc}$ $\|$barns$\|$ | $\sigma_{abs}$ $\|$barns$\|$ | X-rays f$\|10^{-12}$cm$\|$ sinθ = 0 | sinθ/2 = 0.5 Å$^{-1}$ | Atomic number |
|---|---|---|---|---|---|---|---|
| 1 | H | -0.374 | 79.7 | 0.33 | 0.28 | 0.02 | 1 |
| | D | 0.667 | 2.0 | 0.00046 | | | |
| | $^{3}$H | 0.47 | | | | | |
| 2 | $^{3}$He | 0.62 | 1.2 | 5500 | | | 2 |
| | $^{4}$He | 0.30 | 0 | 0 | 0.56 | 0.15 | |
| 3 | Li | -0.214 | 0.7 | 71.0 | 0.84 | 0.28 | 3 |
| | $^{6}$Li | 0.18+0.025i | | 945 | | | |
| | $^{7}$Li | -0.233 | 0.8 | | | | |
| 4 | Be | 0.774 | 0 | 0.010 | 1.13 | 0.39 | 4 |
| 5 | B | 0.54+0.021i | 0.7 | 755 | 1.41 | 0.42 | 5 |
| | $^{10}$B | | | 3813 | | | |
| | $^{11}$B | 0.60 | | | | | |
| 6 | C | | 0 | 0.0033 | 1.69 | 0.48 | 6 |
| | $^{12}$C | 0.665 | | | | | |
| | $^{13}$C | 0.60 | 1.0 | | | | |
| 7 | N | | 0.4 | 1.88 | 1.97 | 0.53 | 7 |
| | $^{14}$N | 0.94 | | | | | |
| | $^{15}$N | 0.65 | | | | | |
| 8 | O | 0.580 | 0 | <0.002 | 2.25 | 0.62 | 8 |
| | $^{17}$O | 0.578 | | | | | |
| | $^{18}$O | 0.600 | | | | | |
| 9 | F | 0.56 | 0 | <0.010 | 2.53 | 0.75 | 9 |
| 10 | Ne | 0.46 | 0.1 | <2.8 | 2.82 | 0.96 | 10 |
| 11 | Na | 0.36 | 1.7 | 0.505 | 3.09 | 1.14 | 11 |
| 12 | Mg | | 0.04 | 0.063 | 3.38 | 1.35 | 12 |
| | $^{24}$Mg | 0.55 | | | | | |
| | $^{25}$Mg | 0.36 | | | | | |
| | $^{26}$Mg | 0.49 | | | | | |
| 13 | Al | 0.35 | 0 | 0.230 | 3.65 | 1.55 | 13 |
| 14 | Si | 0.415 | 0 | 0.16 | 3.95 | 1.72 | 14 |
| 15 | P | 0.51 | 0 | 0.20 | 4.23 | 1.83 | 15 |
| 16 | S | 0.28 | 0.01 | 0.52 | 4.5 | 1.9 | 16 |
| 17 | Cl | 0.96 | 3.5 | 33.6 | 4.8 | 2.0 | 17 |
| | $^{35}$Cl | 1.18 | | | | | |
| | $^{37}$Cl | 0.26 | | | | | |

Table 3.1 (continued)

| Atomic number | Neutrons<br>Element of isotope | b $\|10^{-12}cm\|$ | $\sigma_{inc}$ $\|barns\|$ | $\sigma_{abs}$ $\|barns\|$ | X-rays $f\|10^{-12}cm\|$<br>$\sin\theta = 0$ | $\sin\theta/2 = 0.5$ Å$^{-1}$ | Atomic number |
|---|---|---|---|---|---|---|---|
| 18 | Ar | 0.20 | 0.15 | 0.66 | 5.07 | 2.2 | 18 |
| | $^{36}Ar$ | 2.43 | | | | | |
| 19 | K | 0.37 | 0.4 | 2.07 | 5.3 | 2.2 | 19 |
| | $^{39}K$ | 0.37 | | | | | |
| 20 | Ca | 0.47 | 0.0 | 0.46 | 5.6 | 2.4 | 20 |
| | $^{40}Ca$ | 0.49 | 0.1 | | | | |
| | $^{44}Ca$ | 0.18 | | | | | |
| 21 | Sc | 1.21 | 4.5 | 24.0 | 5.9 | 2.5 | 21 |
| 22 | Ti | -0.34 | 2.7 | 5.8 | 6.2 | 2.7 | 22 |
| | $^{46}Ti$ | 0.48 | | | | | |
| | $^{47}Ti$ | 0.33 | | | | | |
| | $^{48}Ti$ | -0.58 | | | | | |
| | $^{49}Ti$ | 0.08 | | | | | |
| | $^{50}Ti$ | 0.55 | | | | | |
| 23 | V | -0.04 | 4.9 | 4.98 | 6.5 | 2.8 | 23 |
| 24 | Cr | 0.352 | 1.7 | 3.1 | 6.8 | 3.0 | 24 |
| | $^{52}Cr$ | 0.49 | | | | | |
| 25 | Mn | -0.372 | 0.32 | 13.2 | 7.0 | 3.1 | 25 |
| 26 | Fe | 0.95 | 0.21 | 2.53 | 7.3 | 3.3 | 26 |
| | $^{54}Fe$ | 0.42 | | | | | |
| | $^{56}Fe$ | 1.01 | | | | | |
| | $^{57}Fe$ | 0.23 | | | | | |
| 27 | Co | 0.28 | 5.0 | 37.0 | 7.6 | 3.4 | 27 |
| 28 | Ni | 1.03 | 4.7 | 4.8 | 7.9 | 3.6 | 28 |
| | $^{58}Ni$ | 1.44 | | | | | |
| | $^{60}Ni$ | 0.28 | | | | | |
| | $^{61}Ni$ | 0.76 | | | | | |
| | $^{62}Ni$ | -0.87 | | | | | |
| | $^{64}Ni$ | -0.037 | | | | | |
| 29 | Cu | 0.76 | 0.6 | 3.77 | 8.2 | 3.8 | 29 |
| | $^{63}Cu$ | 0.67 | | | | | |
| | $^{65}Cu$ | 1.11 | | | | | |

Table 3.1 (continued)

| Atomic number | Neutrons: Element of isotope | b $\|10^{-12}cm\|$ | $\sigma_{inc}$ $\|barns\|$ | $\sigma_{abs}$ $\|barns\|$ | X-rays f $\|10^{-12}cm\|$: $\sin\theta = 0$ | $\sin\theta/2 = 0.5$ Å$^{-1}$ | Atomic number |
|---|---|---|---|---|---|---|---|
| 30 | Zn | 0.57 | 0.02 | 1.10 | 8.5 | 3.9 | 30 |
| | $^{64}Zn$ | 0.55 | | | | | |
| | $^{66}Zn$ | 0.63 | | | | | |
| | $^{68}Zn$ | 0.67 | | | | | |
| 31 | Ga | 0.72 | 0.2 | 2.80 | 8.8 | 4.1 | 31 |
| 32 | Ge | 0.819 | 0.1 | 2.45 | 9.0 | 4.2 | 32 |
| 33 | As | 0.67 | 0.8 | 4.3 | 9.3 | 4.4 | 33 |
| 34 | Se | 0.80 | 0.3 | 12.3 | 9.6 | 4.5 | 34 |
| 35 | Br | 0.68 | 0.1 | 6.7 | 9.8 | 4.7 | 35 |
| 36 | Kr | 0.74 | | 31 | 10.2 | 4.9 | 36 |
| 37 | Rb | 0.71 | 0.0 | 0.70 | 10.4 | 5.0 | 37 |
| | $^{85}Rb$ | 0.83 | | | | | |
| 38 | Sr | 0.69 | 4.0 | 1.21 | 10.7 | 5.2 | 38 |
| 39 | Y | 0.77 | 0.1 | 1.31 | 11.0 | 5.4 | 39 |
| 40 | Zr | 0.70 | 0.1 | 0.18 | 11.3 | 5.5 | 40 |
| 41 | Nb | 0.71 | 0.05 | 1.15 | 11.5 | 5.7 | 41 |
| 42 | Mo | 0.69 | 0.6 | 2.7 | 11.8 | 5.9 | 42 |
| 43 | Tc | 0.68 | | 122 | | | 43 |
| 44 | Ru | 0.73 | 0.1 | 2.56 | 12.5 | 6.2 | 44 |
| 45 | Rh | 0.59 | 1.2 | 156 | 12.8 | 6.4 | 45 |
| 46 | Pd | 0.60 | 0.3 | 8.0 | 12.9 | 6.5 | 46 |
| 47 | Ag | 0.60 | 0.5 | 63 | 13.3 | 6.7 | 47 |
| | $^{107}Ag$ | 0.83 | 1.3 | | | | |
| | $^{109}Ag$ | 0.43 | 3.7 | | | | |
| 48 | Cd | 0.37+0.16i | | 2450 | 13.6 | 6.9 | 48 |
| | $^{113}Cd$ | -1.5+1.2i | | 20.000 | | | |
| 49 | In | 0.39 | | 196 | 13.9 | 7.1 | 49 |
| 50 | Sn | 0.61 | 0 | 0.625 | 14.2 | 7.3 | 50 |
| | $^{116}Sn$ | 0.58 | | | | | |
| | $^{117}Sn$ | 0.64 | | | | | |
| | $^{118}Sn$ | 0.58 | | | | | |
| | $^{119}Sn$ | 0.60 | | | | | |
| | $^{120}Sn$ | 0.64 | | | | | |
| | $^{122}Sn$ | 0.55 | | | | | |
| | $^{124}Sn$ | 0.59 | | | | | |

Table 3.1 (continued)

| Atomic number | Neutrons: Element of isotope | Neutrons: b $\|10^{-12}$cm$\|$ | Neutrons: $\sigma_{inc}$ $\|$barns$\|$ | Neutrons: $\sigma_{abs}$ $\|$barns$\|$ | X-rays f$\|10^{-12}$cm$\|$: sinθ = 0 | X-rays f$\|10^{-12}$cm$\|$: sinθ/2 = 0.5 Å$^{-1}$ | Atomic number |
|---|---|---|---|---|---|---|---|
| 51 | Sb | 0.56 | 0.2 | 5.7 | 14.5 | 7.4 | 51 |
| 52 | Te | 0.55 | 0.8 | 4.7 | 14.7 | 7.6 | 52 |
| | $^{120}$Te | 0.52 | | | | | |
| | $^{123}$Te | 0.57 | | | | | |
| | $^{124}$Te | 0.55 | | | | | |
| | $^{125}$Te | 0.56 | | | | | |
| 53 | I | 0.53 | 0 | 7.0 | 15.0 | 7.7 | 53 |
| 54 | Xe | 0.49 | | 74 | 15.3 | 8.0 | 54 |
| | $^{135}$Xe | | | $2.7\times10^{6}$ | | | |
| 55 | Cs | 0,54 | 4.6 | 29 | 15.5 | 8.1 | 55 |
| 56 | Ba | 0.52 | 2.5 | 1.2 | 15.8 | 8.3 | 56 |
| 57 | La | 0.83 | 1.6 | 9.3 | 16.1 | 8.4 | 57 |
| 58 | Ce | 0.48 | 0.0 | 0.77 | 16.3 | 8.6 | 58 |
| | $^{140}$Ce | 0.47 | | | | | |
| | $^{142}$Ce | 0.45 | | | | | |
| 59 | Pr | 0.44 | 1.6 | 11.6 | 16.6 | 8.8 | 59 |
| 60 | Nd | 0.78 | 11.0 | 46 | 16.9 | 9.0 | 60 |
| | $^{142}$Nd | 0.77 | | | | | |
| | $^{144}$Nd | 0.28 | | | | | |
| | $^{146}$Nd | 0.87 | | | | | |
| 61 | $^{147}$Pm | | | 90 | 17.2 | 9.2 | 61 |
| 62 | Sm | | | 5600 | 17.5 | 9.3 | 62 |
| | $^{149}$Sm | -1.9+4.5i | | 41.000 | | | |
| | $^{152}$Sm | -0.5 | | 210 | | | |
| | $^{154}$Sm | 0.96 | | 5.5 | | | |
| 63 | Eu | 0.68 | | 4300 | 17.8 | 9.5 | 63 |
| | $^{153}$Eu | | | 450 | | | |
| 64 | Gd | 1.5 | | 49000 | 18.2 | 9.7 | 64 |
| | $^{157}$Gd | 4.3+4i | | | | | |
| | $^{160}$Gd | 0.915 | | 0.77 | | | |
| 65 | Tb | 0.74 | | 46 | 18.5 | 9.8 | 65 |

Table 3.1 (continued)

| Atomic number | Element of isotope | Neutrons b $\|10^{-12}$cm$\|$ | $\sigma_{inc}$ $\|$barns$\|$ | $\sigma_{abs}$ $\|$barns$\|$ | X-rays f$\|10^{-12}$cm$\|$ sinθ = 0 | sinθ/2 = 0.5 Å$^{-1}$ | Atomic number |
|---|---|---|---|---|---|---|---|
| 66 | Dy | 1.71 | | 950 | 18.6 | 10.0 | 66 |
| | $^{160}$Dy | 0.67 | | 55 | | | |
| | $^{161}$Dy | 1.03 | | 585 | | | |
| | $^{162}$Dy | -0.14 | | 200 | | | |
| | $^{163}$Dy | 0.50 | | 140 | | | |
| | $^{164}$Dy | 4.94 | | 2300 | | | |
| 67 | Ho | 0.85 | 4 | 65 | 18.9 | 10.2 | 67 |
| 68 | Er | 0.80 | 7 | 173 | 19.2 | 10.3 | 68 |
| 69 | Tm | 0.70 | | 127 | 19.5 | 10.5 | 69 |
| 70 | Yb | 1.26 | | 37 | 19.8 | 10.7 | 70 |
| 71 | Lu | 0.73 | | 112 | 20.0 | 10.9 | 71 |
| 72 | Hf | 0.78 | | 105 | 20.3 | 11.1 | 72 |
| 73 | Ta | 0.70 | 0.5 | 21 | 20.5 | 11.3 | 73 |
| 74 | W | 0.48 | 2.0 | 19.2 | 20.8 | 11.4 | 74 |
| | $^{182}$W | 0.83 | | | | | |
| | $^{183}$W | 0.43 | | | | | |
| | $^{184}$W | 0.76 | | | | | |
| | $^{186}$W | -0.12 | | | | | |
| 75 | Re | 0.92 | | 86 | 21.1 | 11.6 | 75 |
| 76 | Os | 1.07 | 0.5 | 15.3 | 21.4 | 11.8 | 76 |
| | $^{188}$Os | 0.78 | | | | | |
| | $^{189}$Os | 1.10 | | | | | |
| | $^{190}$Os | 1.14 | | | | | |
| | $^{192}$Os | 1.19 | | | | | |
| 77 | Ir | 1.06 | | 440 | 21.7 | 12.0 | 77 |
| 78 | Pt | 0.95 | 0.7 | 8.8 | 22.0 | 12.1 | 78 |
| 79 | Au | 0.76 | 0.5 | 98.8 | 22.2 | 12.3 | 79 |
| 80 | Hg | 1.27 | 6 | 375 | 22.5 | 12.5 | 80 |
| 81 | Tl | 0.89 | 0.1 | 3.4 | 22.8 | 12.7 | 81 |
| 82 | Pb | 0.94 | 0.0 | 0.170 | 23.1 | 12.9 | 82 |
| 83 | Bi | 0.86 | 0.0 | 0.036 | 23.3 | 13.1 | 83 |
| 90 | Th | 1.03 | 0.0 | 7.56 | 25.3 | 14.4 | 90 |
| 91 | Pa | 1.3 | | 200 | 25.7 | 14.6 | 91 |

Table 3.1 (continued)

| Atomic number | Element or isotope | Neutrons b $\|10^{-12}$cm$\|$ | $\sigma_{inc}$ $\|$barns$\|$ | $\sigma_{abs}$ $\|$barns$\|$ | X-rays f$\|10^{-12}$cm$\|$ $\sin\theta = 0$ | $\sin\theta/\lambda = 0.5\ Å^{-1}$ | Atomic number |
|---|---|---|---|---|---|---|---|
| 92 | U | 0.86 | | 7.68 | 25.9 | 14.8 | 92 |
| | $^{235}$U | 0.98 | | 694 | | | |
| | $^{238}$U | 0.85 | | 2.71 | | | |
| 93 | Np | 1.06 | | 170 | | | 93 |
| 94 | $^{239}$Pu | 0.75 | | 1026 | | | 94 |
| | $^{240}$Pu | 0.35 | | 295 | | | |
| | $^{242}$Pu | 0.81 | | | | | |
| 95 | $^{243}$Am | 0.76 | | 8000 | | | 95 |
| 96 | $^{244}$Cm | 0.7 | | | | | 96 |

b = coherent scattering length for neutrons; (BACON, 1972, 1974; KOESTER, 1977); the coherent scattering cross-section $\sigma_{coh}=4\pi b^2$ can be calculated; $\sigma_{inc}$= incoherent cross-section (WILLIS, 1973; KOESTER, 1977); $\sigma_{abs}$ = nuclear absorption cross-section (SOODAK, 1962); f = atomic form factor for X-rays (BACON, 1962); $\theta$ = half the scattering angle; $\lambda$ = wavelength of the X-ray beam.

### 3.1.1 Elastic and Inelastic Scattering

We would like to recall the scattering function for elastic Bragg scattering

$$S_{Bragg}(\underline{Q}) = \left| \sum_{d}^{\text{unit cell}} \bar{b}_d \exp[-W_d(\underline{Q}) + i\underline{Q}\underline{d}] \right|^2 \tag{3.5}$$

where $\underline{d}$ is the position vector to the equilibrium position of atom d in the unit cell, and $W_d(\underline{Q})$ the exponent of the Debye-Waller factor of atom d. It is assumed that the equilibrium positions have translational periodicity throughout the crystal leading to Bragg scattering only if $\underline{Q}$ coincides with a reciprocal lattice vector $\underline{\tau}$; this is not given explicitly in (3.5). Equation (3.5) gives rather the variation of Bragg intensity for different reflections due to the interference of the atoms in the unit cell.

Generally, the atoms are displaced from their equilibrium position $\underline{d}$ by a vector $\underline{u}_d$. This $\underline{u}_d$ may or may not be time-dependent. In any case it leads to scattering different from Bragg scattering, appearing at $\underline{Q}$'s all over the reciprocal space. For example, inelastic one-phonon scattering is counted among these additional scattering processes. The intensity is taken out of the Bragg intensity. It is the best interpretation for the Debye-Waller factor that this factor decreases the

Bragg intensity by the amount, which goes to processes created by the existence of displacements $\underline{u}_d$.

The formalism for the coherent inelastic scattering function has been reviewed in Chap.1, see (1.96a) or (1.101). We perform the summation over $\underline{\tau}$ and $\underline{q}$ to eliminate the $\delta$-function in these quantities. This simply means that the orientation of $\underline{Q}$ in the reciprocal lattice of our sample is defined. This determines the Brillouin zone under consideration, as $\underline{\tau}$ is the vector to its center, and the phonon wave-vector $\underline{q}$ inside the Brillouin zone, as $\underline{q}$ is the vector pointing from the Brillouin zone center to the end of $\underline{Q}$

$$\left(\frac{d^2\sigma}{d\Omega\, dE'}\right)_{coh} = \frac{k'}{k} \sum_{j}^{\text{all modes}} S_j(\underline{Q},\omega) \quad . \tag{3.6}$$

$S_j(\underline{Q},\omega)$ is the one-phonon scattering function for the mode j with phonon wave-vector $\underline{q}$. Furthermore we can split the scattering function $S_j$ into two components: a $\underline{Q}$-dependent structure factor $G_j$, and an energy $\hbar\omega$ and temperature T depending function $F_j$

$$S_j(\underline{Q},\omega) = |G_j(\underline{q},\underline{Q})|^2 \times F_j\,[\omega,\omega_j(\underline{q}),T] \quad . \tag{3.7}$$

The structure factor $G_j$ (see (1.96))

$$G_j(\underline{q},\underline{Q}) = \sum_{d}^{\text{unit cell}} b_d \exp[-W_d(\underline{Q}) + i\underline{Q}\underline{d}][\underline{Q} \cdot \underline{\sigma}_d^j(\underline{q})]M_d^{-1/2} \tag{3.8}$$

contains the eigenvector $\underline{\sigma}$, which is normalized to unity and describes the pattern of displacements in one unit cell. It has 3r components, where r is the number of atoms per unit cell. The displacements $\underline{u}_d$ are periodic through the lattice (rather, we restrict ourselves to the periodic ones)

$$\underline{u}_d^j = A_d^j \underline{\sigma}_d^j\,(\underline{q}) \exp(i\underline{q}\underline{\ell}) \tag{3.9}$$

where $A^j$ is an amplitude factor, which may depend on time, temperature, and eventually on the position of the unit cell in the crystal, and $\underline{\ell}$ is the vector to the $\ell$th unit cell. The effect of time and temperature dependence of A will be discussed below in connection with the function $F_j[\omega,\omega_j(\underline{q}),T]$. The dependence on position will be discussed in Subsec.3.1.3.

Under the assumption that the undistorted structure is known, i.e., that the positions $\underline{d}$ of the atoms in the unit cell are known and that the Debye-Waller

factor is fairly well determined, then the eigenvector $\underline{\sigma}^j$ can be determined experimentally. Identical $\underline{q}$ vectors appear at least once in each Brillouin zone (in high-symmetric systems several times).

The integrated intensity

$$S_j(\underline{Q}) = \int S_j(\underline{Q},\omega)d\omega = |G_j(\underline{q},\underline{Q})|^2 \times \int F_j[\omega,\omega_j(\underline{q}),T]d\omega \tag{3.10}$$

varies from Brillouin zone to Brillouin zone as given by the structure factor $G_j$. The integral over $F_j$ is the same for all $\underline{Q}$. Thus a measurement of $S_j(\underline{Q})$ for a certain wave-vector $\underline{q}$ at many different $\underline{Q}$ allows us to determine the eigenvector $\underline{\sigma}$ (BROCKHOUSE et al., 1961). It is a structure determination of a periodic distortion in a known lattice. If the distortion is time-dependent, this technique is called a dynamical structure determination.

While $G_j$ describes the intensity variation in reciprocal space, it is the function $F_j$ which determines the inelasticity of a process as e.g., phonon excitation, overdamped mode, diffusive motion. $F_j[\omega,\omega_j(\underline{q}),T]$ is the Fourier transform of the function $f_j[t,\omega_j(\underline{q}),T]$.

$$F_j[\omega,\omega_j(q),T] \sim \int_{-\infty}^{+\infty} \exp(-i\omega t)\; f_j[t,\omega_j(q),T]dt \quad . \tag{3.11}$$

For a well-behaved phonon we find from (1.75)

$$f_j[t,\omega_j(\underline{q}),T] = \exp[-i\omega_j(\underline{q})t]\;\; n_j(T) + \exp[i\omega_j(\underline{q})t][1 + n_j(T)] \quad . \tag{3.12}$$

The Fourier transform yields $\delta$-functions in the energy transfer at $\pm\omega_j(\underline{q})$. The Bose occupation number $n_j(T)$ describes the temperature dependence of the intensity. In classical language $n_j(T)>>1$: we can say that $\sqrt{n_j(T)}$ controls the amplitude $A_d^j$ together with the temperature independent factor $(M_d\omega_j)^{-1/2}$. As long as we can treat the system classically (quantum mechanically at low temperatures or for high frequencies the phonon intensity on the energy gain side is zero and on the energy loss side proportional to $\omega_j^{-1}$), the phonon intensity is proportional to the square of the atomic amplitudes.

$$A_d^{j2} \sim \frac{n_j(T)}{M_d\omega_j} = \frac{1}{M_d\omega_j} \times \frac{kT}{\hbar\omega_j} \sim \frac{T}{M_d\omega_j^{\,2}} \quad . \tag{3.13}$$

Immediately we understand that the amplitudes $A_d^j$ are proportional to $\omega_j^{-1}$. It will be important for the diffuse X-ray scattering that the low frequencies be more visible than the high ones.

In the classical approximation, i.e., $kT \gg \hbar\omega$ energy loss or gain of the neutron has equal probability. But for very low temperatures $n_j$ goes to zero and energy gain processes become impossible. On the other hand, energy loss, which is proportional to $(1 + n_j)$, approaches a constant probability. It is still meaningful to interpret this constant probability or intensity as interaction of the neutron with the zero point amplitudes.

In the case that $f_j$ does not depend on time, the Fourier transform yields a $\delta$-function at $\omega = 0$, and nevertheless an amplitude $A_d^j$, which may or may not depend on temperature T. For example, at a structural phase transformation a certain pattern of displacements condenses into a static distortion. Usually the amplitude of the distortion is proportional to the order parameter. The temperature dependence of the order parameter will be discussed in Subsec.3.2.3.

### 3.1.2 Experimental Technique

*Inelastic Neutron Scattering*

In this section, certain experimental aspects of neutron scattering will be discussed which are of particular relevance for this chapter, and for phonon spectroscopy in general. In this context, a careful distinction is necessary between the "physical" parameters occurring in the scattering law, as $\underline{k}$, $\underline{k}'$, $\underline{E}$, $E'$, $\underline{Q}$, and $\omega$, and the "nominal" parameters, as defined by the setting of the instrument, which we denote, respectively, $\underline{k}_I$, $\underline{k}_F$, $E_I$, and $E_F$, with $\underline{Q}_0 = (\underline{k}_I - \underline{k}_F)$ and $\hbar\omega_0 = E_I - E_F$ which denote the nominal momentum and energy transfer.[1]

In inelastic neutron spectroscopy we have to determine the energy $E_I$ of the neutron before being scattered, and $E_F$ after scattering. There are essentially two methods: 1) the time-of-flight technique, which determines the energy by determining the neutron velocity v. This method will be discussed in Chap.4, 5 and 6. Here we concentrate on 2) the three-axis spectrometers. (Three-axis means one monochromator axis, one sample axis, one analyzer axis, in contrast to two-axis intruments where the analyzer is missing). A recent review of different instrumentation has been given by G. DOLLING (1974).

Figure 3.7 is a schematic drawing of an inelastic neutron scattering experiment. The "black boxes" called collimator and monochromator and collimator and analyzer determine a certain distribution of wave-vectors $k_i$ around the mean or "nominal" $\underline{k}_I$, and of $\underline{k}_f$ around the mean $\underline{k}_F$. These distributions in the reciprocal space volumes $V_I$ and $V_F$ provide intensity (proportional to $V_I \cdot V_F$) and resolution controlled by the folding of the two distributions (B. DORNER, 1972).

---

[1]The distinction between nominal (or experimental) quantities and "true" physical quantities, which is necessary for the discussion presented in Subsec.3.1.2, is not made explicit in the remainder of the book since the distinction is mostly unnecessary in discussing the interpretation of the experimental data reported.

Neutrons with a certain energy are selected by Bragg reflections from large single crystals like copper, pyrolytic graphite,... by determination of $|\underline{k}|$

$$2|\underline{k}| \sin\theta = |\underline{\tau}| \qquad (3.14)$$

where $\theta$ is the Bragg angle. Recent reviews on monochromator crystals are given by FREUND (1975,1976) and by FREUND and SCHNEIDER (1976).

Besides the chosen $|\underline{k}|$, the beam reflected from a monochromator contains additional higher order contamination, arising from the fact that multiples of k may be reflected by multiples of $\tau$, as long as the reactor provides a flux for the larger k's. The higher order k's are generally unwanted and are difficult to suppress. The problem can be solved only in particular areas.

The simplest is to use first order $|\underline{k}|$'s from the maximum of the reactor spectrum (around 40 meV). Then the higher orders come from an energy range where the reactor flux is very low. Quite often, however, one needs lower energies for better resolution.

There are crystal-structures where second-order reflections have vanishing intensity, like [111] in Ge or Si.

Very often filters are used such as polycrystalline Be (VAN DINGENDEN and HAUTECLER, 1963) which only allows neutrons to pass with $E \leq 5.2$ meV, or pyrolytic graphite (MINKIEWICZ and SHIRANE, 1970) which is particularly efficient for E = 13.7 and 14.8 meV.

A curved (better S-type curved) neutron guide (MAIER-LIEBNITZ, 1963; JACROT, 1970) has a natural cut-off due to the curvature. For example, a neutron guide with 2700 m radius of S-type curvature does not transmit $|k|$'s larger than 2 $\text{Å}^{-1}$ (or energies larger than 18 meV).

The higher order problem exists for X-rays as well, and will be discussed in *X-Ray Scattering*.

A three-axis spectrometer is an instrument which uses large single crystals as monochromators and analyzers (see Fig.3.8). Since the first three-axis spectrometer was built by BROCKHOUSE (1961), the basic principle did not change, only the mechanical and electronic engineering were developed so that such instruments can run day and night by computer control.

The three-axis spectrometer IN2 at ILL has a more sophisticated double monochromator (BROCKHOUSE et al., 1968; STEDMAN et al., 1969) which reflects the neutron beam two times from two single crystals having the same lattice spacing. By the two reflections we obtain a very clean beam (less $\gamma$-radiation, less fast neutrons, even less higher order contamination). It allows one to use large Bragg angles up to $2\theta = 165^{\circ}$. This was essential for the work on $CD_4$, as will be discussed in Subsec. 3.2.2. The monochromatic beam is always parallel to the white beam coming out of the reactor. $E_I$ is varied by translating and rotating the two crystals simultaneously.

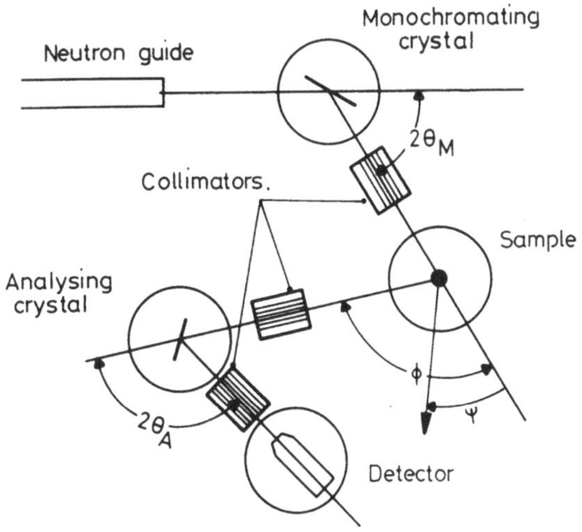


Fig. 3.8 Scheme of a three-axis spectrometer (IN3 at ILL). $\theta_M$ and $\theta_A$ are the Bragg angles of monochromator and analyser, $\phi$ is the scattering angle and $\psi$ the orientation of the sample with respect to the incoming beam. Generally the directions are defined by Soller-slit collimators

The sample table rests at the same place. The scattering angle $\phi$ is defined by the position of the analyzer. Finally the detector has to be positioned to an angle $2\theta_A$ (two times the Bragg angle of the analyzer) with respect to the scattered beam.

There are several modes of operation: constant-$\underline{Q}$-scans where the energy transfer is varied, constant-$\hbar\omega$-scans where $\underline{Q}$ is varied, or any combined scan.[2] In other words, we want to determine an energy $\hbar\omega$ as a function of $\underline{Q}$ or $\underline{q}$. In the experimental plane we have only two $\underline{Q}$ components accessible. Therefore we have 3 unknowns: $\omega$, $Q_x$, $Q_y$ to be determined by the experiment, where we have four variables: $|\underline{k}_I|$, $|\underline{k}_F|$, $\phi$ and the sample orientation in the experimental plane $\psi$. Apparently we can perform any scan in the $\omega$, $Q_x$, $Q_y$ space by keeping one instrumental variable fixed. Most often $|\underline{k}_I|$ or $|\underline{k}_F|$ are kept fixed. A constant-Q-scan with fixed $|\underline{k}_I|$ is shown in Fig.3.9.

Let us come back now to the density distributions $p_i(\underline{k}_i)$ around $\underline{k}_I$ and of $p_f(\underline{k}_f)$ around $\underline{k}_F$ to discuss intensity and resolution problems. We determine volumes in k-space (B. DORNER, 1972), see Fig.3.7,

$$V_I = \int p_i(\underline{k}_i)d\underline{k}_i \qquad V_F = \int p_f(\underline{k}_f)d\underline{k}_f$$
$$V_I = C_M k_I^3 \cot\theta_M \qquad V_F = C_A k_F^3 \cot\theta_A \qquad (3.15)$$

---

[2]By means of peak height and width there is no preferable mode of operation. For technical reasons one sometimes chooses a scan in which the instrumental parameter variations are smallest (i.e., keep the background constant). By means of physical interpretation the constant-Q scan is a distinct one (DOLLING and SEARS, 1973) because the density of states is constant in Q-space.

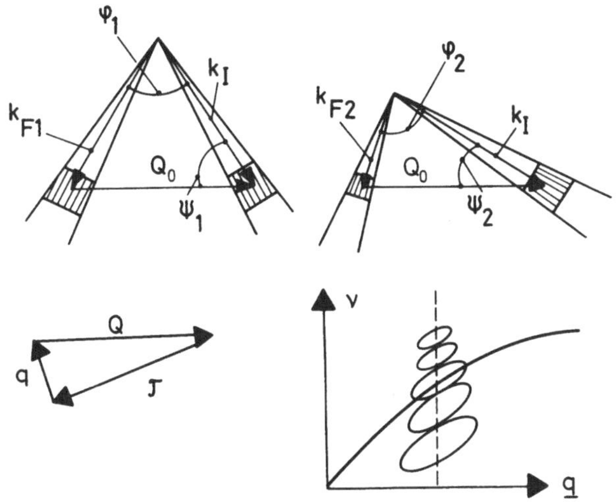


Fig. 3.9 Q-constant-scan with $k_I$ fixed. $\phi$ and $\psi$ are scattering angle and sample orientation. The dashed areas give the distributions of $k_i$ around $k_I$, and of $k_f$ around $k_F$. q is the phonon wave-vector. In $\underline{Q}$-$\omega$ space a constant-Q scan is drawn with varying resolution. (DORNER, 1976)

where $\theta_M$ and $\theta_A$ are the Bragg angles and the constant factors $C_M$ and $C_A$ contain collimator divergencies and mosaic widths of monochromator and analyzer, respectively.

The transmission, or resolution, function R is the folding of the two distributions

$$R(\underline{Q}-\underline{Q}_0,\omega-\omega_0,\underline{Q}_0,\omega_0) = \iint p_i(\underline{k}_i)p_f(\underline{k}_f)\delta[\underline{Q} - (\underline{k}_i - \underline{k}_f)] \times \delta[\omega - \frac{\hbar}{2m}(k_i^2 - k_f^2)]\, d\underline{k}_i d\underline{k}_f \quad . \tag{3.16}$$

Here $\underline{Q}_0$ and $\omega_0$ are the nominal position of the instrument corresponding to $\underline{k}_I$ and $\underline{k}_F$. The normalisation of R is obtained by integration

$$\iint R(\underline{Q}-\underline{Q}_0,\omega-\omega_0,\underline{Q}_0,\omega_0)\, d\underline{Q}\, d\omega = V_I \cdot V_F \quad . \tag{3.17}$$

And the measured intensity $I_{meas}$ is

$$I_{meas}(\underline{Q}_0,\omega_0) = \iint R(\underline{Q}-\underline{Q}_0,\omega-\omega_0,\underline{Q}_0,\omega_0)S_j(\underline{Q},\omega)d\underline{Q}d\omega \quad . \tag{3.18}$$

The actual shape of R in the 4-dimensional $\underline{Q}$-$\omega$-space was first derived by COOPER and NATHANS (1967) using Gaussian approximations for the distributions. Resolution programs are now part of the standard computer libraries at reactor institutes. A graphical method should explain some basic properties of R, which leads to focusing. The distribution of $p(\underline{k})$ being reflected by a single crystal monochromator exhibits a correlation between $|\underline{k}|$ and the angle within the divergence controlled by a Soller slit collimator, see Fig.3.10b. Somewhat simplifying (small mosaic width), the main extension is parallel to the reflecting planes. Regarding a certain path of the neutrons through the three-axis spectrometer (Fig.3.10a) in real space, we can draw Fig.3.10c in reciprocal space showing the $\underline{k}$ distributions around $\underline{k}_I$ and $\underline{k}_F$. As the instrumental transmission or resolution function R is the folding of these two distributions, we have now, with (3.16), to sort out $\underline{k}_i - \underline{k}_f$ combinations for their $\Delta Q_{\parallel}$, $\Delta Q_{\perp}$ and $\Delta\omega$ contributions. For $|\underline{k}_i| - |\underline{k}_f| > |\underline{k}_I| - |\underline{k}_F|$, the $\Delta\omega$ is positive (energy loss of the neutron). Using only a few extreme combinations the qualitative shapes as given in Fig.3.10d can be obtained.

We learn from Fig.3.10d that the function R has a certain inclination in $\Delta\underline{Q}$ - $\Delta\omega$ - space. This inclination depends on $|\underline{k}|$, on the monochromator and analyzer crystals used, and on the path of the neutron, i.e., scattering to the right or to the left at M, S or A.

In the case of neutron scattering one speaks of focusing if the long axis of R is as closely aligned with the dispersion surface as possible, and the component perpendicular is as small as possible. Then a constant-$\underline{Q}$-scan, as shown in Fig.3.9, reveals a narrow peak with high intensity, and one can separate dispersion branches which are close to each other (RAINFORD et al., 1972). Nevertheless, the focusing possibilities are limited; under conventional conditions, it is impossible to obtain a transmission where the long axis is in the $\Delta Q_{\parallel}$ - $\Delta Q_{\perp}$ - plane, yielding a good energy resolution, combined with a relaxed $\underline{Q}$-resolution which is desirable for

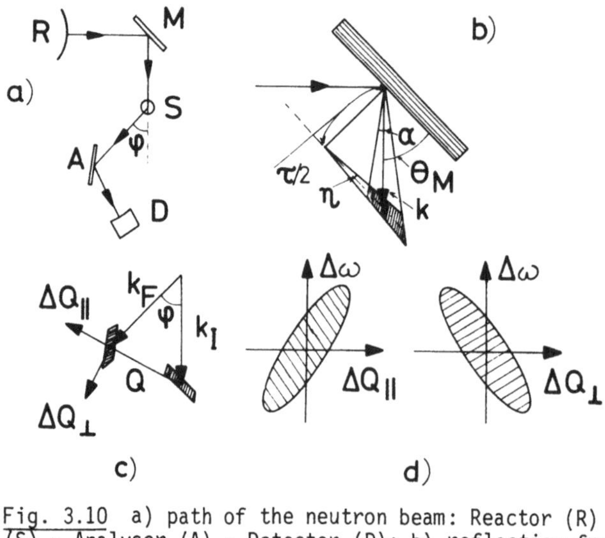


Fig. 3.10 a) path of the neutron beam: Reactor (R) - Monochromator (M) - Sample (S) - Analyser (A) - Detector (D); b) reflection from a single crystal monochromator with mosaic width $\eta$ and Bragg angle $\theta_M$. The dashed area gives the distribution of $k_i$ around $k_I$; c) scattering diagram for (a) in reciprocal space; d) projections of the resolution (DORNER, 1976)

the investigation of flat (optic) modes. Sophisticated instruments such as the backscattering instrumtent (ALEFELD, 1972; HEIDEMANN, 1975) achieve a zero inclination of R using Bragg angles of $90^\circ$ at monochromator and analyzer. A range of energy transfer $\omega$ is provided by moving one crystal to produce a Doppler shift. Another sophisticated idea is the use of horizontally bent cyrstals (MAIER-LEIBNITZ, 1967, 1972; KALUS 1975). By changing the radius of curvature any desired inclination can be obtained. This technique is somewhat cumbersome as the best-suited samples are thin cylinders, but it has no limitation in the transferred energy $\omega$.

During a constant Q-scan (see Fig.3.9) the transmission function varies. With $k_I$ fixed, and varying $k_F$ towards more energy loss, the transmission decreases as $V_F$ decreases. Therefore phonon peaks, which have been obtained with $|\underline{k}_I|$ fixed, have to be corrected for the varying resolutions during the scan. Omitting these corrections can lead to false determinations of peak centers and widths.

A correction, or normalisation, factor $N(\omega_0) = V_I \cdot V_F$ normalizes R. Therefore the corrected intensity $I_{corr}$

$$I_{corr}(\underline{Q}_0,\omega_0) = \frac{I_{meas}(\underline{Q}_0,\omega_0)}{N(\omega_0)} = \int \frac{R(\underline{Q}-\underline{Q}_0,\omega-\omega_0,\underline{Q}_0,\omega_0)}{N(\omega_0)} S(\underline{Q},\omega)d\underline{Q}d\omega \tag{3.19}$$

represents data as if they had been obtained with a constant resolution all along the scan, yet the curve is not unfolded. This means $I_{corr}$ still has a width, which contains the resolution. It is the integral

$$\int I_{corr}(\underline{Q}_0,\omega_0)d\omega_0 = \int S(\underline{Q},\omega)d\omega \tag{3.20}$$

over the pointwise corrected data which gives the integrated phonon intensity without any influence of resolution.

The $N(\omega_0)$ has a very simple form

$$N(\omega_0) = C_M C_A k_I^3 \cot\theta_M \, k_F^3 \cot\theta_A \quad . \tag{3.21}$$

In the case where the resolution is much smaller than the width of the response $F_j$ (for example, a damped phonon) then we can as well describe the data by

$$I_{meas}(\underline{Q}_0,\omega_0) = N(\omega_0)F_f[\omega_0,\omega_j(\underline{q}),T] \cdot |G_j(\underline{q},\underline{Q}_0)|^2 \quad . \tag{3.22}$$

Fig.3.11 shows the response of the damped (not overdamped) soft mode in $Tb_2(MoO_4)_3$ in energy gain and loss with constant $k_I$. The full line is a least squares fit of a $F_j(\omega,T)$ as given in (3.43) modified by $N(\omega_0)$. All the asymmetry is due to resolution and is well taken care of by $N(\omega_0)$. $F_j(\omega,T)$ is symmetric for such low frequencies at such high temperatures.

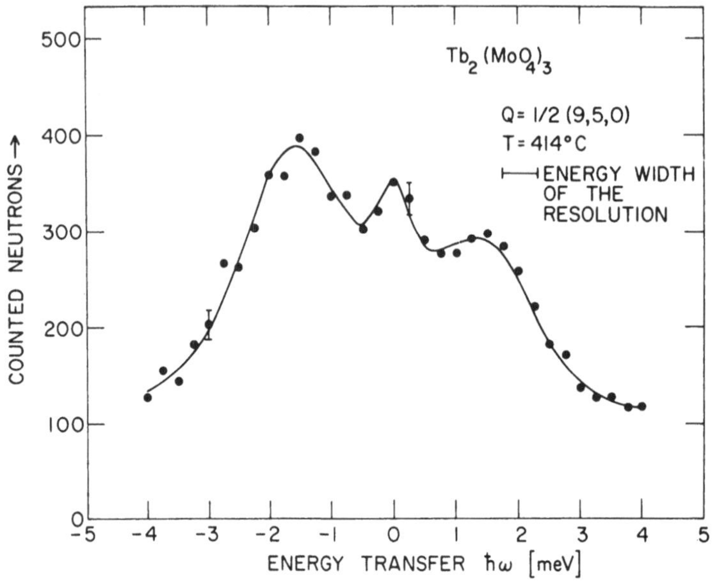


Fig. 3.11 Constant-Q scan through the soft mode at the M-point 255°C above the phase transformation (159°C). The soft mode is heavily damped but not overdamped. The line is a least squares fit of (3.55), which is symmetric in energy gain and loss. The apparent asymmetry arises exclusively from the variation of the instrumental transmission and was accounted for by including a resolution normalization factor in the fit. The little maximum at zero energy transfer is background only (DORNER et al., 1972)

The resolution and normalization have so far been discussed in terms of density distributions $p(\underline{k})$ (3.15) and not in terms of fluxes $k.p(\underline{k})$. This appears to be the appropriate choice as it corresponds to the scattering function $S(\underline{Q},\omega)$(see (3.6)) while fluxes correspond to the cross-section $d^2\sigma/d\Omega dE'$.

When $k_F$ is kept fixed, then a k independent monitor in the monochromatic beam would measure the incoming flux, $K_I \cdot V_I$. But if the monitor has a $1/k_I$ characteristic, as it usually has, then it measures $V_I$. If data are collected, for constant monitor count rate, then $N(\omega_0)$ is taken care of by the monitor. In other words, $I_{corr} \equiv I_{meas}$. Unfortunately it is not always possible to keep $k_F$ fixed. For example, if one uses a pyrolytic graphite (PG) filter in the incoming beam to reduce higher order contamination, then k has to be fixed at 2.57 or 2.67 $Å^{-1}$ because the PG filter is most efficient for these $k_I$ values.

Sometimes one is only interested in the overall energy resolution $\Delta\omega$ of a three-axis spectrometer, for example, to investigate the quasi-elastic scattering from relaxation processes, as will be discussed in Subsec.3.2.2

$$\Delta\omega = (\Delta E_I^2 + \Delta E_F^2)^{1/2} \tag{3.23}$$

with

$$\Delta E_I = 2E_I \cot(\theta_M) \left( \frac{\alpha_0^2\alpha_1^2 + \alpha_0^2\eta_M^2 + \alpha_1^2\eta_M^2}{\alpha_0^2 + \alpha_1^2 + 4\eta_M^2} \right)^{1/2} \tag{3.24}$$

for the incoming beam (KALUS and DORNER, 1973) with $\theta_M$ = Bragg angle of the monochromator $\alpha_0$ and $\alpha_1$ collimation before and after the monochromator and $\eta_M$ the mosaic width of the monochromator. We have three ways of improving the energy resolution.

1) decrease the incoming energy $E_I$. This is not always possible as the necessary momentum transfer sets a lower limit to the energy,

2) use a large $\theta_M$. This depends on the choice of the monochromator crystal and on the mechanical possiblities of the instrument (advantage of a double monochromator).

3) reduce the collimation.

If technically feasible, 2) is superior to 3) since it reduces the intensity less. With 2) the solid angle of the beam is maintained and only the energy width is reduced.

*X-Ray Scattering*

X-ray scattering, can be measured both by counter techniques and by photographic methods. In principle, the counter techniques for the measurement of X-ray scattering are very similar to those used in usual structure analysis, and we shall therefore only stress here some essential differences. The photographic methods will be described in more detail because they are less known, and also because they give on one pattern a general overlook on a whole section of reciprocal space, which is probably more useful as a complementary investigation to neutrons than a counter, point-by-point, analysis which has to be repeated anyway with neutrons if the dynamical properties are to be characterized.

Looking to the future, position-sensitive X-ray detectors should replace the photographic plates in an essentially identical set up, and combine the advantages of the quantitative counter analyses and of a simultaneous investigation of a large section of reciprocal space.

X-ray scattering necessitates measurements of intensities which are in the range of $10^{-3}$ to $10^{-6}$ compared to usual Bragg peaks; the conditions to be satisfied are therefore high intensity beams with low background.

A schematic set-up for X-ray scattering measurements is simply obtained by replacing the analyzer of the triple-axis instrument shown in Fig.3.10 by the detector: this gives a classical 2-axis instrument.

An essential feature seldom used in X-ray diffraction analysis is the monochromator. The X-ray anodes deliver the very intense characteristic lines, but also a white spectrum due to the bremsstrahlung radiation. This white spectrum is about $10^{-3}$ weaker than the characteristic lines; if no monochromator is used, it will

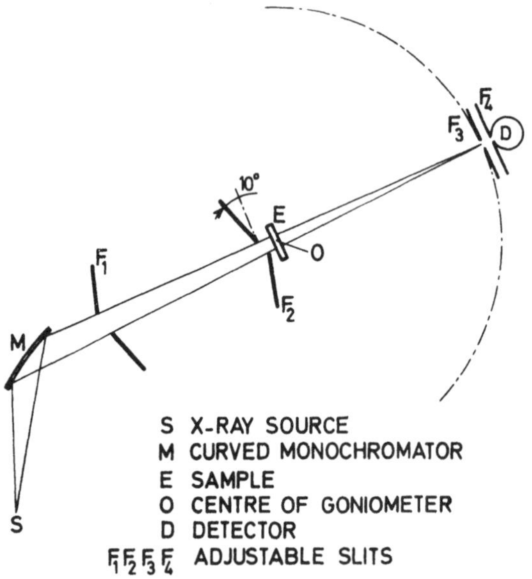


Fig. 3.12 Schematic view of the experimental setup for diffuse X-ray scattering with detector. The monochromator is curved such as to focus onto the detector

be Bragg reflected by the sample,producing an undesirable background which is of the same order of magnitude as usual scattering from phonons making such measurements impossible. Classical X-ray monochromators use the [101] reflection from quartz, the [200] reflection from lithium fluoride, and more recently the [002] reflection of pyrolitic graphite. In order to increase the intensity, collimators are removed and replaced by adjustable slits and the monochromator crystals are bent (see, for example, GUINIER, 1963) in such a way as to focus the X-ray beam on the detector slit (see Fig.3.12). This increase of intensity by relaxed resolution is possible because diffuse scattering of interest, even localized in reciprocal space, varies slowly as a function of Q compared to Bragg peaks.

As with neutrons, higher order reflections of wavelengths $\lambda/2$, $\lambda/3$... from the white spectrum will also be reflected by the monochromators, but they are easily eliminated by the use of proportional counters, or solid state detectors, which have a sufficient resolution in energy. The use of such counters is highly recommended, anyway, to eliminate possible X-ray fluorescence from the sample which is another source of undesirable background. When shorter wavelength characteristic lines are used (MoK$\alpha$,AgK$\alpha$) it is also possible to reduce the anode tension in such a way as to suppress the creation of the $\lambda/2$ radiation.

As can be seen in Fig.3.12, in order to take full advantage of the focusing beams, the sample surface will be larger than for usual diffraction and should best cover the complete beam.

A further reduction of background can be obtained by putting the whole instrument under vacuum or light gas atmosphere (helium); this is useful in particular when measurements have to be performed at relatively small scattering angles, with higher wavelength X-rays, in which conditions scattering from air is most important.

Recent examples of X-ray scattering performed with counter detection can be found in the study of $NH_4Br$ by RENZ et al. (1972) and the study of a one-dimensional metal TTF-TCNQ by KAGOSHIMA et al. (1975).

The photographic method for the measurement of X-ray scattering has been used long ago for investigations on alloys (see for example LAMBOT, 1950; GEISLER et al., 1948; GRAF, 1955) and later extended to studies on radiation damage (LAMBERT, 1957) or doped salts (LEVELUT et al., 1968; SUZUKI, 1961). It is only relatively recently that it has been used for studies on phase transitions (one of the oldest studies is by CANUT et al. (1964) on $NaNo_2$). A schematic setup is shown in Fig.3.13. As the sample is kept fixed during the exposure, this method has often been improperly described as the monochromatic Laue method.

The intensity collected on the photographic film, which can be flat or more generally cylindrical, corresponds then, point by point, to all general points of reciprocal space which simultaneously satisfy the relation

$$\underline{Q}_0 = h\underline{a}^* + k\underline{b}^* + l\underline{c}^* = \underline{k}_I - \underline{k}_F$$

where h,k,l are not restricted to integer values as for usual Bragg diffraction. Fig.3.13 shows a vertical section, containing the incident beam, the locus of the

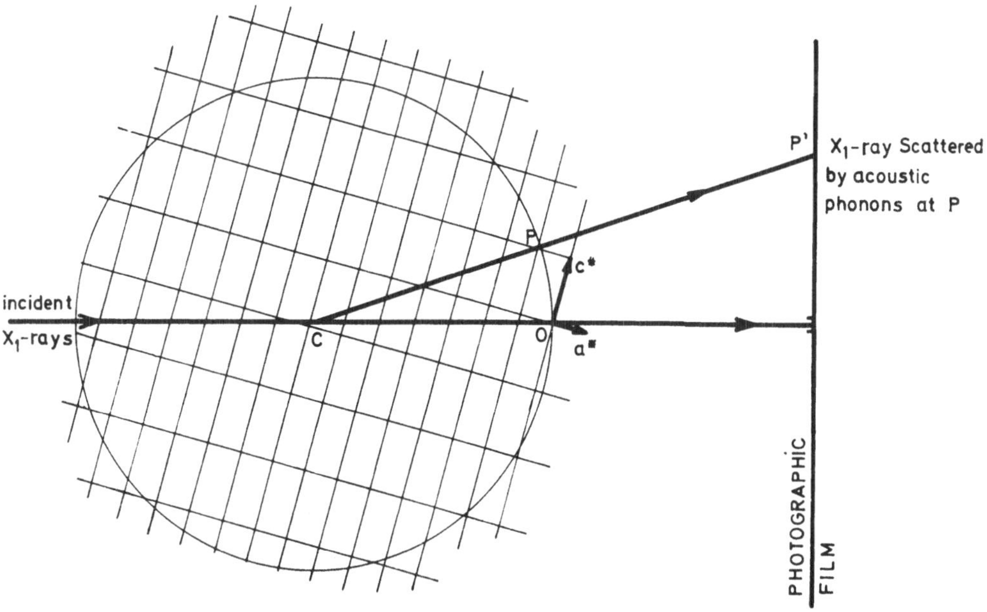


Fig. 3.13 Vertical section containing the incident beam and showing an example of the relative positons of the EWALD sphere, a reciprocal lattice, and the photographic film (COMES et al., 1973b)

points $\underline{k}_I - \underline{k}_F$ which is here a section of a sphere (EWALD sphere) as $|k_I| \approx |k_F|$ for X-rays, a reciprocal space as referred to a particular reciprocal lattice, and the photographic film.

Besides scattering from the acoustic modes ($I \propto 1/\omega^2$) from $\Gamma$ points which are nearly in a reflecting position (on the circle of Fig.3.13), and a few Bragg spots themselves, a pattern made with a standard crystal will show only a slowly varying background due to the phonons of generally slowly varying dispersion surfaces. If, however, some kind of correlation takes place arising from chemical disorder, or from phonons having an exceptionally low frequency in restricted parts of the Brillouin zone, it will to some extent concentrate the scattering in restricted parts of the photographic film. A complete analysis of such extra scattering will need successive patterns for different orientations of the sample.

As an example, Fig.3.14 shows a schematic perspective view describing how one-dimensional correlations, corresponding to intensity localized in reciprocal sheets, will produce a pattern such as shown in Fig.3.3b for orthorhombic $KNbO_3$.

With this photographic technique, if the experimental setup is well built, the intensity requirements can be less tight than for counter measurements. The unwanted background originates almost exclusively in the crystal itself, and high intensity X-ray beams, such as obtained from high power generators with rotating anodes, only reduce the exposure times. This is, of course, convenient, in particular for measurements with variable temperature, but not always essential.

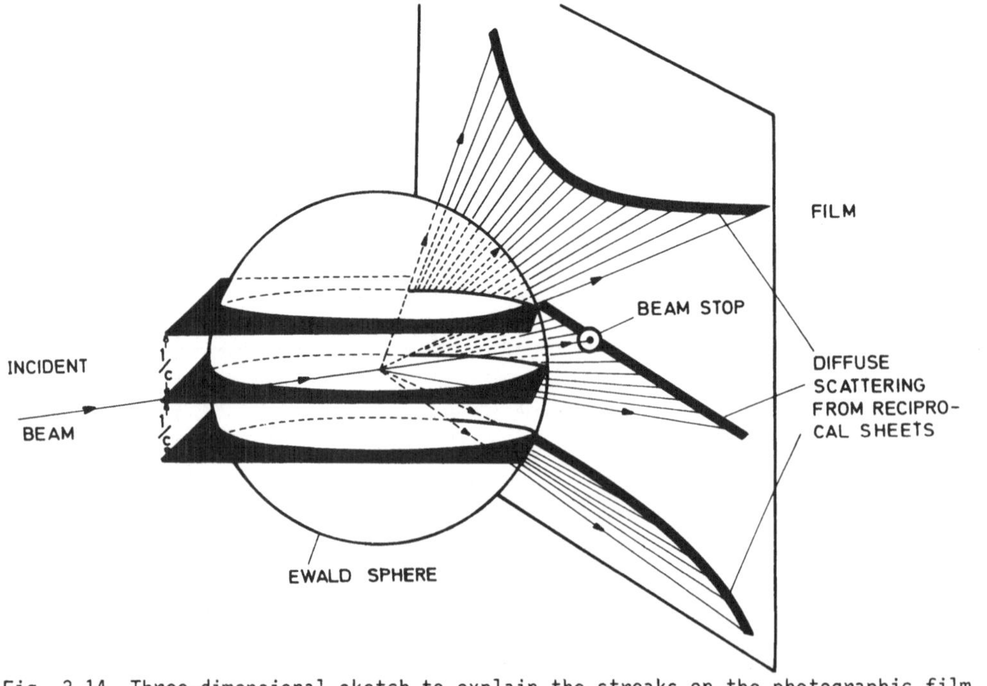


Fig. 3.14 Three-dimensional sketch to explain the streaks on the photographic film arising from planes in reciprocal space

The monochromators best suited for such photographic patterns are double bent crystals which produce beams with approximate point focusing, the horizontal and vertical focusing both occurring at the photographic plate.[3] For this purpose one uses lithium fluoride or pyrolithic graphite monochromators.

Patterns can be obtained both by reflection on a crystal plate, or by transmission. Generally the transmission setup has been most widely used. In this case the crystal thickness has to be optimized as a function of the X-ray radiation which is used, in order to obtain a maximum scattered intensity

$$I \propto x \exp(-\mu\rho x) \qquad (3.25)$$

where x is the thickness

$\mu$ is the mass absorption coefficient of the sample for the wavelength used

$\rho$ is the density of the sample.

The optimum size is then

$$x_0 = \frac{1}{\mu\rho} \quad .$$

The wavelength itself, which is used for such experiments, should be adapted to the sample not only in order to get reasonable values for $x_0$ (of the order of a few tenths of a mm), but also in order to avoid as much as possible short wavelength fluorescence from the constitutive elements of the sample. This is a real difficulty in many cases: $SrTiO_3$, for example, which has been extensively studied with neutrons (RISTE et al., 1971, SHAPIRO et al., 1972) has so far not been studied with X-rays. If $CuK_\alpha$ ($\lambda$ = 1.54 Å) is used, the value of $x_0$ becomes too small for practical purposes. If $MoK_\alpha$ is used ($\lambda$ = 0.710 Å) reasonable values of $x_0$ are obtained, but Sr produces a fluorescence of wavelength $\lambda$ = 0.879 Å ($K_{\alpha 1}$) which is impossible to eliminate by absorbing screens.

The best suited wavelength should be slightly higher than the K absorption edge of strontium ($\lambda_k$ = 0.77 Å) but there is no element which can be practically used to produce an intense characteristic $K_\alpha$ line of this wavelength.

In this respect, the use of X-ray synchrotron radiation should be of interest because the wavelength is continuously tunable.

In conclusion, it is clear that besides the relative low cost of producing X-ray beams, their main advantage compared to neutrons is the possibility of using much smaller samples, and to some extent to be able to make photographic measurements. The main disadvantage is that the scattering cannot be analysed in energy and also, unless counters are used, results will remain in part qualitative.

---

[3] A description of all kinds of X-ray monochromators can be found in the International Tables of Crystallography, Vol. III.

In particular, if measurements are performed as a function of temperature, the relative scaling of the intensities of successive patterns will be very imprecise.

### 3.1.3 Correlations and Dimensionality in Real and Reciprocal Space

To a distribution of scattering particles in real space there corresponds a distribution of intensity (amplitudes) in reciprocal space which is obtained by a Fourier transformation. Let us start with a one-dimensional example, a string of M equidistant identical atoms with spacing a, and use coordinates x and $Q_x$ in real and reciprocal space, respectively. The Fourier sum for the amplitudes gives

$$\sum_{m=1}^{M} e^{iQ_x ma} = \frac{1-e^{iMQ_x a}}{1-e^{iQ_x a}} = e^{i\left(\frac{M-1}{2}\right) Q_x a} \frac{\sin(\frac{1}{2} MQ_x a)}{\sin(\frac{1}{2} Q_x a)} \tag{3.26}$$

and the intensitites are given by the squared modulus

$$I \sim \frac{\sin^2(\frac{1}{2}MQ_x a)}{\sin^2(\frac{1}{2}Q_x a)} \quad . \tag{3.27}$$

We see that the intensity has maxima for $Q_x = h(2\pi/a) = ha^*$ ($a^*$ is the reciprocal spacing and h an integer), but is independent of $Q_y$ and $Q_z$. This defines intensity planes with a width proportional to 1/Ma, the inverse length of the string (the full width at half maximum of such intensity planes is given by $\Delta Q_x = 0.888 \times 2\pi/Ma$).

The response in $Q_x$ is sharper if the string is longer, or in other words if there are more particles. For the infinite string we get $\delta$ functions for each $Q_yQ_z$ intensity plane: successive orders of a one-dimensional Bragg reflection:

$$I \sim \frac{\sin^2(\frac{1}{2} MQ_x a)}{\sin(\frac{1}{2} Q_x a)} \rightarrow \delta(Q_x - ha^*) \text{ for } M \rightarrow \infty \quad . \tag{3.28}$$

In the opposite limit of one single particle the intensity distribution in reciprocal space is constant

$$I \sim \left| \int \delta(x) e^{iQ_x x} dx \right|^2 = \text{constant} \quad . \tag{3.29}$$

In the case of several parallel chains of the same length Ma, in random order (a two-dimensional gas of parallel strings), the total intensity is just the product of the intensity of one single chain and the number of chains.

If the chains have different lengths, we consider an average length $L_x = \langle M \rangle a$. For $\langle M \rangle \gg 1$, we can replace the average over the intensity contributions from different chains by the main value and write

$$I \sim \frac{\sin^2(\frac{1}{2} L_x Q_x)}{\sin^2(\frac{1}{2} a Q_x)} \tag{3.30}$$

and find a width proportional to $1/L_x$.

In conclusion, if we have order in one dimension in real space, we get planes of intensity in reciprocal space.

If the atoms are ordered in two dimensions with extension Ma in x-direction and Nb in y-direction, then the intensity distribution is given by

$$I \sim \frac{\sin^2(\frac{1}{2}MQ_x a)}{\sin^2(\frac{1}{2}Q_x a)} \cdot \frac{\sin^2(\frac{1}{2}NQ_y b)}{\sin^2(\frac{1}{2} Q_y b)} \quad . \tag{3.31}$$

If there are several planes of the same size parallel in x and y but not correlated in z-direction (gas-like), then the number of planes enters as a factor into (3.31).

The intensity distribution in reciprocal space produced by two-dimensional order (x and y) in real space is independent of $Q_z$. There are rods for $Q_x = ha^*$; $Q_y = k(2\pi/b) = kb^*$ (h and k being integers) with widths $\Delta Q_x \sim 1/Ma$ and $\Delta Q_y \sim 1/Nb$.

Finally, if we have P planes ordered in z-direction with a distance c, the intensity in reciprocal space will consist in spots at $Q_x = ha^*$; $Q_y = kb^*$; $Q_z = \ell_c^*$ with a width $\Delta Q_x \sim 1/Ma$; $\Delta Q_y \sim 1/Nb$; $\Delta Q_z \sim 1/Pc$. For $M,N, P \to \infty$, we obtain $\delta$-functions for all three directions, the well-known Bragg peaks (cf. Eq.(1.9) of Chap.1). These $\delta$-functions are factors which have been omitted in (3.5).

In this introductory part we have discussed the intensity distributions due to the basic order of atoms (of their equilibrium positions). Intensity in planes arises from one-dimensional order, as found for germanium atoms in AlMgGe (LAMBOT, 1950); intensity in rods arises from two-dimensional order, as found for lithium platelets in neutron irradiated LiF (LAMBERT, 1957); and intensity in spots from, the usual, three-dimensional order. The widths of the intensity distribution in different directions is proportional to the reciprocal of the extension of ordered atoms in the particular direction.

In the context of structural phase transformations we assume, for the high symmetry phase, that the average or the equilibrium positions are ordered throughout the crystal. But we will apply the above considerations about order to the pattern of displacements $\underline{u}$ from the equilibrium positions. The periodicity of these displacements in the high symmetry phase is limited to domains or clusters; the so-called short-range order. These clusters usually have a lifetime $\tau$, with which they decay and reconstitute themselves elsewhere. We disregard the time dependence for the moment. To derive the scattered intensity we have to sum the scattered amplitudes of all the atoms in the crystal and then to take the absolute value. We start summing up the amplitudes for the nth unit cell

$$F_n = \sum_{d}^{\text{unit cell } n} b_d e^{-W_d[\underline{Q}\underline{u}_d(n)]} e^{i\underline{Q}\underline{d}} \quad . \tag{3.32}$$

This $F_n$ describes the scattering from the displacements of the atoms. The Bragg scattering is already separated and not discussed here. The short range order creates a difficulty, as now $F_n$ is varying from unit cell to unit cell. We have to sum over the whole crystal and cannot use simplifications due to translational symmetry. The scattered intensity is

$$I(\underline{Q}) = \sum_{n}^{\text{crystal}} \sum_{n'} F_n F_{n'}^{*} e^{i\underline{Q}(\underline{x}_n - \underline{x}_{n'})} \tag{3.33}$$

where $\underline{x}_n$ is the vector to the origin of the nth unit cell. We can rewrite (3.33) introducing the distance $\underline{\ell}$ between the unit cells n and n':

$$I(\underline{Q}) = \sum_{\ell}^{\text{crystal}} \left( \sum_{n}^{\text{crystal}} F_n F_{n+\ell}^{*} \right) e^{i\underline{Q}\underline{\ell}} \quad . \tag{3.34}$$

Now we assume that the relative displacement of all the atoms in one unit cell is described by an eigenvector $\underline{\sigma}_d^j$, which is the same for each unit cell, and an amplitude $A_n$, which is varying through the crystal. Furthermore, the eigenvector may vary periodically from cell to cell with a wave-vector $\underline{q}_\sigma$. The $F_n$ reads

$$F_n = A_n \sum_{d}^{\text{unit cell}} b_d e^{-W_d} [\underline{Q}\underline{\sigma}_d^j(\underline{q}_\sigma)] e^{-i\underline{q}_\sigma \underline{x}_n} e^{i\underline{Q}\underline{d}} \quad . \tag{3.35}$$

It follows that

$$\sum_{n}^{\text{crystal}} F_n F_{n+\ell}^{*} = |G_j(\underline{q},\underline{Q})|^2 e^{-i\underline{q}_\sigma \underline{\ell}} \sum_{n}^{\text{crystal}} A_n A_{n+\ell} \quad .$$

Here we introduced again the structure factor $G_j$ as already defined in (3.8). The sum over the amplitudes

$$\sum_{n}^{\text{crystal}} A_n A_{n+\ell} = V(\underline{\ell}) < A(0)A(\underline{\ell})> \tag{3.36}$$

leads us to the correlation function $<A(0)A(\underline{\ell})>$ of the amplitudes. The correlation function describes the probability to find the amplitude A in the unit cell at $\underline{\ell}$, if there exists the amplitude A at $\underline{\ell} = \underline{0}$. $V(\underline{\ell})$ is a function describing the periodicity of the unit cells in the crystal as used by GUINIER (1963).

Now we find for the intensity

$$I(\underline{Q}) = |G_j(\underline{q},\underline{Q})|^2 \sum_{\ell}^{\text{crystal}} V(\underline{\ell}) < A(0)A(\ell)> e^{i(\underline{Q}-\underline{q}_\sigma)\underline{\ell}} \quad . \tag{3.37}$$

The sum over $\ell$ is a Fourier transformation of a product of two functions depending on $\underline{\ell}$. The transformation yields the convolution of the two functions transformed separately. For the infinite crystal the Fourier transform of $V(\underline{\ell})$ gives $\delta[\tau - (Q-q_\sigma)]$ and the convolution with the Fourier transform of the correlation function $F[<A(0)A(\ell)>]$ defines the center of $F[<A(0)A(\underline{\ell})>]$ to be each $\underline{Q} = \underline{\tau} + \underline{q}_\sigma$. This means that the scattered intensity arising from the short-range order is centered around positions which get superlattice Bragg positions for long range order (in the low symmetry phase).

For the analytic shape of the correlation function one can assume many different expressions (GUINIER, 1963). In the context of phase transformations the ORNSTEIN-ZERNIKE approximation has been found to be very useful. Following ORNSTEIN-ZERNIKE the correlation function for the isotropic case reads

$$<A(0)A(|\underline{\ell}|)> = k_B T \frac{1}{|\underline{\ell}|} \exp(-\frac{|\underline{\ell}|}{\xi}) \tag{3.38}$$

where $\xi$ is the correlation length. After the Fourier transformation we find for the intensity

$$I(\underline{Q}) = |G_j(\underline{q},\underline{Q})|^2 \frac{k_B T}{\frac{1}{\xi^2} + q^2} \tag{3.39}$$

where each $\underline{\tau} + \underline{q}_\sigma$ serves as an origin for $\underline{q}$.

In the anisotropic case one writes

$$I(\underline{Q}) = |G_j(\underline{q},\underline{Q})|^2 \frac{k_B T}{\frac{1}{\xi^2} + \frac{\xi_x^2}{\xi^2} q_x^2 + \frac{\xi_y^2}{\xi^2} q_y^2 + \frac{\xi_z^2}{\xi^2} q_z^2} \quad . \tag{3.40}$$

Again the widths of the intensity distribution in directions x,y and z are proportional to $1/\xi_x$, $1/\xi_y$ and $1/\xi_z$, respectively. Taking as above a one-dimensional example in which the intensity distribution is independent of $Q_y$ and $Q_z$, the full width at half maximum of the intensity plane in x-direction is given by $\Delta Q_x = 2/\xi_x$. Note here that the correlation length $\xi_x$ as deduced from (3.40) is by a factor $\pi \times 0.888$ smaller than the length $L_x = <M>a$ resulting from (3.30), for the same $\Delta Q_x$ which is the measured quantity. In conclusion we see that the absolute values of the sizes ($\xi_x$ or $L_x$) are dependent on the type of analysis which is used. This $I(\underline{Q})$ is the intensity measured by diffuse X-ray scattering.

Finally we discuss briefly the time dependence of the clusters. Usually the decay is described by $\exp(-t/\tau)$ with the relaxation time $\tau$. The Fourier transform of this expression gives a normalized function of frequency. We obtain an $I(\underline{Q},\omega)$ which can be determined by inelastic neutron scattering (if $\tau$ is short enough to produce an energy distribution large enough compared to the instrumental resolution)

$$I(\underline{Q},\omega) = |G_j(\underline{q},\underline{Q})|^2 \ F_j(\omega,\underline{q},T) \quad . \tag{3.41}$$

In comparison with (3.7) there is no dispersion $\omega_j(\underline{q})$ any more, only a $\underline{q}$-dependence is left with a pattern $\underline{\sigma}^j$ in real space.

$$F_j(\omega,q,T) = \frac{\omega}{1-\exp(-\frac{\hbar\omega}{k_BT})} \frac{1}{\frac{1}{\xi^2} + \frac{\xi_x^2}{\xi^2} q_x^2 + \frac{\xi_y^2}{\xi^2} q_y^2 + \frac{\xi_z^2}{\xi^2} q_z^2} \frac{1/\pi\tau}{\frac{1}{\tau^2} + \omega^2} \quad . \tag{3.42}$$

Generally $\tau$ is a function of T and q. This will be discussed in Subsec.3.2.2.

The introduction of a relaxation time $\tau$ in connection with correlations in space may seem obvious but it is not necessarily. This means correlations are produced not only by relaxing clusters but also by well-behaved phonons which have a dispersion in reciprocal space. In the introduction we have explained that phonon dispersion gets visible in diffuse X-ray scattering due to the fact that the intensity, for $k_BT \gg \hbar\omega$, is proportional to $1/\omega^2$. Now we will discuss the diffuse X-ray intensity coming from phonons by means of correlations in $\underline{u}$. From (1.89) and (3.9) we understand that the displacement $\underline{u}(\underline{\ell})$ of a particular atom in the $\ell$th unit cell is described by the sum over all phonons, that is by a sum over $\underline{q}$ and j. We then can write for the atoms of species d (we omit the index d)

$$\langle u_\alpha(\underline{\ell})u_\beta(0)\rangle = \sum_{j,\underline{q}} A^{j^2}(\underline{q})\sigma_\alpha^j(\underline{q})\sigma_\beta^j(\underline{q}) \exp(i\underline{q}\underline{\ell}) \quad . \tag{3.43}$$

Mixed products in j do not contribute to the correlation function since, different modes are orthogonal, and mixed products in $\underline{q}$ vanish if we interpret the brackets $\langle\ \rangle$ to be a time average or, just as well, by an ensemble average at constant time. We prefer the latter, as we talk about constant time correlations.

The contributions of different phonons, represented here by plane waves with wave-vector $\underline{q}$, are controlled by the exponential phase factor. Waves with $\underline{q}\perp\underline{\ell}$ have correlated amplitudes for all $\underline{\ell} \to \infty$ (plane wave). As soon as the angle between $\underline{q}$ and $\underline{\ell}$ deviates from $\pi/2$ the positive correlation is vanishing at about $\underline{q}.\underline{\ell} = \pi$ for the contribution by the particular $\underline{q}$. When $A^j(\underline{q})$ is, for a particular j, the same for all $\underline{q}$ there is no correlation between the amplitudes of the atoms, from this mode. But if there is a $\underline{q}$-dependence of $A^j(\underline{q})$, for example with $A^j(\underline{q})$ large at a particular $\underline{q}$ and rapidly decreasing as $\underline{q}$ goes away from this position, then we find a correlation. The plane waves with large amplitudes produce a correlation in the $\underline{u}$'s, which cannot be averaged away by the other waves because they have smaller amplitudes. For $k_BT \gg \hbar\omega$, the $A^j(\underline{q})$ are proportional to $1/\omega_j(\underline{q})$.

In Fig.3.15 a) we approximate a minimum of dispersion at $\underline{q}_M$ (M is a symmetry point at the Brillouin zone boundary) by a constant frequency for $0 \le q_\perp \le 1/6|\underline{q}_M|$ and an extremely steep rise at $q_\perp = 1/6\ |\underline{q}_M|$. The corresponding amplitudes for two waves ($q_\perp = 0$ and $q_\perp = 1/6|\underline{q}_M|$) are sketched in real space. These two waves extinguish their amplitudes at $|\underline{\ell}| = \xi$ in the direction $q_\perp$. All other waves between the two extreme ones produce extinctions of amplitudes at larger $|\underline{\ell}|$. Therefore a

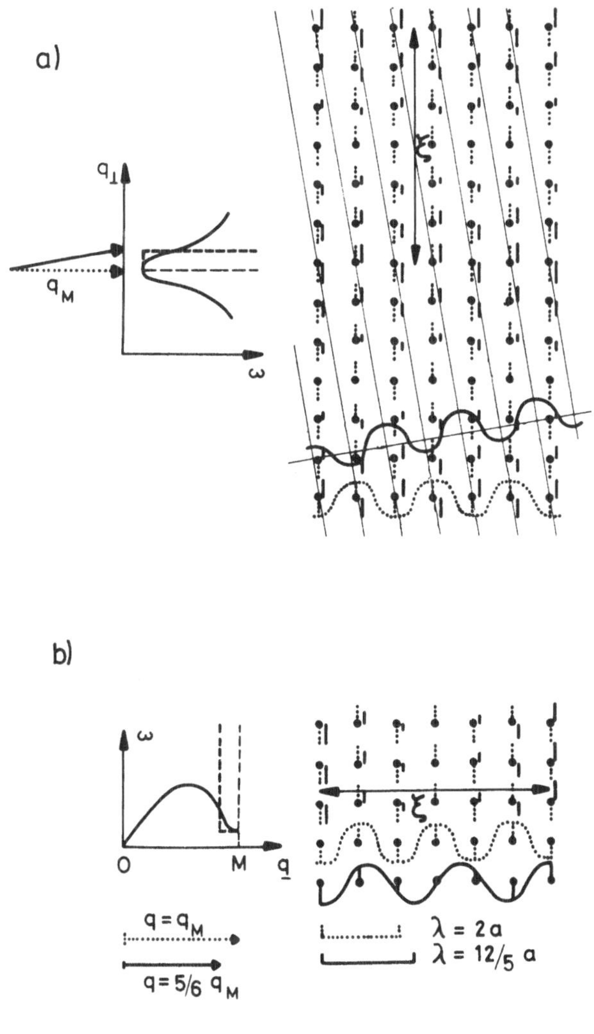


Fig. 3.15a,b Correlation of atomic amplitudes in real space due to a dispersion in reciprocal space. A minimum at $q_M$ is approximated by a constant frequency around $q_M$, which goes abruptly to high frequencies at a distance $\Delta q = 1/6\ q_M$. The dots give the equilibrium positions of the atoms and the solid lines the amplitudes due to $q_M$. The dotted lines are the amplitudes due to: a) $q_\perp = 1/6\ |\underline{q}_M|$ and b) $\underline{q} = 5/6\ \underline{q}_M$

correlation of amplitudes exists only for $|\underline{\ell}| < \xi$. Waves for $q_\perp > 1/6\ |\underline{g}_M|$ cannot destroy the correlation as they have vanishing amplitudes within our approximation.

Fig.3.15 b) explains a correlation parallel to $\underline{g}_M$ in the same approximation. Again all waves $5/6|\underline{g}_M| < |\underline{g}| < |\underline{g}_M|$ extinguish the correlation of amplitudes for $|\underline{\ell}| > \xi$ in the direction $\underline{g}_M$.

A discussion of constant time correlations in the amplitudes of phonons was given by HÖLLER (1969) to interpret the results of diffuse X-ray scattering.

This understanding of correlations in amplitudes allows us to reinterpret the idea of Einstein oscillators associated with flat dispersion branches (usually, optic modes).

If the $A^j(\underline{q})$ are, for one mode j, constant in three dimensions (constant frequency) then the intensity (disregarding the structure factor $G_j$) is constant everywhere. We conclude that there is no correlation between identical atoms in neighbouring unit cells in this mode j of oscillation. The atoms in one unit cell then form an Einstein oscillator, which oscillates in the averaged field of the neighbouring cells.

If the intensity is confined to planes, we have correlation in chains perpendicular to the planes. In this case, the Einstein oscillators consist of chains. We expect an increasing frequency for $\underline{q}$-components parallel to the chains as found in $KNbO_3$ (CURRAT et al., 1974).

If the intensity appears like rods in reciprocal space, correlation exists in planes perpendicular to the rods. These planes may be called Einstein oscillators as found in NaOH (BLEIF et al., 1971) in $CD_4$ (HÖLLER and PRESS, 1972) and in Na Nb $O_3$ (DENOYER et al. 1970, 1974, 1976).

## 3.2 Fluctuations and Order

The static long-range order in single crystals leads to symmetries and symmetry operations, which transform each atom into an identical one. If we apply symmetry operations to an atom d, which has an actual displacement $\underline{u}_d$, then the atom is transformed into an identical one d' but the new $\underline{u}_{d'}$ can be positive or negative to $\underline{u}_d$ (neglecting possible phase factors) depending on the symmetry operation. With the help of symmetry operations, we can group the displacement patterns j (modes j) into different irreducible representations in a space group (KOVALEV, 1965; ZAK, 1969). If there is no degeneracy, the number of irreducible representations is equal to the number of symmetry operations. In the case of degeneracy, the sum of the squared degeneracies is equal to the number of symmetry operations. Modes which belong to the same representation can couple to each other, but their dispersion surfaces cannot intersect either by varying $\underline{q}$ or by varying T (modes getting soft).

Phase transformations are usually accompanied by a change in symmetry. As this can be found in the textbook of LANDAU and LIFSHITZ (1969) we will only briefly outline some conditions. For a first order phase transformation to occur there are no symmetry requirements, it is only necessary for the free energy of the two phases to be the same at the transition temperature. In a second order, or continuous, phase transition the properties of the crystal change gradually, and only one phase exists at any temperature. This is only possible if certain symmetry conditions are satisfied.

We describe the positions of atoms d in the unit cell $\ell$ by a vector $\underline{R}_0(\ell,d)$. The distribution of atoms, described by the vector $\underline{R}_0$, is invariant under the operations of the space groups (MARADUDIN and VOSKO, 1968).

If the crystal changes continuously, the new structure can be described by a vector,

$$\underline{R}(\ell,d) = \underline{R}_0(\ell,d) + \underline{u}(\ell,d) \quad . \tag{3.44}$$

The continuous change implies that the symmetry group of $\underline{u}$ must be a subgroup of $\underline{R}_0$ (first condition).

The condensed vector $\underline{u}$ which produces the new structure, of lower symmetry than the structure described by the vectors $\underline{R}_0$, belongs to only one irreducible representation of the system $\underline{R}_0$ (second condition). This irreducible representation has characters +1 for symmetry operations which exist in both phases $\underline{R}_0$ and $\underline{R}$. The characters for symmetry operations not present in $\underline{R}$ are -1. This explains why the soft mode in the less symmetric phase is always in the totally symmetric representation ($A_1$). The vector $\underline{u}$ within the irreducible representation of dimension i (i-fold degenerate modes) can be written as

$$\underline{u}(\ell,d) = \sum_i c_i \underline{\phi}_i(\ell,d) \quad . \tag{3.45}$$

The change of the crystal corresponds, of course, to a change of the free energy. For small values of the coefficients $c_i$ one can write the free energy F as a power series in $c_i$. Substituting $c_i = \eta\gamma_i$, with $\sum_i \gamma_i^2 = 1$, and $\eta$ being the order parameter one obtains

$$F = F_0 + \frac{1}{2}A\eta^2 + \frac{1}{3}Bf^{(3)}\eta^3 + \frac{1}{4}Cf^{(4)}\eta^4 + \dots \quad . \tag{3.46}$$

The coefficients A,B,C, etc., are functions of the temperature, pressure, field, concentration, etc. Here we restrict ourselves to the temperature dependence only. $f^{(\ell)}$ is a homogeneous function of order $\ell$ in the coefficients $\gamma_i$. As the free energy is not changed by the symmetry operations of the structure $\underline{R}_0$, only functions $f^{(\ell)}$ occur which are invariant under all operations of the group $\underline{R}_0$. There is no term linear in $\eta$ because first order invariants exist only for the identity representation. Also $f^{(2)} = 1$ since only one quadratic invariant exists for any representation.

The stable state of the crystal is found by minimising F with respect to $\eta$ and $\gamma_i$. The functions $f^{(n)}$ are discussed for ferroelectrics of the $BaTiO_3$ type by HAAS (1965), and for the doubly degenerate soft mode in $Tb_2(MoO_4)_3$ (DORNER et al., 1972). From the conditions for stability, $\partial F/\partial\eta = 0$ and $\partial^2 F/\partial\eta^2 > 0$, one finds that the state $\eta = 0$ is stable for $A > 0$, whereas for $A < 0$ the stable state must have $\eta \neq 0$. Therefore, a phase transition from the state of high symmetry, $\eta = 0$, to a state with a lower symmetry could occur at the point where $A = 0$.[4] However, for the crystal to be stable at the point where $A = 0$ and $\eta = 0$, F must increase both for small positive and negative changes of $\eta$. This cannot be the case if $Bf^{(3)} \neq 0$. Therefore, a continuous phase transition is possible only if third-order terms in the free energy are zero, i.e., the function $f^{(3)}$ should vanish by symmetry (third condition).

These three conditions are necessary symmetry conditions for a continuous phase transformation to occur. But even if all three conditions are fulfilled, the transition will occur as a first order phase transition of $C < 0$. The lattice is then stabilized by higher order terms in $\eta$. The free energy versus $\eta$ for different temperatures is presented in Fig.3.16 for positive and negative C. As already pointed out, C may depend on other physical parameters as well. Varying these parameters C may pass from negative values to positive ones, thus changing the character of the phase transformation from first to second order. For $C = 0$ the phase transition is continuous at the temperature where $A = 0$, and the first stabilising term in the free energy expansion is of 6th order in $\eta$. Such a point is called a tricritical point. $ND_4Cl$ is a system which exhibits a tricritical point under pressure (YELON et al., 1974).

Each of the basic vectors in (3.45) can be written in the form

$$\underline{\phi}_i(\ell,d) = \underline{\sigma}_d^i(\underline{q})\exp(i\underline{q}\underline{\ell}) \tag{3.47}$$

where $\underline{\sigma}_d^j(\underline{q})$ has the periodicity of the lattice, so that the irreducible representations are characterised by vectors $\underline{q}$ in reciprocal space, representing the properties for translations. LANDAU and LIFSHITZ give a fourth condition which in effect states that only such values of $\underline{q}$ are allowed as can be written as simple fractions of a reciprocal lattice vector. Consequently the unit cell of the crystal below the transition point is a simple multiple of the original unit cell.

[4]Often the high symmetry phase appears at higher temperatures than the low symmetry phase, but it may be inverted as in NaOH (BLEIF et al., 1971).

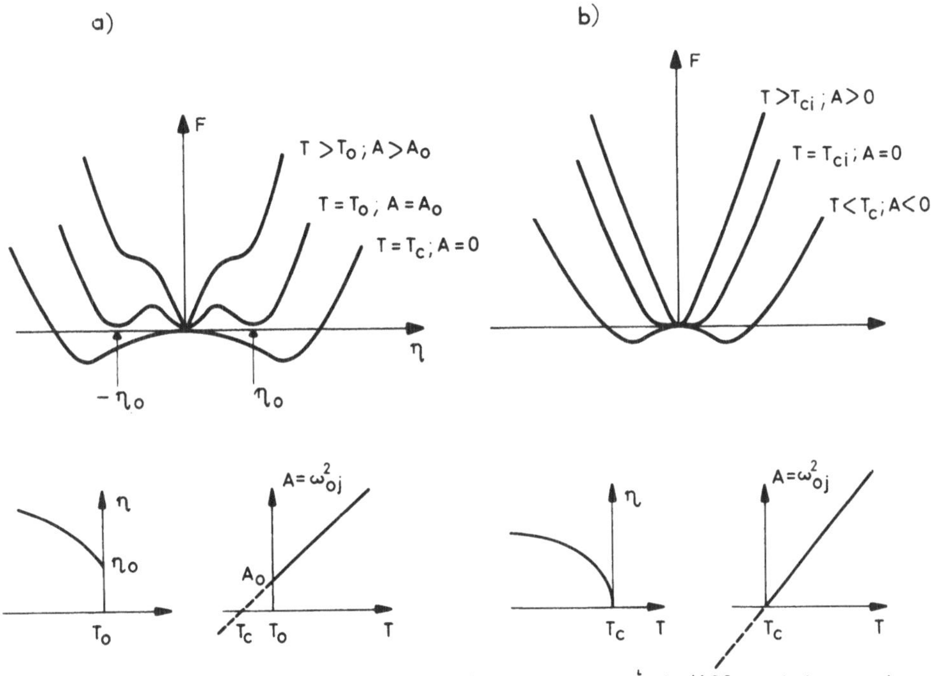


Fig. 3.16a,b Free-energy versus the order parameter η at different temperatures: a) for a first-order phase transformation; b) for a second-order phase transformation. Below each figure the corresponding temperature dependence of the order parameter η and the square of the soft mode frequency $\omega_{0j}^2$ are shown

The fourth condition is invalid since, for example, continuous transformations have been found with spin spiral magnetic structures, where the spiral periodicity is incommensurate with that of the lattice, and the symmetry cannot even be described by a three-dimensional space group. Superstructures in the atomic positions have been found in TTF TCNQ by COMES et al. (1976) and $TaSe_2$ by MONCTON et al. (1975), which are incommensurate with the high temperature lattice periodicity. In both cases the phase transformation appears to be due to a strong electron-phonon interaction.

A small, but significant, extension of the second condition has been pointed out by COCHRAN (1971). In (3.45) the i denotes the dimension of the representation, but even for dimension i = 1 the vector $\underline{\phi}$ consists of j orthogonal eigenvectors $\underline{\phi}^j$, where j is the number of modes within the representation depending on the number of atoms in the unit cell. For Cochran it is very improbable that more than one mode out of the j will produce a new structure at a particular temperature.

The inverse susceptibility associated with the order parameter is $\partial^2F/\partial\eta^2|_{\eta=0} = A$, for η = 0, in the high symmetry phase. In the case of an order-order phase transformation $A = \omega_0^2$, where $\omega_0$ is the soft mode frequency discussed in Subsecs. 1.3.2 and 3.2.1. For a disorder-order transition the susceptibility is related to the correlation length of ordered clusters, as discussed in Subsec.3.2.2. The particular eigenvector $\phi_i^j$ or with (3.47) $\sigma_d^{ij}(\underline{q})$ will be related to symmetry operations in Subsec.3.2.3.

### 3.2.1 Soft Modes

If a crystal undergoes a structural phase transformation from one ordered structure to another ordered structure, then we have an order parameter $\eta$, which is zero in one-phase (the higher symmetric) and non-zero in the lower symmetric phase. $\eta$ describes a pattern of static displacements of the equilibrium positions of the ions. As already explained in Subsec.3.1.1, this pattern can be described by an eigenvector $\sigma_d^j$ close to the transition temperature $T_0$. Going away from $T_0$ the eigenvector of the static displacements may change due to a coupling of the order parameter to other distortions,as described by elastic strain or other modes. This will be discussed in Subsec.3.2.3.

Now we assume that in the phase where $\eta$ is zero, there exists a mode with the same eigenvector $\underline{\sigma}_d^j$. It is then plausible that the restoring forces for this particular mode will decrease on approaching the phase transformation and consequently the frequency of this mode will decrease and get soft in the high symmetry phase. In the low symmetry phase there is a soft mode as well. Its eigenvector may differ slightly from the eigenvector in the high temperature phase. In general, the eigenvector is the derivative of the actual displacements at a certain temperature with respect to the order parameter (see Subsec. 3.2.3). We will first discuss the soft mode in the phase ($\eta = 0$) and later together with $\eta$ in the phase ($\eta \neq 0$).

Historically, the concept of a soft mode (COCHRAN, 1959; ANDERSON, 1960) was introduced from the temperature dependence of the dielectric constant at para-ferroelectric transformations. Several examples were known, where the static dielectric constant $\varepsilon_0$ followed a Curie-Weiss law (cf. Subsec.1.3.2).

$$\varepsilon_0 \sim \frac{1}{T-T_c} \quad . \tag{3.48}$$

The connection of $\varepsilon_0$ to polar optic modes $\omega_{TO,n}$ is given by the Lyddane-Sachs-Teller relation which reads for n polar modes in a cubic system:

$$\frac{\varepsilon_\infty}{\varepsilon} = \prod^n \left( \frac{\omega_{TO,n}^2}{\omega_{LO,n}^2} \right) \quad . \tag{3.49}$$

Here $\varepsilon_\infty$ is the dielectric constant at frequencies high above the lattice frequencies. If there is only one polar mode we get

$$\frac{1}{\varepsilon_0} \sim \omega_{To}^2 \sim (T - T_c) \tag{3.50}$$

The soft mode idea was generalised later for all order-order structural phase transformations which are (at least almost) second order. This temperature dependence of the soft mode is, perhaps, difficult to understand. In Chap.1, (1.104), it was explained in terms of an anharmonic positive frequency shift $\Delta$, which is linear in temperature. This $\Delta$ is stabilising an unstable frequency $\omega_o^2 < 0$.

We will follow a somewhat different approach (AXE, 1971). Instead of expanding the potential energy V of the lattice in terms of displacements $\underline{u}_d$, we expand immediately in the amplitudes $A^j$ of the normal modes j. We neglect the third order term which, if it exists, contributes like the fourth order term, and write

$$V = \sum_j \frac{1}{2} \omega_{o,j}^2 A^{j2} + \sum V^{(4)}_{j1j2j3j4} \; A^{j1}A^{j2}A^{j3}A^{j4} + \dots \quad . \tag{3.51}$$

This is a quite complicated expression which is simplified by taking the mean potential experienced by the jth normal mode.

$$\langle V\rangle = \frac{1}{2}\omega_{o,j}^2 (A^j)^2 + \sum_{j_1} V^{(4)}_{j_1 j_1 jj} \langle (A^{j1})\rangle^2 \; (A^j)^2$$
$$+ \sum_{j_1} V^{(4)}_{j_1 j_1 jj} \left( A^{j_1 2} - \langle A^{j_1 2}\rangle^2 \right) (A^j)^2 + \dots \quad . \tag{3.52}$$

The cross products between normal coordinates of different modes drop out since normal modes are independent, at least in lowest order. The frequency $\omega_{oj}$ of mode j is then replaced by a quasiharmonic frequency $\omega_j$ given by

$$\omega_j^2 = \omega_{oj}^2 + \sum_{j_1} V^{(4)}_{j_1 j_1 jj} \; [2n(\omega_{j_1}) + 1] + \dots \tag{3.53}$$

and the last term of (3.52) expresses a fluctuation in frequency which introduces a damping $\Gamma_j$.

For the thermal average of the amplitude $A^j1$, of the mode $j_1$, we use the Bose occupation factor n. This identity is strictly true only in the harmonic case. A better calculation has to be self-consistent, meaning that the fluctuations and the anharmonic contributions from other modes should be included.

For nonpathological modes, $\omega_{oj}^2$ is positive and quite large, so that the anharmonic contribution of the second term is relatively small. But it might be that the second term is larger $\omega_j^2$, and $\omega_{oj}^2$ is negative, which means unstable. The fact that $\omega_j$ is real is due to the anharmonic coupling to the averaged amplitudes of all the other modes. If we now assume that the main anharmonic contribution comes from modes for which $\hbar\omega_{j_i} \ll k_BT$, then the anharmonic part is linear in temperature and it follows that

$$\omega_j^2 \sim (T - T_o) \quad . \tag{3.54}$$

As T reaches $T_o$ from above, the mode j becomes unstable or better would become unstable if the lattice distortion of the new phase did not stabilize it again.

The soft mode investigated in $Tb_2(MoO_4)_3$ ($T_o$ = 159°C (DORNER et al., 1972)) at the M-point (see Fig.3.17) at high temperatures be described by the damped harmonic oscillator:

$$F_j[\omega,\omega_j(\underline{q})T] = \frac{\omega}{1 - \exp(-\frac{\hbar\omega}{k_B T})} \quad \frac{\Gamma_j}{[\omega^2 - \omega_j^2(\underline{q})]^2 + \omega^2\Gamma_j^2} \tag{3.55}$$

(see Fig.3.11). Below 400°C the mode becomes overdamped ($\Gamma_j^2 \geq 2\omega_j^2(\underline{q})$), as seen in Fig.3.18. At temperatures above 400°C, $\omega_j$ was determined by a least squares fit to the shape of the response, fitting $\Gamma$ and $\omega_j$ simultaneously; see Fig.3.19. Below 400°C this was no longer possible. The data near $T_o$ were obtained from the integrated intensity, (3.13), which holds for the damped and overdamped harmonic oscillator as well.

Apparently the Curie-Weiss law is well fulfilled. We might call $1/\omega_j^2$ the structural susceptibility of the system corresponding to the order parameter $\eta$ (see end of Sec.3.2). $Tb_2(MoO_4)_3$ is one of the rare examples where the soft mode moves out of the overdamped regime with increasing temperature and becomes easily resolvable at high temperatures. In this system, as in many others, the damping constant $\Gamma$

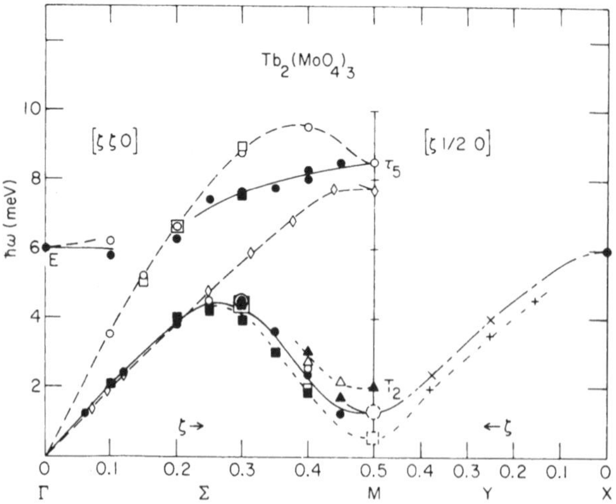


Fig. 3.17 Some low frequency dispersion curves exhibiting the soft mode at the M-point. The Δ at M corresponds to the phonon displayed in Fig.3.11. For lower temperatures the phonon was overdamped and the quasiharmonic frequency was derived from the integrated intensity (DORNER et al., 1972)

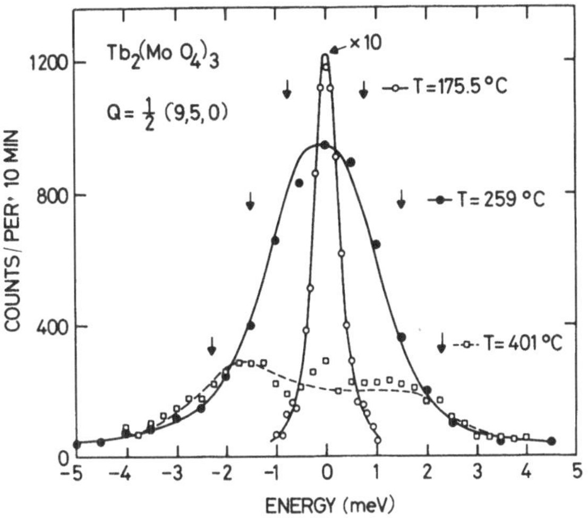


Fig. 3.18 Energy profile of scattering at a superlattice point in the PE phase, showing the evolution from underdamped phonon to strongly overdamped critical scattering as the temperature approaches $T_C$. The curves through the points are of the form of (3.55) modified by an approximate instrumental correction to account for asymmetry. The arrows indicate the values of $\omega_M(T)$ estimated by this fitting procedure (AXE et al., 1971)

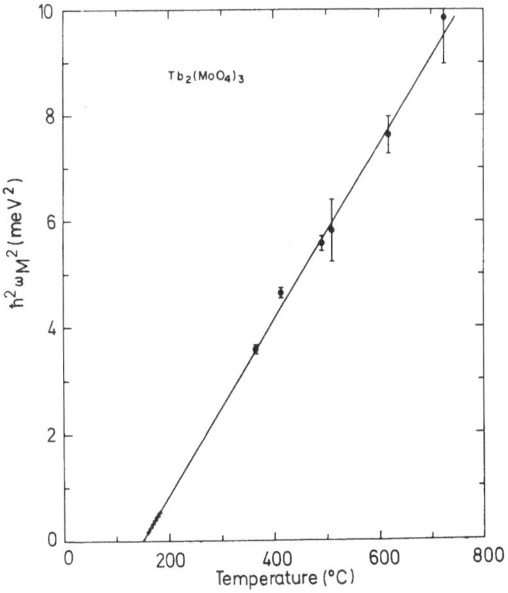


Fig. 3.19 The square of the soft mode frequency versus temperature. For temperatures above 400°C, $\omega_M$ was determined directly, while the crosses near the phase transformation represent inverse integrated intensities times temperature. The slope was fitted to the high temperature results (DORNER et al., 1972)

was found to be constant for all $T > T_0$; the overdamped character of the soft mode when approaching $T_0$ is therefore due to the decreasing frequency (at about 200°C above $T_0$, $\omega_j^2$ becomes smaller than $\Gamma^2/2$).

As a concluding remark, it should be pointed out that up to now the soft mode description of displacive phase transitions has been mainly a phenomenological approach (THOMAS, 1969, 1971, 1976). The microscopic origin of the unstable mode ($\omega_{0j}^2$ negative in (3.52)), except for the low dimensional metals (see Subsec.3.3.3) has been seldom discussed. In the so-called "vibronic theory" of ferroelectric phase transitions (KRISTOFEL et al., 1973) a predominant role is assigned to another type of electron-phonon interaction under the assumption of a band analogy of the Jahn-Teller effect (BERSUKER et al., 1968; BIRMAN, 1962). A clear example of a cooperative Jahn-Teller effect was discussed by BIRGENEAU et al. (1974) in their report on the 151 K phase transition of $PrAlO_3$.

### 3.2.2 Critical Slowing Down

In the preceding subsection we noted that a soft mode usually appears with a structural order-order phase transformation. The decreasing frequency has two effects:
1) the amplitudes of the atoms get larger, and
2) the motion of the atoms becomes slower.
The latter means that a time-dependent distortion pattern given by the eigenvector, once it is established, lasts longer. All this is included in the soft mode-overdamped mode picture describing order-order phase transformations.

For order-disorder phase transformations the case is different. The most studied continuous order-disorder phase transformations are magnetic ones, where the disordered phase is the paramagnetic one. Magnetic phase transitions will be discussed briefly in Chap.7. Examples of structural order-disorder phase transformations are provided by crystals, which have a disorder in the orientation of the molecules, such as $CD_4$, $ND_4Cl$ or $ND_4Br$. (The deuterated samples are especially suited for neutron investigation as the incoherent cross-section of deuterium is about 20 times smaller than for hydrogen. For the present discussion the incoherent scattering produces only unwanted background). We restrict ourselves to orientational disorder and do not talk about translational disorder, like β-brass, or the ionic conductors such as AgI.

In solid methane the centers of mass of the randomly oriented molecules form a fcc lattice. The centers of mass of $ND_4$ in $ND_4Cl$ and $ND_4Br$ occupy body-centered positions in a cubic framework of the halides. In the solid disordered phase, well above the phase transformation temperature, the molecules tumble around by rotational diffusion independently from their neighbors. This rotational diffusion is rather continuous for $CD_4$, and more like jumps for the $ND_4$ molecules within the Cl-cage. This single molecule reorientation gives rise to a quasielastic scattered intensity around $\omega = 0$ with an energy width proportional to the inverse of the single particle relaxation time $\tau_s$. The $|\mathbf{Q}|$-dependence is described by Bessel functions appropriate to the symmetry of the molecules.

Approaching the phase transformation by lowering the temperature, neighbouring molecules start to build clusters of ordered orientation (short-range order). These clusters disappear and are reconstituted elsewehre within a cluster relaxation time $\tau_c$. On lowering the temperature these clusters grow larger (correlation length $\xi$) and they live longer (increasing $\tau_c$). The expression "critical slowing down" comes from this increase in lifetime of the clusters. Furthermore, it expresses the fact that, near the phase transition, the system becomes extremely slow in reaching thermal equilibrium after a temperature change. The basic origin for all critical effects is believed to be the divergence of the correlation length.

While the lifetime $\tau_c$ of the clusters is diverging at the phase transition, the single particle reorientation does not show a critical behavior. This means that any particular molecule, even in a cluster, diffuses away from the proper orientation and comes back within a time $\tau_s$.

Above the phase transition one discusses the "critical" region $q \gg 1/\xi$ (q large enough to investigate correlations within one cluster) and the "hydrodynamic" region $q \ll 1/\xi$ (q small enough to study correlations larger than the clusters). In magnetic systems (see for example TUCCIARONE et al., 1971) one has found collective excitations in the "critical" region, in other words spin waves with a wavelength much smaller than the correlation length, so that they can propagate within a cluster. In the "hydrodynamic" region only diffusive motion can be found.

Many structural phase transformations are slightly of first order; consequently the "critical" region remains inaccessible because the phase transformation sets in too early. In $CD_4$, for example, the fluctuations never get large enough to cause non-mean-field-like behavior (KROLL and MICHEL, 1977). Therefore we restrict ourselves only to the "hydrodynamic" region.

We write the scattering law as given in (3.7), and point out that there exists a structure factor $G_j$ as well for relaxation processes. The structure factor stays essentially the same as in (3.8), where $\underline{\sigma}$ is now the eigenvector of the new structure with orientationally ordered molecules. For orientational order it is more appropriate to replace the mass in (3.8) by the moment of inertia and to describe $\underline{\sigma}$ by spherical harmonics, i.e., rigid molecules. This technique was developed and applied by PRESS and HÜLLER (1973) and PRESS (1973). Without going into further details we state that the precursers in intensity above the phase transformation can be described by an eigenvector as well for relaxation type modes as for soft modes. The order (partial order) in phase II of $CD_4$ is shown in Fig.3.20.

In the "hydrodynamic" region $F_j$ (see (3.7) and (3.42)) can be written

$$F_j\,[\omega,\omega_j(\underline{q}),T] = F_j(\omega,\underline{q},T) = f(T)\chi(\underline{q},\xi)C(\omega,\tau_c)$$

with

$$f(T) = \frac{\omega}{1-\exp\left(-\frac{\hbar\omega}{k_BT}\right)}\ ;\ \chi(\underline{q},\xi) = \frac{1}{q^2+\frac{1}{\xi^2}}\ ;\ C(\omega,\tau_c) = \frac{\frac{1}{\pi\tau_c}}{\omega^2+\frac{1}{\tau_c^2}} \qquad (3.56)$$

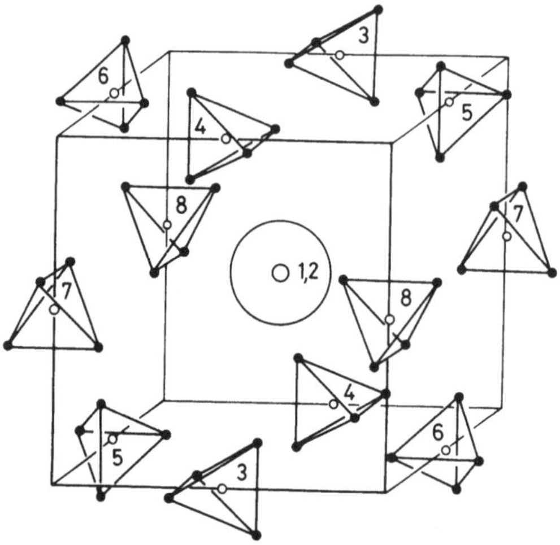


Fig. 3.20 Arrangement of the eight sublattices in $CD_4$ II. The figure shows the "octahedral" symmetry in phase II (space group Fm3c), with a disordered molecule in the symmetry center. (PRESS, 1972)

Here, $\chi(\underline{q},\xi)$ is the static susceptibility and $\chi(\underline{q},\xi) \cdot C(\omega,\tau_c)$ the dynamic susceptibility, with $C(\omega,\tau_c)$ normalised to unity (cf. (1.184)). The correlation length $\xi$ and the relaxation time $\tau_c$ depend on temperature T and $\underline{q}$. Every superstructure position is an origin for $\underline{q}$.

In $CD_4$ (PRESS et al., 1974) critical slowing down of orientational fluctuations on approaching the disorder-order transition has been detected for the first time. The relaxation time $\tau_c$ (about $10^{-11}$s, 11 K above the phase transformation) is already large for an inelastic neutron scattering experiment. To separate the critical scattering which has an energy width of $1/\tau_c$ from the elastic incoherent background very good energy resolution $\Delta\omega$ was needed. Another complication was that the nearest superlattice position (L-point) with sufficient intensity was 1/2 (5,1,3) having a $Q_L = 3.18$ Å$^{-1}$, and thus demanding a relatively high neutron energy.

For the $CD_4$ experiment best resolution $\Delta\omega = 0.025$ THz was achieved using $E_I = 3.3$ THz and a Cu (220) monochromator yielding $2\theta_M = 145°$ (solution ii, as discussed in Subsec.3.1.2). Fig.3.21 shows a constant Q scan for $CD_4$ at the L-point (q = 0) 11° above the phase transformation. Fig.3.22 shows $1/\tau_c$ versus temperature. One sees clearly the small discontinuity at $T_0 = 27.0$ K due to the first order phase transformation. The experimental data of Fig.3.22 were fitted by a power law

$$\frac{1}{\tau_c(q = 0,T)} \sim \left(\frac{T - T_c}{T_c}\right)^{\rho} \tag{3.57}$$

with $T_c = 26.2 \pm 0.2$ K, and $\rho = 1.13 \pm 0.13$. This latter value is slightly above the mean field value $\rho = 1$. Recently KROLL and MICHEL (1976) pointed out that this can be explained by other enhanced fluctuations which do not get critical but comparable to the critical fluctuations several degrees above $T_0$. In other words, more than one relaxation time is involved.

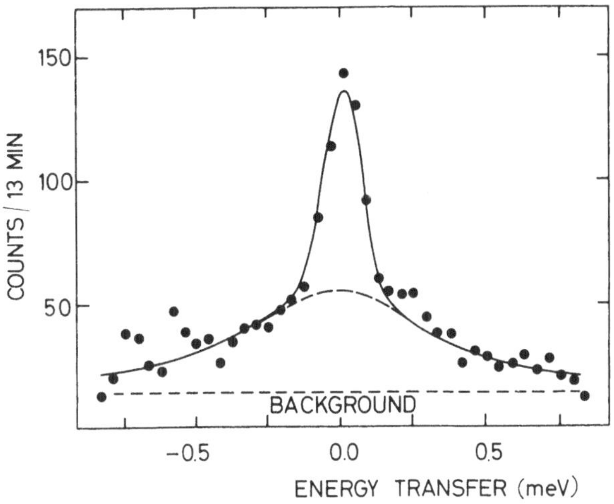


Fig. 3.21 Constant-Q-scan in $CD_4I$ (T = 38.2 K) at the L point [$Q_L = (5\bar{1}\bar{3})/2$]. The incoherent elastic scattering and the critical scattering are well separated ($E_I$ = 14.8 meV). The former is Gaussian shaped with instrumental width; the latter has a Lorentzian shape with a half width $\Gamma(q = 0, T = 38.2\ K) = 0.38 \pm 0.05$ meV. The Lorentzian is shown dashed in the vicinity of E = 0 (PRESS et al., 1974)

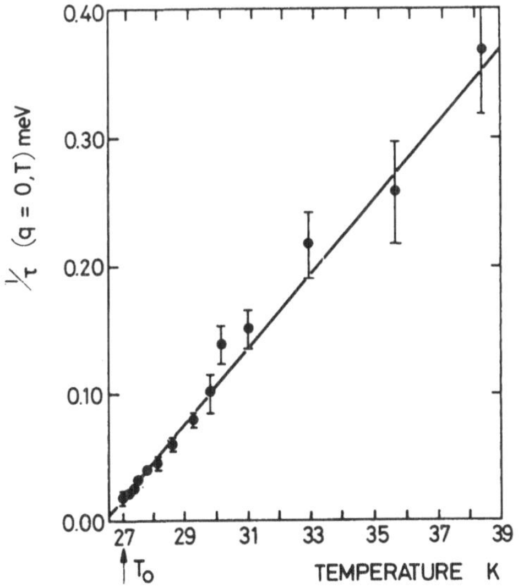


Fig. 3.22 Inverse relaxation time $1/\tau$ (q=0,T) of orientational fluctuations in $CD_4I$ at $Q = Q_L$. The error bars shown include the uncertainty due to the statistical error of the instrumental resolution (PRESS et al., 1974)

The correlation in $CD_4$ was found to be anisotropic, i.e., quasi-two-dimensional (HÜLLER and PRESS, 1972). The static susceptibility then reads

$$\chi(\underline{q},\xi) = \frac{1}{\frac{1}{\xi^2} + \frac{\xi_{\parallel}^2}{\xi^2} q_{\parallel}^2 + \frac{\xi_{\perp}^2}{\xi^2} q_{\perp}^2} \tag{3.58}$$

$\xi_\perp/\xi_\parallel \approx 4$ at 0.6 K above the phase transformation. This ratio is expected to be temperature independent as found for NaOH by BLEIF et al. (1971), where it was 1.9.

For $CD_4$ HÜLLER and PRESS (1972) could explain qualitatively the anisotropy in planes by the next nearest neighbor octupole-octupole interaction. But, contrary to two-dimensional magnetic systems (see Chap.7) where strong correlations occur in one set of planes only, one has to consider four sets of equivalent planes (rather disks) perpendicular to the (111) directions in the case of methane.

The temperature dependence of the static susceptibility in $CD_4$ was found to show the Curie-Weiss behavior at $|\underline{q}| = 0$ (PRESS and HÜLLER, 1974), by measuring the intensity $I(q = 0,T)$ integrated over energy by means of very relaxed resolution

$$\frac{I(q = 0,T)}{T \cdot |G_j(\underline{Q})|^2} \sim \xi^2 \sim (T - T_c)^{-1} \qquad (3.59)$$

where again the extrapolated Curie temperature $T_c$ was found to be 0.4 K below $T_0$. The temperature factor $f(T)$ in (3.56) for $\hbar\omega \ll k_BT$ is proportional to T.

The $\underline{q}$-dependence of the cluster relaxation time $\tau_c(q)$ near $T_0$ was investigated for $|\underline{q}| > 1/\xi$ under the assumption that (3.56) is still applicable, although one is nearer to the critical than to the hydrodynamic region. This is justified because collective excitations of librons are not expected in the disordered phase of $CD_4$ as they could not even be observed below the phase transformation (PRESS et al., 1974). Thus the frequency dependence will be of the diffusive type as described by a Lorentzian in (3.56).

Near the phase transformation it was found for $CD_4$

$$\frac{1}{\tau_c(\underline{q},T)} = \frac{1}{\tau_c(q = 0,T)} + \left(q_\perp^2 + \frac{\xi_\perp^2}{\xi_\parallel^2}\, q_\parallel^2\right) D \quad . \qquad (3.60)$$

This means that the characteristic relaxation time decreases with increasing $|\underline{q}|^2$, or that smaller clusters (contributing to the intensity at larger $|\underline{q}|$) decay faster than larger clusters. One can notice, moreover, that an anisotropic relaxation time corresponds to the anisotropy in $\xi$. This means the short-range order of the correlated "disks" decays faster for directions perpendicular to the disk than parallel to the disk.

The existence of an eigenvector in the high temperature phase describing the order below the phase transformation, and given by the structure factor $|G_j(\underline{q},\underline{Q})|^2$, was established clearly from the fact that the critical scattering intensity $I(q_\parallel) \neq I(-q_\parallel)$.

The ammonium halides are studied extensively. At room temperature $ND_4Br$ and $ND_4Cl$ have the cubic CsCl structure, where the $ND_4$ molecules occupy body centered positions.

Their orientation is disordered over two possible orientations of the $ND_4$ tetrahedron within the cubic frame of the halide ions. $ND_4Br$ undergoes two phase transformations into an antiparallel order below $T_c$ = 215 K and then into parallel order below $T_0$ = 167 K, while $ND_4Cl$ exhibits only one phase transformation at $T_c$ = 249 K into a ferro-order directly (EGERT et al., 1971).

With the diffuse X-ray scattering technique diffuse rods of intensity have been found in the Γ-M direction at room temperature for both systems (non-deuterated) by RENZ et al. (1972). Approaching the phase transformation the intensity on the rods concentrated on the M and Γ points (RENZ et al., 1972; COUZI et al., 1974). This indicates that the fluctuations sense the two possible ordered configurations

1) antiparallel with superlattice reflections at the M-points, or
2) parallel at the Γ-point without a volume change,

which compete with each other. Underlying these are competing interactions

1) the direct $ND_4$-$ND_4$ interaction favors a parallel order, while
2) the indirect interaction via the polarizable halide ions leads to antiparallel order.

The microscopic theory for the ammonium halides (HÜLLER, 1972; YAMADA et al., 1972; HÜLLER and KANE, 1974) is sufficiently advanced to give quantitative agreement for the structures, the transition temperatures (even differences between deuterated and non-deuterated samples) and the rotational potentials.

The dynamical critical phenomena in $ND_4Br$ have been studied by inelastic neutron scattering (YAMADA et al., 1974). $ND_4Br$ is interesting as the ordering of the $ND_4$ molecules is connected to a collective displacement of the $Br^-$ ions. Critical scattering at the M-point has been observed only for the relaxational motion of the $ND_4$ molecules and not for the $TA_2$ mode, which has the eigenvector of the $Br^-$ displacements. The $TA_2$ mode which condenses out below $T_c$ does not decrease in frequency approaching $T_c$ from above. This is explained by YAMADA et al. (1974) by the slow relaxation of the $ND_4$ molecules. The neighboring $Br^-$ ions perform about 50 oscillations before a particular $ND_4$ molecule relaxes into another orientation. Therefore, for the high frequency phonons, the relaxational fluctuations can be simply considered as static disturbance, by which the equilibrium positions of the $Br^-$ ions are displaced from the high symmetry site, see also the theory by SOKOLOFF and LOVELUCK (1973). The relaxation time for the $ND_4$ molecules was too long to study the critical slowing down.

The phase transformation in $ND_4Cl$ at 249 K is slightly of first order at normal pressure, indicating that the coefficient C in the expansion equation (3.46) is small and negative. As already mentioned, C may be pressure dependent. For $ND_4Cl$ C increases with increasing pressure, thus going through zero at 128 bar and becomes positive for higher pressures. In other words, the phase transformation is of second order for pressures above 128 bar. An excellent neutron diffraction study of the effects has been performed by YELON et al. (1974).

The phase transformation point in a p - T phase diagram, where C = 0, is called a tricritical point (GRIFFITHS, 1970), at T = 250.2 K for $ND_4Cl$. Inelastic neutron investigations have not yet been performed, because the slow fluctuations demand extremely high energy resolution.

As a concluding remark to Subsec.3.2.1 and 3.2.2, where we have made a clear-cut distinction between order-order (soft mode) and disorder-order (relaxation) phase transitions, it may be useful to show that a heavily overdamped harmonic oscillator gives the same response as a diffusion fluctuation with a relaxation time $\tau$. For $k_BT \gg \hbar\omega$ we write for the damped harmonic oscillator, see (3.55).

$$F_j[\omega,\omega_j(\underline{q}),T] = k_BT \frac{\Gamma_j}{(\omega^2 - \omega_j^2(\underline{q}))^2 + \omega^2\Gamma_j^2} \quad . \tag{3.61}$$

The denominator has a minimum away from $\omega = 0$ only as long as $\Gamma_j^2 < 2\omega_j^2(\underline{q})^2$. The response $F_j$ of the damped harmonic oscillator then has two maxima, one for energy loss and one for energy gain.

As soon as $\Gamma_j^2 \geq 2\omega_j^2(\underline{q})$ the harmonic oscillator is overdamped and the response has only one maximum at $\omega = 0$. Now we rewrite $F_j$ for $\Gamma_j^2 \gg 2\omega_j^2$

$$\begin{aligned} F_j[\omega,\omega_j(\underline{q}),T] &= \frac{k_BT}{\omega_j^2(\underline{q})} \frac{\omega_j^2(\underline{q}) \cdot \Gamma_j}{\omega^4 + \Gamma_j^2\omega^2 + \omega_j^4} \\ &= \frac{k_BT}{\omega_j^2(\underline{q})} \frac{\omega_j^2(\underline{q})/\Gamma_j}{(\omega^2/\Gamma_j^2 + 1)\omega^2 + [\omega_j^2(\underline{q})/\Gamma_j]^2} \quad . \end{aligned} \tag{3.62}$$

Writing $\omega_j^2(\underline{q})/\Gamma_j = 1/\tau(\underline{q})$ and considering only $\omega^2 \ll \Gamma_j^2$ we obtain

$$F_j[\omega,\omega_j(\underline{q}),T] = \frac{k_BT}{\omega_j^2(\underline{q})} \frac{1/\tau(\underline{q})}{\omega^2 + [1/\tau(\underline{q})]^2}$$

If $\omega_j$ is the frequency of a soft mode, then in first approximation it has a dispersion[5] in its neighborhood, for the isotropic case

$$\omega_j^2(q) = \omega_{jo}^2 + \alpha q^2 \tag{3.63}$$

The phonon wave-vector $\underline{q}$ counts now from the position of the condensing soft mode or in other words from the center of the critical scattering. Then $F_j$ reads

[5] For the anisotropic case it should read $\omega_j^2(\underline{q}) = \omega_{j0}^2 + \underline{q}\,\underset{\sim}{\alpha}\,\underline{q}$, where $\alpha$ is a 3 by 3 matrix. In rectangular crystal structures it reduces to $\omega_j^2(\underline{q}) = \omega_{j0}^2 + \alpha_{11}q_1^2 + \alpha_{22}q_2^2 + \alpha_{33}q_3^2$, with $q_1$ to $q_3$ pointing in orthogonal symmetry directions.

where the reciprocal correlation length $1/\xi^2 = \omega_{j0}^2/\alpha \approx (T-T_c)$, which is to be compared with (3.59).

As discussed already, the relaxation time depends on q and T. Here we find:

$$\frac{1}{\tau(q,T)} = \frac{\omega_j^2(\underline{q})}{\Gamma_j} = \frac{\omega_{j0}^2}{\Gamma_j} + \frac{\alpha}{\Gamma_j} q^2 = \frac{1}{\tau(q=0,T)} + \frac{\alpha}{\Gamma_j} q^2 \quad . \tag{3.65}$$

With the help of footnote[5] we again find (3.60). We also find (compare (3.57))

$$\frac{1}{\tau(q=0,T)} \approx (T-T_c) \quad . \tag{3.66}$$

For this derivation, $\Gamma_j$ was assumed to be independent of the temperature as found so far for all soft modes. By the foregoing argument we have shown that a heavily overdamped soft mode has the same response as a system which is diffusive at all temperatures and controlled by a correlation length $\xi$ and a relaxation time $\tau$. Unless a really underdamped soft mode is observed, it is quite difficult to decide which type of description is applicable to certain systems.

To complete the discussion we have to talk about the assumption that $\Gamma_j$ is independent of T. This implies

$$\tau(q=0,T) \sim \xi^2(T) \quad . \tag{3.67}$$

Although this is found experimentally for $CD_4$ (PRESS et al., 1974), the relation is different for other fluctuating systems.

In general, the relaxation time of a macroscopic variable undergoing critical slowing down is written as

$$\frac{1}{\tau} \sim \frac{L}{\chi} \tag{3.68}$$

where L is the Onsager kinetic coefficient and $\chi$ the susceptibility associated with the macroscopic variable (KAWASAKI, 1968). In our case $L^{-1}$ is the damping constant of the soft mode and $\chi \sim \omega_j^{-2}(q)$, see (3.63). In the conventional theory of critical slowing down, L is assumed to be finite and the anomaly in the relaxation is attributed only to the enormous increase in $\chi$ near the critical point. This is not always

$$F[\omega,\omega_j(\underline{q}),T] = F_j(\omega,\underline{q},T) = \frac{1}{\alpha} \frac{k_BT}{\frac{1}{\xi^2} + q^2} \frac{1/\tau(\underline{q})}{\omega^2 + [1/\tau(\underline{q})]^2} \tag{3.64}$$

so (KAWASAKI, 1968; HALPERIN and HOHENBERG, 1969). For isotropic Heisenberg ferromagnets the theory predicts that L also has a singularity and therefore

$$\tau \sim \xi^{5/2} \tag{3.69a}$$

and for isotropic Heisenberg antiferromagnets

$$\tau \sim \xi^{3/2} \tag{3.69b}$$

as was found for $RbMnF_3$ (TUCCIARONE et al., 1971).

### 3.2.3 Eigenvector and Order Parameter

As pointed out in the preceding subsections, structural phase transitions, as long as they involve a symmetry change, can always be described by an eigenvector which is the order parameter $\eta$. In the higher symmetry phase $\langle\eta\rangle$ always equals 0. We will call it the $\eta = 0$ phase because a relation to temperature is not appropriate inasmuch as the $\eta = 0$ phase may be the phase above or below the transition temperature. As discussed before, the time dependence of fluctuations with a pattern of this eigenvector may be oscillatory (soft mode) as well as diffusive (relaxation) in the $\eta = 0$ phase. The order parameter, and thus the eigenvector connected to the structural change, is always defined in terms of the symmetries of the $\eta = 0$ phase. If the space groups of both phases are known, it is easy to find the group theoretical representation in which the eigenvector has to transform. Depending on the number of atoms per unit cell, there will be several modes (say n) in the same representation. As all the n modes have to be orthogonal, there must be n parameters in each eigenvector which are not determined by symmetry.

In the phase ($\eta = 0$) the eigenvector of the soft mode stays the same at all temperatures in contrast to the phase ($\eta \neq 0$). A well-investigated example is the $\alpha$ (low symmetry)-$\beta$ (high symmetry) phase transformation in quartz. In the $\alpha$-phase there are 4 $A_1$ modes which correspond to $1A_1$ and $3B_1$ modes in the $\beta$-phase (see Fig. 3.23). The soft mode in $\beta$ is a linear combination of the 3 $B_1$ modes; in $\alpha$ the soft mode is generally a linear combination of all the 4 $A_1$ modes. But approaching the phase transformation temperature ($T_0$ = 573°C) from below, the coupling to the 4th $A_1$ mode, which stays $A_1$ in $\beta$, has to go to zero. As explained in Subsec.3.1.1, an eigenvector determination at room temperature by AXE and SHIRANE (1970) found that the contribution from the 4th $A_1$ mode was about 20% of the others. Thus the eigenvector of the soft mode in $\alpha$ is varying with temperature. The same holds for the pattern of the lattice distortion. A model (GRIMM and DORNER, 1975) which reproduces the quartz-structure in both phases by rigid $SiO_4$ tetrahedra, was able to explain the necessity of coupling to the 4th $A_1$ mode. The order parameter is the tilt angle $\delta$ of the $SiO_4$-tetrahedra around 2-fold axes. The eigenvector of the soft mode

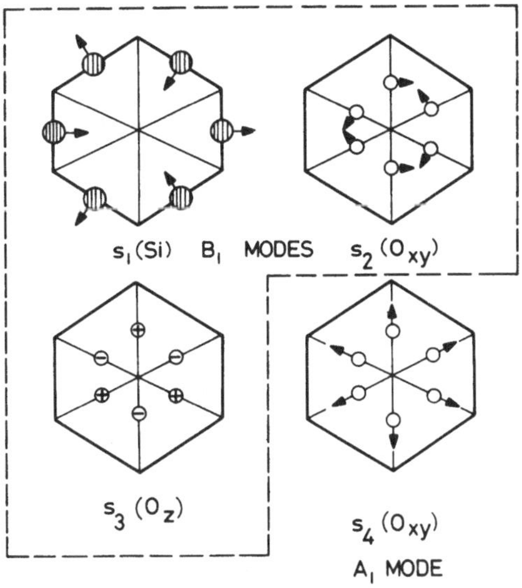


Fig. 3.23 Symmetry mode vectors used to describe the displacements associated with the α-β quartz transformation (AXE and SHIRANE, 1970)

varies with the variation of the eigenvector of the lattice distortion. The free energy expansion for quartz, including a 6th order term describing the first order phase transformation, reads

$$F = \frac{1}{2}\omega_0^2(B_1)f^{(2)}(B_1)\delta^2 + \frac{1}{4}C\,f^{(4)}(B_1)\delta^4 + \frac{1}{6}D\,f^{(6)}(B_1)\delta^6 + \ldots$$
$$+ \frac{1}{2}\omega_0^2(A_1)u^2(A_1) + g\,u(A_1)\,\delta^2 + \ldots \quad . \qquad (3.70)$$

Here $\omega_0(B_1)$ is the soft mode frequency in the β-phase, $\omega_0(A_1)$ is the temperature independent frequency and $u(A_1)$ the amplitude of the fourth $A_1$ mode.

In the α-phase we get from the equilibrium condition $\partial F/\partial u(A_1) = 0$ that $|u(A_1)| = -g/[\omega_0^2(A_1)]\delta^2$. This means that the condensed out lattice distortion $\underline{u}$ (phase α) contains $B_1$ and $A_1$ components

$$\underline{u}\ (\text{phase } \alpha) \sim f(B_1)\delta + g(A_1)\delta^2 \quad . \qquad (3.71)$$

The eigenvector σ of the soft mode in α is not proportional to the distortion itself, but to the derivative of the distortion with respect to the order parameter

$$\underline{\sigma}_{\text{soft mode}}(\text{phase } \alpha) = \frac{\partial u}{\partial \delta} \sim f(B_1) + 2g(A_1)\delta \quad . \qquad (3.72)$$

We see that the soft mode eigenvectors in α and β quartz are not identical. Approaching the phase transformation from below the eigenvector becomes more and more similar to the eigenvector in β-quartz, because the order parameter δ is decreasing.

In the $\alpha$-phase neither the lattice distortion nor the soft mode eigenvector has exclusively the symmetry of the order parameter $\delta$, which in both phases remains defined by $B_1$ symmetry (of the $\beta$-phase).

The contribution of $A_1$ to $\underline{\sigma}_{soft\ mode}$ is in first approximation proportional to $\delta$. More accurately it is proportional to $\tan(\delta)$ as derived from the model by GRIMM and DORNER (1975). The results of the model have been compared to the room temperature lattice distortion (YOUNG, 1962) and to the room temperature eigenvector (AXE and SHIRANE, 1970). The agreement is excellent.

This coupling of the order parameter to other physical parameters such as to another mode, to an elastic deformation, to a ferroelectric polarization, etc., is often referred to as coupling of the primary-order parameter to secondary ones. The secondary-order parameters may belong to a representation different from the one for the primary-order parameter, as discussed above. The only requirement for this higher-order coupling is that the free energy expansion remains invariant under all symmetry operations.

$Tb_2(MoO_4)_3$ is an example of a material in which the primary-order parameter corresponds to a zone-boundary mode at the M-point (1/2,1/2,0) (DORNER et al., 1972), while ferroelectric polarization is a secondary-order parameter.

This material, and the isomorphous $Gd_2\,(MoO_4)_3$, provides the first definite examples of induced ferroelectricity. Simplified to bare essentials the appropriate free energy is of the form

$$F = \frac{1}{2}\,\omega_{j0}^2\eta^2 + \frac{1}{4}C\eta^4 + \frac{1}{6}\,D\eta^6 + \frac{1}{2}\,\chi_{33}^{\sigma-1}\,P_z^2 + \frac{1}{2}\,g\,P_z\,\eta^2 + \dots \quad . \tag{3.73}$$

In this expression all functions $f^{(n)}(\gamma_i)$ for the doubly degenerate mode are set equal to unity. The true mechanism to produce the ferroelectric polarization $P_z$ is a coupling of the order parameter $\eta$ to a shear strain $u_{xy}$ and the strain in turn couples piezoelectrically to the spontaneous polarization. The free energy as written above is already minimized with respect to the strain. Therefore the dielectric susceptibility at constant stress $\chi_{33}^{\sigma}$ appears instead of the susceptibility at constant strain $\chi_{33}^{u}$. The relation is

$$\frac{1}{\chi_{33}^{\sigma}} = \frac{1}{\chi_{33}^{u}} - \frac{a_{36}^2}{C_{66}^{P}} \tag{3.74}$$

where $a_{36}$ is a piezoelectric coupling constant and $C_{66}{}^{P}$ the elastic constant at constant polarization.

The secondary-order parameter $P_z$ belongs certainly to another group theoretical representation than the order parameter, as $P_z$ appears at the Brillouin zone center ($\Gamma$-point) and the soft mode at the zone boundary (M-point). The lowest order of possible coupling is therefore $P_z\eta^2$ involving two soft mode wave-vectors $\underline{q}_M$ of opposite direction to couple to $P_z$ at $\underline{q} = 0$.

The dielectric susceptibility $\chi_{33}^{\sigma}$ therefore does not exhibit an anomaly in the $\eta = 0$ phase (CROSS et al., 1968). Critical behavior is confined to the soft mode only. This consideration shows that Landau's theory is incomplete rather than incorrect, as his conclusions remain true for what we term primary-order parameter. Throughout this chapter we mean the primary-order parameter, when we simply talk about the order parameter.

The order parameter which is condensing out in the $\eta \neq 0$ phase has a certain temperature dependence. For a continuous phase transformation we find from (3.46) with $\partial F/\partial\eta = 0$ and $A < 0$

$$\eta \sim A^{1/2} \sim (T_c - T)^{1/2} \quad . \tag{3.75}$$

The exponent 0.5 appears in the Landau theory (molecular field approximation) and holds only if $\langle\eta\rangle^2 \gg \langle\eta^2\rangle$. Near the phase transformation the fluctuations play a dominant role and the exponent is close to 1/3. A beautiful experiment on the temperature dependence of the order parameter in $SrTiO_3$ (MÜLLER and BERLINGER, 1971) shows clearly this change in the temperature dependence.

For a first-order phase transformation it is always dangerous to use power laws, as the quantities do not usually extrapolate to zero. In addition, fluctuations will be less important—even unobservable—because the discontinuity in $\eta$ may immediately fulfill the molecular field condition given above. For a first-order phase transformation, the free energy has to be expanded to the 6th order term in $\eta$ with $C < 0$ and $D > 0$.

$$F = \frac{1}{2} A\eta^2 + \frac{1}{4} C\eta^4 + \frac{1}{6} D\eta^6 + \dots \quad . \tag{3.76}$$

The reciprocal susceptibility is $A \approx (T - T_c)$ and C and D are assumed to be independent of T. Then the first-order phase transformation appears at a $T_0$, where the reciprocal susceptibility reaches $A_0$. The low symmetry phase establishes itself at $T_0$ by a finite order parameter $\eta_0$.

$$A_0 = \frac{3}{16}\frac{C^2}{D} , \quad \eta_0^2 = -\frac{3}{4}\frac{C}{D} \quad . \tag{3.77}$$

Note that $T_0 > T_c$ if the $\eta = 0$ phase is the high temperature phase. The temperature dependence of $\eta$ below $T_0$ then reads

$$\eta^2 = \frac{2}{3}\eta_0^2\left[1 + \left(1 - \frac{3}{4}\frac{T-T_c}{T_0-T_c}\right)^{1/2}\right] \quad . \tag{3.78}$$

This expression was successfully applied to the primary and to two secondary-order parameters in $Tb_2(MoO_4)_3$ (DORNER et al., 1972) and recently to quartz (DORNER et al., 1974; GRIMM and DORNER, 1975; BACHHEIMER and DOLINO, 1975; BANDA et al., 1975). The example of quartz should be discussed in some detail, as there is some misleading interpretation in the literature. In his excellent review SCOTT (1974) gives reference to a paper which was never published in the cited form and in fact the interpretation is incorrect. As finally, and correctly, stated by BANDA et al., (1975) the lattice expansion in quartz couples to the square of the order parameter and thus can be described by (3.78). A power law extrapolating $\eta$ to zero at some temperature $T_c$* above $T_0$ can be compared to (3.78) by adjusting first and second derivatives. It was pointed out to us by BLUME in 1974 that a power law

$$\eta \sim (T_c^* - T)^{1/6} \tag{3.79}$$

has a very similar temperature dependence near $T_0$ as (3.78). This is discussed in detail and published by BACHHEIMER and DOLINO (1975)(see.Fig.3.24). We would like to state that the interpretation of experimental data below a first-order phase transformation by a power law is not meaningful as long as molecular field theory and (3.78) can describe the data.

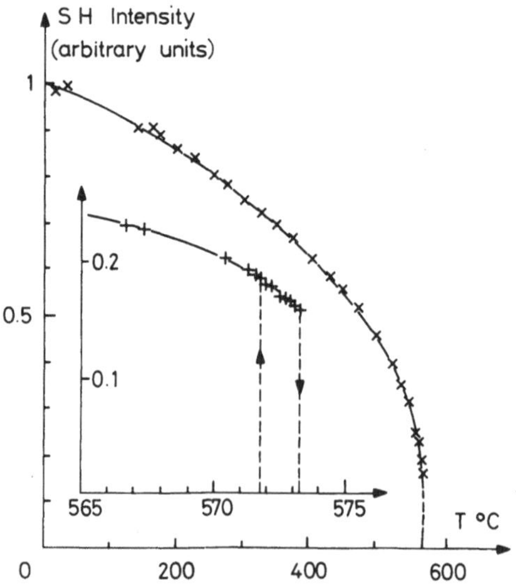


Fig. 3.24 Variation of the second harmonic intensity (for an incidence angle of + 15°) as a function of temperature. The transition region is expanded to show the hysteresis of about 1.5°C (BACHHEIMER and DOLINO, 1975)

## 3.3 Experimental Results

Several systems have been discussed already in the foregoing sections as they served for examples. We have touched upon the different $\underline{q}$-vectors which might be connected with a soft mode. With Fig.3.25 we now would like to go into more detail. A structural phase transformation, which does not change the volume[6] of the unit cell, will be called a ferrodistortive phase transformation. All identical atoms are displaced in phase. This implies that the wavelength of the distortion is $\lambda = \infty$ ($\underline{q} = 0$). The

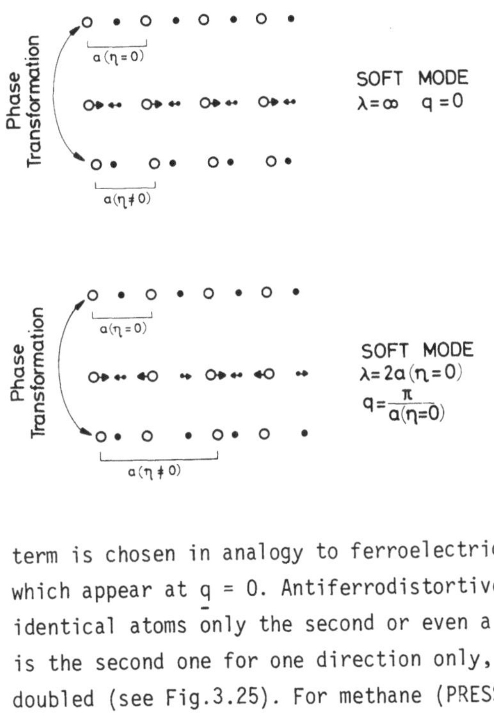


Fig. 3.25 Scheme of a phase transformation without volume[6] change ($q = 0$, ferrodistortive) and with volume[6] change ($q = \pi/a$, antiferrodistortive). The arrows indicate the eigenvector of the soft mode

term is chosen in analogy to ferroelectric or ferromagnetic phase transformations, which appear at $\underline{q} = 0$. Antiferrodistortive, to the contrary, means that among the identical atoms only the second or even a further one is displaced in phase. If it is the second one for one direction only, then the volume of the unit cell is doubled (see Fig.3.25). For methane (PRESS, 1972) the identity period is doubled in all 3 cubic directions and thus the volume of the unit cell changes by a factor of 8 (see Fig.3.20). Many real systems just double the volume of the unit cell going from cubic to tetragonal such as $SrTiO_3$ or from tetragonal to orthorhombic such as $Tb_2(MoO_4)_3$. In Fig.3.26 we show the volume change for $Tb_2(MoO_4)_3$, where the z-axis stays unchanged. A shear deformation in the x-y plane leads to new orthorhombic $a \neq b$ axes. Undergoing the phase transformation the single crystal (tetragonal) will not transform into one orthorhombic domain (monodomain) but into several domains where a and b are interchanged. These domains exhibit a severe limitation for the investigation of the $\eta \neq 0$ phase. Sometimes this difficulty can be overcome by

---

[6]By volume we understand here the unit cell content which is constant in the ferrodistortive case and a multiple in the antiferrodistortive case; we do not mean the small changes ($\approx$ 1%) always observed.

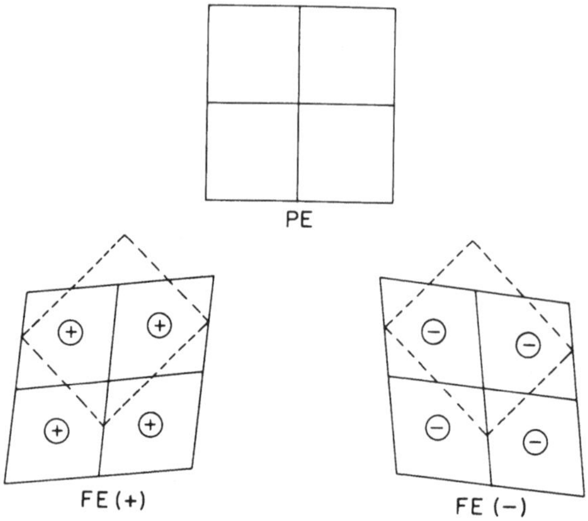


Fig.3.26 Paraelectric (PE)-Ferroelectric (FE) phase transformation in $Tb_2(MoO_4)_3$ from a tetragonal to an orthorhombic unit cell with doubling of the volume. The new orthorhombic a- and b-axes are rotated by 45° with respect to the tetragonal axes. In the orthorhombic ferroelectric phase two domains are possible which have positive or negative polarisation. In this substance the domains can be switched from one to the other by applying stress along an a- or b-axes (DORNER et al., 1972)

applying an electrical or magnetic field or axial stress, while cooling the sample below the transformation to obtain a monodomain.

In general the inelastic neutron investigation of the $\eta \neq 0$ phase is more difficult than the $\eta = 0$ phase, because

1) the symmetry is lowered,
2) there may be domains, and
3) if the volume of the unit cell is increased, the Brillouin zone size is reduced and as a consequence there are more dispersion branches per $\underline{q}$-vector. Therefore there exist many more results for $\eta = 0$ in the various systems (SHIRANE, 1974).

### 3.3.1 Ferrodistortive Transformations

In Table 3.2 we list some examples of substances which exhibit structural phase transformations in the Brillouin zone center that have been investigated by inelastic neutron scattering, and/or diffuse X-ray scattering.

### 3.3.2 Antiferrodistortive Transformations

In Table 3.3 we list some examples of substances which undergo phase transitions accompanied by an increase in the volume of the unit cell, as explained before.

Table 3.2 Examples for ferrodistortive phase transformations with references to inelastic neutron scattering and to diffuse X-ray scattering investigations

| | Neutrons | X-rays |
|---|---|---|
| $BaTiO_3$ | SHIRANE (1967)<br>YAMADA et al. (1969)<br>SHIRANE et al. (1970)<br>HARADA et al. (1971) | LAMBERT (1964)<br>KOMATSU and TERAMOTO (1966)<br>HARADA and HONJO (1967)<br>COMES et al. (1968) |
| Boracites $Me_3B_7O_{13}$ | | FELIX et al. (1974) |
| GeTe | PAWLEY (1969) | |
| $KD_2PO_4$ | PAUL et al. (1970)<br>SKALYO et al (1970)<br>ZEYEN et al. (1976)<br>ZEYEN and MEISTER (1976) | |
| $KH_2PO_4$ | GRIMM et al. (1970)<br>ARSIC-ESKINJA et al. (1971) | |
| $KNbO_3$ | NUNES et al. (1971)<br>CURRAT et al. (1974) | COMES et al. (1970) |
| $K(Nb_xTa_{1-x})O_3$ KTN | YELON et al. (1971)<br>ZACCAI and HEWAT (1974) | ZACCAI and HEWAT (1974) |
| $KTaO_3$ | AXE et al. (1970)<br>COMES et al. (1972) | COMES et al. (1971) |
| $NaNO_2$ | DOLLING et al. (1970) | SHIBUYA (1961)<br>TANISAKI (1963)<br>CANUT and MENDIOLA (1964)<br>CANUT and HOSEMANN (1964) |
| NaOH | | BLEIF et al. (1971) |
| $Nb_3Sn$ | SHIRANE and AXE (1970, 1971a, 1971b)<br>AXE and SHIRANE (1973) | |
| $V_3Si$ | SHIRANE et al. (1971) | |
| $ND_4Cl$ | YELON et al. (1974) | |
| $NH_4Cl$ | | RENZ et al. (1972)<br>COUZI et al. (1974) |
| $PbTiO_3$ | SHIRANE et al. (1970) | |
| SbSI | PIERREFEU et al. (1976) | |
| $SiO_2$ (quartz) | AXE and SHIRANE (1970)<br>DORNER et al. (1974) | ARNOLD (1965) |
| $SrTiO_3$ | YAMADA et al. (1969) | |
| TBBA | HERVET et al. (1974)<br>DIANOUX et al. (1975)<br>HERVET et al. (1975)<br>VOLINO et al. (1976) | LEVELUT and LAMBERT (1971)<br>de VRIES (1973)<br>DOUCET et al. (1973)<br>LAMBERT et al. (1974)<br>LEVELUT et al. (1974) |
| $TbVO_4$ | HUTCHINGS et al. (1975) | |
| Thiourea | RUSH (1967)<br>GOSH (1972)<br>McKENZIE (1975) | SCHIOZAKI (1971) |

Table 3.3 Examples for antiferrodistortive phase transformations with references to inelastic neutron scattering and to diffuse X-ray scattering investigations

| | Neutrons | X-rays |
|---|---|---|
| $CD_4$ | PRESS et al. (1974) | |
| $CsPbCl_3$ | FUJII et al. (1974) | |
| $KMnF_3$ | MINKIEWICZ et al. (1969)<br>SHIRANE et al. (1970)<br>GESI et al. (1972)<br>SHAPIRO et al. (1972) | MINKIEWICZ et al. (1970)<br>COMES et al. (1971) |
| KCP<br>($K_2Pt(CN)_4Br_{0.30}$-$3D_2O$) | RENKER et al. (1973)<br>RENKER et al. (1974)<br>LYNN et al. (1975)<br>COMES et al. (1975)<br>CARNEIRO et al. (1976) | COMES et al. (1973a)<br>COMES et al. (1973b) |
| $LaAlO_3$ | AXE et al. (1969)<br>KJEMS et al. (1973) | |
| $NaNbO_3$ | DENOYER et al. (1976) | DENOYER et al. (1970)<br>ISHIDA et al. (1971)<br>ISHIDA et al. (1973)<br>DENOYER et al. (1974) |
| $NaNO_2$<br>(anti ferroelectric phase) | DOLLING et al. (1970) | YAMADA et al. (1963)<br>HOSHIMO and MOTEGI (1967) |
| $NbO_2$ | SHAPIRO et al. (1974)<br>AXE et al. (1976) | |
| $V_{0.90}Nb_{0.10}O_2$ | | COMES et al. (1974) |
| $ND_4Br$ | YAMADA et al. (1974) | |
| $PrAlO_3$ | BIRGENEAU et al. (1974) | |
| $RbCaF_3$ | ROUSSEAU et al. (1976) | |
| SnSe, SnS, SnTe | PAWLEY (1966)<br>PAWLEY (1969) | |
| $SrTiO_3$<br>(105 K transition) | COWLEY et al. (1964)<br>SHIRANE et al. (1969)<br>RISTE et al. (1971)<br>STIRLING et al. (1972,1976)<br>SHAPIRO et al. (1972)<br>CURRAT et al. (1977) | |
| $TaSe_2$<br>IT - $TaS_2$ | MONCTON et al. (1975)<br>ZIEBECK et al. (1976) | SCRUBY et al. (1975) |
| $Tb_2(M_oO_4)_3$ | AXE et al. (1971)<br>DORNER et al. (1972) | |
| TTF-TCNQ | COMES et al. (1975)<br>SHIRANE et al. (1976)<br>MOOK et al. (1976)<br>ELLENSON et al. (1976) | DENOYER et al. (1975)<br>KAGOSHIMA et al. (1975) |
| ZrNb alloys | AXE et al. (1975) | |

### 3.3.3 Modulated Lattices and Low Dimensional Systems

So far we have considered examples of phase transitions in which the unit cell below the transition temperature is an integer multiple of the high temperature original unit cell. In reciprocal space this yields critical scattering, or softening of phonon modes at $\underline{q}$ values which can be written as simple fractions of the reciprocal lattice vector of the high temperature phase. In these cases, both the high and low temperature phases can be described within the usual space group formalism.

Looking at the low temperature phase as an average structure, essentially identical to the high temperature phase, plus a condensed phonon mode, cases should exist in which the wave-vector of the condensed mode cannot be written as a simple fraction of the reciprocal lattice vectors, as already mentioned in Sec.3.2. The unit cell below the phase transition is then not an integer multiple of the high temperature unit cell. Such structures are referred to as modulated structures.

Modulated structures have been known for a long time, in particular in the case of alloys where one deals with composition modulations, or in the case of polytypism where one deals with periodic stacking faults. More recently, displacive modulations have been observed in various systems, and we shall describe the case of low dimensional metals, in which the coupling of the electrons to the $2k_F$ wave-vector phonons leads to particular metal-insulator transitions, since they provide an example where the microscopic-origin of the unstable mode is understood. In these cases, the low temperature insulating phase has a modulated structure.

It is known that the motion of ions in metals is screened by the conduction electrons. Such screening results, in the simplest descriptions, in anomalies on the longitudinal phonon dispersion. If $\Omega_q$ denotes the unrenormalized phonon frequency, $\lambda$ describes the strength of the electron-phonon coupling and $\chi(q)$ the electron gas susceptibility, the renormalized phonon frequency can be written to a good approximation,

$$\omega^2(q) = \Omega_q^2 \, [1 - \lambda\chi(q)] \tag{3.80}$$

where the electron gas susceptibility

$$\chi(q) = \sum_k \frac{f(E_k) - f(E_{k+q})}{E_{k+q} - E_k} \quad . \tag{3.81}$$

Here $E_k$ is the energy of the electronic state k, and $f(E_k)$ is the Fermi distribution function.

It is seen that an important contribution to $\chi(q)$ originates from energy conserving scattering processes which take an electron from a state $E_k$ to an empty state $E_{k+q}$ such as $\Delta E = 0$. These are scattering processes between states on the Fermi surface with a momentum change of $q = 2k_F$ which give rise to the well-known Kohn anomaly (KOHN, 1959). Of particular interest is the case of a 1-D metallic system in which the Fermi surface consists of 2 planes at $+ k_F$ and $- k_F$; energy conserving scattering takes place exclusively between these two planes with a momentum change of $2k_F$ which leads to a pronounced singularity in $\chi(q)$ at $q = 2k_F$ (AFANASEV and KAGAN, 1963). For a free electron system, in 1-D and T = 0, the integration of $\chi(q)$ yields:

$$\chi(q) = \frac{2Lm}{\pi h^2 q} \ln\left(\frac{2k_F + q}{2k_F - q}\right) \tag{3.82}$$

where L is the length of the metallic chain. As a consequence of the divergence of $\chi(q)$, satellite Bragg peaks, characteristic of a modulated structure, appear at the wave-vector $q = \pm 2k_F$. This opens a gap in the conduction band, and a "Peierls" insulator is created (PEIERLS, 1964). In a strictly 1-D system such a metal-insulator, Peierls transition is expected only at T = 0, with a giant Kohn anomaly at finite temperature. In real systems, which are only quasi one dimensional, because of possible charge compensations between adjacent chains (BARISIC and SAUB, 1973) or slight non-planarity of the Fermi surface (HOROWITZ, GUTFREUND and WEGER, 1974), a real phase transition is expected at finite temperatures between a high temperature conducting state and a low temperature insulating state. One reaches then a picture, similar to that of a structural phase transition, in which the $2k_F$ giant Kohn anomaly provides a high temperature soft mode (in the metallic state) which condenses to form a low temperature modulated structure (in the insulating state).

Two real systems of this type have been investigated recently, both with X-ray diffuse scattering and with neutron inelastic scattering.

$K_2Pt(CN)_4Br_{0.30}$-$3H_2O$ (abbreviated KCP) is built with platinum chains separated by about 10 Å and which the Pt-Pt spacing is only slightly larger than in pure platinum (2.77 Å). In this compound the 1-D metallic properties along the Pt chains arise from the additon in non-stoichiometric proportion of the halogen which removes a few electrons from the last filled band (platinum $d_z^2$ orbitals) of the originally insulating $K_2Pt(CN)_4$ (KROGMANN, 1969). This gives a $2k_F$ value of 0.30 c* (c* = reciprocal lattice constant).

X-rays provided the first observation of the corresponding 1-D scattering in the form of satellite diffuse sheets perpendicular to the c axis and the wave-vector 0.30 c* (COMES et al. 1973a) and the evidence of a low temperature 3D-ordering (COMES et al., 1973b). The corresponding pattern at 300 K is shown in Fig.3.27. At low temperature the platinum chains are modulated sinusoidally in such a way that neighbouring chains are in opposition of phase. Such a coupling had been earlier predicted in a 1-D approach of the phase transition in A 15 compounds (BARISIC, 1972).

Inelastic neutron scattering studies (RENKER et al., 1973) indeed revealed that most of the 1-D scattering observed with X-rays was of inelastic nature at room

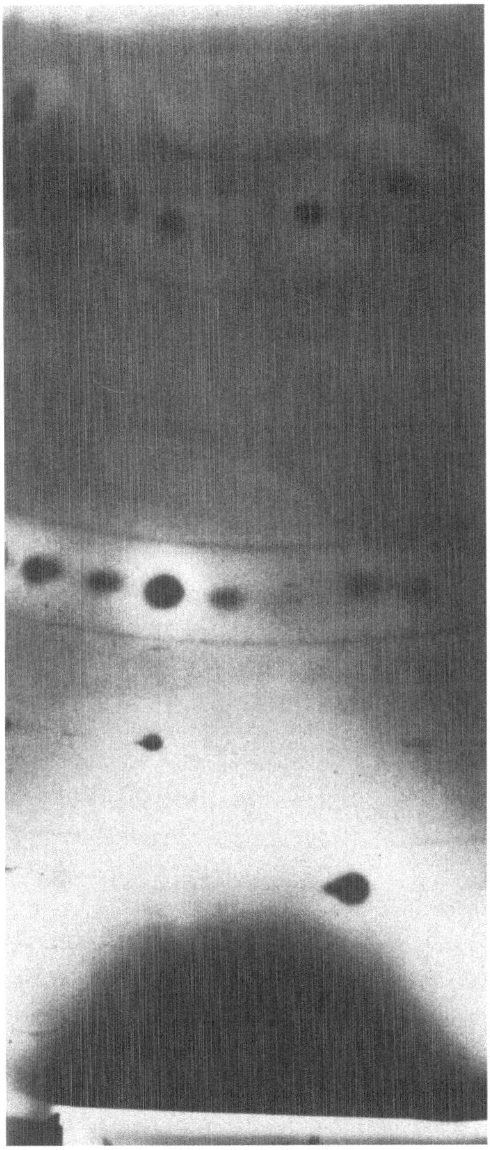

Fig. 3.27 Diffuse X-ray scattering from $K_2Pt(CN)_4Br_{0.3}\cdot 3H_2O$ at 300 K. The streaks corresponding to planes at 2 $k_F$ are clearly visible; the broad spots are created from acoustic phonons near the Bragg positions

temperature. Figure 3.28 shows a pronounced dip in the phonon dispersion surface in a plane perpendicular to the z-axis and at a distance of $2k_F$ from the center of the Brillouin zone. This giant Kohn anomaly in the longitudinal dispersion parallel to the chain direction reflects the 1-D character of the electron wave functions. The low temperature phase transition was incomplete in the sense that transverse order (coupling between chains) at the lowest temperatures (5 K) extends to about 3 Pt chains (RENKER et al., 1974). Further, very detailed neutron scattering experiments (LYNN et al., 1975) showed that not only the Pt chains, but the whole $Pt(CN)_4$ stacks

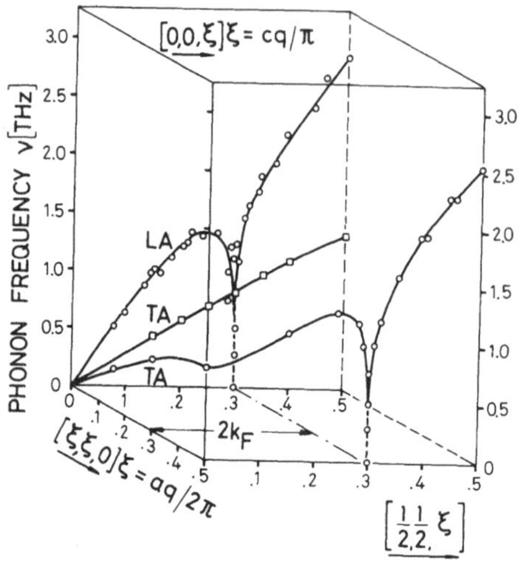


Fig. 3.28 Phonon dispersion in $K_2Pt(CN)_4Br_{0.3}\cdot 3H_2O$ at 300 K. The dip in the dispersion corresponds to a giant Kohn anomaly which appears in planes perpendicular to the z-axis and at a distance of 2 $k_F$ from the origin. The planar dip corresponds to the streaks in Fig. 3.27 (RENKER et al., 1975)

were modulated accounting for the asymmetric line shapes of the low temperature "satellites" in a direction perpendicular to the chain direction, but confirmed the incomplete character of the phase transition.

The reason for the absence of long-range order for the interchain coupling is very likely related to the built-in disorder of this compound (0.6 Br per unit cell). Other difficulites encountered for a clear determination of the dispersion around $2k_F$ motivated the very careful measurements of the $(\omega,q)$ intensity contours through the Kohn anomaly and lead to a new way of displaying them as shown in Fig.3.29 (CARNEIRO et al., 1976). The complexity of such results limits a quantitative comparison between experiment and the existing theory, but the general features agree with the theoretical expectations.

Tetrathiofulvalene-tetracyanoquinodi-methane (TTF-TCNQ) is another real system displaying very exceptional 1-D metallic properties (COLEMAN et al., 1973; FERRARIS et al., 1973). Here the 1-D metallic properties are due to a charge transfer between 1-D electron bands arising from stacks of respectively TTF and TCNQ molecules. The evidence of phase transitions to low temperature modulated structures was given again both by X-rays (DENOYER et al., 1975; KAGOSHIMA et al., 1975), and by further neutron scattering experiments (COMES et al., 1975); despite the extremely difficult experimental conditions (complex structure and extremely small crystals) the precurser pronounced Kohn anomaly in the metallic state could also be measured (MOOK and WATSON, 1976; SHIRANE et al., 1976). Without going here into details about this second example yet incompletely studied, TTF-TCNQ shows in its modulated phases an additional and new interesting feature. While the modulation below 38 K is found to be $4a \times 3.4b \times c$, there is an intermediate phase between 38 and 49 K where the *modulation in a direction* varies continuously as a function of temperature such as to be $2a \times 3.4b \times c$ around 49 K (Fig.3.30). This modulation then remains unchanged up to 54 K (BAK and EMERY, 1976; ELLENSON et al., 1976) temperature above which the

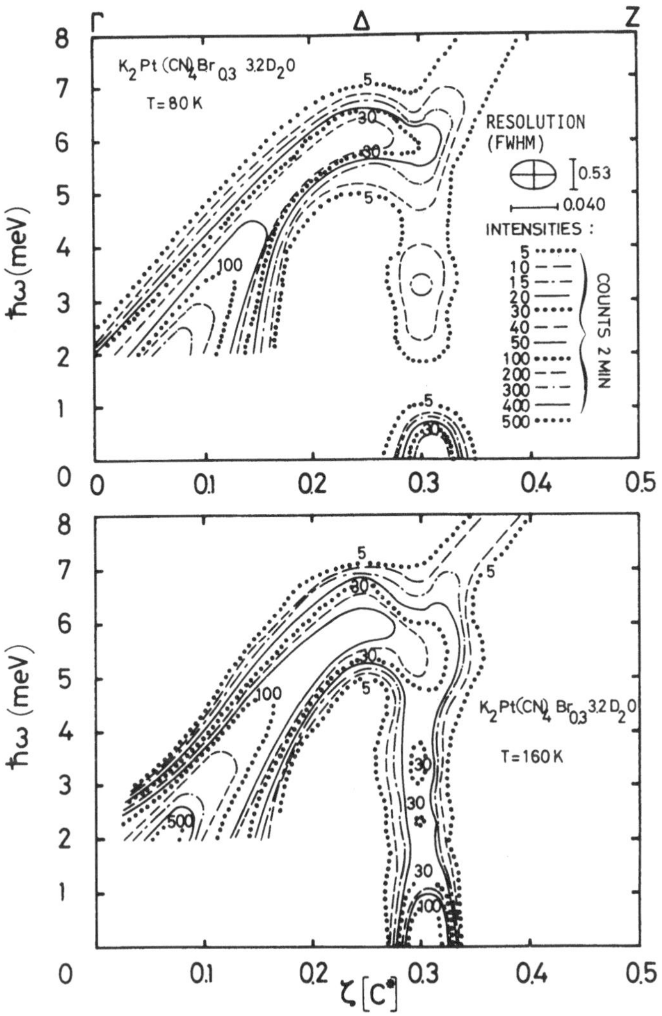


Fig. 3.29 Normalized intensity contours of the neutron scattering from the excitations of wave-vector $\underline{q} = \zeta\underline{c}^*$ at T = 80 and 160 K, measured with an average resolution as indicated. The elastic incoherent scattering and the inelastic background have been subtracted. The intensity unit counts/(2 min) is not to be taken as representative of the actual counting time (CARNEIRO et al., 1976)

structure is unmodulated and the crystal metallic. This is one of the rare examples of temperature dependent modulations, other recent cases were found for layered compounds (WILSON et al., 1975; MONCTON et al., 1975) in relation with electron-phonon coupling in 2-D metals. Older unexplained cases include $\gamma$-$Na_2CO_3$ (DUBBELDAM and DEWOLF, 1969) and some alkali molybdates and tungstates (VAN DEN BERG et al., 1973) and ferroelectric compounds such as $NaNO_2$ (HOSHINO and MOTEGI, 1967) and thiourea (SCHIOZAKI, 1971; McKENZIE, 1975). See the recent review by AXE (1976).

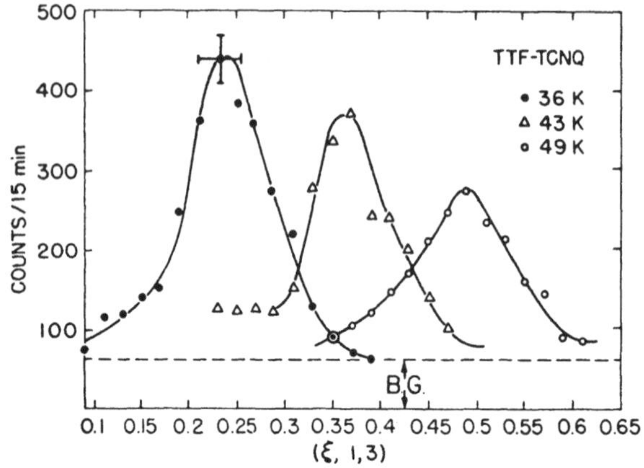


Fig.3.30 Scans along $\underline{a}^*$ at three different temperatures. These scans were performed with a manual point-by-point adjustment of the sample tilt and yield, therefore a less precise positioning in reciprocal space, as shown with the horizontal error bars (COMES et al., 1975b)

## References


Afanasev, A.M., Kagan, Y. (1963): Sov. et Phys. JETP 16, 1030
Alefeld, B. (1972): Kerntechnik 14, 15
Anderson, P.W. (1960): *Fizika dielektrikov,* ed.by G.I. Skanavi (Acad, Nauk SSSR, Moscow) p. 290
Arnold, H. (1965): Z. Krist. 121, 145
Arsic-Eskinja, M., Grimm, H., Stiller, H. (1971): 1st European Conference on Condensed Matter, Summaries, Florence, Italy, 14-17 Sept. 1971, p.60, Geneva, Switzerland: European Phys. Soc.
Axe, J.D., Dorner, B., Shirane, G. (1971); Phys. Rev. Lett. 26, 519
Axe, J.D., Harada, J., Shirane, G. (1970): Phys. Rev. B1, 1227
Axe, J.D., Keating, D.T., Moss, S.C. (1975); Phys. Rev. Lett. 35, 530
Axe, J.D., Shirane, G. (1970): Phys. Rev. B1, 342
Axe, J.D., Shirane, G. (1973): Phys. Rev. Lett. 30, 214
Axe, J.D., Shirane, G. Müller, K.A.(1969): Phys. Rev. 183, 82o
Axe, J.D., Pynn, R., Thomas, R. (1976): to be published
Axe, J.D. (1976): Proceedings of the Conference on Neutron Scattering Gatlinburg Tenn., U.S. Department of Commerce, Springfield, Virginia 22161
Bachheimer, J.P., Dolino, G. (1975): Phys. Rev. B11, 3195
Bacon, G.E. (1962): *Neutron Diffraction* (Oxford University Press, London)
Bacon, G.E. (1972): Acta Cryst. A28, 357
Bacon, G.E. (1974): Acta Cryst. A30, 847
Bak, P., Emery, V.J. (1976): Phys. Rev. Lett. 36, 978
Banda, E.J.K.B., Craven, R.A., Parks, R.D., Horn, P.M., Blume, M. (1975): Solid State Commun. 17, 11-14
Barisic, S. (1972): Phys. Rev. B5, 941
Barisic, S., Saub, K. (1973): J. Phys. C 6, L367
Birgeneau, R.J., Kjems, J.K., Shirane, G., van Uitert, L.G. (1974): Phys. Rev. B10, 2512
Birman, T. (1962): Phys. Rev. 125, 1959; Phys. Rev. 127, 1093
Bersuker, I.B., Vekhter, B.G. (1968): Soviet Phys. 9, 2084
Bjerrum Møller, H., Riste, T. (1975): Phys. Rev. Lett. 34, 996
Bleif, H., Dachs, H., Knorr, K. (1971): Solid State Commun. 9, 1893-7
Brockhouse, B.N. (1961): *Inelastic Scattering of Neutrons in Solids and Liquids* (IAEA., Vienna) pp. 133-151
Brockhouse, B.N., Becka, N.L., Rao, K.R., Woods, A.D.B. (1961): *Inelastic Scattering of Neutrons in Solids and Liquids* (IAEA., Vienna) p. 113
Brockhouse, N.B., De Witt, G.A., Hallman, E.D., Rowe, J.M. (1968): *Neutron Inelastic Scattering,* Vol. II (IAEA., Vienna) p. 259
Canut, M., Hosemann, R. (1964): Acta Cryst. 17, 973
Canut, M.L., Mendioza (1964): Phys. Stat. Solids 5, 313
Carneiro, K., Shirane, G., Werner, S.A., Kaiser, S. (1976): Phys. Rev. B13, 4258
Cochran, W. (1959): Phys. Rev. Lett. 3, 412
Cochran, W. (1971): *Structural Phase Transitions and Soft Modes,* ed. by Samuelsen, E.J., Andersen, E., Feder, J. (Universitetsforlaget Oslo-Bergen-Tromsö).

Coleman, L.B., Cohen, M.J., Sandman, D.J., Yamagashi, F.G., Garito, A.F., Heeger, A.J. (1973): Solid State Commun. 12, 1125
Comes, R., Denoyer, F., Deschamps, L., Lambert, M. (1971): Phys. Lett. 34A, 65
Comes, R., Denoyer, F., Lambert, M. (1971): J. de Phys. France 32, C52-195
Comes, R., Lambert, M., Guinier, A. (1968): C.R. Acad. Sc. Paris 266, 959
Comes, R., Lambert, M., Guinier, A. (1970): Acta Cryst. A26, 244
Comes, R., Lambert, M., Zeller, H.R. (1973a): Phys. Status Solidi (b) 58, 587
Comes, R., Lambert, M., Launois, H., Zeller, H.R. (1973b): Phys. Rev. B8, 571
Comes. R., Renker, B., Pintschovius, L., Currat, R., Gläser, W., Scheiber, G. (1975a): Phys. Status Solidi 71, 171
Comes, R., Shapiro, S.M., Shirane, G., Garito, A.F., Heeger, A.J. (1975b): Phys. Rev. Lett. 35, 1518; (1976) Phys. Rev. B14, 2376
Comes, R., Shirane, G. (1972): Phys. Rev. B5, 1886
Cooper, M.J., Nathans, R. (1967): Acta Cryst. 23, 357
Couzi, M., Denoyer, F., Lambert, M. (1974): J. de Phys. (Paris) F. 35, 753
Cowley, R.A. (1963): Phys. Rev. 134, A981
Cross, L.E., Fouskova,A., Cummins, S.E. (1968): Phys. Rev. Lett. 21, 812
Currat, R., Comes, R., Dorner, B., Wiesendanger, E. (1974): J. Phys. C 7, 252
Currat, R., Mueller, K.A., Denoyer, F., Berlinger, W. (1977): to be published.
Denoyer, F., Comes, R., Garito, A.F., Heeger, A.J. (1975): Phys. Rev. Lett. 35, 445
Denoyer, F., Comes, R., Lambert, M. (1970): Solid State Commun. 8, 1979
Denoyer, F., Comes, R., Lambert, M., Guinier, A. (1974): Acta Cryst. A30, 423.
Denoyer, F., Lambert, M., Comes, R., Currat, R. (1976): Solid State Commun. 18, 441
De Vries, A. (1973): Proceedings of the Bangalor International Conference on Liquid Crystals.
Dianoux, A.J., Volino, F., Heidemann, A., Hervet, H. (1975): J. de Phys. Lett. 36, L-275
Dolling, G. (1974): *Dynamical Properties of Solids*, Vol. 1, p. 541, ed. by G.K. Horton and A.A. Maradudin (North-Holland, Amsterdam)
Dolling, G., Sakurai, J., Cowley, R.A. (1970): J. Phys. Soc. Japan, Proceedings of 2nd IMF 1969
Dolling, G., Sears, V.F. (1973): Nucl. Instr. Methods 106, 419
Dorner, B. (1972): Acta Cryst. A28, 319-27
Dorner, B. (1976): *Molecular Spectroscopy of Dense Phases* ed. by M. Grosmann, S.G. Elkomoss and J. Ringeisen ( Elsevir Scientific Amsterdam)
Dorner, B., Axe, J.D., Shirane, G. (1972): Phys. Rev. B 6, 1950-63
Dorner, B., Bauer, K.H.W., Jagodzinski, H., Grimm, H. (1974): Ferroelectrics 7, 291
Doucet, Y., Levelut, A.M., Lambert, M. (1973): Mol. Cryst. 24, 317
Dubbeldam, G.C., De Wolf, P.M. (1969): Acta Cryst. B25, 2665
Egert, G., Jahn, I.R., Renz, D. (1971): Solid State Commun. 9, 775-8
Ellenson, W.D., Comes, R., Shapiro, S.M., Garito, A.F., Heeger, A.J. (1976): Solid State Commun. 20, 53
Felix, P., Lambert, M., Comes, R., Schmid, H. (1974): Ferroelectrics 7, 131
Ferraris, J.P., Cowan, D.D., Walatka, V.V., Perlstein, J.E. (1973): J. Am. Chem. Soc. 95, 948
Freund, A. (1975): Nucl. Instr. Meth. 124, 93
Freund, A. (1976): Proceedings of the Conference on Neutron Scattering Gatlinburg, Tenn. U.S.A 6-10 June 1976, U.S. Department of Commerce, Springfield, Virginia 22161
Freund, A., Schneider, J.R. (1976): Proceedings of the "Dreiländer-Jahrestagung über Kristallwachstum und Kristallzüchtung" Jülich, Germany 17-19 Sept. 1975, JÜL-CONF-18; and ILL Report 75F236S
Fujii, Y., Hoshino, S., Yamada, Y., Shirane, G. (1974): Phys. Rev. B9, 4549
Geisler, A.H., Hill, J.K. (1948): Acta Cryst. 1, 238
Gesi, K., Axe, J.D., Shirane, G., Linz, A. (1972): Phys. Rev. B5, 1933
Ghosh, R.E. (1972): Thesis (Oxford University)
Graf, R. (1955): Thèse Université Paris
Griffiths, R.B. (1970): Phys, Rev. Lett. 24, 715
Grimm, H., Dorner, B. (1975): J. Phys. Chem. Solids 36, 407
Grimm, H., Stiller, H., Plesser, Th. (1970): Phys. Stat. Solids 42, 207
Guinier, A. (1963): *X-Ray Diffraction in Crystals, Imperfect Crystals and Amorphous Bodies* (Freeman, San Francisco) p. 155
Haas, C. (1965): Phys. Rev. 140A, 863-8
Halperin, B.I., Hohenberg, P.C. (1969): Phys. Rev. 177, 952
Harada, J., Axe, J.D., Shirane, G. (1971): Phys. Rev. B4, 155

Harada, J., Honjo, G. (1967): J. Phys. Soc. Japan 22, 45
Heidemann, A. (1975): Z. Physik B20, 385
Hervet, H., Volino, F., Dianoux, A.J., Lechner, R.E. (1974): J. de Phys. Lett. 35, L151
Hervet, H., Volino, F., Dianoux, A.J., Lechner, R.E. (1975): Phys. Rev. Lett. 34, 451
Horowitz, B., Gutfreund, H., Weger, M. (1974): Phys. Rev. B9, 1246
Hoshino, S., Motegi, H. (1967): Japan. J. Appl. Phys. 6, 708
Hüller, A. (1969): Solid State Commun. 7, 589-91
Hüller, A. (1972): Z. Physik 254, 456
Hüller, A., Kane, J.W. (1974): J. Chem. Phys. 61, 3599-609
Hüller, A., Press, W. (1972): Phys. Rev. Lett. 29, 366-9
Hutchings, M.T., Scherm, R., Smith, S.H., Smith, S.P.R. (1975): J. Phys. C 8, L-393
Ishida, K., Honjo, G. (1970): J. Phys. Soc. Japan 30, 899
Ishida, K., Honjo, G. (1973): J. Phys. Soc. Japan 34, 1279
Jacrot, B. (1970): *Instrumentation for Neutron Inelastic Scattering Research* (IAEA., Vienna) p. 225
Kagoshima, S., Anzai, H., Kayimura, K., Ishigoro, T. (1975): J. Phys. Soc. Japan 39, 1143
Kalus, J. (1975): J. Appl. Cryst. 8, 361
Kalus, J., Dorner, B. (1973): Acta Cryst. A29, 526-28
Kawasaki, K. (1968): Progr. Theoret. Phys. 39, 285
Kjems, J.K., Shirane, G., Müller, K.A., Scheel, H.J. (1973): Phys. Rev. B8, 1119
Koester, L. (1977): Springer Tracts in Mod. Phys. 80
Kohn, W. (1959): Phys. Rev. Lett. 2, 393
Komatsu, K., Teramoto, K. (1966): J. Phys. Soc. Japan 21, 1152
Kovalev, O.V. (1965): Irreducible Representations of the Space Groups (Gordon and Breach, New York)
Kristoffel, N., Konsin, P. (1973): Ferroelectrics 6, 3
Krogmann, K. (1969): Ang. Chem. Intern Ed. Engl. 8, 35
Kroll, D.M., Michel, K.H. (1977): Phys. Rev. B15, 1136
Lambert, M., Guinier, A. (1957): Compt. Rend. Acad. Sci. Paris 245, 526
Lambert, M. (1964): Bull. Soc. Sci. Bretagne (France) 39, 93
Lambert, M., Levelut, A.M. (1974): *Anharmonic Lattices, Structural Transitions and Melting*, ed. by Ristet (Noordhoff, Leiden)
Lambot (1950): Rev. Métallurgie 47, 709
Landau, L.D., Lifshitz, E.M. (1969): *Statistical Physics* (Addison-Wesley, London)
Levelut, A.M., Doucet, J., Lambert, M. (1974): J. Phys. 35, 773
Levelut, A.M., Lambert, M. (1971): Compt. Rend. Acad. Sci. Paris 272, 1018
Levelut, A.M., Lambert, M., Guinier, A. (1968): Acta Cryst. A24, 459
Lynn, J.W., Izumi, M., Shirane, G., Werner, S.A., Saillant, R.M. (1975): Phys. Rev. B12, 1154
Maradudin, A.A., Vosko, S.H. (1968): Rev. Mod. Phys. 40, 1-37
McKenzie, D.R. (1975): J. Phys. C8, 1607
Maier-Leibnitz, H., Springer, T. (1963): J. Nucl. Energy 17, 217
Maier-Leibnitz, H. (1967): Ann. Acad. Sci. Ennicae Ser. AVI, 267
Maier-Leibnitz, H. (1972): *Neutron Inelastic Scattering* (IAEA, Vienna) P. 681
Meister, H., Skalyo, J., Frazer, B.C., Shirane, G. (1969): Phys. Rev. 184, 550
Minkiewicz, V.J., Shirane, G. (1969): J. Phys. Soc. Japan 26, 674
Minkiewicz, V.J., Shirane, G. (1970): Nucl. Instr. Methods 89, 109
Minkiewicz, V.J., Fujii, Y., Yamada, Y. (1970): J. Phys. Soc. Japan 28, 443
Moncton, D.E., Axe, J.D., Disalve, F.J. (1975): Phys. Rev. Lett. 34, 634
Mook, H.A., Watson, C.R. (1976): Phys. Rev. Lett. 36, 801
Müller, K.A., Berlinger, W. (1971): Phys. Rev. Lett. 26, 13
Nunes, A.C., Axe, J.D., Shirane, G. (1971): Ferroelectrics 2, 291
Paul, G.L., Cochran, W., Buyers, W.J.L., Cowley, R.A. (1970): Phys. Rev. B2, 4603
Pawley, G.S., Cochran, W., Cowley, R.A., Dolling, G. (1966): Phys. Rev. Lett. 17, 753
Pawley, G.S. (1968): J. Phys. Suppl. 29, C4, 145
Peierls, R.E. (1964): *Quantum Theory of Solids* (Clarendon, Oxford)
Pierrefeu, A., Dorner, B., Steigmeier, E.F. (1976): Ferroelectrics, 12, 125
Press, W. (1972): J. Chem. Phys. 56, 2597-609
Press, W. (1973): Acta Cryst. A29, 257

Press, W., Hüller, A. (1973): Acta Cryst. A29, 252-56
Press, W., Hüller, A. (1974): *Anharmonic Lattices, Structural Transitions and Melting* ed. by T. Riste (Noordhoff,Leiden)
Press, W., Hüller, A., Stiller, H., Stirling, W., Currat, R. (1974): Phys. Rev. Lett. 32, 1354
Rainford, B.D., Houmann, J.G., Guggenheim H.J. (1972): *Neutron Inelastic Scattering* (IAEA, Vienna) p. 655
Renker, B., Pintschovius, L., Gläser, W., Rietschel, H., Comes, R. (1975): *Lecture Notes in Physics*, Vol. 34, 53 (Springer, Berlin, Heidelberg, New York)
Renker, B., Pintschovius, L., Gläser, W., Rietschel, H., Comes, R., Liebert, L., Drexel, W. (1974): Phys. Rev. Lett. 32, 836
Renker, B., Rietschel, H., Pintschovius, L., Gläser, W., Brüesch, P., Kuse, D., Rice, M.J. (1973): Phys. Rev. Lett. 30, 1144
Renz, D., Breutner, K., Dachs, H. (1972): Solid State Commun. 11, 879-83
Riste, T., Samuelsen, E.J., Otnes, K. (1971): Solid State Commun. 9, 1455
Rousseau, M., Nouet, J., Almairac, R., Hennion, B. (1976): J. de Phys. Lettres 37, L-33
Rush, J.J. (1967): J. Chem. Phys. 47, 4278
Schiozaki, Y. (1971): Ferroelectrics 2, 245
Scott, J.F. (1974): Rev. Mod. Phys. 46, 83
Scruby, C.B., Williams, P.H., Parry, G.S. (1975): Phil. Mag. 31, 255
Shapiro, S.M., Axe, J.D., Shirane, G., Riste, T. (1972): Phys. Rev. B6, 4332
Shapiro, S.M., Axe, J.D., Shirane, G., Raccah, P.M. (1974): Solid State Commun. 15, 377
Shibuya, I. (1961): J. Phys. Soc. Japan 16, 490
Shirane, G. (1974): Rev. Mod. Phys. 46, 437
Shirane, G., Axe, J.D. (1970): Phys. Rev. Lett. 30, 214
Shirane, G., Axe, J.D. (1971a): Phys. Rev. B4, 2957
Shirane, G., Axe, J.D. (1971b): Phys. Rev. Lett. 27, 1803
Shirane, G., Axe, J.D. (1970): Phys. Rev. Lett. 30, 214
Shirane, G., Axe, J.D., Birgeneau, R.J. (1971): Solid State Commun. 9, 397
Shirane, G., Axe, J.D., Harada, J., Linz, A. (1970): Phys. Rev. B2, 3651
Shirane, G., Axe, J.D., Harada, J., Remeika, J.P. (1970): Phys. Rev. B2, 155
Shirane, G., Frazer, B.C., Minkiewicz, V.J., Leake, J.A. (1967): Phys. Rev. 19, 234
Shirane, G., Minkiewicz, V.J., Linz, A. (1970): Solid State Commun. 8, 1941
Shirane, G., Shapiro, S., Comes, R., Garito, A.F., Heeger, A.J. (1976): Phys. Rev. B14, 2325
Shirane, G., Yamada, Y. (1969): Phys. Rev. 177, 858
Skalyo, ., Frazer, B.C., Shirane, G. (1970): Phys. Rev. B1, 278
Sokolof, J.B., Loveluck, J.M. (1973): Phys. Rev. B7, 1644-5o
Soodak, H. (1962): *Reactor Handbook* ed. by H. Soodak (Interscience Publishers, New York, London)
Stedman, R., Almquist, L., Raunio, G., Nilsson, G. (1969): Rev. Sci. Instr. 40, 249
Stirling, W.G. (1972): J. Phys. C 5, 2711
Stirling, W.G., Currat, R. (1976): J. Phys. C9, L 519
Suzuki, K. (1961): J. Phys. Soc. Japan 16, 67
Tanisaki, S. (1963): J. Phys. Soc. Japan 18, 1181
Thomas, H. (1967): Proceedings of the International School of Physics, Course "Local Properties and Phase Transitions" Varenna, 1973
Thomas, H. (1969): IEEE Transactions on Magnetics MAG-5, 874
Thomas, H. (1971): *Structural Phase Transitions and Soft Modes* ed. by E.J. Samuelsen E. Andersen and J. Feder (Universitetsforlaget Oslo,) p. 15
Tucciarone, A. Lau, H.Y., Corliss, L.M., Delapalme, A., Hastings, J.M. (1971): Phys. Rev. B4, 3206-45
Van den Berg, A.J., Tuinstra, F., Warczewski, J. (1973): Acta Cryst. B29, 586
Van Dingenden, Hautecler (1963): C.E.N.-S.C.K., Mol, BLG260
Volino, F., Dianoux, A.J., Hervet, H. (1976): Solid State Commun 18, 453
Walker, C.B. (1956): Phys. Rev. 103, 547
Werner, S.A., Pynn, R. (1971): J. Appl. Phys. 42, 4736
Willis, B.T.M. (1973): *Chemical Application of Thermal Neutron Scattering*, ed. by B.T.M. Willis (Oxford University Press) p. 296
Wilson, J.A., Disaluo, F.J., Mahajan, S. (1975): Adv. Phys. 24, N2, 117
Yamada, Y., Mori, M., Noda, Y. (1972): J. Phys. Soc. Japan 32, 1565

Yamada, Y., Noda, Y., Axe, J.D., Shirane, G. (1974): Phys. Rev. B9, 4429-38
Yamada, Y., Shibuya, I., Hoshino, S. (1963): J. Phys. Soc. Japan 18, 1594
Yamada, Y., Shirane, G. (1969): J. Phys. Soc. Japan 26, 296
Yamada, Y., Shirane, G. Linz, A. (1969): Phys. Rev. 177, 848
Yelon, W.B., Cochran, W., Shirane, G., Linz, A. (1971): Ferroelectrics 2, 261
Yelon, W.B., Cox, D.E., Kortmann, P.J., Daniels, W.B. (1974): Phys. Rev. B 9, 4843-56
Young, R.A. (1962): Defence Documentation Centre, rep. N° AD276235, Washington 25 D.C.
Zaccai, G., Hewat, A.W. (1974): J. Phys. C7, 15
Zak, J. (1969): *The Irreducible Representations of Space Groups* (W.A. Benjamin, Inc. New York)
Zeyen, C.M.E., Meister, H., Kley, W. (1976): Solid State Commun 18, 621
Zeyen, C.M.E., Meister, H. (1976): Ferroelectrics, to be published in the Proceedings of the "Third European Meeting on Ferroelectricity EMF-3, Zürich Sept. 1975"
Ziebeck, K.R.A., Dorner, B., Stirling, W.G., Schoellhorn, R. (1977): J. Phys. C, in press

## Recent References with Titles

Damien, J.C., Lefebvre, J., More, M., Hennion, B., Currat, R., Fouret, R. (1977): Lattice vibrations in the plastic phase of deuterated adamantane. J. Phys. C (to be published)
Denoyer, F., Lambert, M., Comès, R., Currat, R. (1976): Inelastic neutron scattering study in cubic $NaNbO_3$. Solid State Comm. 18, 441
Di Salvo, F.J., Moncton, D.E., Waszczak (1976): Electronic properties and superlattice formation in the semimetal $TiSe_2$. Phys. Rev. B14, 4321
Halperin, B.I., Varma, C.M. (1976): Defects and the central peak near structural phase transitions. Phys. Rev. B14, 4030
Heidemann, A., Lasseques, J.C., Lechner, R., Schlaak, M. (1976): Molecular Reorientations in Crystalline $(CH_3)_3$ NHCl. In *Molecular Spectroscopy of Dense Phases*, ed. by Grosmann et al. (Elsevier Scientific Publishing Company) p. 327
Kagoshima, S., Ischiguro, T., Anzai, H. (1976): X-Ray scattering study of phonon anomalies and superstructures in TTF-TCNQ. J. Phys. Soc. Jap. 41, 2061
Koehler, T.R., Bishop, A.R., Krumhansl, J.A., Schrieffer, J.R. (1975): Molecular dynamics, simulation of a model for (one-dimensional) structural phase transitions. Solid State Comm. 17, 1515
Migoni, R., Bilz, H., Bäuerle, D. (1976): Origin of raman scattering and ferroelectricity in oxidic perovskites. Phys. Rev. Lett. 37, 1155
Mook, H.A., Watson, C.R.Jr. (1976): Neutron inelastic scattering study of Tetrathiafulvalene-Tetracyanoquinodimethane (TTF-TCNQ). Phys. Rev. Lett. 36, 801
Pouget, J.P., Khanna, S.K., Denoyer, F., Comès, R., Garito, A.F., Heeger, A.J. (1976): X-ray observation of 2 $k_F$ and 4 $k_F$ scatterings in Tetrathiafulvalene-Tetracyanoquinodimethane (TTF-TCNQ). Phys. Rev. Lett. 37, 437
Shapiro, S.M., Shirane, G., Garito, A.F., Heeger, A.J. (1977): Transverse acoustic modes in TTF-TCNQ (Tetrathiafulvalene-Tetracyanoquinodimethane). Phys. Rev. B15, 2413
Shapiro, S.M., Moss, S.C. (1977): Lattice dynamics of face-centered-cubic $Co_{0.92}Fe_{0.08}$. Phys. Rev. B15, 2726
Scherm, R., Schedler, E., Wagner, V., Dolling, G., Teuchert, W. (1977): A variable curvature analyser crystal for three-axis-spectrometers. Nucl. Instr. Meth. (in press)
Aubry, S., Pick, R. (1974): Dynamical behavior of a coupled double-well system. Ferroelectrics 8, 471
Bak, P., Emery, V.J. (1976): Theory of the structural phase transformations in Tetrathiafulvalene-Tetracyanoquinodimethane (TTF-TCNQ). Phys. Rev. Lett. 36, 978
Schultz, A.J., Stucky, G.D., Williams, J.M., Koch, T.R., Maffly, R.L. (1977): X-ray diffuse scattering of one-dimensional tetracyanoplatinate salts. Solid State Comm. 21, 197
Currat, R., Mueller, K.A., Berlinger, W., Denoyer, F. (1977): Neutron scattering study of $SrTiO_3$ under [111] uniaxial stress. Phys. Rev. (to be published)

# 4. Dynamics of Molecular Crystals, Polymers, and Adsorbed Species

J.W. White

With 35 Figures

Elastic and other properties of molecular solids are strongly influenced by the difference in magnitude of the forces between atoms in the same molecule and between the molecules themselves. On the whole covalently bound atoms have force constants at least an order magnitude greater than the intermolecular forces, and though many of the effects of condensation into a crystal or upon a surface can be observed by optical and other spectroscopic methods, neutron scattering offers a unique insight into the intermolecular forces—through the determination of phonon dispersion curves—and permits a study of the mutual interaction between internal and external modes.

In this chapter neutron scattering studies of simple molecular crystals, crystalline polymers and adsorbed species are reviewed with reference to models for the intermolecular forces and the effects of the external fields on intramolecular excitations. For polymers these questions have immediate practical importance for interpreting both the elasticity of polymer crystals and the eleastic properties of normal bulk polymers where crystalline and amorphous regions are blended. The measurements so far on the structure and dynamics of adsorbed and intercalated molecules are germane to the subject of heterogeneous catalysis. Most work has been done on simple molecular crystals so far, but the approximations tested and the methods evolved there are immediately applicable to the situation for polymers and adsorbed molecules.

## 4.1 Molecular Crystals

Because intermolecular forces are weak compared to those between chemically bound atoms, the normal modes of vibration for a molecular solid generally separate into two distinct types; the external modes related to displacements of and rotations about the molecular centre of mass, and the internal modes, of much higher frequency, related to vibrations with respect to the centre of mass. To model such complicated solids assumptions have been necessary in the theory, and the thrust of experimental work so far has been to define the limits of validity of the necessary approximations.

An approximation which has been widely used is that the molecular framework can be treated as rigid for calculating the external modes. The extent to which this is practicable is clearly of interest from the point of view of simplifiying calculations. Again, the weakness of the intermolecular forces generally means that the phonon dispersion curves are of low frequency in relation to $k_BT/\hbar$* at normal temperatures. Molecular crystals are not only soft—that is to say have a relatively low Debye temperature— but also anharmonic, and this should always be taken into consideration explicitly in descriptions of the dynamics at temperatures above approximately 77 K. This has been known for a long time and shows up in large temperature variations of the Raman active lattice frequencies (ITO et al., 1968). Calculations which take anharmonicity into account have been performed, e.g., for solid argon (KLEIN and HOOVER, 1971), but a much simpler method—the extrapolation method (HAHN, 1962; COWLEY, 1968)—is promising. The degree to which harmonic calculations are applicable for different molecular systems is a question of considerable importance. Finally comes the question of the way in which the intermolecular potential is best represented in molecular systems where the number of atoms in a unit cell is usually so great that traditional methods, such as the Born-von Karman approach, require too many disposable parameters, thereby obscuring physical aspects of the potential.

Some theoretical approaches to modeling the dynamics of molecular solids have recently been reviewed (VENKATARAMAN and SAHNI, 1970) and so here we will avoid the mathematical formulation, reviewing instead experiments which have tested the validity or otherwise of the above physical approximations, and which have suggested ways to extend current models of the intermolecular potential.

### 4.1.1 Models for the Lattice Dynamics in Molecular Crystals

Because of the difficulty of using the Born-von Karman model, "intermolecular models" have been developed to calculate an effective potential, valid for motion of the molecular centre of mass, and with respect to it. One of the earliest of these was by COCHRAN and PAWLEY (1964). For molecules of high symmetry this method works well but generally requires too many parameters. An approach which has been quite successful in fitting dispersion curves has been to expand the crystal potential as a Taylor series in the translational and librational displacements of the molecules from their equilibrium positions and orientations. For small displacements the intermolecular force constants are the second derivatives of the potential with respect to these displacements and a tensor force field with a minimum number of parameters to fit the observed dispersion curves can be developed (DOLLING and POWELL, 1970), (DOLLING et al., 1973). An alternative, and equally successful, vector model has

---

*$\hbar = h/2\pi$ (normalized Planck's constant)

been proposed by RAFIZADEH and YIP (1970). The use of such tensor and vector models has been adequately demonstrated for hexamethylenetetramine (see Sec.4.4) and allows the prediction of branches in the phonon dispersion curves which are unobservable by neutron scattering as well as giving a good basis for calculating thermodynamic properties.

A further method evolved by PAWLEY (1967, 1972) is to develop the intermolecular pseudo-potential as a sum of pairwise interactions between the separate atoms in neighbouring molecules. A simple analytic form such as the Buckingham potential

$$V_{ij} = \frac{-A_{ij}}{r_{ij}^{6}} + B_{ij}\ \exp-(C_{ij}r_{ij}) \tag{4.1}$$

has been used following the success of this approach (WILLIAMS, 1966 and 1967) for predicting the structures of a wide variety of molecular crystals. The parameters of this potential, $A_{ij}$ $B_{ij}$ $C_{ij}$, were derived by KITAIGORODSKII (1961), KITAIGORODSKII and DASHEVSKII (1968) and WILLIAMS (1966, 1967) using information from selected sets of crystal structures, heats of sublimation and other thermodynamic data. Although the parameters are often strongly correlated because of the method by which they have been derived, they have been found transferable to a large variety of molecular substances.

The procedure followed in using this approach is first to calculate the thermal equilibrium crystal structure for an assumed set of parameters. In practice, interatomic distances greater than about 5.5 Å need not be included as they produce an error in the subsequent frequencies of the order of only 1%, which is much less than other systematic errors arising from the assumption of molecular stiffness or harmonic forces. To find the stable crystal structure, minima in the total potential energy of the crystal are sought by exploring the potential energy hypersurface with small displacements of the molecular centre of mass and small rotations about it. The general condition for equilibrium (KITAIGORODSKII, 1966) is that the first derivatives of the total potential energy, $\Phi(u_i,u_j..)$, with respect to the appropriate normal coordinates should be zero ((4.2))

$$\sum_j \left(\frac{\partial\Phi}{\partial u_i}\right) = 0 \quad \text{etc.} \tag{4.2}$$

As pointed out by PAWLEY (1972) this is not a sufficient condition and one must ensure that there is no strain in the crystal at the equilibrium position found by the calculation. For this to be true the sum of all forces and couples on a molecule must also be zero. With this condition the "self-terms" in the crystal energy can be found. The best criterion to ensure that the self-terms have been adequately calculated is that the calculated crystal structure is the same as the observed, and if this is not so small, changes must be made in the force constants or in the molecular orientations.

Under the conditions that one has a strain-free crystal at equilibrium, it is possible to take the second differentials of the crystal potential with respect to displacements as the appropriate components of a tensor force field. Then follows calculation of the phonon dispersion curves in the harmonic approximation, as a function of wavevector (BORN and HUANG, 1966). Below we give two examples where this procedure has been adopted:

1) β-paradichlorobenzene, where hydrogen-chlorine and chlorine-chlorine interactions are important, as well as those between hydrogen itself and carbon.

2) hexamethylenetetramine, where there are only three atomic contacts of importance, hydrogen-hydrogen, hydrogen-carbon, and hydrogen-nitrogen.

These two examples illustrate the methods by which parameters for the atom-atom model may be derived, the degree of reliance to which such a model may be held for molecules in which intermolecular electrostatic interactions are significant, and the likely defects in the calculations arising from anharmonicity and the breakdown of the rigid molecule approximation.

### 4.1.2 The Lattice Dynamics of β-paradichlorobenzene

The chemical formula and crystal structure of the β phase of fully deuterated paradichlorobenzene are shown in Fig.4.1. This crystal may be prepared from a single crystal of the α phase by cooling through the phase transition at 30.4°C and is an attractive one to study from the point of view of coherent inelastic neutron scattering despite the fact that it has a triclinic unit cell (a = 7.34 Å, b = 5.82 Å, c = 3.90 Å, α = 90.9, β = 112.9, γ = 92,4 at 80 K). There is only one molecule per unit cell and the Brillouin zone is larger than that for any known rigid hydrocarbon. This confers the advantage of only a few well-spaced excitations which may by assigned without ambiguity.

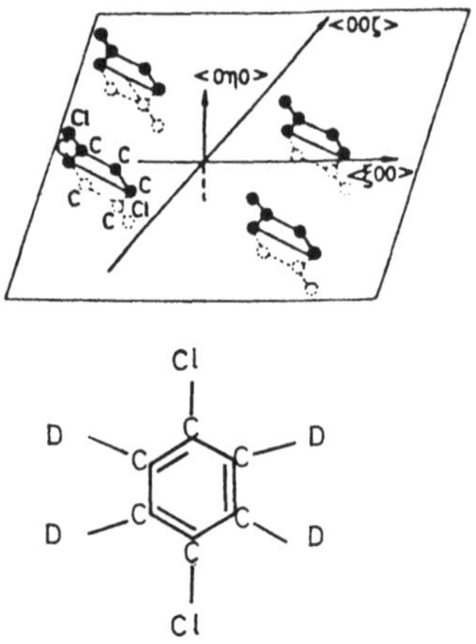


Fig. 4.1 Relative orientations of the reciprocal axes and the molecules (REYNOLDS et al., 1974)

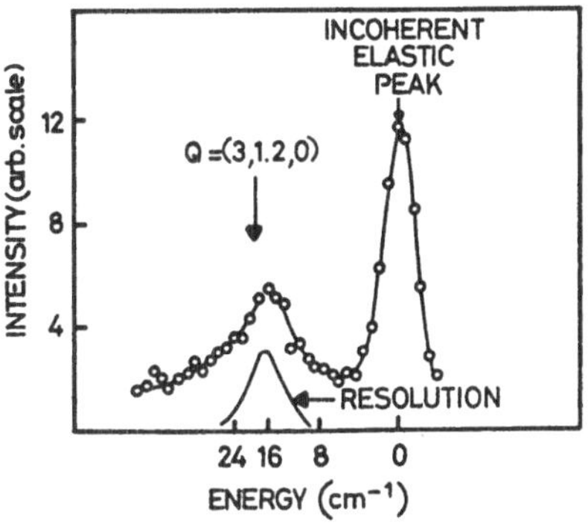


Fig. 4.2 An example of an acoustic neutron group for β-phase p-$C_6D_4Cl_2$ at 295 K (REYNOLDS, KJEMS and WHITE, 1974)

In this work the rigid molecule approximation was adopted and the aims of the experiment were to test the validity of the pseudo-harmonic approximation, to see how far the atom-atom method could be taken for interatomic interactions where no previously known parameters for the Buckingham potential were available, and to investigate the consequences of intermolecular electrostatic interactions and anisotropic molecular polarisability.

From the point of view of the rigid molecule approximation, the intramolecular frequencies and force field are very well known (SCHERER, 1967, 1968) and the lowest internal mode has about twice the frequency of the highest frequency external excitation. The phonon dispersion curves were determined at 295 K and at 90 K (REYNOLDS et al., 1974) using the MARX spectrometer at A.E.K., Risø. Well-separated phonon groups were obtained and a representative example is shown in Fig.4.2. It was even found possible, using a very small resolution ellipsoid and a defocused scan, to avoid contamination from elastic Bragg scattering, to observe phonons between approximately 0.1 and 0.6 THz from fully hydrogenous crystals of β p-$C_6H_4Cl_2$. Fig.4.3 shows the phonon dispersion curves measured at 90 K for β p-$C_6D_4Cl_2$. Phonon groups were checked by observations in more than one zone and the curves are the best estimated fit to the points. Assignments to a given branch were made by qualitative comparison with the inelastic structure factors calculated from the atom-atom potential model. The simple crystal structure made it in general unnecessary to use a quantitative comparison.

The three librational and three translational branches in any given direction can be clearly distinguished in Fig.4.3. The highest librational mode is flat and could just conceivably arise from incoherent scattering, although precautions were taken to minimise this. The general appearance of the diagram shows that there are three approximately flat librational branches, through which pass three sine-curve-like acoustic modes with mixing preventing actual cross-overs. The most dispersed of the librational branches is that at lowest energy. In the (ξ00) direction it almost halves its energy between the zone centre and the zone edge. The widths of the excitations are of the order of 0.2 THz even at 90 K and this large value was generally the limitation on the accuracy of the measurements.

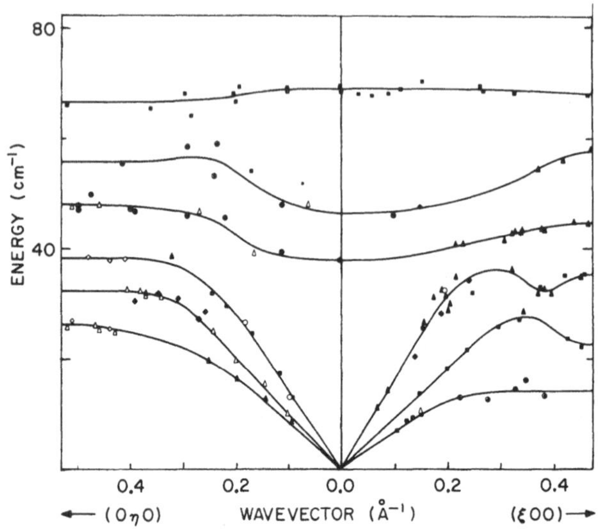


Fig. 4.3 Experimental dispersion curves for β-phase p-$C_6D_4Cl_2$ at 295 K (REYNOLDS et al., 1974)

In order to test the model calculation, dispersion curves measured at 295 K and 90 K were used in a frequency extrapolation to derive the dispersion curves appropriate to 0 K which should result from a calculation based on the harmonic part of the real anharmonic potential. This may be compared with the harmonic component of the model anharmonic atom-atom potential produced by the calculation described in Sec.4.2.

*The Crystal Potential Model*

It was the aim of this model calculation to assess the most serious physical differences between a model crystal and the real crystal, rather than to produce a best fit of the data. The extrapolation method hopefully avoids errors due to misplaced assumptions about the anharmonicity in a crytal, but in using the atom-atom potential method it is tacitly assumed that:

1) the two-body intermolecular potential is decomposable into pairwise atomic interactions;

2) the atomic interactions are represented analytically by the Buckingham potential;

3) either to a first approximation, the many body intermolecular interactions may be neglected, or are included in the two-body parameters because of the method used to derive them.

4) neglect of interactions due to static electric moments

For the β-paradichlorobenzene crystal there are eighteen A,B and C parameters for the pairs of interacting atoms, C-C, C-H, H-H, Cℓ-Cℓ, Cℓ-C and Cℓ-H. This number may be reduced significantly by additional approximations. First, it may be reliably assumed that the cross-terms in the attractive A parameter are described by

$$A_{XY} = (A_{XX}A_{YY})^{1/2} \qquad (4.3)$$

which has a theoretical basis in the London equation for the dispersion energy and is also true to a good approximation in other formulae for this quantity (PITZER, 1959). The nine C-C. C-H and H-H parameters were those due to WILLIAMS (1967) and had been derived from the widest spread of observed quantities whilst using the above approximation for determing A. Because the parameters B and C are closely correlated when obtained from equilibrium properties of the crystal, it is possible to choose values of C with the complementary B which, to some extent, balances errors (REYNOLDS et al., 1974). Thus the values for the chlorine-chlorine C was chosen to be 3.65 $Å^{-1}$, intermediate between the value known for argon-argon and neon-neon. This value was chosen to simplify the choice of $A_{XY}$ cross-terms since it is very similar to that for H-H and C-C. The values for the C parameters associated with Cℓ-H and Cℓ-C were chosen by an arithmetic mean, *thereby leaving only four parameters,* $A_{Cℓ\text{-}Cℓ}$, $B_{Cℓ\text{-}Cℓ}$, $B_{Cℓ\text{-}C}$, $B_{Cℓ\text{-}H}$, to be fitted. These quantities were fitted to 34 observables independent of β p-$C_6D_4Cℓ_2$ and β p-$C_6H_4Cℓ_2$. The fit obtained was tested against predicted Raman and infrared frequencies as well as the sublimation energy for hexachlorobenzene and paradichlorobenzene, and found to be quite good.

Table 4.1 Parameters for the crystal potential in chlorobenzenes

| | A(kcal $Å^{-6}$) | B(kcal) | C($Å^{-1}$) |
|---|---|---|---|
| H-H | 27.3 | 2 654 | 3.74 |
| H-C | 125 | 8 766 | 3.67 |
| C-C | 568 | 83 630 | 3.60 |
| C-Cl | 1140 | 180 000 | 3.62 |
| H-Cl | 250 | 18 000 | 3.70 |
| Cl-Cl | 2300 | 426 000 | 3.65 |

*Comparison of Experimental and Model Dispersion Curves*

Figure 4.4 shows a comparison between the extrapolated dispersion curves (solid lines) and the calculated theoretical curves (dashed lines). For a model with so few parameters the correspondence is very encouraging. The average frequency is neither too high nor too low, but the agreement is by no means as good as can be obtained using many parameter force fields,and for molecular crystals containing only carbon and hydrogen contacts (see Sec.4.4). The degree of agreement reached in the dynamics should be contrasted with the great success of the model potential in predicting the unit cell lengths, extrapolated to 0 K, to within 0.05 Å and the molecular orientation to within $2^{o}$ of that observed at room temperature. The model also predicts the sublimation energy to be 57 kJ $mole^{-1}$ compared to the observed value of 63 kJ $mole^{-1}$, a fact which indicates that perhaps the attractive part of the potential is not quite strong enough. That the dificiencies cannot be due to the attractive part of the potential is almost sure, but to ensure that it is the model which is deficient and not the potential parameters, the observed dispersion curves and other properties of β p-$C_6D_4C\ell_2$ were used to obtain a refined set of parameters. Little improvement in the dispersion curve fit was found, which strongly indicates that the potential proposed was *unique* within the scope defined by the initial assumptions.

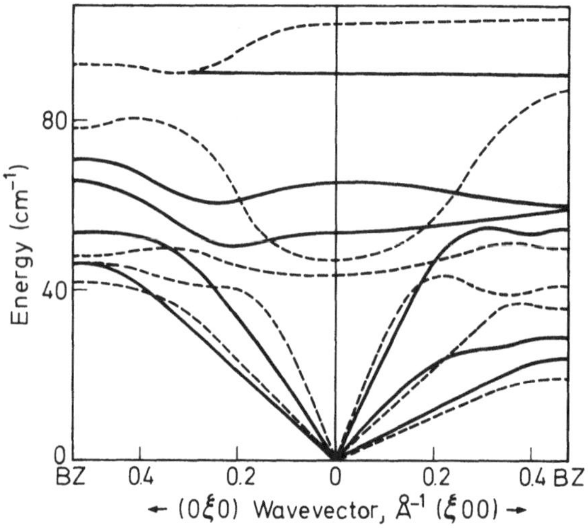


Fig. 4.4 Comparison of extrapolated experimental dispersion curves for deutero β paradichlorobenzene at 0 K (full lines) with predictions from the harmonic part of the interatomic force field (dotted lines) derived from atom-atom potentials (REYNOLDS et al., 1974)

*Improvements to the Model Potential*

There are several obvious inadequacies in the atom-atom potential used above. The first is that the multicentre energy has been neglected. Attempts to meet this objection and thereby improve the repulsive part of the potential for simple molecules have been made with some success (CATLOW and HAYNS, 1974; CATLOW et al., 1975). In this approach small molecules like hydrogen, methane, and hydrogen fluoride were used. The molecular wave functions were calculated by *ab initio* methods and the total potential energy of one molecule with respect to the other, as a function of molecular orientation, was determined under two approximations, namely when the molecular dipole moments (where applicable) were assumed constant as a function of the intermolecular separation, and secondly allowing for electronic rearrangement as the molecules approached more closely. In each case the electrostatic contribution to the total potential energy was subtracted to give an effective Born-Meyer term for the intermolecular repulsion. This method of simulating the repulsive part of the interatomic interaction is interesting since it minimises the inclusion of multicentre repulsions in determining the Born-Meyer parameters. It therefore gives some independent check of the degree to which multicentre effects have been included in empirically determined quantities.

Another inadequacy in the above calculations is the attractive part of the potential by the tacit assumption of a spherically symmetrical molecular polarisability. In fact, for such a molecule as paradichlorobenzene, where the first electronic transitions are strongly polarised with respect to the long molecular axis, the polarisability must necessarily be anisotropic. For β-paradichlorobenzene the consequences of this were calculated (REYNOLDS et al., 1974). The third obvious inadequacy of the above calculation was its failure to include electrostatic forces since the paradichlorobenzene molecule has almost certainly an appreciable electric quadrupole moment. *Ab initio* calculations suggest (MOROKUMA et al., 1963) that the charge distribution may correspond to about 1/4 electron on the chlorine and -1/4 electron on the adjacent carbon.

When the effects due to the anisotropic polarisability and the electric quadrupole moment were included in the calculation with summation out to about 26 nearest neighbours from the central molecule, significant changes in the fitting of the crystal structure, the sublimation energy, and the molecular dynamics were observed. The extra anisotropic polarisability has rather a small effect on the dynamics but improves the fit to unit cell lengths. The sublimation energy is also increased by 11.8 kJ mole$^{-1}$, making it slightly larger than the experimentally observed value which is, however, somewhat uncertain. The effect of introducing the C-Cl bond dipole electrostatic forces is to slightly decrease the sublimation energy whilst the effect on the dynamics is again small, except for the two lowest energy branches at (1/2, 0,0). The 1.0 THz branch is shifted to 0.39 THz. Since this is a purely translational mode mainly in the c direction, and the C-Cl dipoles lie almost in the a-b plane, the effect is not unexpected. The introduction of other small electric dipoles such as in the C-D bond produces large effects; for example, the translational mode at (1/2, 0,0) became imaginary. The size of this effect indicates the need to include electrostatic effects in detailed studies, even of hydrocarbons, but the general conclusion is that electrostatic effects on the vibrational properties of they crystal are very restricted.

### 4.1.3 The Dynamics of Hexamethylenetetramine

The high symmetry of the hexamethylenetetramine molecule (HMT), I$\bar{4}$3m ($T_d^3$), allows the substance to crystallise in a body centred cubic, one molecule per unit cell lattice, space group I43m (a = 7.021 Å at 298 K). Although the number of atoms is rather larger than in paradichlorobenzene, the simplification in the dynamics produced by the higher crystal symmetry, and the fact that the interatomic contacts are confined to the first two rows of the Periodic Table, reduces the severity of the approximations involved in the atom-atom potential method and allows the consequences of a failure of the rigid body approximation to be studied. The molecule is shown in Fig.4.5 and a thorough study of the crystal and molecular dynamics has been made (DOLLING and POWELL, 1970; DOLLING et al., 1973).

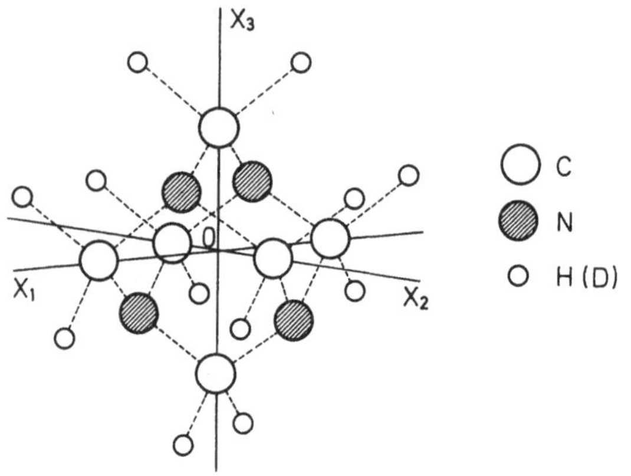


Fig. 4.5 The structure of the hexamethylenetetramine molecule (DOLLING and POWELL, 1970)

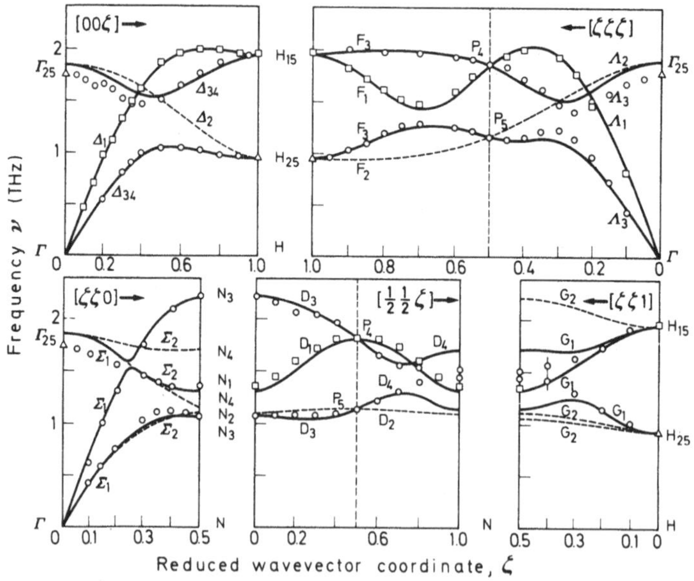


Fig. 4.6 Comparison of experimental intermolecular or lattice mode frequencies in DHMT at 298K and those calculated from model I. The experimental results (DP) shown as squares, triangles and circles represent pure translational, pure librational and mixed modes respectively. The calculated dispersion curves are shown as solid and dashed lines, the latter representing pure librational branches which are experimentally unobservable (DOLLING, PAWLEY and POWELL, 1973)

For the atom-atom potential method interactions between H-H, H-C and H-N alone need to be considered, but to reduce the number of parameters it was assumed as a first approximation that the H-C and H-N interactions were the same. The dispersion curves were determined at 298 K by triple axis spectrometry for the fully deuterated hexamethylenetetramine molecule and are shown for some major symmetry direction in Fig.4.6. The experimental points are shown with designated assignments to translational (squares), pure librational (triangles), of mixed (circles) modes, respectively. The solid lines are a fitting of experimental points given by a seven parameter tensor force field of the kind described in Sec.4.1 and the dotted lines show experimentally unobservable modes calculated using the same tensor force field. It can be seen that the fit is extremely good. A physically interesting feature which appears from this fit is that the shortest D-N bond is found to be the dominant bond from the point of view of choosing the best self-consistent force field.

The atom-atom potential method was also applied to the system assuming the equivalence of carbon and nitrogen and the parameter set shown in Table 4.2.

Table 4.2 The constants of the 6-exponential interatomic pair potential functions

| Atom pair | A $J/(\text{bond } m^6)$ | B J bond | $\frac{C}{nm^{-1}}$ |
|---|---|---|---|
| C...C | $249 \times 10^{-80}$ | $29\ 181 \times 10^{-20}$ | 35.8 |
| C...H | $107 \times 10^{-80}$ | $29\ 181 \times 10^{-20}$ | 41.2 |
| H...H | $40 \times 10^{-80}$ | $29\ 181 \times 10^{-20}$ | 48.6 |

It was found that the A and C parameters for the deuterium-deuterium interaction were the most important in determining the crystal dynamics and so the fitting of the observed dispersion curves to those calculated was used iteratively to obtain improved values for these parameters. Unlike the situation in β-paradichlorobenzene this did produce an improvement in the fit, and the recalculated dispersion curves are shown as the top lines of the banded dispersion curves in Fig.4.7. Good qualitative agreement with the experiment was found—generally the frequencies are within 5 or 10% of those observed—but the model is not as good as that from the tensor force field. The refined atom-atom potentials are, however, of value for proceeding to the next stage of this discussion of the dynamics of molecular crystals, namely the effects of the rigid molecule approximation.

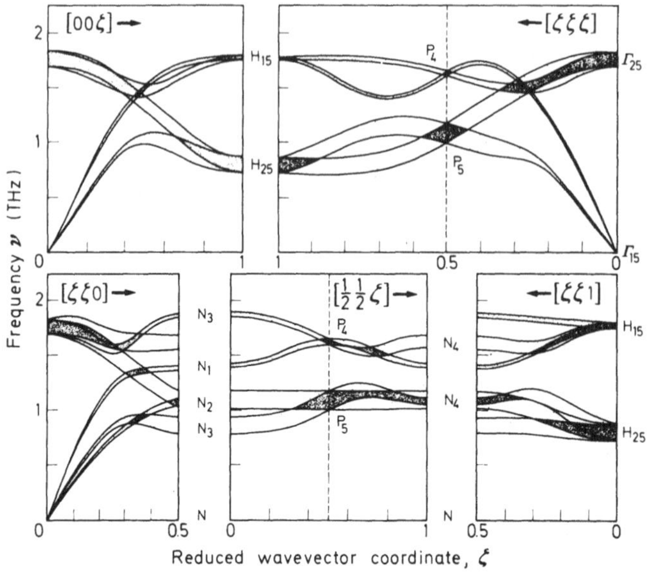


Fig. 4.7 Calculated intermolecular mode dispersion curves when molecular distortions due to intramolecular mode vibrations are included. The top of each band is a dispersion curve calculated assuming the rigid molecule approximation. The bottom of the band is the corresponding result when molecular distortion is included (DOLLING, PAWKY and POWELL, 1977)

*The Rigid Molecule Approximation*

A procedure for assessing the effects of molecular distortion on the external modes of vibration for molecular crystals was developed by PAWLEY and CYVIN (1970) and was used to calculate the dispersion curves by the atom-atom potential method for deutero-hexamethylenetetramine. In principle, to include the intramolecular excitations, the dynamical matrix must be extended to include the additional coordinates required to specify the relative motion and all atoms in the unit cell. For hexamethylenetetramine, with 22 atoms per molecule, this requires an additional 60 coordinates so that the final dynamical matrix is 66 × 66. The intramolecular force field was available from gas phase optical measurements of the molecular vibrational frequencies (ELVEBREDD and CYVIN, 1972) using the harmonic approximation. The dispersion curves calculated using this large dynamical matrix are shown by the lower bounds of the curves in Fig.4.7. To obtain the rigid molecule situation (the upper bound curves) all internal force fields were increased by a factor of 100 and the calculation repeated. It is clear that the effects of intramolecular distortion are quite pronounced, even though the lowest internal mode for hexamethylenetramine is approximately five times the frequency of the highest external mode of the crystal. The widths of the bands are greater than the normal errors involved in phonon frequency measurements, and it must be concluded that a failure to admit molecular distortion to a discussion of the dynamics of molecular crystals, of this kind, is a serious error.

Few general conclusions can be made at this stage about the effects of molecular distortions. The external modes at some points in the Brillouin zone are more strongly reduced in frequency than at others and this depends markedly on the symmetry of the internal and external excitations concerned. As a general rule in HMT the largest shifts were always associated with modes of purely or largely librational character. In addition to the frequency shifts produced by the non-rigid molecule interacting with the lattice modes, there seems also to be an indication that electrostatic effects between molecules are not insignificant for hexamethylenetetramine, as was also found for paradichlorobenzene.

### 4.1.4 Incoherent Neutron Scattering, Phonon Density of States, and Intramolecular Vibrations

Unlike phonon dispersion curve measurements, which for hydrogenous materials require fully deuterated single crystals, measurements of incoherent scattering spectra may be done simply with powder samples of the hydrogenous substance. In the case of some polymers where crystals are not obtainable this method was for a long time the only route to the crystal dynamics. The methods have been extentively discussed (WHITE, 1972) and provide some information about the density of phonon states and rather more about the excitation spectrum of the molecular modes. The intramolecular vibrations are, on the whole, weakly dispersed. This is evident from the non-rigid molecule calculations for naphthalene and hexamethylenetetramine (PAWLEY and CYVIN, 1970; DOLLING et al., 1973), as well as from extensive calculations which have been made in the past on polymers (SHIMANOUCHI et al., 1962; PISERI and ZERBI, 1968). Such modes can be conveniently observed by incoherent neutron scattering spectroscopy (WHITE, 1972). For some molecules there are distinct advantages to the neutron method compared to optical techniques because of the cross-section and eigenvector weighting in neutron spectroscopy. In this section we illustrate the study of intramolecular modes by reference to paradichlorobenzene, hexamethylenetetramine, and some polymeric materials, and discuss the relationship between the density of phonon states observed by neutron scattering and the density of phonon states seen in connection with optical fluorescence and absorption spectra of molecular crystals.

*Paradichlorobenzene*

The incoherent differential scattering cross-section measures $\partial^2\sigma/\partial\Omega\partial E$ the intensity of neutrons scattered per fractional solid angle, $\partial\Omega$, and fractional energy change $\partial E = \hbar\partial\omega$, and is given by

$$\left(\frac{\partial^2\sigma}{\partial\Omega\partial E}\right)_{inc} = \frac{k'}{2k}\sum_{d}\exp[-2W_d(\underline{Q})]\,\overline{|b_d-\bar{b}_d|^2}\cdot\frac{1}{M_d}\sum_{j,\underline{q}}|\underline{Q}\cdot\underline{\sigma}_d^j(\underline{q})|^2\cdot\frac{1}{\omega_j(\underline{q})}\left[n_j(\underline{q})\pm\frac{1}{2}+\frac{1}{2}\right]\delta[\omega\pm\omega_j(\underline{q})] \tag{4.4}$$

where the symbols are as defined in Chap.1 (Eq.(1.83) et seq.). The important features of this cross-section are that the intensity of scattering for a given mode, j, is directly proportional to the squares of the momentum transfer, $\underline{Q}$, and the cartesian eigenvector, $\underline{\sigma}_d^j(\underline{q})$, of the scattering atom, d, excited in the mode. The cross-section is also weighted by the squared "incoherent scattering length" of the atom, $\overline{|b_d-\bar{b}_d|^2}$. For light atoms the amplitude of vibration is large and so they contribute disproportionately to the intensity of scattering. These effects are immediately apparent in the incoherent scattering spectrum of paradichlorobenzene at 295 K shown in Fig.4.8. The low frequency region of the spectrum between 0 and about 110 $cm^{-1}$ ($1cm^{-1} = 3 \times 10^{10}Hz$) may be interpreted with reference to the dispersion curves for β-paradichlorobenzene shown in Fig.4.3. The two peaks at approximately $47cm^{-1}$ and $108cm^{-1}$ arise from maxima in the density of states associated with the Brillouin zone boundary translational excitations and the librational branches, respectively. In particular, the maximum at about $108cm^{-1}$ comes mainly from the highest frequency librational mode for which the librational axis is approximately through the chlorines Further measurements exist on related halobenzene crystals (REYNOLDS and WHITE, 1969; REYNOLDS et al., 1972).

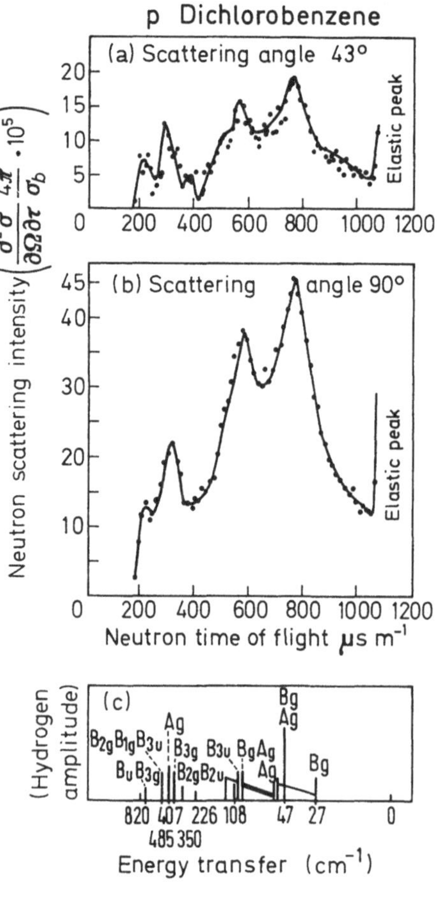


Fig. 4.8a-c Neutron spectrum of paradichlorobenzene at room temperature; scattering angles 43 and 90°. (c) optical data weighted by the hydrogen amplitude squared (REYNOLDS and WHITE, 1969)

At frequencies above $200 cm^{-1}$, three well-defined excitations can be seen in the time-of-flight spectra, and with the higher resolution provided by measurements with the beryllium filter technique (HAYWOOD, 1969) many more modes can be distinguished (REYNOLDS et al., 1972). The number of molecular modes observed and their relative intensities are different from that seen, for example, with infrared spectroscopy (SCHERER, 1963). As in the case of small molecules (WHITE, 1972) the neutron spectra are simplified because only those modes with large hydrogen displacements appear strongly as expected from the vibrational amplitude and cross-section weighting of (4.4).

A more subtle weighting phenomenon may also enter into density of states measurements by incoherent neutron scattering, even in the absence of light atoms, when the unit cell is not cubic. This arises because the dispersion curves may not be uniformly sampled by the scattering experiment. In a typical cold neutron experiment the range of momentum transfer available may extend only up to about 2 $Å^{-1}$ and so for crystals with moderately large Brillouin zones there may be only partial sampling of the dispersion surfaces. The problem is most acute when the sample contains a large proportion of coherent scattering nuclei and when the Brillouin zones for different directions are of very different size (asin the case of some polymers). *Mutatis mutandis* in such cases the phenomenon can be of use for qualitative assignment of singularities in the excitation spectrum since it can be shown that the amplitude weighting does not appreciably shift the frequencies at which singularities occur in the observed spectra.

The effect was investigated for deutero-naphthalene (PAWLEY et al., 1971) by a model calculation using the atom-atom potential method. Both the density of states $g(\omega)$ and the deuterium or hydrogen amplitude weighted density of states $G^{inc}(\omega)$ were computed (under the rigid molecule approximation). By even sampling at 2000 points in the asymmetric unit of the Brillouin zone, it was first shown that $g(\omega)$

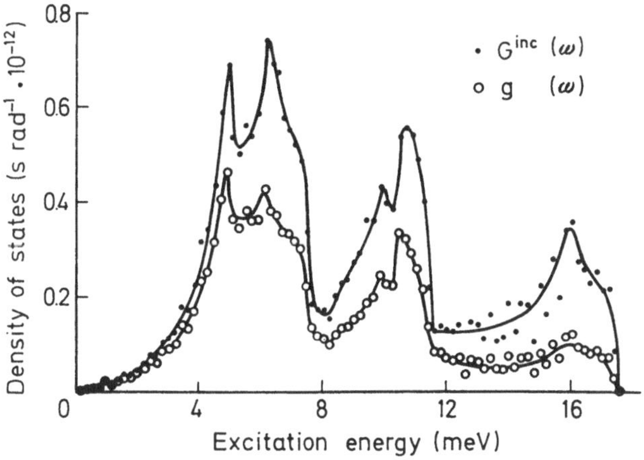


Fig. 4.9 The density of states, $g(\omega)$ and hydrogen amplitude-weighted density of states, $G_H^{inc}(\omega)$, for hydrogenous naphthalene (PAWLEY et al., 1971)

and $G^{inc}(\omega)$ have singularities at the same frequencies for hydrogenous naphthalene (Fig.4.9).

The $(\omega,q)$ space was again evenly sampled in spheres of radius 2.0 Å$^{-1}$ and 2.5 Å$^{-1}$ to simulate possible experimental conditions for perdeuteronaphthalene and the results are shown in Fig.4.10. Allowing for the statistics associated with the small number of sampling points, the curves for the two cases are both quite similar to the true density of states. Discrepancies in detail are most noticeable for the librational modes above 8.0 meV (64 cm$^{-1}$) and faintly in the translational density of states. The frequency distribution shows the strong weighting of librational modes against translational modes expected from the larger amplitudes of hydrogen displacements and by comparision the effects of uneven sampling are small.

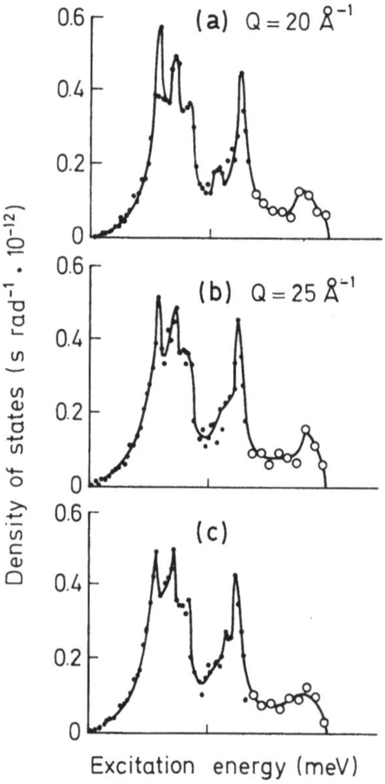


Fig. 4.10 (a) The density of states sampled with a sphere of radius 2.0Å$^{-1}$ in the reciprocal lattice of deuterated naphthalene (b) The density of states sampled with a sphere of radius 2.5Å$^{-1}$ in the reciprocal lattice of deuterated naphthalene (c) The density of states for deuterated naphthalene (PAWLEY et al., 1971)

### *Molecular Modes in Hexamethylenetetramine*

Insofar as almost quantitative agreement has been found between the observed intensities of incoherent neutron scattering from molecular vibrational modes and the intensities calculated using the intramolecular harmonic force fields available from optical spectroscopic studies (REYNOLDS and WHITE, 1969), it seems desirable to take the next step and see whether such force fields can be improved by using neutron spectra to constrain the eigenvectors. Some steps towards achieving this have recently been made through an accurate comparison of the intensities of scattering from molecular modes in hexamethylenetetramine, with model calculations (THOMAS and GHOSH, 1975). To make such a comparison they gave explicit attention to the Debye-Waller factor in (4.4) by including information from crystal diffraction and the known temperature dependence of the phonon dispersion curves. In the isotropic approximation the hydrogen contribution to the incoherent cross-section (4.4) becomes

$$\left(\frac{\partial^2\sigma}{\partial\Omega\partial E'}\right) = \frac{3N}{2}\,\frac{k'}{k_0}\,\frac{\sigma_H^{inc}}{4\pi M_H}\cdot\sum_d\left\{\exp-\left[\frac{Q^2\langle U_d^2\rangle}{3}\right]\right\}\,Q^2\int d\omega'\langle[U_d(\omega)^2\rangle\,\frac{Z(\omega')}{\omega'}\,[n(\omega')+1]\delta(\omega-\omega') \tag{4.5}$$

wether $\langle U_d^2\rangle$ is the hydrogen mean square amplitude.

In making such measurements it has been found desirable to work at as low temperatures as possible. Fig.4.11 shows the observed and calculated spectra for hexamethylene-tetramine at 100 K as well as a simulated spectrum produced by convoluting the experimental resolution function of the beryllium filter spectrometer with δ functions appropriate to the molecular mode frequencies and intensities. It is obvious that, to a first approximation, the agreement between calculated and observed spectra is very good, although at high frequencies there are systematic deviations-the observed intensities always being some 20 or 30% greater than calculated. The exact reasons for this are not at present clear. The difference between

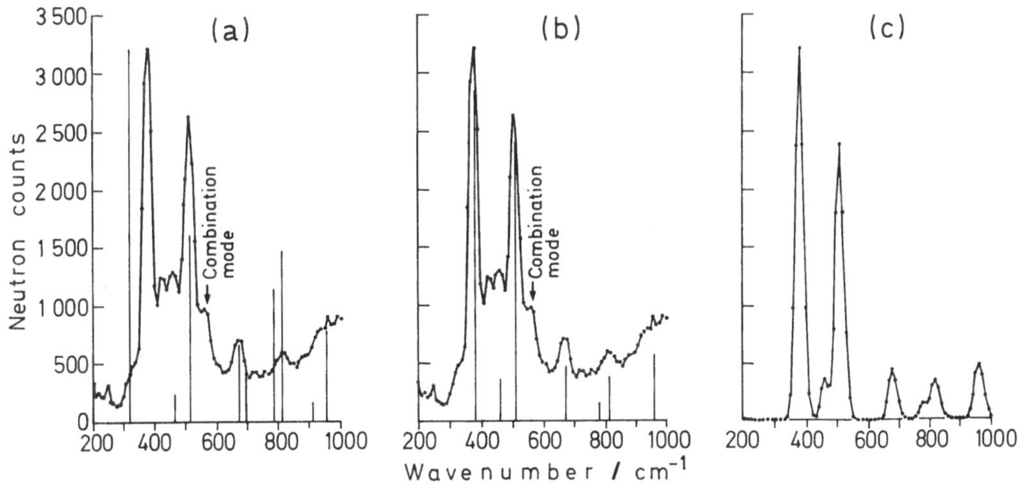


Fig. 4.11a-c Experimental and calculated neutron spectra for hexamethylene-tetramine (a) Experimental spectrum at 100K. Spikes represent the position and relative amplitudes of the internal modes based on the calculation of Elvebredd and Cyvin (b) same as (a) but with spikes based on the optical data of Mecke and Spieseche. The two-phonon shoulder at 550 $cm^{-1}$ is identified by an arrow. (c) Calculated one-phonon spectrum with instrumental resolution included, based on optical data of Mecke and Spieseche (THOMAS and GHOSH, 1975)

the two force fileds for hexamethylene-tetramine (ELVEBREDD and CYVIN, 1972; MECKE and SPIESECHE, 1955) is also very apparent. Clearly the neutron scattering from this substance has resolved the difference in assignments given by these authors. The question of an extension of this work to a detailed comparison of the force fields derived from frequency measurements alone and from frequency measurements plus eigenvector measurements should continue to be investigated.

### 4.1.5 Molecular Torsional Motions and Tunneling Spectroscopy

The lack of optical selection rules in neutron inelastic scattering is advantageous for observing excitations which are forbidden or very weak in optical spectroscopy, and a notable case of these has been the torsional vibrations of methyl groups on small molecules (ALDRED et al., 1967; WHITE, 1972) and on polymers (HIGGINS et al., 1972; LONGSTER and WHITE, 1968; SAFFORD and NAUMANN, 1967; TREVINO and BOUTIN, 1968). The value of observing these motions is that the hindering potential for the methyl group is frequently very sensitive to the molecular conformation and so this latter can be studied from a knowledge of the torsion frequency with some assumptions about the shape of the potential. The torsion frequency in polypropylene oxide has been identified by incoherent neutron scattering in a comparison between the spectra for polypropylene oxide itself and deuteromethylpolypropylene oxide. Because the scattering cross-section of the deuterium is of the order of 10 times less than that for hydrogen, the inelastic cross-section is proportionately reduced and implies a torsional assignment for the mode at 228 $cm^{-1}$ (Fig.4.12). Recently, another characteristic of the potential around methyl groups has been observed using the very high resolution available from the backscattering spectrometers at Jülich and Grenoble (ALEFELD and KOLLMAR, 1976; BATLEY et al., 1977). Tunneling splittings for the methyl torsions have been observed. The tunneling splitting is of the order

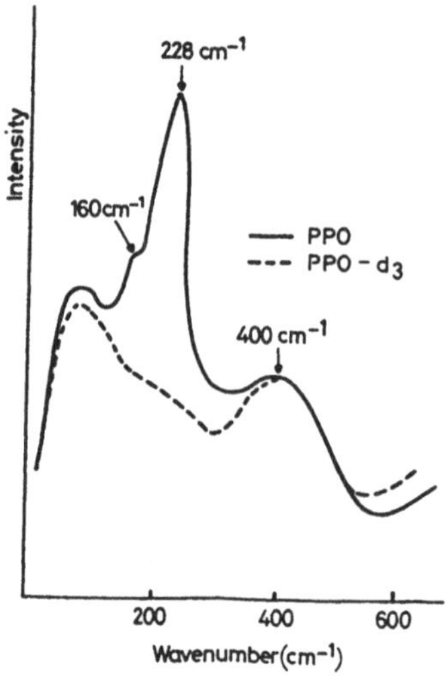


Fig. 4.12 Incoherent inelastic neutron scattering spectra of poly (propylene oxide) $\{CH(CH_3)-CH_2-O\}_n$ and of poly (propylene oxide)-$d_3$ $\{CH(CD_3)-CH_2-O\}_n$ (HIGGINS et al., 1972)

of $10^{-7}$ eV and is observable only at temperatures below about 30 K. At high temperatures modulation of the potential by the intermolecular dynamics is sufficiently severe, and activated motions between levels sufficiently fast, to destroy the tunneling splitting and give rise to a quasielastic component associated with rotary motion of the methyl group. A fuller treatment of these phenomena will be given in Chap.5.

### 4.1.6 Phonon Density of States Measurements by Optical Spectroscopy

In the optical absorption and emission spectra of molecular crystals it is frequently found at low temperatures that a density of phonon states appears as a side band on electronic and vibronic transitions (HOCHSTRASSER and PRASAD, 1972; BROUDE et al., 1967; KOPELMAN et al., 1972). The singularities in these density of states are often very sharp, particularly in the case of phosphorescence and other weak electronic transitions. The physical origin of these bands is that upon excitation or de-excitation the intermolecular forces within the crystal are changed. In the case of a localised electronic excitation this can be thought of as the need for neighbouring molecules to relax their positions as a consequence of the changed molecular size at the centre of excitation. More generally, one invokes the coupling between the excitons and phonons in the crystal and much attention has been given to the diagnostic use of phonon side bands for estimating the strength of this coupling (WILSE-ROBINSON, 1969; SHEKA and DOLGANOV, 1973; SHEKA, 1975; RASHBA, 1966).

In principle, a great deal can be known about the density of phonon states for the molecular crystal from optical spectroscopic measurements of this kind (KOPELMAN, et al., 1972) but it has been pointed out (REYNOLDS, 1973) that these optically produced density of states suffer a distortion of a complementary kind to that introduced by the hydrogen eigenvector weighting in incoherent neutron scattering. This arises because the molecular electronic transitions have often a well-determined electronic polarisation with respect to the molecular axes. In such cases the allowed transitions into the exciton band are also strongly polarised and, as a direct consequence, there is distinct anisotropy in the strain produced by excitation. Phonons of well-defined polarisation result and hence only a limited density of states is seen. The result is very easy to see in the phosphorescence spectrum of pyrazine, dichloro- and dibromobenzene.

In conclusion, whereas neutron scattering preferentially emphasises the density of states from librational excitations, the density of states from optical spectroscopy more usually emphasises translational modes, particularly along directions associated with the transition dipole moment for the vibronic excitation.

## 4.2 Polymers

For polymers the question of how to model the intermolecular force field assumes practical as well as academic interest. Synthetic bulk polymers have the general formula $-(CXY)_n-$ which implies the repetition of a certain group of atoms along a chain possibly many millions of units in extent. The properties of the materials, as produced, depend upon the dynamical properties within the chain and between neighbouring chains, as well as on the microscopic texture associated with the way in which the chains arrange themselves in the material. Typically, bulk polymeric material contains both amorphous and crystalline regions. The chains are ordered with respect to one another, for example by repeated chain folding (KELLER, 1968) in the crystalline regions and a three-dimensional structure arises because of close packing. These crystalline regions are of limited extent (typically several hundred angstroms to several thousand angstroms on edge) and are interspered by amorphous regions where the crystals have not been able to form because of chain tangling. In a typical material the crystalline to amorphous fraction may be of the order of 50%. To understand mechanical properties such as the modulus of elasticity as well as relaxation phenomena, it is essential to have some model of the way in which the properties of the two components of this composite interact. Various models have been tried (HOLLIDAY and WHITE, 1971; BUCKLEY et al., 1969; HADLEY et al., 1969; HADLEY and WARD, 1975) but a severe limitation to this work has been the lack of any knowledge either of the isotropic modulus in the amorphous region or of the anisotropic modulus appropriate to the crystalline fractions.

There have been a number of theoretical attempts to calculate the elastic properties of polymer crystals and the situation for polyethylene, the simplest polymer crystal, is typical. Because the chain is much stiffer than the lateral interactions the dynamics of the isolated, elongated chain were worked out first of all using a valence or Urey-Bradley (UREY and BRADLEY, 1930) force field. Early attempts at a three-dimensional calculation considered no interactions between chain and interchain modes, the latter being calculated from various nearest neighbour force fields (ENOMOTO and ASAHINA, 1964; MIYAZAWA and KITAGAWA, 1964; ODAJIMA and MAEDA, 1966). Quite different results were obtained from the different force fields, whose diversity arose either from the need to simplify the dynamical calculation or the desire to use available data on the external vibrations of the closely related n-paraffins. Improved force fields for the interchain modes were proposed by TASUMI and KRIMM (1967) and KITAGAWA and MIYAZAWA (1970) and a detailed study of expected critical points based on a further calculation has been made by MARTIN and MARSH (1972).

Neutron inelastic scattering measurements have been definitive in testing the force fields for polyethylene by density of states studies, "average" phonon dispersion curve measurements on bulk or partially ordered material, and by measuring phonons in highly ordered samples (MYERS and RANDOLPH, 1968; TWISLETON and WHITE, 1972; REYNOLDS et al., 1977). Polymers of greater complexity such as polytetrafluoroethylene and polyoxymethylene are now susceptible to study (TWISLETON and WHITE, 1972; ANDERSON et al., 1977) and several interesting differences from the dynamics of simple molecular crystals have been revealed which are associated with their inherent disorder and the composite texture. Studies on rubbers and polymer solutions (ALLEN, 1973), particularly by neutron quasielastic scattering, appear to be most promising to the complementary problems raised by the amorphous fraction but are beyond the scope of this review.

### 4.2.1 Polyethylene-the Isolated Chain and Incoherent Scattering Studies

Polyethylene is produced by high pressure or catalytic linking of ethylene ($CH_2 = CH_2$), and under suitable conditions an unbranched chain $-(CH_2)_n-$ can be produced. It is the dynamics of this material and its fully deuterated analogue which will be discussed here. Very small polyethylene crystals may be obtained by recrystallisation from xylene and in the normal crystal the chain conformation is a planar zig-zag. This is packed in an orthorhombic unit cell, two chains per unit cell, space group $D_{2h}^{16}$ or Pnam whose dimensions are a = 7.155 Å, b = 4.889 Å and c = 2.547 Å at 77 K (TEARE, 1959). At higher temperatures the amplitudes of the torsional motions about the chain axis become large and the a/b ratio approaches $\sqrt{3}$ although no hexagonal form of the polymer has yet been positively identified. The orthorhombic structure is shown in Fig.4.13 along with nearest neighbour contacts used to develop the lateral force field by the atom-atom method (Subsec.4.1.1).

The dispersion curves for hydrogenous polyethylene as an isolated chain and with an intermolecular force field are compared in Fig.4.14a,b. The effect of the second molecule in each cell is to double each branch but the weakness of the inter-

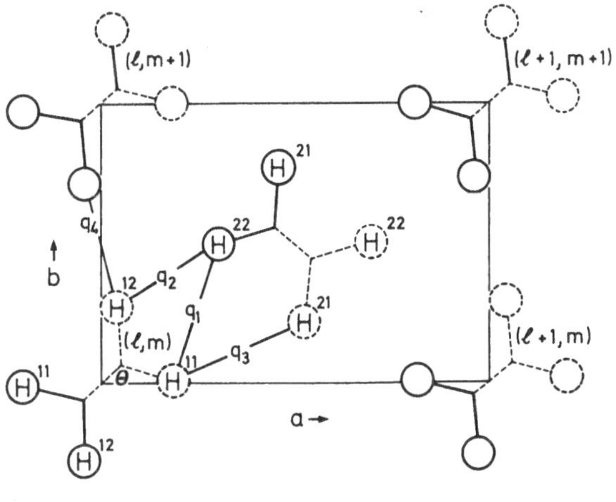


Fig. 4.13 Cross-section of the unit of crystalline polyethylene (TASUMI and KRIMM, 1961)

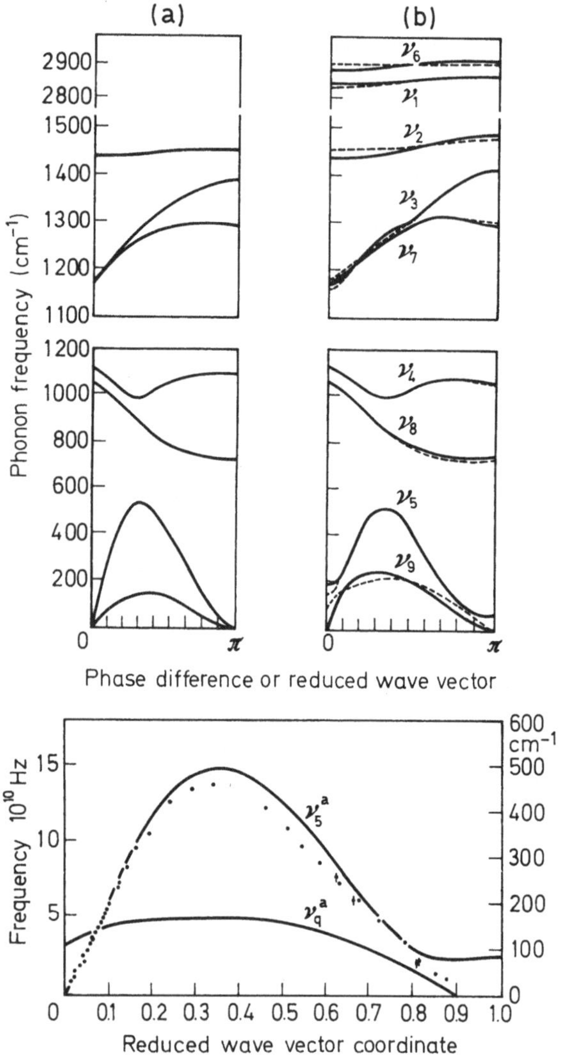


Fig. 4.14a-c Comparison of theoretical dispersion curves for chain-axis modes in polyethylene (a) calculated using the isolated chain and valence force field approximation (b) calculated using a Urey-Bradley molecular force field and four intermolecular H...H constants (c) comparison of experiment with calculated $\nu_9^a$ and $\nu_9^b$ dispersion curves (After PIZERI and ZERBI, 1968; TASUMI and KRIMM, 1967 and FELDKAMP et al., 1968)

chain field means that the splittings are usually small. Hitherto, measurements by optical spectra of these splittings at the zone centre and of the *new* zone centre frequencies due to interactions within $\nu_5$ and $\nu_9$ (longitudinal acoustic and transverse acoustic) branches were the only available information on the intermolecular force field.

Because it is impossible to get large single crystals of polythylene, the first neutron scattering measurements were made by incoherent scattering on the normal material (MYERS and RANDOLPH, 1968) and indicated the approximate cut-off frequencies associated with the longitudinal acoustic and transverse acoustic excitations in the chain. Such density of states measurements were again used (TWISLETON and WHITE, 1972) to reveal some of the maxima associated with the interchain modes but these studies represented the limits to which neutron incoherent scattering could be taken.

The first coherent neutron inelastic scattering measurements on polyethylene were performed by FELDKAMP et al. (1968) who used fully deuterated polyethylene to maximise the coherent scattering, and a sample which had been elongated 8 times by stretching to produce orientation of the chain direction. With this specimen longitudinal acoustic phonons in the chain direction were observed which, at low momentum transfers, were in good agreement with the isolated chain model. At higher frequencies and momentum transfers the experimental points are about 7% lower than calculated but the cut-off frequency is in the expected ratio to that determined by MYERS and RANDOLPH (1968) for hydrogenous polythylene. This result indicates a slight error in the force field of TASUMI and KRIMM (1967). The experiment (points) and theory (curves) are compared in Fig.4.14c.
Attempts to observe the transverse polarised $\nu_9$ mode were unsuccessful because of the almost random relative orientation of crystallites about the c axis. No modes associated with vibrations of the chains relative to one another were observable for the same reason.

*Selective Observation of Interchain Vibrations in Polyethylene*

For polycrsytalline or partially oriented materials it has been found possible to observe some branches of the phonon dispersion curves using cold neutron scattering by working at momentum transfers up to that associated with the first Debye-Sherrer ring of the diffraction pattern. Constrast between the coherent scattering from the crystalline regions and the incoherent scattering of the amorphous fraction arises from the form of the coherent inelastic structure factor ((4.6))

$$G_j(\underline{Q}) = \sum_d \left(\frac{\hbar}{2M_d\omega}\right)^{1/2} b_d\,[\underline{Q}\cdot\underline{\sigma}_d^j(q)]\,\exp(i\,\underline{Q}\cdot\underline{\alpha}_d)\,\exp(-W_d) \qquad (4.6)$$

where the sum is over atoms d of scattering length $b_d$ at positions $\underline{\alpha}_d$ within a unit cell. The vibrational eigenvector in the mode j at wave-vector $\underline{q}$ is $\underline{\sigma}_d^j(\underline{q})$ and the atomic mass is $M_d$. This function is strongly peaked in $(Q,\omega)$ space because of the structure factor

$$b \sum_d \exp(i\,Q\cdot\underline{\alpha}_d) \quad .$$

By comparison, the cross-section for incoherent scattering is a slowly varying function in Q space. The use of this method for polymers was first tried for polytetrafluoroethylene (TWISLETON and WHITE, 1972) where the excitations are relatively simple since the molecules are haxagonal close packed. The technique was extended to polyethylene (TWISLETON and WHITE, 1972) and has subsequently been used to measure dispersion curves in pyrolytic graphite in the bulk (ROSS, 1973) and as microscopic particles (GAMLEN and WHITE, 1976). The results for polyethylene illustrate the procedure.

Since the c axis in polyethylene is about 2.5 Å long and the orthorhombic a and b axes 7.16 and 4.89 Å, respectively, the chief excitations accessible by coherent inelastic scattering from deuteropolyethylene, at low momentum transfers, are those, in the basal plane of the crystal. The neutron scattering spectra for incident 4.5 Å neutrons at a number of different angles of scattering are shown in Fig.4.15. In the experiment, excitations along, a, b and (ξξ0) directions were seen simultaneously and, in order to assign the coherent neutron scattering spectrum as a function of the scattering angle, calculations using the atom-atom potential method were

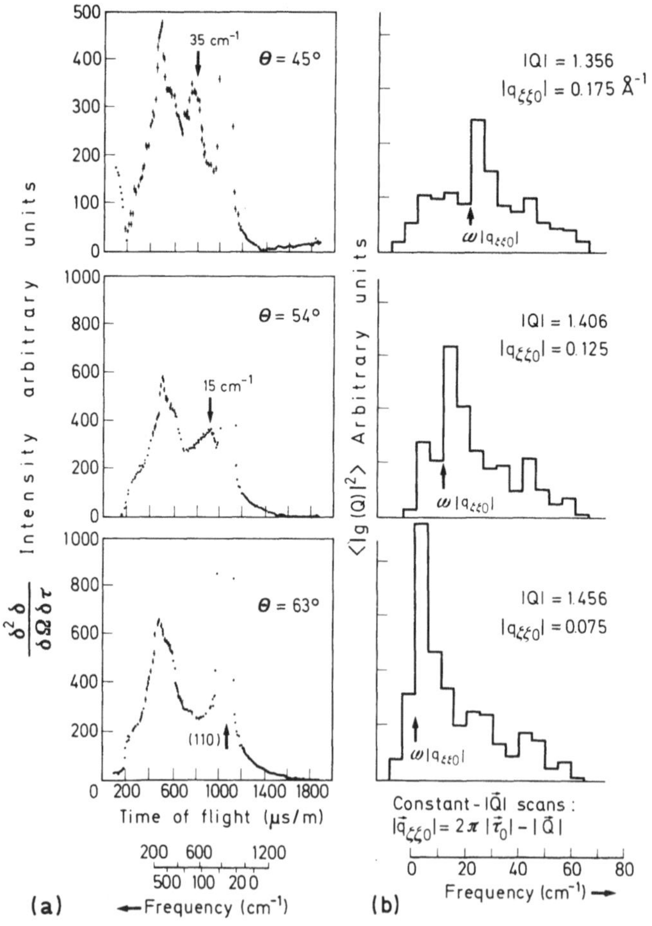


Fig. 4.15 Angular dependence of the inelastic scattering from deutero-polyethylene at 298 K. The momentum transfer dependence of the polycrystalline phonon group is contrasted with that calculated using Shimanouchi's force field for constant Q scans at Q values appropriate to the momentum transfer at the peaks in the experiment (TWISLETON and WHITE, 1972)

used to generate a model spectrum which is shown compared to the experimental results. When a polycrystalline sample was used there was a contribution to the spectrum from low frequency excitations polarised parallel to the chain axis. It was found, by using stretch oriented material, that these corruptions could be minimised and that the dispersion curve for excitations along the (ξξ0) direction was obtained unambiguously and in good agreement with that calculated from the Williams set VII atom-atom parameters. The data points found in this way are shown in Fig.4.16, along with those found using the three-dimensionally oriented specimen and three axis spectrometry as discussed in the next section.

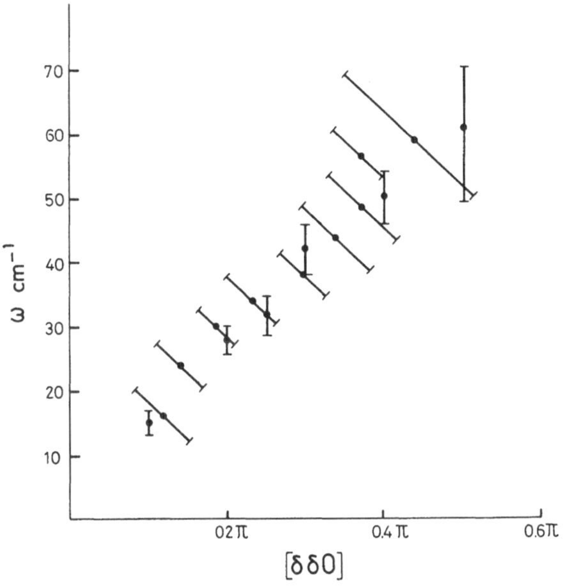


Fig. 4.16 Experimental LA phonon groups along ζζ0 from time-of-flight experiments (⟍) on c-oriented D.P.E. compared with those derived from triple axis constant Q scans ( I ) on single crystal texture D.P.E., both at 298 K (TWISLETON, 1973)

*Measurements on Highly Oriented Polyethylene*

Specimens of deuteropolyethylene with well-developed a,b and c axes have recently been produced by the controlled annealing, under high pressure, of highly drawn material (YOUNG et al., 1973; TWISLETON, 1973). These specimens offered for the first time the chance to measure phonon dispersion curves perpendicular to the polymer chain direction by conventional methods, and the hope to observe transverse excitations along and perpendicular to the chain. To facilititate these studies an atom-atom potential model for polyethylene was developed. The crystal structure was calculated using Williams set IV and VII potential parameters (WILLIAMS, 1966, 1967) with summation limits of 5.0, 5.5 and 6.0 Å for separations between H-H, C-H and C-C non-bonded atom pairs, respectively. The lattice parameters were reproduced within an error of at most 0.2% and the setting angle, θ, between the plane containing the C-C bonds of a chain and the b-c plane was found to be 46.5°. There is a variety of experimentally found setting angles between 42.3 and 48.6° which arises to some extent because of a correlation between the thermal parameters and the setting angle in structure refinements of polyethylene, and so the quality of the fit to the setting angle is undetermined. Polyethylene is, in effect, a two-dimensional molecular crystal since one of the crystallographic axes coincides with the axis of an extended molecular chain. Consequently the four "external normal modes" involving librations about the a-b axes are, in fact, intrachain deformations. So there are only eight "genuine" external chain modes for the two molecules per unit cell structure. In the lattice dynamical calculation for the external modes the chain was treated as being discontinuous at the cell boundaries with the underlying assumption that dynamical properties derived form a single $C_2H_4$ unit pertain to the extended chain. The assumption was tested by a series of calculations for $C_{21}H_{41}$ segments up to $C_{20}H_{40}$, all of which yielded the same phonon frequencies and molecular eigenvectors. To do this the inertial axes for the segments were redefined to take account of their incorporation within the infinite chain by introducing two dummy atoms of very large mass above and below each segment and along the crystallographic c axis (REYNOLDS et al., 1977). The dispersion curves were calculated by the method of PAWLEY (1972) and are shown in Fig.4.17.

The best deuteropolyethylene specimen produced by the stretching and annealing technique had diffraction peaks in the a,b and c directions, whose widths at half height were approximately 17°, 17° and 5°, respectively. The specimen was made from the same specimen used for the c axis studies by FELDKAMP et al. (1968). Because of the mosaic spread it was expected, and found, that the phonon groups were rather

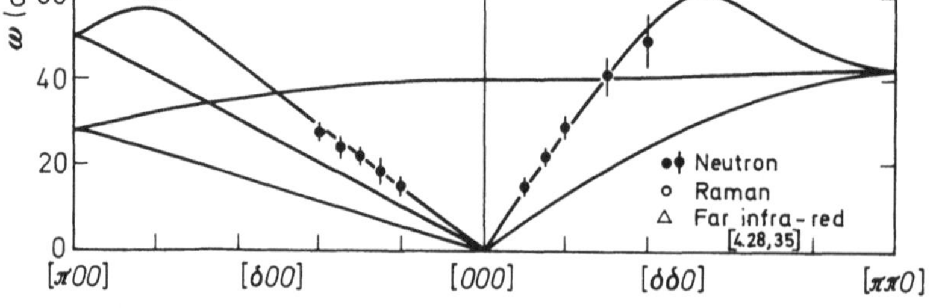

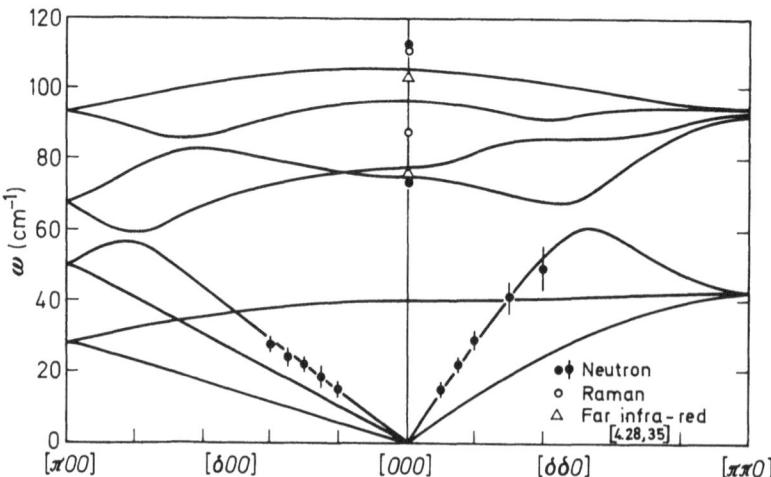

Fig. 4.17 Phonon dispersion curves for D.P.E. along ζ00 and ζζ0 calculated from Williams set VII potential parameters together with the available experimental data at 77 K (TWISLETON, 1973)

broad. In fact the groups were approximately twice the widths of the resolution function of the three-axis spectrometer whose resolution had been relaxed using monochromator and analyser graphite crystals of mosaic 0.5 and 1.2$^{\circ}$, respectively. Well-defined phonons were observed for the longitudinal acoustic excitations in the ($\xi$00) and the ($\xi\xi$0) directions. No transverse excitations could be seen but two zone centre frequencies of optical branches were observed. The highly oriented specimen was also used to determine Raman and far infrared active zone centre frequencies. The measurements therefore give seven experimental parameters of the interchain force field of polyethylene which were used to test the available force models for this crystal. The experimental points are shown plotted on the dispersion curves of Fig.4.17 and agree with this rigid chain model calculation within an average error of 5%. It is, therefore, reasonable to assume, as a first approximation, that the inherent stiffness of the polyethylene chain satisfies the conditions for the rigid chain model and that elastic moduli, and other properties, may be calculated from it using the parameters of our best force field.

In this way values of the elastic moduli along the b and a crystal directions were found to be 8 and 9 $\times$ $10^{10}$ dyne cm$^{-2}$, in close agreement with an average interchain modulus derived from bulk measurements (BUCKLEY et al., 1969). This is in considerable contrast to the predictions of precious theoretical models which indicated a factor of two difference in modulus for the a and b directions. These studies by no means close the subject of the dynamics of polyethylene; major questions such as the force field and even the spectroscopic assignments for the rotatory optic branches are unresolved and are most relevant to macroscopic properties such as crystal melting and phase transitions under pressure. It is expected that the neutron intensities at high flux beam reactors, such as the Institut Laue-Langevin in Grenoble, may allow considerable further progress.

### 4.2.2 Polyoxymethylene

Polyoxymethylene has the chemical formula $-(CH_2O)_{\overline{n}}$. The molecule normally forms in a 9/5 helical conformation (9 monomer units in 5 turns being the repeating unit of the crystal). Chains of this conformation pack in a pseudo-hexagonal lattice. Raman scattering measurements have been made on this modification (PISERI and ZERBI, 1968) as well as on the orthorhombic form for which the chains are in an extended zig-zig conformation. Macroscopic samples with apparent crystalline form up to sizes of several millimetres were produced by topotactic polymerisation of trioxane (UCHIDA and TADAKORO, 1967; CHATANI et al., 1968). This technique has been perfected to produce centimetre sized single domain specimens of fully deuterated material, thereby permitting coherent neutron inelastic scattering measurements (ANDERSON et. al., 1977; WHITE, 1976). The reported space group for this crystal is $C_3^1$ or $C_3^2$ depending upon the chirality of the helix. Figure 4.18 is a schematic diagram of the unit cell whose dimensions are c = 17.39 Å, a = b = 4.46 Å with one chain per unit cell at 298 K.

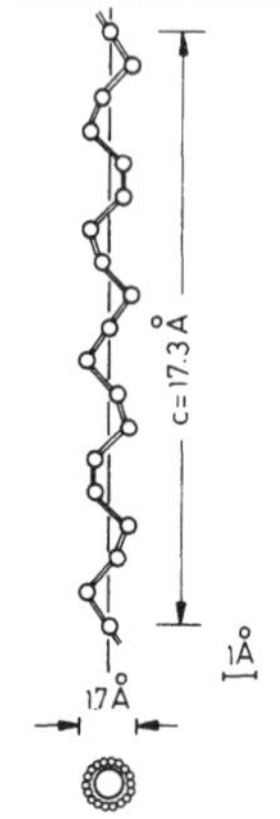


Fig. 4.18 Molecular structure of polyoxymethylene (ANDERSON et al., 1977)

The large number of atoms in the unit cell and the helical structure greatly complicate calculation of the three-dimensional lattice dynamics. The only model available uses a valence force field to give the modes for the isolated chain (PISERI and ZERBI, 1968). As with other helical polymers it is convenient to present the dispersion curves using an extended Brillouin zone based upon the primary repeat unit (in this case $CD_2O$). The problem of representing the dynamics of a polymer of this complexity is also complicated by the existence of inherent disorder within and between the microfibrils of this material.

There is evidence from the diffraction pattern (WHITE, 1976; ANDERSON and WHITE, 1976) that some disorder occurs at the molecular level due to a lack of registry between the chains in neighbouring unit cells. In addition, specimens produced by topotactic polymerisation have a fibrillar texture due to the contraction in volume between the mother crystal of trioxane and the product crystal of polyoxymethylene. This fibrillar texture leads to some disorder between neighbouring fibrils (typically about several hundred angstroms in diameter). Despite all of this, well-defined reflections whose width is usually never greater than about $5^{\circ}$ may be observed in the diffraction pattern, thereby permitting three-axis phonon dispersion curve measurements.

Both longitudinal acoustic excitations along and perpendicular to the polymer chain, and transverse acoustic excitations, have been seen in studies with this crystal. The dispersion curves for some crystallographic directions are shown in Fig.4.19. The great anisotropy between the chain direction and basal plane direction is obvious, the stiffness constants for the longitudinal acoustic branches being in the ratio of about 5:1. These measurements are unique for any polymer and give enough information to compile the stiffness matrix, thereby for the first time allowing a detailed analysis to begin of the role of crystalline strains in the elastic properties of a bulk polymer.

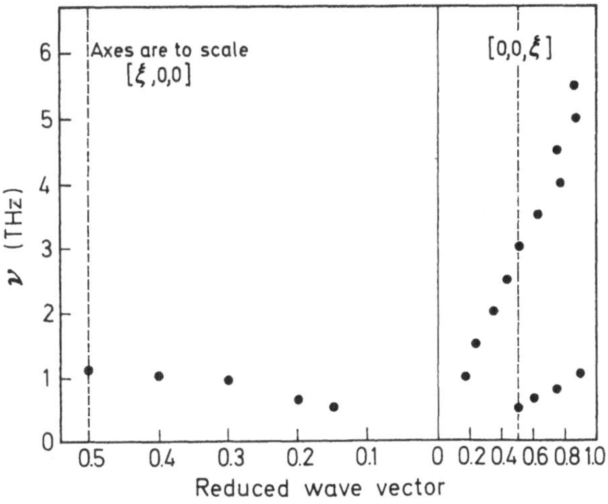


Fig. 4.19 Part of three branches of the phonon dispersion curves for deutero-polyoxymethylene measured at 296 K. The extreme anisotropy of the phonons in this oriented polymer is obvious from the slopes. (ANDERSON et al., 1977)

Additionally, the simpler polymer crystals, because of their low dimensionality, are most appropriate systems to further scrutinise the approximations of molecular crystal lattice dynamical models. For them also the question of long-range electrostatic interactions is of particular interest since often there are large electric dipole moments along or perpendicular to the chain arising, for example, from polar side groups. At present the importance of these dipoles for determining the conformation and crystalline habit is unknown. The problem may be of particular interest in connection with biological macromolecules where, in addition, the existence of strong hydrogen bonds between chains directly affects the physical and chemical properties.

### 4.2.3 Biopolymers

Fibrous proteins and similar biological macromolecules have many features in common with the synthetic fibres already discussed, except that there is an increase in the complexity of the molecular subunits and often the presence of more or less bulky side groups. The chief difference, however, between biopolymers and synthetic macromolecules is that the formula cannot be represented as the n-fold repeat of one single unit—biological macromolecules in fact have a pseudo-random sequence whereby a number of basic units are permuted along the chain. In an attempt to bridge the gap between this situation and the more simple one in commercial synthetic fibres, pseudobiological polymers such as polyglycine and polyalanine have been studied. These have regularly repeating amino-acid subunits and only approximate to the biological materials insofar as they evoke the same types of conformation (the α helix) and hydrogen bonding patterns. The second obvious difference between most biological macromolecules and materials such as polyethylene or polyalanine is that in living tissue the fibre is always closely associated with a second component, usually water or an aqueous solution, and this greatly influences its elastic properties and dynamics. Because of the strong isotopic contrast between hydrogen and deuterium, neutron scattering offers unique possibilities for studying the essential role of water in these composite systems. Below we will refer to some of the questions which arise for the connective tissue collagen, and for muscle, but first we consider models which have been made for the dynamics of materials such as polyalanine as a guide to the broader situation in real biological fibres.

*The Dynamics of Polypeptides*

The problems of treating the full lattice dynamics of dry polypeptides is even more complicated than that for polyoxymethylene and as yet no such attempt has been made. Calculations on the basis of the isolated chain model have been made for polyglycine (MIYAZAWA, 1967; GUPTA et al., 1968) and poly-L-alanine (FANCONI et al., 1971; FANCONI and PETICOLAS, 1971). The most straightforward way of making such calculations is to use a valence force field for interatomic interactions within the chain. The results of such calculations have been tested by Raman and infrared measurements. Neutron incoherent scattering (GUPTA et al., 1968) has been used to obtain the eigenvector weighted density of states for polyglycine but there has, as yet, been no attempt at coherent inelastic scattering measurements. Most work has been concerned with developing models whereby the very complicated force field for such polymers can be simplified and the work of FANCONI et al. (1971) demonstrates one technique. They first calculated the phonon dispersion curves for the isolated $\alpha$ helix of polyalanine

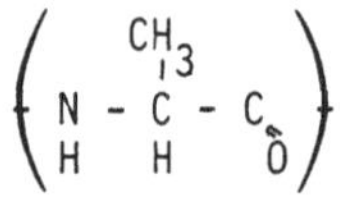


by the FG matrix method initially developed by HIGGS (1953) (the helix has an 18/5 conformation with a 5.4 Å pitch). FANCONI et al. (1971) then compared the calculations with two simplified models which were based upon either treating the whole alanine molecule as a single mass unit or treating it as a two-mass unit; hydrogen and a group of all atoms except the hydrogen bonded hydrogen. The mass of hydrogen was used as the mass at the amide hydrogen position and the remaining mass of the alanine residue was used for the second mass which was placed at the oxygen position with bonding between nearest neighbour residues through the oxygen position masses. The dispersion curves for the single-mass model were rather grossly different from those using the seven-mass model (whereby all atoms except the methyl hydrogens were included separately in the calculation). The two-mass model was a fair approximation to the more detailed calculation and this can be seen in Fig.4.20 where the two results are compared. Most of the salient features appear in both sets of curves.

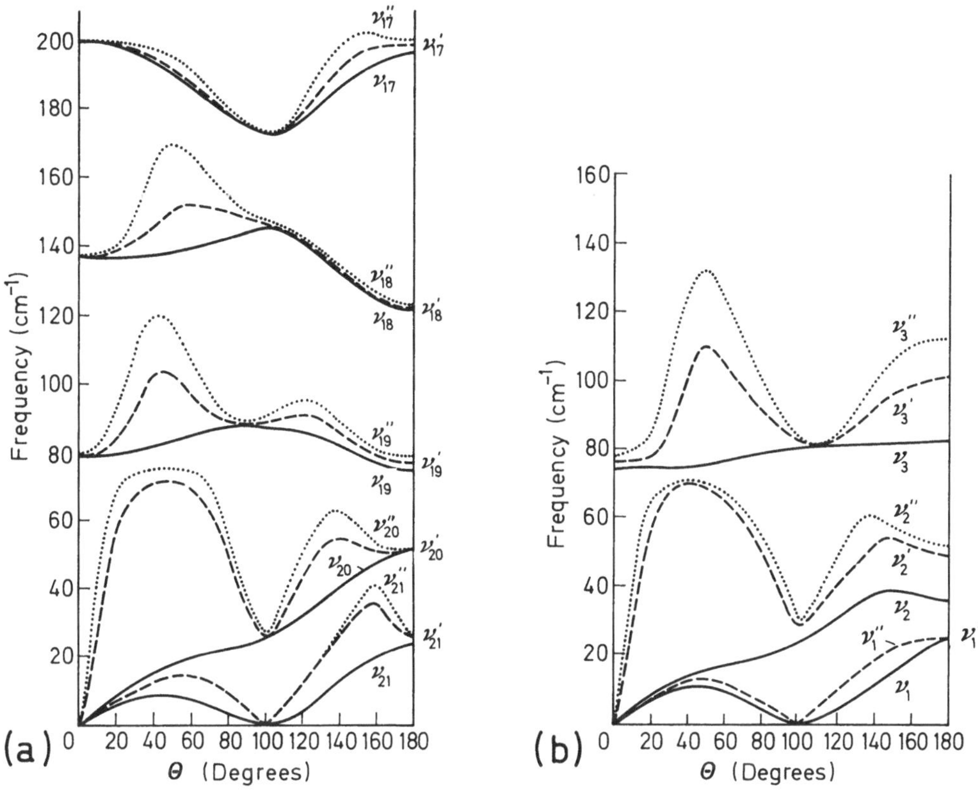


Fig. 4.20 (a) Low frequency phonon dispersion curves of the seven-mass model of α-poly-L-alanine for three values of the hydrogen-bond force constant (—)$F_{N-H\cdot\cdot O}$=0.0; (--)$F_{N-H\cdot\cdot O}$=0.15 mdyn/Å; (···)$F_{N-H\cdot\cdot O}$=0.3 mdyn/Å (b) Low frequency phonon dispersion curves for the two mass model of α-poly-L-alanine for three values of the hydrogen-bond force constant: (—)$F_{N-H\cdot\cdot O}$=0.0; (--)$F_{N-H\cdot\cdot O}$=0.15 mdyn/Å; (··)$F_{N-H\cdot\cdot O}$=0.3 mdyn/Å (FANCONI and PETICOLAS, 1971)

The initial slopes and the dip at a phase angle of 100° arise directly from the assumption of nearest neighbour and third nearest neighbour interactions. It would clearly be advantageous to compare experimental dispersion curves with the calculations to further improve these models since they offer the only chance at present to discuss the dynamics and hence the force fields in actual biological substances. For the moment it can be concluded that models similar to the two-mass model for polyalanine would be worth trying in more complicated systems such as collagen and muscle.

*Collagen and Muscle*

All of the polymers so far considered have consisted of isolated strands of the fibrous molecule packed with one or two molecules per unit cell. In collagen the amino-acid sequence in each molecule is almost completely determined (HULMES et al., 1973). The molecules arrange themselves in triple stranded ropes which then pack together to form the microfibril possibly on a five-fold basis. The microfibril is the equivalent to the unit cell for ths crystalline polymers described above. The exact packing of the fibres and the microfibril is not yet known but neutron scattering measurements of the diffraction pattern have given promising new information (MILLER et al., 1976). Collagen is of widespread occurrence in biology as a connective tissue, its elastic properties are of fundamental biological importance but little understood. The problems are similar to those posed several years ago for synthetic macromolecules. Macroscopic measurements of the elastic properties give only part of the story since the fibre lengths are only of the order of thousands of angstroms. The elastic properties are extremely dependent on temperature and humidity and it is, therefore, clear that the aqueous component plays an important part (WHITE, 1976). A study of these elastic properties can be attempted using the methods and experience surveyed in this chapter but it is also clear that force fields will have to be tested against experiment using low frequency Raman and Brillouin scattering because of present technical limitations on neutron elastic and quasielastic scattering.

In the case of muscle the situation is more complex than for collagen since several different proteins are involved in the fibrillar structure as well as an extensive aqueous medium. Neutron diffraction has again given novel insights into the structure of muscle, and the way in which energy is released into contractile processes through the interplay of the polymer conformations and the aqueous medium still remains mysterious.

## 4.3 Structure and Dynamics of Adsorbed Species

The interaction between neutrons and matter is weak compared to the Coulomb interactions between electrons or other charged species and most substances. Typically, the cross-sections for neutron scattering are about $10^{-8}$ of those for electron scattering and so it is both natural that neutrons have been used to study the dynamics and structure of bulk solids and liquids, and paradoxical that they should also find a use in studies of the dynamics and structure of adsorbed species. Because the cross-sections are low, studies of adsorbed monolayers on single crystal surfaces, which are currently possible with low energy electron diffraction (LEED) (PENDRY and GARD, 1975; VENABLES, 1975) are at present impossible with neutrons. For some systems it has been shown that this difficulty is outweighed by the insights given by the neutron method which derive from the simpler scattering theory for neutrons compared to electrons (MARSHALL and LOVESEY, 1971).

Table 4.3 gives a list of critical parameters for a number of possible adsorbates and substrates. The coherent cross-sections for the molecules are calculated in the long wavelimit appropriate for the first few orders of diffraction. The spherical Bessel funtion of the rotational form factor (SEARS, 1966) $J_0^2(Q_\ell r)$ has been evaluated at $Q = 1.71\ \mathrm{\AA}^{-1}$ which corresponds to the diffraction peak of the $\sqrt{3} \times \sqrt{3}$ regis-

Table 4.3 Neutron scattering and capture cross-sections of some gases and substrates for which measurements have been done or are in progress

| Coherent scattering | | | | | Incoherent scattering |
|---|---|---|---|---|---|
| Gas | $4\pi(\Sigma_d b_d)^2$ $\times 10^{-28}M^2$ | $r \times 10^{-9}$ | $4\pi(\Sigma_d b_d)^2 . j_0^2(Qr_d)$ $\times 10^{-28}M^2$ $(Q = 1.7\ \text{Å}^{-1})$ | $\sigma_a$ $\times 10^{-28}M^2$ | $\sum_d \sigma_{inc,d}$ $\times 10^{-28}M^2$ |
| $^4He$ | 1.13 | - | 1.13 | - | - |
| Ne | 2.66 | - | 2.66 | 0.032 | 0.24 |
| $^{36}A$ | 74.20 | - | 74.20 | 0.005 | - |
| Kr | 6.88 | - | 6.88 | 25.0 | - |
| Xe | 2.90 | - | 2.90 | 24.5 | - |
| $H_2$ | 7.03 | 0.037 | 6.13 | 0.66 | 159.4 |
| $D_2$ | 22.36 | 0.037 | 19.49 | 0.001 | 4.02 |
| $N_2$ | 44.41 | 0.0547 | 32.75 | 3.7 | 0.6 |
| $O_2$ | 16.90 | 0.0604 | 11.64 | 0.001 | 0.2 |
| $Cl_2$ | 46.13 | 0.0994 | 15.47 | 66.0 | 6.84 |
| $CD_4$ | 139.6 | 0.109 | 36.3 | 0.002 | 8.0 |
| $CH_4$ | 8.68 | 0.109 | - | 1.32 | 318 |
| $ND_3$ | 108.7 | 0.094 | 43.1 | 1.85 | 6.33 |
| | | 0.038 | 93.0 | | |
| $NH_3$ | 0.42 | 0.094 | - | 2.8 | 239.4 |
| | | 0.038 | | | |
| $D_2O$ | 46.0 | 0.076 | 25.8 | | |
| | | 0.059 | 34.4 | 0.002 | 4.03 |
| | | 0.096 | 16.0 | | |
| $H_2O$ | 0.36 | 0.076 | - | 0.66 | 159.5 |
| | | 0.059 | | | |
| | | 0.096 | | | |

| Substrate | $4\pi(\Sigma b_d)^2$ $\times 10^{-28}M^2$ | Space group | Lattice parameter | Surface area per gram $(M^2)$ | $\Sigma\sigma_{inc,d}$ $\times 10^{-28}M^2$ |
|---|---|---|---|---|---|
| Graphite,c | 5.56 | | a=0.247nM,c=0.693M | 20-90 | - |
| $MgBr_2$ | 44.4 | | | ~10 | 0.9 |
| $FeI_2$ | 50.76 | | | ~ 4 | 1.0 |
| $PbI_2$ | 50.3 | | | ~ 7 | 0.84 |
| $SiO_2$ | 31.4 | | | 20-500 | - |
| $Al_2O_3$ | 74.8 | | | c. 200 | - |
| Zeolite 13X | | | | c. 1000 | - |
| Ni | 13.33 | | | c. 200 | 4.7 |
| Pt | 11.34 | | | c. 200 | 0.6 |

tered phase in graphite (KJEMS et al., 1976). For molecules with low symmetry the value for each axis is calculated. The nuclear scattering lengths, $b_j$, are taken from SHULL (1972) BACON (1976) and the radii of gyration, r, from HERZBERG (1950).

The problems to which neutron scattering measurements can be addressed are similar to those already encountered for molecular crystals and liquids. First they can be used to study collective effects associated with aggregated molecules in two dimensions such as the structure of adsorbed monolayers and multilayers, their dynamics and underlying interatomic and molecule surface interactions. Table 4.3 shows that coherent and incoherent scattering offer complementary approaches to such problems. For expamle, there should be important differences in the coherent elastic scattering cross-sections for molecules depending on the ordering of molecular axes rotating phases. The coherent inelastic scattering from such systems is usually too weak to observe but incoherent inelastic scattering is feasible. A case in point which has been studied is ammonia on graphite (GAMLEN and WHITE, 1976).

Concerning the collective phenomena, it can be envisaged that techniques such as the atom-atom potential method (see Subsec.4.1.1) will be useful for modeling purposes and already there exists a wealth of thermodynamic evidence pointing towards phase changes in adsorbed monolayers and multilayers. Fig.4.21 shows the adsorption isotherms for methane on exfoliated graphite at a number of different temperatures determined by THOMY and DUVAL (1970). The curves resemble those for monotomic adsorbates where at least two different states of the adsorbed species are indicated by points of inflexion in ths isotherms. In the past it has only been possible to suggest that these may be solid or liquid-like by comparisons between the measured isosteric enthalpies of adsorption and known values of transition enthalpies for three-dimensional materials. Low energy electron diffraction from single crystals (SUZANNE et al., 1973) has greatly assisted such interpretations, but the limitations of LEED to elastic scattering measurements necessarily limit the distinguishability between, for example, a disordered solid and a liquid phase. The relationship between critical temperatures found by thermodynamic measurements and

Table 4.4 Critical and triple point temperatures for two- and three-dimensional phases

| Adsorbate | $T_1$(3D) (K) | $T_c$(3D) (K) | $T_{c1}$(2D) (K) | $T_{c1}$(2D) (K) | $T_{c1}$(2D) / $T_c$(3D) | $T_{c1}$(2D) Literature values |
|---|---|---|---|---|---|---|
| Neon | 24.6 | 44.4 | - | - | 0.36 | $\leftarrow$16.1[5] |
| Argon | 84 | 151 | - | 65 $\rightarrow$ | 0.43 | 67[5]<br><64[6] |
| Krypton | 116 | 209 | 77 | 87 $\rightarrow$ | $0.45_5$ | 86.5[5]<br>82[7] |
| Xenon | 161 | 290 | 99 | 117 | 0.40 | 120[8] |
| Methane | 91 | 191 | - | 75 $\rightarrow$ | 0.39 | |

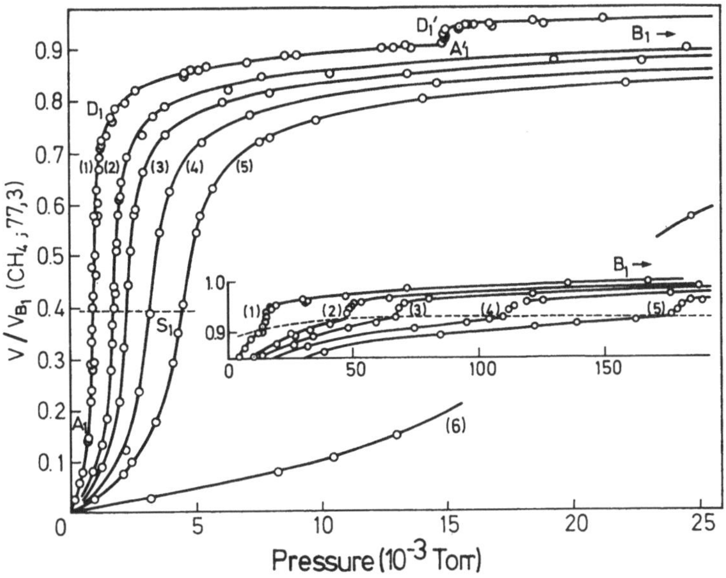


Fig. 4.21 Adsorption isotherms for methane on exfoliated graphite. Formation of the first layer. (1) 77.3 K; (2) 80.0 K; (3) 80.9 K; (4) 82.3 K; (5) 83.5 K; (6) 90.1 K (THOMY and DUVAL, 1970)

various transition temperatures for the bulk material, for some rare gases and methane, can be seen from Table 4.4 (THOMY and DUVAL, 1970).

In addition to systematic studies of neutron diffraction and inelastic scattering experiments from adsorbed molecules as a function of the strength and symmetry of the adsorption potential, there are two natural areas of contact which arise with the problem of heterogeneous catalysis. The first is that since neutron measurements must perforce be done upon macroscopic specimens of high surface area to maximise the ratio of the scattering from the adsorbed molecule to the scattering from the substrate, one is led to study questions of percolation and capilallary condensation between fine grains. The second and possibly more important area embraces the finely divided substrate itself which not only needs adequate chacaterisation from a morphological point of view but may have electronic and vibrational states, different from the bulk material, on account of the small particle size and as a consequence of extensive chemisorption.

### 4.3.1 Substrates

The neutron scattering cross-sections per atom/molecule of most suitable substrates for adsorption experiments varies between about 5 and 80 barns ($10^{-28} m^2$). To achieve reasonable signal to noise for scattering from the adsorbed material compared to the substrate, the adsorbant needs to have at least the same scattering cross-section per atom/molecule as the substrate and the surface area of the substrate should be greater than about 10 $m^2$ per gram. Various forms of graphite have been extensively used for physisorption studies because of the low scattering cross-section for carbon and because materials can be produced which either have large surface areas (e.g., graphon-surface area 90 $m^2$ per gram) or have relatively high crystallographic orientation (e.g., grafoil–surface area approximately 20 $m^2$ per gram). For such material selective orientation of the scattering vector with respect to the planes on which mos of the adsorption occurs is possible. Materials such as molecular sieve zeolites, which are extensively used as industrial catalysts, are alumino-silicate cage structures and may have surface areas up to 1000 $m^2$ per gram since the cages are interconnected by channels of molecular dimensions. These materials have the disadvantage that their surface is less homogeneous than the graphites and they frequently contain hydroxyl groups which must be deuterated to minimise the background scattering due to the substrate (EGELSTAFF et al., 1966). Closely related to them are the clay minerals, vermiculite and montmorillonite which have a much more homogeneous adsorption surface. In the domain of chemisorption the adsorption enthalpy is usually greater than 80 kJ $mole^{-1}$ and examples of suitable materials where neutron measurement have been made are Raney nickel (STOCKMEYER et al., 1975) (surface area 185 $m^2 gm^{-1}$) and platinum black (WADDINGTON et al., 1976).

The characteristics of microscopic particles such as graphon, Raney nickel and platinum black, whose particle sizes may be as small as 50 Å on edge, may be defined by high and low angle diffraction of X-rays or neturons and by measurements of the neutron incoherent and neutron coherent scatterin spectra (GAMLEN and WHITE, 1976). For graphon it has been shown that the phonon dispersion curve for longitudinal acoustic excitations perpendicular to the sheets is measurable using the method developed for polycrystalline polymers with highly anisotropic lattice parameters (see Selective Observation of Interchain Vibrations in Polyethylene). Such measurements may be an effective probe of changes in the electronic structure of the substrate following chemisorption since the particles are small enough that monolayer coverage adds a not insignificant number of atoms and electrons to the system. The method could clearly be extended to cubic crystals using average phonon calculations of the kind described by DE WETTE and RAHMAN (1968).

### 4.3.2 Physisorption

When adsorption occurs with an enthalpy change of less than about 80 kJ $mole^{-1}$, the binding between the adsorbate and the substrate arises by the van der Waals forces, and the models applicable to the external modes in the dynamics of molecular crystals are applicable for a discussion of the structures which arise and for modelling the phase transitions between them. Neutron diffraction measurements have now been made from adsorbed monolayers and multilayers of nitrogen, argon, hydrogen, oxygen, nitric oxide and ammonia on grafoil or graphon (Table 4.3). These systems give reasonable signal to noise for the diffraction pattern of the adsorbed phase since the coherent scattering lengths are large in the low resolution approximation appropriate to measurements of the first few orders of diffraction. Here the molecular scattering cross-section $(4\pi(\Sigma b_d)^2)$ more nearly represents the scattering power than that of the separate atoms $\Sigma 4\pi_d^2$. Other advantages and disadvantages of coherent neutron scattering for surface studies have been analysed by KJEMS et al. (1976).

Where the coherent scattering length is sufficiently large to give good constrast with respect to the substrate, phonon dispersion curve measurements are possible (TAUB et al., 1975) but Table 4.3 shows that incoherent scattering from hydrogenous species may be more generally applicable for studies of the inelastic scattering because the large incoherent scattering cross-section nearly always gives good contrast between hydrogenous adsorbates and the substrate.

*Diffraction*

Here the neutron diffraction from nitrogen, hydrogen (deuterium) and argon adsorbed on grafoil are compared. The systems show rather different features because of the relative strengths of the intermolecular interaction and the molecule-surface potentials. For the experiments on adsorbed nitrogen, KJEMS et al.(1976) used a 60 gram specimen of grafoil, the orientation of whose basal planes had a rocking curve, determined by neutron diffraction, of $\pm$ 30°. The sample was held in a aluminium container connected to a gas measuring system and the adsorption isotherms at various temperatures determined volumetrically. The isotherm for 78 K is shown in Fig.4.22.

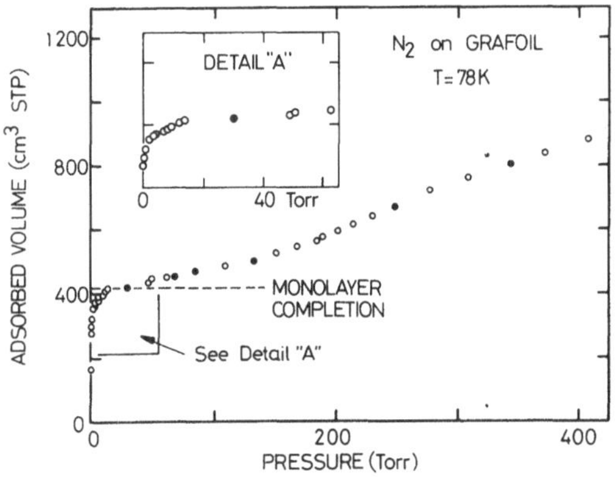


Fig. 4.22 Adsorption isotherms for $N_2$ on grafoil(KJEMS et al., 1976)

The curve is strongly concave downwards indicating that nitrogen wets the surface and one can expect monolayers and multilayers to form. The plateau region corresponds to one monolayer and this occurred when 415 $cm^3$ of gas at STP had been adsorbed. The diffraction pattern from rather less than one monolayer (volume adsorbed = 360 $cm^3$ STP) is show in Fig.4.23 where the strong diffraction peak at $2\theta \approx 78^\circ$ due to the (002) reflection in graphite has been substrated. The nitrogen diffraction peak at 78 K is very broad, peaking at about 69° in $2\theta$. As the temperature was lowered (Fig.4.23b,c,d) to 50 K this peak progressively sharpened without changing its position appreciably, indicating an increase in size of the scattering "crystal".

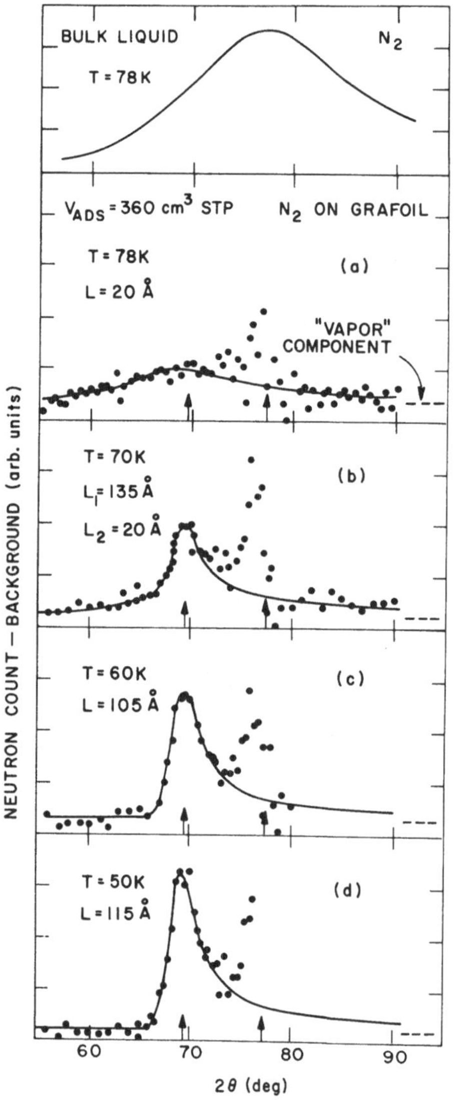


Fig. 4.23a-d Neutron diffraction from one mono-layer of nitrogen on grafoil at different temperatures compared with the diffraction pattern for liquid nitrogen at 78 K (KJEMS et al., 1976)

The peak shapes for the adsorbate diffraction pattern were asymmetric with a long "tail" towards higher values of $2\theta$, and this is characteristic of the diffraction pattern for a layer lattice. This problem was treated by WARREN (1941) who showed that for incident wavelength $\lambda$, the diffraction pattern from randomly oriented lamellae is given by

$$I_{hk} \sim \frac{N.m_{hk}.|F_{hk}|^2 \ \exp(-2W)}{(\sin\theta)^{3/2}} \left(\frac{L}{\pi^{1/2}\lambda}\right)^{1/2} \cdot F(a) \quad . \qquad (4.7)$$

where L is the "size" of the array in the direction (hk), $2\theta$ is the scattering angle, $m_{hk}$ the multiplicity of the hk th reflexion and exp(-2W) a two-dimensional Debye-Waller-factor. The function F(a) is defined by

$$F(a) = \int_0^\infty \exp[-(x^2 - a)^2]dx \tag{4.8}$$

where $a = 2\pi^{1/2}L/\lambda\ (\sin\theta - \sin\theta_{hk})$ and the 2D spacing for the hkth reflexion $d_{hk}$ is $(2d_{hk})^{-1} = \sin\theta_{hk}/\lambda$. If a is large (4.7) reduces to a simpler form

$$I_{hk} \sim N.\ \frac{n_{hk}\cdot|F_{hk}|^2 .\ f^2(\theta)\ \exp(-2W)}{\sin\theta(\sin^2\theta - \sin^2\theta_{hk})^{1/2}}\ . \tag{4.9}$$

The molecular form factor $f(\theta)$ has been introduced to cover the case of a molecular layer and is given by the spherical Bessel function,

$$f(\theta) = 2j_0(Qr) \tag{4.10}$$

whose argument is the product of the molecular radius of gyration, r, and the momentum transfer, Q, defined by $Q = 4\pi\ \sin\theta/\lambda$. It can readily be seen that (4.9) leads to an asymmetric line shape whose width is governed by the parameter "a" and hence the layer lattice extent. KJEMS et al. (1976) showed how the effects of preferred orientation may be included in the treatment above. Such a diffraction pattern also arises for grafoil itself and microcrystalline graphite (GAMLEN and WHITE, 1976). The crystal sizes for the adsorbed nitrogen monolayers were calculated using Warren's theory and a fitting program to the observed data. The specific lengths, L, are shown on Fig.4.23.

By comparing the diffraction pattern to that for bulk liquid nitrogen at 78 K it can be seen that the intermolecular spacing is rather larger in the adsorbed state. Indeed, the layer lattice peaks index well, on the assumption that the nitrogen molecules form a $\sqrt{3} \times \sqrt{3}$ structure in registry with the graphite basal plane. This type of structure has been seen in low energy electron diffraction from xenon on graphite (SUZANNE et al., 1973).

When further nitrogen gas was added to the sample, the peak associated with the $\sqrt{3} \times \sqrt{3}$ structure remained almost unchanged until a little over one monolayer had been completed, thereupon there was a progressive development of a peak at $2\theta \approx 76^\circ$, so long as measurements were made at temperatures rather less than 50 K. By the time the second monolayer had been completed the peak associated with the $\sqrt{3} \times \sqrt{3}$ structure was no larger visible and all the diffraction intensity occurred at $76^\circ$. This peak at higher angles shows that the nitrogen molecules pack more closely upon formation of the second layer and, as a result, come out of registry with the basal plane of the graphite. It is clear that the intermolecular forces, rather than the molecule-surface forces, become dominant in determining the structure as a second layer is formed. It is also clear that, on heating a system in which there is rather more than one monolayer, there arises a sharp transition in structure above approximately 54 K; the peak $2\theta = 76^{\circ}$ drops sharply in intensity to be replaced by the peak at $2\theta = 69^{\circ}$ characteristic of the $\sqrt{3} \times \sqrt{3}$ structure. KJEMS et al. (1976) followed such behaviour as a function of surface coverage and temperature, and were able to propose a partial phase diagram for the nitrogen/grafoil system up to two layers. Briefly, below 55 K and for less than approximately 0.065 molecules per square angstrom, there is a registered phase plus vapour. Above this coverage there is a registered phase plus a dense phase and at coverages of the order of 0.08 molecules per angstrom squared a dense solid. Above approximately 55 K the coexistence of a registered phase and a liquid is proposed but inelastic scattering measurements are required to differentiate between these.

The adsorption of hydrogen and deuterium on grafoil has been studied by NIELSEN and ELLENSON (1975). Diffraction from both species may be measured so long as pure para-hydrogen is used and the incident neutron energy is less than approximately 20 meV, the energy needed for orthopara conversion. At low temperatures both molecules form a solid-like $\sqrt{3} \times \sqrt{3}$ structure in registry with the graphite. Upon the addition of further gas, the monolayer is compressed into a non-commensurate triangular structure and the lattice parameter is proportional to the square root of the gas filling (Fig.4.24). This behaviour ceases when the number of molecules added is such that the densest possible monolayer has been produced, whereupon a second layer starts to be formed. For both hydrogen and deuterium the $\sqrt{3} \times \sqrt{3}$ structure melts at 20 K but the dense monolayer may persist up to about 30 K. The continuous variation of the dense monolayer lattice parameter contrasts with the abrupt changes found for adsorbed nitrogen, and has important consequences for the inelastic scattering from the adsorbed phase (see below).

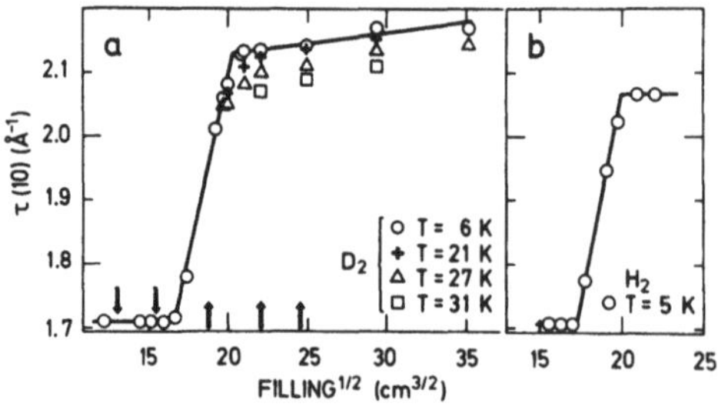


Fig. 4.24 The reciprocal lattice parameter, $\tau$, for adsorbed deuterium (a) and hydrogen (b) on grafoil as a function of temperature and coverage (NIELSEN and ELLENSON, 1975)

In contrast to both the nitrogen and hydrogen, the adsorption of $^{36}Ar$ on grafoil at 5 K is non-commensurate. $^{36}Ar$ is the strongest coherent scattering nucleus and with it TAUB et al. (1975) were able to observe three peaks in the diffraction pattern for an adsorbed monolayer (Fig.4.25). It can be seen that these peaks all have the characteristic layer lattice shape expected from Warren's theory. They index as the (10) (11) and (20) Bragg reflections from a two-dimensional triangular lattice with the nearest neighbour spacing of 3.88 Å. The measurements were made with the surface covered to approximately 90% of a full monolayer. The nearest neighbour distance in the adsorbed film is only 2% larger than the nearest neighbour distance in three-dimensional solid argon at 20 K, and so it may be inferred that the attractive force between argon atoms is chiefly responsible for establishing the two-dimensional structure seen and that, because of the size of the argon atoms, there is no registration with the rather longer periodicity of the graphite basal plane. It can be shown by simple calculations using the Lennard-Jones potential, for example, that the argon-argon interaction is considerably stronger than the variation of the argon-carbon potential from point to point on the substrate. In this case, therefore, the surface potential does not dominate in establishing the monolayer structure as must be inferred to be the case for nitrogen and hydrogen for coverage up to about one monolayer.

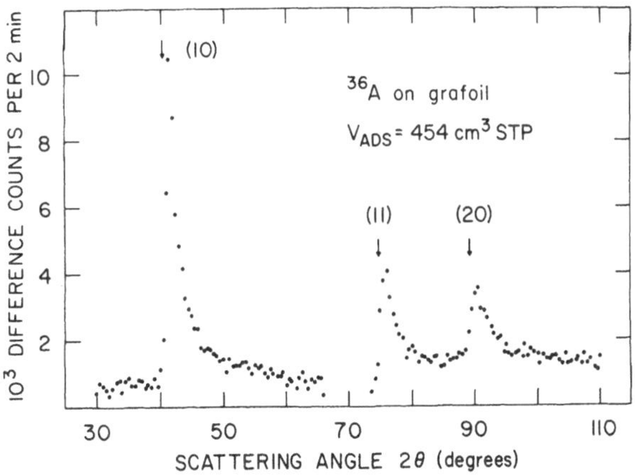


Fig. 4.25 The diffraction pattern from adsorbed $^{36}$Ar on grafoil at 5 K (TAUB et al., 1975)

*Phonon Measurements*

For $^{36}$Ar, hydrogen and deuterium the scattering cross-sections were large enough to allow a follow-up to the diffraction experiments with inelastic scattering studies. Because of the incomplete orientation of the two-dimensional planar polycrystalline layers, which arises because of the lack of complete orientation in the substrate, it is necessary to have a model for the expected excitations in order to interpret the experimental results. The coherent one-phonon dynamical structure factor, $S_{coh}(Q,\omega)$

$$[S_{coh}(\underline{Q},\omega)]_{poly} = \exp(-2W)[n(\omega) + 1](\hbar Q^2/2M\omega)\left\langle \sum_{d=1}^{2} [\underline{Q} \cdot \underline{\sigma}_d(\underline{Q})]^2 \, \delta[\omega - \omega_j(\underline{Q})]\right\rangle , \qquad (4.11)$$

was computed using the known dimensions of the triangular lattice and the Lennard-Jones (6-12) potential for argon to obtain the phonon eigenvectors, $\sigma_d(Q)$. In the expression exp(-2W) is the Debye-Waller factor, $n(\omega)$ is the phonon occupation number, and M is the mass of $^{36}$Ar atoms, whilst $\underline{Q}$ is the usual vector momentum transfer. The square brackets indicate that the two-dimensional polycrystalline average has had to be taken for Q with respect to the crystal directions.

Well-defined excitations were observed for the argon film at 5 K and are shown in Fig. 4.26. The shape of these excitations is quite different at the two different momentum transfers shown, 1.6 $Å^{-1}$ and 2.6 $Å^{-1}$ The theoretical frequency distribution at these momentum transfers, obtained from the model calculations setting the Debye-Waller factor to unity, are in good agreement with the observed results. The shaded distribution is a convolution of the predicted spectra with the spectrometer resolution. The calculations also predict a well-defined structure at momentum transfers of 3.5 and 4.3 $Å^{-1}$ and this has been verified experimentally. These distributions are not single phonon groups because the polycrystalline average associated with the disorder in the film causes sampling of the whole dispersion surface at the momentum transfer appropriate to the experiment. As a result, there are strong peaks at momentum transfers near Brillouin zone edges where the density of states is large. The difference between the spectra seen at $\underline{Q}$ = 1.6 $Å^{-1}$ and Q = 2.6 $Å^{-1}$ arises because of the eigenvector dependence of the cross-section; at the lower momentum transfer the phonon polarisation vector is nearly perpendicular to $\underline{Q}$ near the zone boundary and, therefore, most of the response comes from the transverse acoustic branch at approximately 3.5 meV. For the higher momentum transfer both the longitudinal and transverse modes contribute to the scattering.

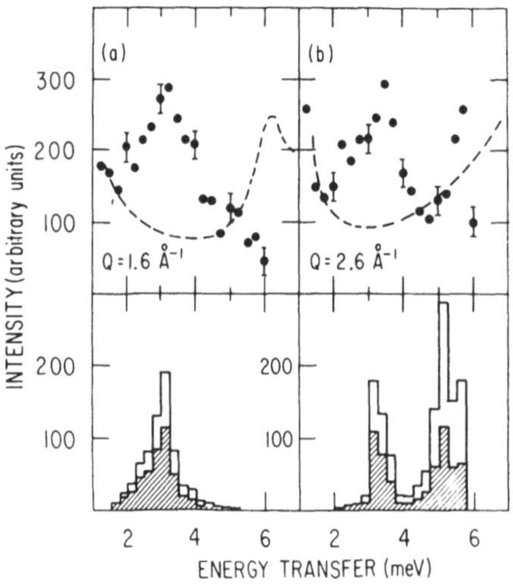


Fig. 4.26 Neutron inelastic scattering spectra from an adsorbed $^{36}$Ar monolayer at two different values of the momentum transfer, Q, compared with predictions for the lattice dynamics of a 2-dimensional volume based on a Lennard-Jones potential between the Argon atoms and a negligible potential energy due to adsorption (TAUB et al., 1975)

In general the agreement between experiment and the simple theoretical model proposed is good and supports a model that the adsorbed argon film has a structure which is primarily determined by the argon-argon interactions and that, to a first approximation, the interaction between argon and the graphite surface can be neglected. No measurements were possible of the excitations perpendicular to the graphite surface, which would bear strongly on the accuracy of this assumption, and the data are not yet of sufficient quality to warrant a more rigorous analysis of the lateral forces than that given by the simple Lennard-Jones potential.

The good contrast between scattering from $^{36}Ar$ and the graphite surface in the previous example arose from the large nuclear cross-section of the adsorbant; for adsorbed hydrogen the inelastic cross-section is also large, both because of the nuclear scattering and because the amplitudes of the vibrational motions are themselves large due to the low mass. Therefore, as a corollary to the diffraction measurements reported above, it was possible to measure some of the excitations for adsorbed hydrogen and deuterium films as a function of the coverage. Extensive phonon dispersion curve measurements have already been made on crystalline solid hydrogen and deuterium (NIELSEN, 1973) and so there exists a good basis for the interpretation of the dynamics of the adsorbed layers. Phonon groups from adsorbed hydrogen and deuterium, as a function of the percentage covered, referred to a single monolayer, are shown in Fig.4.27. For coverages less than one monolayer discrete excitations were seen which were independent of the scattering vector. For deuterium this peak is at 4 meV (96 THz) and has a width of 0.3 meV (0.07 THz). The wave-vector independence indicates a localised excitation of the adsorbed molecule relative to the surface with very little interaction between neighbouring molecules. When the filling is greater than one monolayer these sharp excitations disappear and only a broad spectrum can be observed which indicates that, associated with the disorder already seen in diffraction, the excitations are strongly damped.

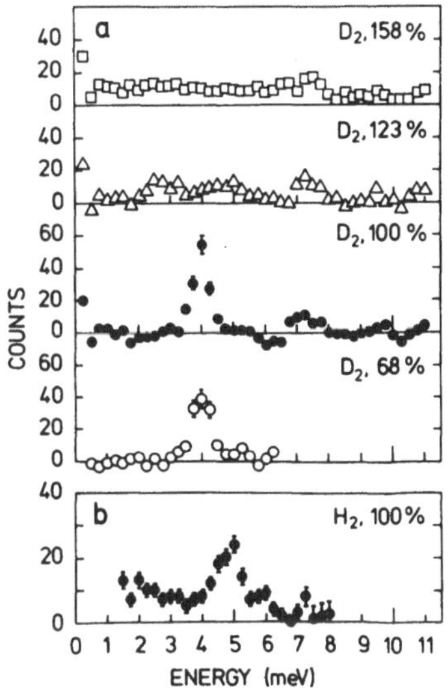


Fig. 4.27 Neutron inelastic scattering from adsorbed hydrogen and deuterium as a function of the percentage of one monolayer coverage. The measurements were made at a temperature of 4.2 K (NIELSEN and ELLENSON, 1975)

Insofar as localised vibrational modes for the adsorbed hydrogen and deuterium were observable for the commensurate surface structure, we have a measure of the perpendicular component of the surface potential at the hydrogen molecule. It has also been possible to observe (NIELSEN, 1976) the excitation of the ortho-para transition at 14.2 meV for hydrogen. The frequency is almost the same as that in bulk hydrogen and no evidence for Stark splitting of the J = 1 rotational state by the surface field was detectable. This evidence allows an upper limit to be set to components of the surface potential along and perpendicular to the adsorption site.

### 4.3.3 Incoherent Scattering

The observability of the ortho-para transition for adsorbed hydrogen on graphite is assisted because that transition carries the full incoherent cross-section (~80 barns per atom) of hydrogen. This cross-section is many times larger than the coherent cross-section available for the other experiments described above and since many molecules of interest, from the point of view of heterogeneous catalysis, have at least one hydrogen atom, incoherent scattering is an attractive method for studying the dynamics of adsorbed molecules. Numerous experiments based upon this have been carried out (TODIREANU and HAUTECLER, 1972; WHITE, 1972). Many of these studies have involved high surface area materials such as zeolites but with the availability of isotherms for graphites (THOMY and DUVAL, 1970; BEZUS et al., 1964) it has been attractive to study the structure and dynamics of molecules such as methane in the physisorbed state. Good signal to noise can be obtained for the adsorbed methane in respect of the background scattering for temperatures up to about 170 K (the boiling point of methane is 109 K) and as an example, the incoherent scattering spectrum from 0.2 monolayers of methane on graphon at 176 K is shown in Fig.4.28 (GAMLEN and WHITE, 1976). At temperatures above 109 K the spectra still show many of the characteristics expected for a liquid. There is a quasieleastic peak whose position is independent of momentum transfer and whose width increases as would be expected for liquid-like diffusion. The theory for neutron scattering from a two-dimensional liquid has been worked out by STOCKMEYER (1976). The scattering is qualitatively different from that expected from a two-dimensional gas and is in accord with the adsorption isotherm which shows that the methane wets the carbon surface and that the methane-carbon interaction is of comparable magnitude to the interaction between methane molecules in the monolayer and submonolayer regions.

In addition to the quasielastic scattering, there is well-developed inelastic scattering of the kind normally seen for relatively free rotational motion. It is not possible to distinguish two-dimensional rotation from three-dimensional rotation with the spectrometer resolution at which these experiments were done, but there is little evidence to suspect that no rotation about the axes parallel to the surface is possible. It is more likely that the rotation about these axes is hindered slightly compared to that about the axis perpendicular to the surface.

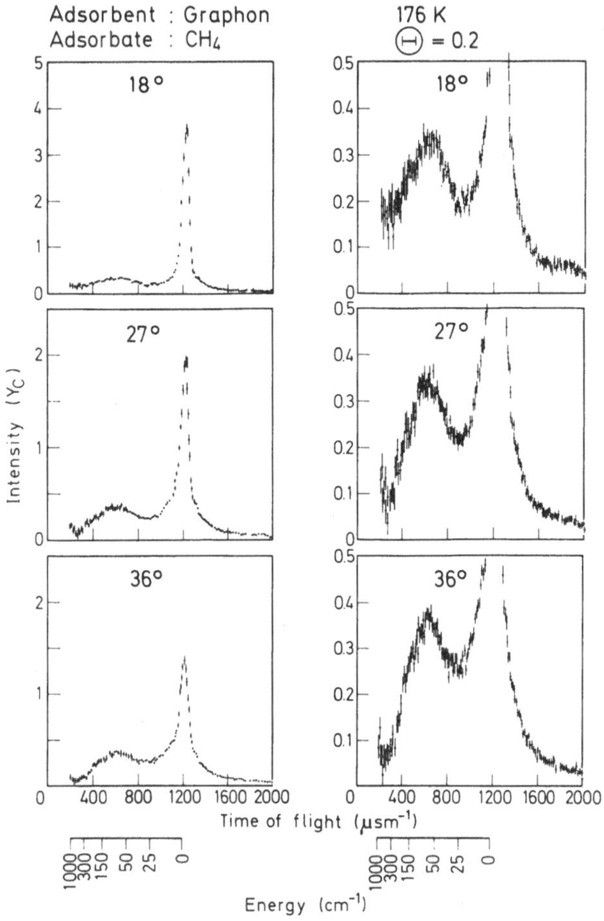


Fig. 4.28 The effect of the periodicity of the graphon surface on the translation and rotation of adsorbed methane (GAMLEN and WHITE, unpublished)

## 4.4 Chemisorption

When atoms or molecules are chemisorbed at surfaces the heats of adsorption are usually large because covalent or ionic bonds are formed between the adsorbate and the substrate The strength of such bonds depends upon the electronic properties of both species and the final electronic state of the system is usually markedly different from that of the reagents. This can be seen by photoelectron and Auger spectroscopy on single crystal metal surfaces (LINNETT et al., 1975; ROBERTS et al., 1974; THOMAS et al., 1974) and sometimes by infrared and Raman spectroscopic measurements on the adsorbate (SHEPPARD et al., 1960; PRITCHARD, 1974; HENDRA, 1975). For optical measurements there must exist a window in the spectrum of the substrate at the absorption or scattering frequency of the adsorbed molecule, or one must use reflection techniques.

For Raman work, fluorescence from the substrate must be weak or absent. With complicated substrates such as zeolites, or for metals such as Raney nickel and platinum black, optical spectroscopic methods are restricted for these reasons and neutron scattering measurements of the intramolecular excitations, the external modes and diffusion are uniquely valuable. Here we consider two systems, differing in the strength of the adsorption potential (GENOT, 1974).

### 4.4.1 Hydrocarbons Adsorbed on Zeolites

Molecular sieve zeolites consist of sodium or potassium alumino-silicate cage structures (BARRER, 1964) whose cations may be exchanged to produce catalysts of high surface area and variable specificity. An extremely stable form of zeolite capable of being outgassed at 600°C without destruction of the lattice is produced by exchanging the $Na^+$ ions in the crystal with $La^{3+}$. This material is a strong Lewis acid catalyst due to the high charge on the lanthanum. Alternatively, the sodium may be replaced with silver, thereby giving a catalyst for which there is a specific interaction between the silver and double bonds in, for example, ethylene. The properties of methane adsorbed in $La^{3+}$-Y-zeolite at room temperature are intermediate between those seen for physisorption and chemisorption. The adsorption isotherm is strongly concave downwards and at normal pressures the structure fills to the extent of about 12 methane molecules per zeolite cage. Even for surface coverages of about 0.1 monolayer the neutron scattering spectrum can be readily separated into the scattering from the adsorbed gas and the lattice. This is done by subtracting the spectrum of the substrate after outgassing at 600°C to less than $10^{-6}$ mm Hg. For a coverage of 0.3 at 245 K the spectra so obtained are shown in Fig.4.29 at four different scattering angles. Qualitatively the scattering resembles that from physisorbed methane on graphite. There are clear indications of a broad inelastic and narrow quasielastic component; the inelastic components resemble the scattering from dense methane gas or liquid methane and the spectrum can be fitted by the superposition of a liquid methane spectrum and the spectrum of a gas where only one molecular rotational axis is available (GRIFFING, 1961). Qualitatively the spectra suggest multilayer formation even at formally less than one monolayer filling, and hence preferential adsorption sites. It can be concluded that at low temperatures the methane molecule is partly chemisorbed and partly physisorbed and that the extent of the process can be estimated from the temperature dependence of the quasielastic scattering intensity.

For methane on the zeolite the hindered rotational modes of the near inelastic spectrum overlie an enhanced spectrum of the lattice due to the firmly bound species. This enhancement of the lattice spectrum by the chemisorbed adsorbant has been noticed for a number of molecular sieves and zeolites (EGELSTAFF et al., 1976).

For the case of ethylene adsorbed to silver exchanged zeolite (HOWARD et al., 1976) the enhanced lattice modes of the substrate are again seen and in addition at least

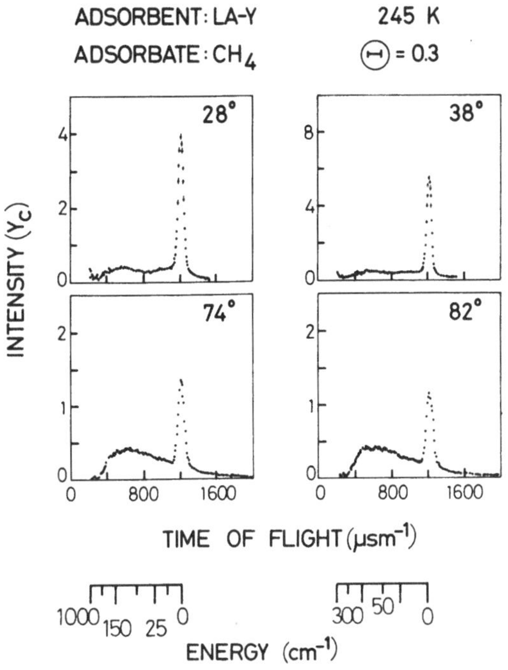


Fig. 4.29 Scattering from methane adsorbed on La-Y zeolite at 176 K (GAMLEN and WHITE, unpublished)

one excitation at low frequency due to the librational motion of the specifically attached ethylene molecule. Experiments on model compounds between ethylene and chloroplatinic acid have allowed these librational modes to be assigned (GHOSH et al., 1975).

### 4.4.2 Hydrogen Adsorbed on Metals

The adsorption enthalpies for hydrogen on nickel and platinum are much greater than those for hydrocarbon molecules on zeolites and, typically, the bond strengths between the hydrogen atoms and the surface may be of the order of 0.5 eV. No case has yet been reported where the hydrogen is undissociated on the metal surface and so the enthalpy of formation of the chemisorbed state must be certainly greater than the bond energy of the hydrogen molecule (approximately 420 kJ $mole^{-1}$). Incoherent neutron scattering has been used to study both hydrogen on nickel (STOCKMEYER et al., 1975) and on platinum black (HOWARD et al., 1976).

For both systems the adsorption of hydrogen produced an enhancement of the density of phonon states for the metal substrate. The tight binding between the hydrogen and its small mass allows the atom to move essentially in phase with the surface metal atoms and, insofar as all these reflect the motions of the bulk material, a true density of states is produced. At frequencies of the order of 0.15 eV (36 THz) modes have been observed for the relative motion of the hydrogen and the surface atom. On nickel the mode was at 140 meV (1120 $cm^{-1}$) and was 56 meV FWHM broad. This mode shifted down in frequency to 100 meV for adsorbed deuterium, as expected for a monatomic bending or stretching frequency. The extreme breadth of this mode indicates either considerable heterogeneity of the adsorption site or very strong lateral interactions between hydrogens adsorbed at neighbouring nickel atoms. Evidence for such strong interactions has recently been presented from studies of the lattice dynamics of hydrogen absorbed in palladium and niobium (ROWE, 1975), where the optic mode associated with the dynamics of the hydrogen atom cannot be thought of as an Einstein oscillator. The dispersion of this excitation in $PdD_{0.63}$ along principal symmetry directions may be as great as 34 meV (272 $cm^{-1}$, 8.16 THz) compared to the lowest frequency of 36 meV.

The dynamics of hydrogen adsorbed on platinum black parallels roughly the presently known situation for hydrogen on nickel and, as yet, no case has been found for adsorption at room temperature where the diffusion coefficient of the hydrogen atoms, measured by the momentum transfer dependence of the quasileastic scattering, exceeds about $10^{-7}$ $cm^2$ $s^{-1}$. The related situation for non-stoichiometric metal hybrides will be discussed in Chap.5.

## 4.5 Intercalation Compounds

Intercalation compounds are formed by a wide variety of molecules and crystals. Many thousands of examples of this behaviour are known but typical are the compounds formed by graphite which can absorb halogens, metal halides and alkali metals to form stoichiometric compounds of the type $C_8X$, $C_{24}X$, etc., or a variety of non-stoichiometric species (BARRER, 1964). The included molecules are arranged in a regular crystallographic relationship to the graphite lattice for the stoichiometric compounds but at high temperature they become mobile and disordered. Another very important class of intercalation compounds is formed by the clay minerals which are again sheet-like structures of similar habit to the micas. The zeolites also belong to this class. The stable structures formed by intercalating crystals are of intrinsic interest but are also extremely relevant to the phenomenon of heterogeneous catalysis since many good catalysts have this property. The intercalation compounds are extremely good systems in which to study physisorption and chemisorption since the surface area available to the intercalating molecules is much greater than can be obtained even with finely divided powders and the intercalated molucules suffer well-defined crystal fields.

In this section we show how incoherent inelastic scattering has been used to study two different types of intercalation compound-tantalum disulphide ammonia and vermiculite-water. In the first of these problems the translational motion of the ammonia molecules is relatively slow compared to the rotational motion and we may study the crystal potential at the ammonia molecule using the librational modes as a probe. In the second case, the effect on molecular translational diffusion of proximity to the alumino-silicate surface may be studied as a function of the water layer thickness since the clay water complexes can be swollen uniformly in the range of water layer thicknesses between one monolayer and approximately 100 layers.

### 4.5.1 Tantalum Disulphide-Ammonia

Like graphite, tantalum disulphide is a layer structure but there exist many allotropic forms. The coordination of the chalcogen about the metal atom can be either trigonal prismatic or octahedral. In the usual notation for hexagonal packing the trigonal prismatic case can be represented by AbA and the octahedral by AbC (where the capital letters refer to the chalcogen and the lower case to the metal atoms). The various crystal structures arise by stacking these trigonal and/or octahedral layers, the simplest when octahedral layers are stacked together leading to the sequence AbCAbC, etc. This polytype is designated 1T. Other polytypes may be designated by a uniform system of nomenclature where the number indicates the number of layers per unit cell and the letter the unit cell symmetry (T = trigonal, H = Hexagonal, R = rhombohedral). A third lower case letter is used when the first two are insufficient to uniquely identify the polytype.

The 2H, $TaS_2$ is the stable polytype at room temperature and has a space group $D_{6h}^4$ - $P6_3/mmc$ with lattice parameters a = 3.315 Å, c = 12.10 Å (JELLINEK, 1962). Numerous structural phase transitions occur between the different polytypes as a function of temperature, and the superconducting properties of some of the polytypes of tantalum disulphide and their intercalation compounds have been investigated (GAMBLE et al., 1971; GEBALLE et al., 1971). For example, the 2H polytype is a superconductor with a critical temperature of 0.8 K. The structure of the 2H and the 1T polytypes are shown in Fig.4.30. Upon intercalation it is found that the electrical anisotropy of the material increases and also the critical temperatures for superconductivity increase. For example, when pyridine molecules are introduced between the layers the critical temperatures are usually in the range 1-5 K. For reasons such as this the electronic and lattice dynamical state of such layer compounds is of interest.

Extensive studies of the crystal structure for a number of intercalation compounds of ammonia and deuteroammonia with tantalum disulphide have been made using neutron diffraction (RIEKEL, 1975). Also studied were the rates of translational diffusion of the ammonia into the crystal by following the time evolution of the crystal diffraction pattern and by using quasielastic neutron scattering at very

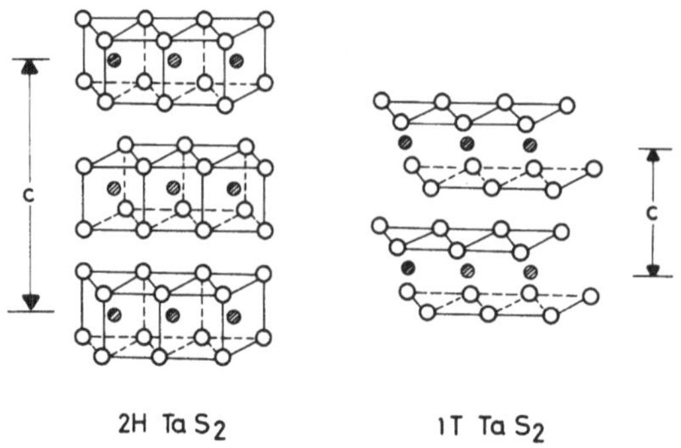


Fig. 4.30 Schematic diagram showing the crystal structures of the 2H and 1T forms of tantalum disulphide

high resolution (RIEKEL and STIRLING, 1976). This work and nuclear magnetic resonance studies now indicate that the ammonia molecule is arranged in the lattice with the nitrogen lone pair electrons approximately parallel to the sheets. In addition to the translational motions. The molecule may rotate and these motions may be studied by a combination of inelastic and quasielastic incoherent neutron scattering, since there is very good contrast between the scattering from the intercalated molecule and the tantalum disulphide. Fig.4.31 shows the inelastic scattering from the stoichiometric tantalum disulphide-$NH_3$ compound as a function of temperature (TAYLOR et al., 1977). At low temperatures a well developed double peak arises at energy transfers of approximately 350 $cm^{-1}$, which becomes extremely broad as the temperature is raised to 300 K. A reasonable assignment of the two peaks is that the one at

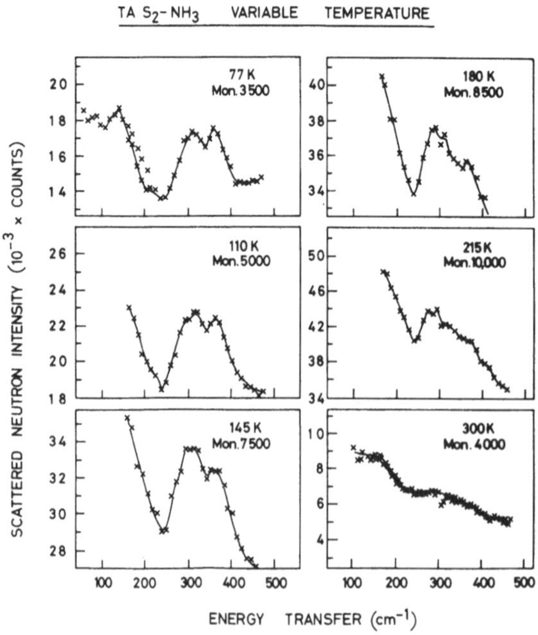


Fig. 4.31 Incoherent neutron inelastic scattering spectra from tantalum disulphide ammonia intercalation compound as a function of temperature showing peaks attributed to the librational motion of the ammonia species

approximately 300 $cm^{-1}$, whose frequency and width are less dependent on temperature, can be associated with the lattice dynamics of the intercalate–possibly a Brillouin zone edge frequency–when neighbouring layers are moving out of phase and where the protons in the ammonia molecule are accordingly displaced. This is, however, strongly mixed with the other mode, centred at approximately 360 $cm^{-1}$ at 77 K, and which may be associated with the torsional motion of the ammonia molecule in the site potential provided by the tantalum disulphide layers. At high temperatures activated motion over the barrier provided by the hindering potential is sufficiently fast and the potential sufficiently anharmonic due to the expansion fo the c axis lattice parameter that the peak becomes very broad.

At the same time as the inelastic scattering spectrum becomes diffuse, there develops a quasielastic scattering which cannot be assigned to translational motion of the ammonia molecules on the basis of the known translational diffusion constant. The spectrum of this rotational quasielastic scattering has not been fully studied due to limitations on the spectrometer resolution. It, and the vibrational spectra, are manifestations of the long time and short time parts of the proton velocity autocorrelation function which, because of the simple crystal structure, may be susceptible to modelling by molecular dynamics calculations (PANGALI, 1975). In this way model potentials for the intercalation compounds may be tested through the means of the included ammonia molecule which acts as a probe.

### 4.5.2 Translational Diffusion in Alumino-Silicate Minerals

The crystal structure of the layer lattice clay minerals, vermiculite and montmorillonite, is shown in Fig.4.32. The structures are built up from silicate tetrahedra and aluminium octahedra, and have a cation exchange capacity which depends upon substitution of lower valent atoms for aluminium in the lattice. The sheets are negatively charged because of this and between the layers there always exist balancing cations. Water is intercalated by most of these minerals but when the surface charge density is high, swelling ceases when only one or two layers of water have been included. For lithium and n-butyl ammonium vermiculites and for sodium montmorillonite the swelling may continue beyond the "crystalline" swelling region (POSNER and QUIRK, 1964) into the "osmotic" swelling region where the layer lattice separation is a simple function of the mass of water available for swelling. Reproducible and reversible swelling can be achieved either by equilibrating the mineral with the vapour from saturated salt solutions at a known temperature or by allowing thermodynamic equilibrium between external, dilute salt solutions and the included water.

Using these methods the incoherent scattering from the included water in vermiculites and montmorillonites has been studied (HUNTER et al., 1971; OLEJNIK and WHITE, 1972) and the dynamics of included probe ions such as the tetramethyl ammonium ion followed in a manner similar to that described for the ammonia in tantalum disulphide

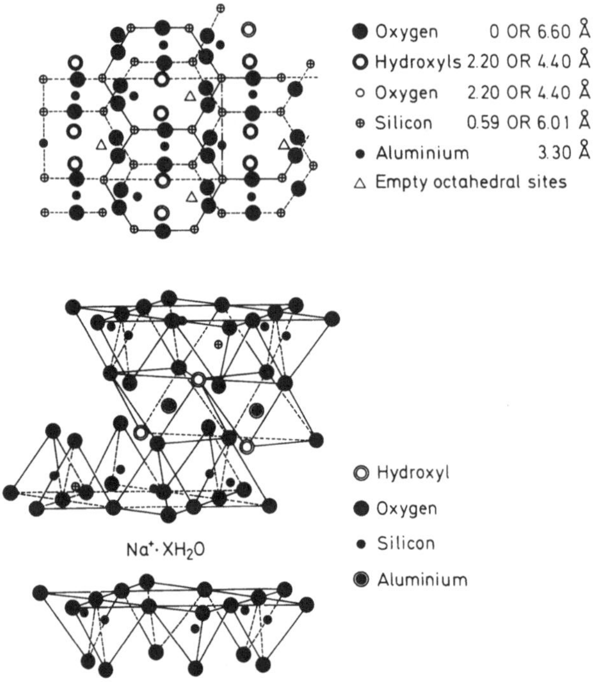


Fig. 4.32 (a) Projection of the layer lattice of montmorillonite onto the basal plane. The heavy small circles represent oxygen or aluminium and the open circles are silicon or dydroxyl (b) Views of the layers with the c axis vertical showing the alumino-silicate lamellae and the interlayer space which contains the balancing cations and water

(OLEJNIK and WHITE, 1973; HAYES and ROSS, 1975). The constrast between the included water and the clay mineral is very good when there are more than two layers of water included, and measurements can be easily made down to one layer. The quasielastic and inelastic scattering spectra from lithium montmorillonite as a function of the c lattice parameter (c = water layer thickness plus 10.9 Å) are shown in Fig.4.33 compared to the scattering spectrum at the same scattering angle for liquid water at 25°C. Pronounced effects due to the water layer thickness are observed in the quasielastic widths and in the near inelastic scattering (0 to 100 $cm^{-1}$). In the dry clay the quasielastic width is of the resolution width and the intensity of the scattering in the near inelastic region is very low—almost at background level. As the number of water layers is increased, the quasielastic width also increases due to diffusion broadening and at the same time the near inelastic spectrum grows.

The quasielastic spectra may be analysed to determine the diffusion coefficient of water between the layers. In the approximation of simple diffusion the quasielastic width ($\Delta E$) is given by $\Delta E = 2\hbar DQ^2$ where D is the molecular translational diffusion

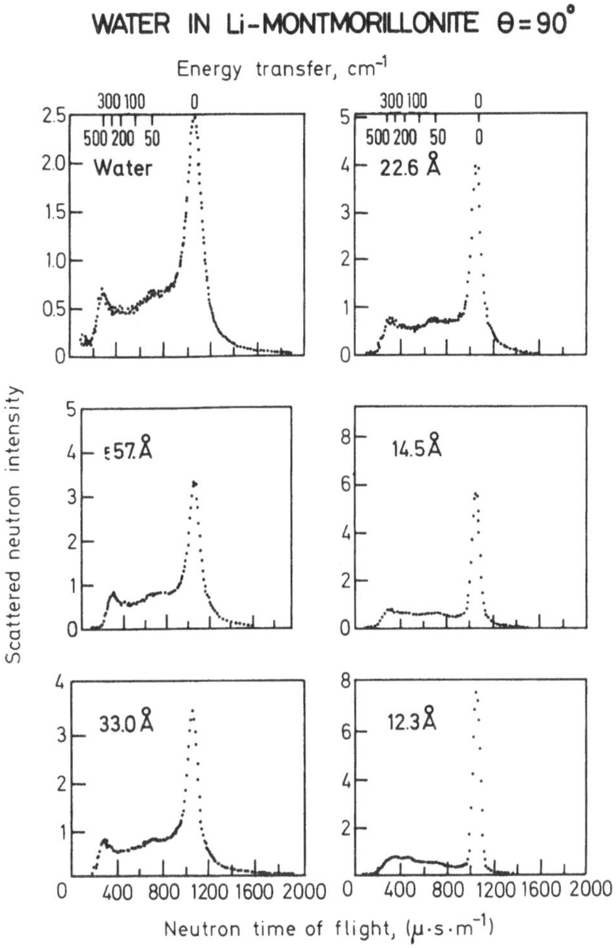


Fig. 4.33 Incoherent neutron quasi-elastic and inelastic scattering from water and from water in lithium montmorillonite as a function of the c axis lattice parameter of the clay (the water spacing is given by subtracting 10.6Å from the c axis length) (OLEJNIK and WHITE, 1972)

coefficient and Q is the momentum transfer. This analysis of the diffusion is shown in Fig.4.34 and it can be seen that the simple diffusion law is obeyed by the more highly swollen samples up to momentum transfer of the order of approximately 1 $Å^{-1}$. The diffusion coefficients are markedly less than the diffusion coefficients for liquid water at the same temperature (dotted line) and for small layer separations the diffusion plots have much lower slopes and show marked curvature and flattening out at high momentum transfers. This is indicative of the smaller diffusion constants and the flattening is consistent with rotational quasielastic scattering which is probably an important contribution to the cross-section for these samples.

The data for a number of different clay minerals with various layer spacings are summarised in Fig.4.35. This figure shows that the diffusion coefficient is log-

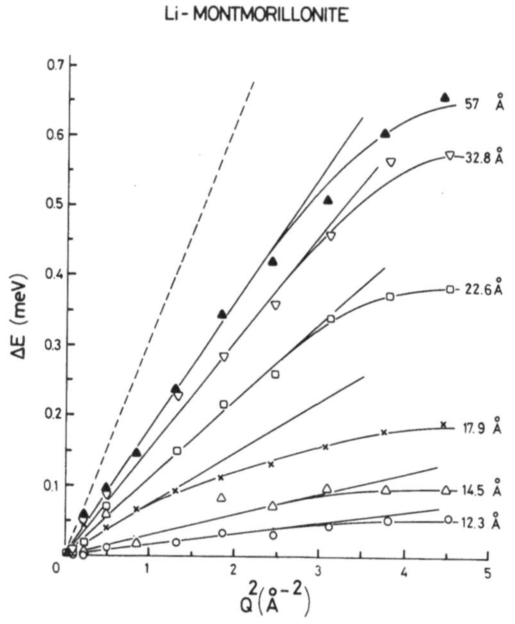


Fig. 4.34 Quasi-elastic energy broadening as a function of the squared momentum transfer for water in lithium montmorillonite clay specimens. The c axis lattice parameters are shown attached to each curve (OLEJNIK and WHITE, 1972)

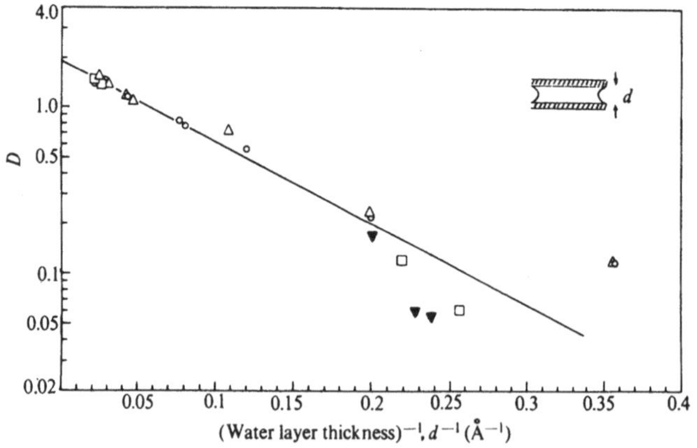


Fig. 4.35 Logarithm self diffusion constant for water (D) in: (△) sodium montmorillonite; (O) lithium montmorillonite; (□) lithium vermiculite, and (▼) sodium vermiculite as a function of inverse water layer thickness. The full line shows a theoretical curve derived from the Kelvin equation using a model for the clay water film shown by the inset (OLEJNIK and WHITE, 1972)

arithmically related to the inverse water layer thickness at water layer thicknesses greater than about 5 Å. The curve extrapolates almost to the diffusion coefficient of bulk water for infinite thickness and we may conclude that the effect of the surface in modifying the translational diffusion has a specific relaxation range of about 15 Å. The straight line relationship can be rationalised (OLEJNIK and WHITE, 1972) on the basis of the Kelvin equation (4.12) which connects the macroscopic change in free energy $\Delta G_i$ for a water layer of thickness, d, to the water surface tension, $\sigma$, and the molar value $V_m$

$$\Delta G_i = \frac{2\sigma V_m}{d} \quad . \tag{4.12}$$

This excess free energy arises because of the proximity of water to the surface and the presence of the counter ions in dilute solution. The change in free energy shows up as a lowering of the water vapour pressure as well as changes in thermodynamically related quantities such as the diffusion coefficient. As a first approximation the Eyring relationship between the free energy and the diffusion constant was used. This gave almost the correct slope for the line in Fig.4.35 when the usual surface tension of water and molecular volume were used, thereby suggesting that other evidence that water suffers long range ordering at alumino-silicate surfaces may be fallacious.

## 4.6 Conclusions

Because of their well-defined structure, ordered molecular crystals are susceptible to accurate theoretical modelling. In the "plastic" phases of molecular crystals and polymers this is progressively less possible and our understanding is more limited to qualitative points, though it seems clear that, by an elaboration of the methods available for simpler crystals, progress is possible. This is also a reasonable expectation for the ordered phases in physisorption and chemisorption where only small modifications of a Schrödinger approach may be needed to formulate effective descriptions of the structure and dynamics. From the basis of our knowledge of ordered materials it is possible now to propose suitable force fields to allow the alternative approach of molecular dynamics to be used for disordered systems. Then the appropriate space-time correlation functions can be generated and it is to be anticipated that the success, already obvious, for simple and molecular liquids will be repeated.

References

Aldred, B.K., Eden, R.C., White, J.W.(1967): Discussions Faraday Soc. No. 43, 169
Alefeld, B., Kollmar, A. (1976): Phys. Lett. 57A, 289
Allen, G. (1973): *Chemical Applications of Thermal Neutron Scattering*, Chap. 5, ed. by Willis, B.T. (Oxford University Press)
Anderson, M.R., White, J.W. (1976): Unpublished data on neutron diffraction from perdeutero and hydrogen polyoxymethylene
Anderson, M.R., Currat, R., Harryman, M.B.M., Steinman, D.K., White, J.W. (1977): Proc. Roy. Soc. (to be published)
Bacon, G.E. (1976): *Neutron Diffraction* (Oxford University Press)
Barrer, R.N. (1964): *Non-stoichiometric compounds*, ed. by Mandlecorn, L. (Wiley-Interscience, New York)
Batley, M., Overs, A.H., Thomas, R.K., White, J.W., Mraw. S., Staveley, L.A.K. (1977): J. Chem. Soc. Faraday II
Bezus, A.G., Dreving, V.P., Kiselev, A.V. (1964): Russ. J.Phys. Chem. 38, 1589
Born, M., Huang, K. (1966): *Dynamical Theory of Crystal Lattices*, (Oxford University Press)
Broude, V.L., Rashba, E.I., Sheka, E.F. (1967): Phys. Stat. Sol. 19, 395
Buckley, C.P., Gray, R.W., McCrum, N.G. (1969): Polymer, Letters 7, 835
Carneiro, K., Nielsen, M., McTague, J.P. (1973): Phys. Rev. Lett. 30, No. 11, 481
Catlow, R., Hayns, M.R. (1972): J. Phys. C5, L237
Catlow, R., Harker, A.H., Hayns. M.R. (1975): J.Chem. Soc. Faraday II, 71, 2
Chatani, J., Uchida, T., Tadakoro, H., Hayashi, K., Nishii, M., Okamura, S. (1968): J. Macromol. Sci. Phys. B2, 567
Cochran, W., Pawley, G.S. (1964): Proc. Roy. Soc. A280, 1
Cowley, R.A. (1968): Rep. Progr. Phys. 31, 123
De Wette, E.W., Rahman, A. (1968): Phys. Rev. 176, 784
Dolling, G., Powell, B.M. (1970): Proc. Roy. Soc. London, A319, 209
Dolling, G., Pawley, G.S., Powell, B.M. (1973): Proc. Roy. Soc. London A.333, 363
Egelstaff, P., Stretton-Downes, J., Rainey, V., White, J.W., (1966): Phys. Rev. Lett. 17, 533
Egelstaff, P., Stretton-Downes, J., White, J.W. (1966): Proc. International Conference on Molecular Sieve Zeolites. J. Appl. Chem.
Elvebredd, I., Cyvin, S.J. (1972): *Molecular Structures and Vibrations*, ed. by S.J. Cyvin Chap. 17 (Elsevier, Amsterdam)
Enomoto, S., Asahina, M. (1964): J. Polymer Sci. A2, 3523
Fanconi, B., Peticolas, W.L. (1971): Biopolymers 10, 2223
Fanconi, B., Small, E.W., Peticolas, W.L. (1971): Biopolymers 10, 1277
Feldkamp, L.A., Venkataraman, G., King, J.S. (1968): *Neutron Inelastic Scattering* (IAEA, Copenhagen) Vol. 2, 159
Gamble, F.R., Osiechi, H.J., Di Salvo, F.J. (1971): J. Chem. Phys. 55, 3525
Gamlen, P.H., White, J.W. (1976): Faraday Transactions of Chemical Society II, No. 2, 446
Geballe, T.H., Menth, A., Di Salvo, F.J., Gamble, F.R. (1971): Phys. Rev. Lett. 27, 314
Genot, B. (1974): J. Chem. Phys. 71, 62
Ghosh, R.E., Waddington, T.C., Wright, C.J. (1973): J. Chem. Soc. Faraday II, 69, 275
Griffing, G.W. (1961): Phys. Rev. 124, 1489
Gupta, V.D., Trevino, S., Boutin, H. (1968): J. Chem. Phys. 48, 3008
Hadley, D.W., Ward, I.M. (1975): Rep. Progr. Phys. 38, 1143
Hadley, D.W., Pinnock, P.R., Ward, I.M. (1969): J. Mater Sci. 4, 152
Hahn, H. (1962): Proc. Inelastic Neutron Scattering, Chalk River (IAEA, Vienna) p. 37,
Hayes, M., Ross, D.K. (1976): Private communication
Haywood, B.C. (1969): Discussions Faraday Soc. 48, 163
Hendra, P.J. (1975): *Raman Spectroscopy*, (Wiley-Interscience, New York)
Herzberg, G. (1950): *Molecular Spectra and Molecular Structure, I Spectra of Diatomic Molecules*, 2nd ed., (D. Van Nostrand and Co. Inc. New York).

Higgins, J.S., Allen, G., Brier, P.N. (1972): Polymer 13, 157
Higgs, P. (1953): Proc. Roy. Soc. London A220, 472
Hochstrasser, R.M., Prasad, P.N. (1972): J. Chem. Phys. 56, 2814
Holliday, L., White, J.W. (1971): Pure Appl. Chem. 26, 545
Howard, J., Waddington, T.C., Wright, C.J. (1975): J. Chem. Soc. Chem. Comm. 775
Howard, J., Waddington, T.C., Wright, C.J. (1976): J. Chem. Phys. 64, 3897
Hulmes, D.J.S., Miller, A., Parry, D.A.D., Piez, K.A., Woodland-Galloway, J. (1973): J. Mol. Biol. 79, 137
Hunter, R.J., Stirling, G.C., White, J.W. (1971): Nature (Physical Sciences) 230, 192
Ito, M., Suzuki, M., Yokoyama, T. (1968): *Excitons, Magnons and Phonons in Molecular Crystals*, ed. by A.B. Zahlan (Cambridge University Press) p. 1
Jellinek, F. (1962): J. Less-Common Metals 4, 9
Keller, A. (1968): Rep. Progr. Phys. 31, 623
Kitagawa, T., Miyazawa, T. (1970): Japan Polymer J. 1, 471
Kitaigorodskii, A.I. (1961) Tetrahedron 14, 230
Kitaigorodskii, A.I. (1966) J. Chem. Phys. 63, 6
Kitaigorodskii, A.I., Dashevskii, V.G. (1968): Tetrahedron 24, 5917
Kjems, J.K., Passell, L., Taub, H., Dash, J.G., Novaco, A.N. (1976): Phys. Rev. 13, 1446
Klein, M.L., Hoover, W.G. (1971): Phys. Rev. 46, 539
Kopelmann, R., Ochs, F.W., Prasad, P.N. (1972): J. Chem. Phys. 57, 5409
Linnett, J.W., Perry, D.L., Egelhoff, W.F., Jr. (1975): Chem. Phys. Lett. 36, 331
Longster, G.F., White, J.W. (1968): J. Chem. Phys. 48, No. 11, 5271
Marshall, W., Lovesey, S. (1971): *Theory of Thermal Neutron Scattering* (Oxford University Press)
Martin, D.H., Marsh, D.I. (1972): J. Phys. C. 5, 2310
Mecke, R., Spieseche, H. (1955): Chem. Ber. 88, 1997
Miller, A., Hulmes, D., Ibel, K., Doyle, B., Haas, J., Timmins, P., White, J.W. (1976): Neutron Diffraction on Collagen. In *Neutron Scattering for the Analysis of Biological Structures*, ed. by B.P. Schoenborn (Brookhaven Natnl. Lab. BNL 50453)
Miyazawa, T. (1967): *Polyamino Acids*, ed. by G.D. Fasman, (Dekker, New York)Chap. 2.
Miyazawa, T., Kitagawa, T. (1964): J. Poly. Sci. B2, 395
Morokuma, K., Fukui, K., Yonezawa, T., Kato, H. (1963): Bull, Chem. Soc. Japan. 36,47
Myers, W.R., Randolph, P.D. (1968): J. Chem. Phys. 49, 1043
Nielsen, M., Ellenson, W.D. (1975): Annual Report Physics Dept., A.E.K., Risø. Report No. 334, p. 44
Odajima, A., Maeda, T. (1966): J. Polymer Sci. C15, 55
Olejnik, S., White, J.W. (1972): Nature Physical Science 236, No. 62, 15
Olejnik, S., White, J.W. (1973): Unpublished work
Pangali, C. (1975): Part II thesis, Oxford University,
Pawley, G.S. (1967): Phys. Stat. Sol. 20, 347
Pawley, G.S. (1972): Phys. Stat. Sol. (b) 49, 475
Pawley, G.S., Cyvin, S.J. (1970): J. Chem. Phys. 52, 4073
Pawley, G.S., Reynolds, P.A., Kjems, J.K., White, J.W. (1971): Solid State Communications (Pergamon Press, New York) p. 1353.
Pendry, J.B., Gard, P. (1975): J. Phys. C. 8, 2048
Piseri, L., Zerbi, G. (1968): J. Chem. Phys. 49, 3840
Pitzer, K.S. (1959): Advan. Chem. Phys. 2, 59
Posner, A.M., Quirk, J.P. (1964): Proc. Roy. Soc. 278A, 35
Pritchard, J. (1976): Experimental Methods in Catalytic Research 3, 281, (Ed. R.B. Anderson and P.T. Davidson, Academic Press).
Rafizadeh, H.A., Yip. S. (1970): J. Chem. Phys. 53, 315
Rashba, E.I. (1966): Zh, Eksperim. Teor. Fiz. 50, 1164
Reynolds, P.A. (1973): Chem. Phys. Lett. 22, 177
Reynolds, P.A., White, J.W. (1969): Discussions Faraday Soc. No. 48, 131
Reynolds, P.A., Kjems, J.K., White, J.W. (1972): J. Chem. Phys. 56, No. 6. 2928
Reynolds, P.A., Kjems, J.K., White, J.W. (1974): J. Chem. Phys. 60, 824
Reynolds, P.A., Twisleton, J., White, J.W. (1977): Proc. Roy. Soc. (to be published)
Riekel, C. (1975): A Neutron Diffraction Study on the Intercalation of Ammonia into Tantalum Disulphide, ed. by J.S. Higgins, Chemistry Information Meeting, Grenoble, 1975, p. 75
Riekel, C., Schöllhorn, R. (1976): Mat. Res. Bull., 11, 365

Riekel, C., Fender, B.E.F., Stirling, G. (1976): Annual Report Inst. Laue-Langevin, Part II, 370, 1977
Roberts, M.W., Atkinson, S.J., Brundue, C.R. (1974): Chem. Phys. Lett. 24, 175
Rowe, J.M. : Private communication
Safford, G.J., Naumann, A.W. (1967): Advan. Polymer Sci. 5, 1
Safford, G.J., Naumann, A.W. (1967): Advan. Polymer Sci. 5, 1
Scherer, J.R. (1963): Spectrochim Acta 19, 1739
Scherer, J.R. (1967): Spectrochim Acta A23, 1489
Scherer, J.R. (1968): Spectrochim Acta A24, 747
Sears, V.F. (1966): Canad. J. Phys. 44, 1279, 1299
Sheka, E.F. (1975): Mol. Cryst. Liq. Cryst. 30, 239
Sheka, E.F., Dolganov, V.K. (1973): Fiz. Tverd. Tela 15, No. 3. 836
Sheppard, N., Little, L.H., Yates, D.J.C. (1960): Proc. Roy. Soc. (London) A259, 242
Shimanouchi, T., Tasumi, M., Miyazawa, T. (1962): J. Mol. Spectr. 9, 261
Shull, C.G. (1972): Coherent scattering amplitudes of bound atoms. *Coherent Neutron Scattering Amplitudes*
Stockmeyer, R. (1976): Report at Meeting on Neutron Scattering, Gatlinburg
Stockmeyer, R., Conrad, H.M., Renouprez, A., Fouilloux, P. (1975): Surface Science 49, 549
Suzanne, J., Coulomb, J.P., Bienfait, M. (1973): Surface Sci. 40, 414
Tasumi, M., Krimm, S. (1967): J. Chem. Phys. 46, 755
Taub, H., Passell, L., Kjems, J.K., Carneiro, K., Mctague, J.P., Dash, J.G. (1975): Phys. Rev. Lett. 34, No. 11, 654
Taylor, A.D., Thomas, R.K., Trewern, T., White, J.W. (1977): J. Chem. Soc. Faraday II, (to be published)
Teare, P.W. (1959): Acta Cryst. 12, 294
Thomas, J.M., Ikemoto, I., Ishii, K., Kuroda, H. (1974): Chem. Phys. Lett. 28, 55
Thomas, M.W., Ghosh, R.E. (1975): Mol. Phys. 29, No. 5, 1489
Thomy, A., Duval, X. (1970): J. Chim. Physique 67, 1101
Todireanu, S., Hautecler, S. (1972): Phys. Rev. Lett. 43A, 189
Trevino, S., Boutin, H. (1968): J. Macromol. Sci. A1, 723
Twisleton, J.F. (1973): D. Phil, Thesis, Oxford University Press
Twisleton, J.F., White, J.W. (1972): *Neutron Inelastic Scattering* (IAEA Grenoble) p. 301
Twisleton, J.F., White, J.W. (1972): Polymer 13, 40
Uchida, T., Tadakoro, H. (1967): J. Polymer Sci. A2, 5, 63
Urey, H.C., Bradley, C.A. (1930): Phys. Rev. 38, 1969
Venables, J.A. (1975): Phys. Bull. 26, 359
Venkataraman, G., Sahni, V.C. (1970): Rev. Mod. Phys. 42, 409
Warren, B.E. (1941): Phys. Rev. 59, 693
White, J.W. (1972): *Proc. Grenoble conference on Neutron Scattering* (IAEA Vienna) p. 315
White, J. (1972): *Polymer Science*, ed. by A.D. Jenkins, Chap. 27, 1744.
White, J.W. (1972): *Chemical Applications of Thermal Neutron Scattering*, ed. by B.T. Willis (Oxford University Press) Chap. 3, pp. 31-48
White, J.W. (1976): *Neutron Scattering for the Analysis of Biological Structures*, ed. by B.P. Schoenborn (Brookhaven Natnl. Lab. BNL 50453) p. VI - 3
Williams, D.E. (1966): J. Chem. Phys. 45, 3770
Williams, D.E. (1976): J. Chem. Phys. 47, 4680
Wilse-Robinson, G.W., Colson, S.D., Kopelman, R. (1967): J. Chem. Phys. 47, 27 and Colson, S.D., Hanson, D.M., Kopelman, R., Robinson, G.W. (1968): J. Chem. Phys. 48, 2215 (Supplementary References)
Young, R.J., Bowden, P.B., Ritchie, J.M., Rider, J.G. (1973): J. Mat. Sci. 8, 23

# 5. Molecular Rotations and Diffusion in Solids, in Particular Hydrogen in Metals

T. Springer

With 21 Figures

This chapter deals mainly with problems related to physical chemistry; namely, Sec.5.1 with diffusion in solids, in particular of hydrogen in metals, Sec.5.2 with rotational motions of hydrogeneous molecules in crystals, and Sec.5.3 with rotation and diffusion in liquids and in liquid crystals. A common aspect in these problems is that such motions are responsible for energy transfers which are small compared to $k_BT$, or to the Debye energy $k_B\theta$, and typical neutron spectra to be discussed appear in energy ranges of a few $10^{-6}$ to a few $10^{-3}$ eV, or $10^{-5}$ to $10^{-2}cm^{-1}$. We restrict ourselves essentially to the discussion of scattering from the protons which is predominantly incoherent ($\sigma_{inc}$ = 79.7 barns; $\sigma_{coh}$ = 1.79 barns). This means that we deal only with the motions, or excitations, of individual protons or protonic groups, ignoring collective excitations and interference.

Fig.5.1 and 5.2 show the qualitative shape of the incoherent neutron spectra for rotating and diffusing molecules or protons. First, let us consider a proton in a molecule which is bound in a *solid* crystal, and which rotates randomly about its equilibrium position. In this case the spectrum consists of a purely elastic line $\delta(\omega)$, and a *quasielastic spectrum* centered at energy transfer *$\hbar\omega = 0$, whose width is related to the characteristic time of the rotation. In the case of periodic rotations, for instance nearly free rotations or tunneling motions, side-peaks will appear, as also shown in Fig.5.1. The side-peaks may have a finite width due to the lifetime of the corresponding quantum state, whereas the central line has always a zero width whatever the motion of the molecule is (this point has been subject of controversy, but it is settled now; see LARSSON, 1973). It is easy to recognize that the purely elastic line has the same physical origin as the Mössbauer line in gamma ray interactions. It comes from processes which leave the lattice without excitations, and which transfer momentum to the lattice as a whole. In terms of the classical autocorrelation function $G_s(\underline{r},t)$ for the scattering proton, this can be understood as follows. We write

$$G_s(\underline{r},t) = G_\infty(\underline{r}) + G'(\underline{r},t) \quad . \qquad (5.1)$$

*$\hbar = h/2\pi$ (normalized Planck's constant)

$G'(\underline{r},t)$ decays to zero for $t \to \infty$ with the correlation time of the random motion. The asymptotic probability distribution $G_\infty(\underline{r}) = \int \langle n(\underline{r}'-\underline{r})n(\underline{r}')\rangle d\underline{r}'$ is finite if the scattering proton is restricted to a finite volume in space[1], e.g., in the case of molecular rotations the proton cannot leave the spherical shell over which it rotates. The Fourier transform of $G_\infty$ leads to the $\delta$-function discussed before. Recognition of the existence of this line, as well as its careful experimental separation, has brought considerable progress in the investigations of random molecular rotations.

If the molecule, or the proton, undergoes *translational diffusion* the $\delta$-function disappears and we obtain a distribution as in Fig.5.2a. The width, due to the translational motion, is related to the self-diffusion constant and the characteristic time of the diffusive step. For molecular liquids the separation of the superimposed rotational and translational components of the quasielastic spectrum is a difficult task which cannot always be solved. This aspect will be discussed in more detail in Sec.5.3.

In the subsequent sections we will assume that the motion of the scattering particle can be described in terms of classical physics, with exception of the tunneling rotations treated in Subsec.5.2.4. We consider $G_s(\underline{r},t)$ to be a real function, which has the meaning explained in Subsec.1.3.4. The corresponding scattering law is symmetric in energy, in contrast to the requirement of detailed balance. It can be expected that the classical description is a good approximation in the region of energy transfers small compared to $k_BT$, or for times

$$t \gtrsim \hbar/k_BT \simeq 3.10^{-13}\ \text{s} \quad . \tag{5.2}$$

With respect to *neutron spectroscopy* at small energy transfers, important progress was achieved a few years ago with the so-called backscattering spectrometer. The method is based on the application of Bragg reflection for energy selection, using a Bragg angle of $\theta = 90^\circ$. Under these circumstances, according to Bragg's law, the neutron wavelength depends on $\theta$ only in second order. Consequently, the wavelength spread due to the finite beam collimation is extremely small, and the resolution is determined by second order effects from the width of the Bragg angle, and by extinction. Spectrometers of this type are in operation at the Jülich and the Grenoble reactor (ALEFELD et al., 1969; HEIDEMANN, 1974), both having energy resolutions between 0.3 and 1 $\mu$eV. The scanning of the spectra is carried out by a Doppler motion of the monochromator crystal. The energy transfer which can be covered in this way is of the order $\pm$ 20 $\mu$eV. The resolution of such spectrometers is about a hundred times better than the best values obtained by the other crystal spectrometers.

---

[1] $n(\underline{r})$ is the asymptotic density distribution of the proton $n(\underline{r},t \to \infty)$. The bracket means the thermal average (see (1.46)). The Fourier transform of $G_\infty$ is often called the "incoherent structure factor", which gives the elastic intensity.

The intensity advantage of the statistical chopper has been discussed extensively for many years (e.g., HOSSFELD et al., 1970). This method has been applied successfully at the Argonne National Laboratory for a number of quasielastic scattering problems to be discussed in the subsequent sections (PRICE, 1970). Another technique to be applied for high resolution spectroscopy is, perhaps, the spin-echo method (MEZEI, 1972) being developed at the Grenoble reactor.

## 5.1 Diffusion in Solids, in Particular Hydrogen in Metals

### 5.1.1 General Aspects

The fast diffusion of hydrogen in certain metals is a subject of great theoretical and practical interest. This diffusion has been investigated by many methods; namely, 1) by permeation experiments, 2) by the Gorski-effect (the use of anelastic relaxation caused by diffusion), 3) by the field-gradient spin-echo method, 4) by proton spin-lattice relaxation measurements, yielding $\tau_c$, the correlation time of the diffusing proton, and finally, 5) by incoherent neutron scattering which determines the self-correlation function of the diffusing proton.

As will be shown, neutron experiments at small scattering vectors Q (say $< 1$ Å$^{-1}$) yield the self-diffusion constant D. Consequently, such measurements are complementary to the "macroscopic" methods as long as these deal with the motion of an *individual* proton, as in the case of 3). As will be discussed in Subsec.5.1.3, the neutron results are not necessarily identical with 2). From experiments on single crystals at *larger* Q (2 - 4 Å$^{-1}$) information on the *single step* of the diffusion is obtained, e.g., the jump vectors, and the space distribution of the proton over the interstitial sites.

By means of conventional spectrometers (resolution $\delta E \simeq 10^{-4}$ eV) the self-diffusion can be easily studied for D in the region of $10^{-5}$ to $10^{-4}$ cm$^{-2}$/s, and for characteristic times of the diffusion of $10^{-11}$ to $10^{-12}$ s. Backscattering spectrometers have extended this region to values of the order $10^{-7}$ cm$^2$/s and $10^{-9}$ s. Limits are imposed on the hydrogen concentration: For vanadium ($\sigma_{inc}$ high) and for tantalum ($\sigma_a$ high) it is difficult to work at concentrations below $\sim$10 atomic %. Due to its very small incoherent cross-section and its weak absorption, hydrogen in niobium can be investigated at concentrations as small as a few 0.1 atomic % ($\sigma_{inc}$ = 6.3 mbarn, SCHMATZ, 1975).

So far, most of the neutron scattering experiments were carried out on hydrogen in palladium, and in the disordered phases of hydrogen in the bcc metals Nb, Ta, and V (see Table 5.1). These hydrogen-metal experiments will be described in detail, and a few remarks will be added on the quasielastic scattering from diffusion in other solids. (For a comprehensive review comparing the data from the various methods for the investigation of hydrogen diffusion see VÖLKL and ALEFELD, 1975).

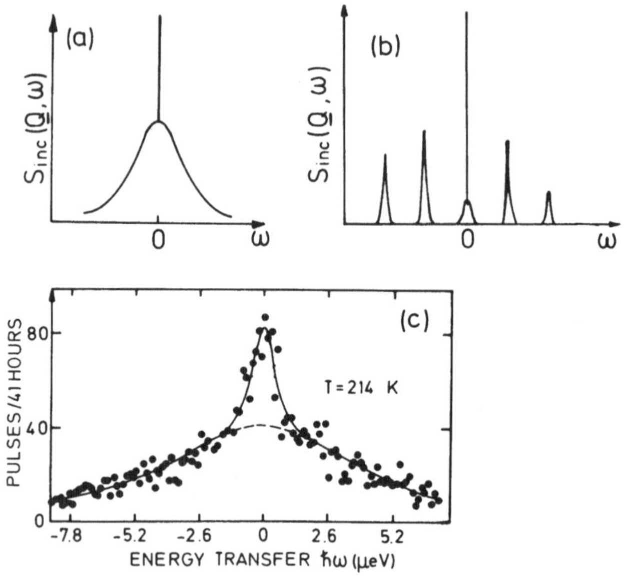


Fig. 5.1a-c Incoherent scattering law as a function of energy transfer $\hbar\omega$ for a scattering proton in a molecule which is bound in a solid, performing (a) random rotations,(b) quantum rotations. In any case a purely elastic line appears at $\hbar\omega = 0$.(c) Experimental spectrum corresponding to case (a), showing the influence of resolution broadening (1μeV): Scattering on adamantane (ALEFELD and STOCKMEYER, 1973). Experimental examples for case (b) are shown in Figs.5.16/5.17

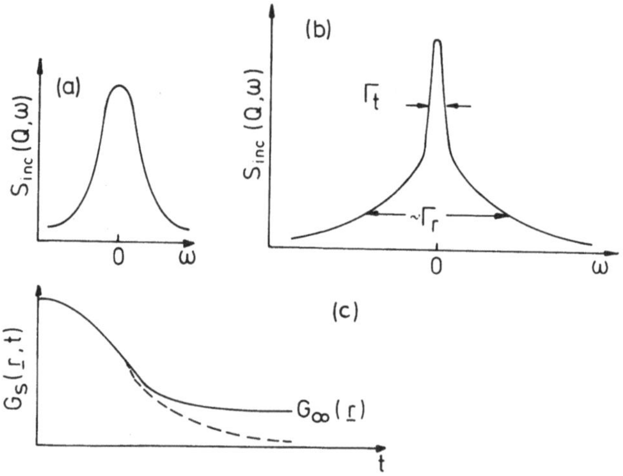


Fig. 5.2a-c Incoherent scattering law (a) for a single diffusing proton, and (b) for a proton in a rotating molecule which diffuses in a liquid. A case is shown where rotational and translational components of the quasielastic spectrum can be separated. (c) Self-correlation function $G_s(\underline{r},t)$ for case (a) in Fig.5.1 (solid line) and for case (b) of this figure (dashed)

Table 5.1 Typical parameters for hydrogen diffusion in metals as determined by quasielastic scattering

| Sample | T (K) | D($10^{-5}$) ($cm^2/s$) | $\tau$ ($10^{-12}s$) | $E^{(act)}$ | $\hbar\omega_{loc}$ | $<u^2>$ | References for $<u^2>$, D, and $\tau$ |
|---|---|---|---|---|---|---|---|
| $\alpha$-$PdH_{0.04}$ | 704 | | | 0.23 eV[a)] | 0.066 eV[b)] | 0.13 Å$^2$ | SKÖLD and NELIN, 1967 |
| | 620 | 3.5 | 2.8 | 0.16 eV[b)] 0.3 eV[e)] | | 0.12 Å$^2$ | CARLILE and ROSS, 1974 |
| $\beta$-$PdH_{\sim 0.6}$ | 470 | 2.8 | 4.7 | | | 0.07 Å$^2$ f) | BEG and ROSS, 1970 |
| $NbH_{0.033}$ | 580 | 2.7 | 2.0 | 0.10 eV[a)] | 0.11 and 0.18 eV[c)] | 0.18 Å$^2$ | GISSLER et al., 1970 |
| $\alpha$-$VH_{0.2}$ | 480 | 8.0 | 0.7 | 0.05 eV[a)] | 0.12 0.17 eV[c)] (very broad) | 0.18 Å$^2$ | ROWE et al., 1971 |
| $\alpha$-$TaH_{0.15}$ | 520 | 1.7 | 2.4 | 0.14 eV[a)] | 0.12 and 0.17 eV | 0.024 Å$^2$ | RUSH et al., 1973 |

a) VÖLKL and ALEFELD, 1975
b) SKÖLD and NELIN, 1967
c) VERDAN et al., 1968
d) RUSH et al., 1973
e) NELIN and SKÖLD (1975) by combining various neutron scattering experiments at different concentrations
f) From recent work of NELIN and SKÖLD (1975) on $\beta$-$PdH_{0.48}$ at 533 K one can calculate $<u^2> \simeq 0.65$ Å$^2$;
g) RUSH and FLOTOW (1968)

$<u^2>$ = mean square amplitude of hydrogen oscillation; from $I_{qe}$ at small Q.
$\tau$ = mean rest time; D = self-diffusion constant; $E_{act}$ = activation energy of $\tau$ or D.
$\omega_{loc}$ = frequency of localized mode (center of line).

[1]Note added in proof: The anomalous values of $< u^2 >$ follow from the large slope of $\ln I_{qe}$ vs. $Q^2$. It is not yet clear if, at least part of the slope, is caused by the difference between $I_{qe}$ and $I'_{qe}$ which increases with increasing Q (see Eqs. 5.17 and 5.18) (LOTTNER, pers. comm.).

## 5.1.2 Experiments on Hydrogen in Pd, a fcc-Lattice

In a fcc lattice the scattering law is particularly simple since, for the hydrogen atoms on tetrahedral or octahedral sites, the interstitial positions form a simple Bravais lattice (Fig.5.3). In this case, the self-correlation function $G_s(\underline{r},t)$ can be calculated from a rate equation

$$\partial P(\underline{r},t)/\partial t = -(1/\tau)P(\underline{r},t) + (1/m\tau)\sum_{k=1}^{m} P(\underline{r} + \underline{s}_k,t) \tag{5.3}$$

where

$$G_s(\underline{r},t) \equiv P(\underline{r},t) \quad \text{if} \quad P(\underline{r},0) = \delta(\underline{r}) \quad . \tag{5.4}$$

$\underline{s}_k$ (k = 1,.. m) is the set of jump vectors connecting an interstitial site $\underline{r}$ with its m neighbour sites. $\tau$ is the average time between successive jumps which are

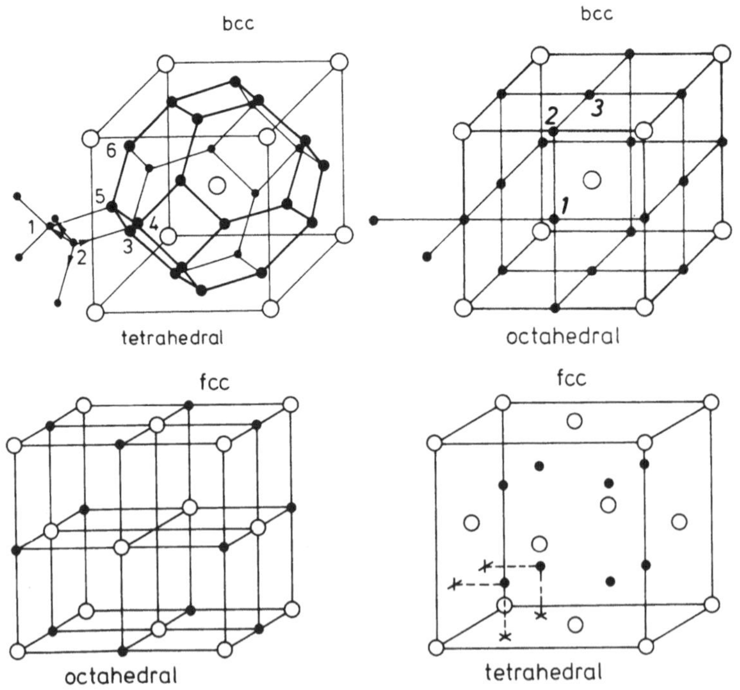


Fig. 5.3 Tetrahedral and octahedral interstitial sites in fcc and bcc host lattices (full circles). Open circles: host lattice atoms. The non-equivalent interstitial sites for the bcc lattice are labeled with 1,2,3 (octahedral) and 1, .. 6 (tetrahedral)

assumed to be uncorrelated. Furthermore, the assumption is made that $\tau$ is large compared to the time needed for the jump from site to site, $\tau_j$, which is of the order of $10^{-13}$ s (5.11). Introducing the intermediate scattering law for diffusion in the classical form

$$F_s^D(\underline{Q},t) = \exp[-tf(\underline{Q})] \tag{5.5}$$

into (1.48) one obtains

$$f(\underline{Q}) = (1/m\tau) \sum_{k=1}^{m} [1 - \exp(-i\underline{Q} \cdot \underline{s}_k)] \quad . \tag{5.6}$$

The corresponding incoherent scattering law is the time Fourier transform of the intermediate scattering law, namely

$$S_s^D(\underline{Q},\omega) = \frac{f(\underline{Q})/\pi}{\omega^2 + f^2(\underline{Q})} \quad , \tag{5.7}$$

which is a normalized Lorentzian with a full width at half maximum $\Gamma = 2f(\underline{Q})$. These relations were derived first by CHUDLEY and ELLIOTT (1961). For $Q^2 s_k^2 \ll 1$, (5.6) yields, in the case of a primitive cubic lattice,

$$\Gamma = 2Q^2(s^2/6\tau) = 2Q^2D \tag{5.8}$$

where s is the jump distance, and D is the self-diffusion constant of the hydrogen. The relation $\Gamma = 2Q^2D$ holds for *any* kind of lattice. Obviously, the quasielastic width in (5.6) is periodic in reciprocal space. In particular, the width is zero if $\underline{Q}$ approaches a vector of the reciprocal lattice of the interstitials. Consequently, the height of the quasielastic line would approach infinity at such $\underline{Q}$-values. This can be understood physically in terms of a Bragg *diffraction* of the neutron wave during its coherency time, by a *single* diffusing proton which is distributed over a finite region of the interstitial lattice; the coherency time is given by $\hbar/\Delta E$, where $\Delta E$ is the energy resolution of the spectrometer applied for the observation of the diffusing proton (SPRINGER, 1972).

Experiments on *$PdH_x$ polycrystals* in the fcc $\alpha$-phase by SKÖLD and NELIN (1967) showed, for the first time, that the quasielastic width follows the predictions of the Chudley-Elliott-model described before. A decision was obtained in favour of single-jump diffusion over an octahedral, rather than a tetrahedral, interstitial lattice, in agreement with neutron diffraction work. Figures 5.4 and 5.5 show typical quasielastic spectra and widths from more recent work on single crystals by ROWE et al. (1972) on $PdH_{0.03}$ for different orientations of $\underline{Q}$ which again confirm the agreement with the octahedral model. The experiments were carried out on a time-of-flight

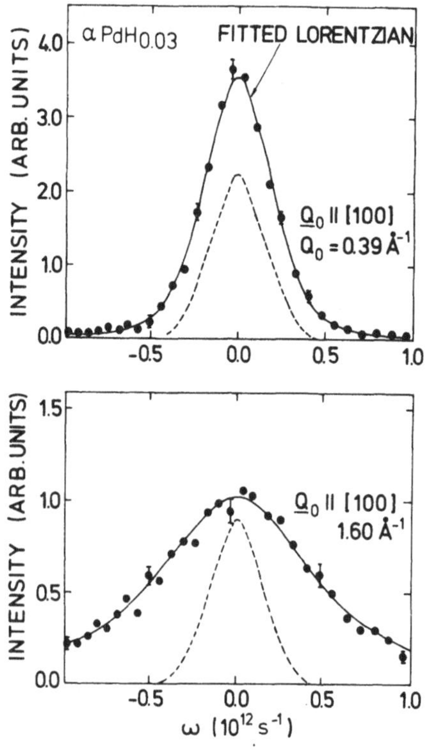


Fig. 5.4 Typical quasielastic spectra for incoherent scattering from protons in $\alpha$-$PdH_{0.03}$ at 350 °C vs. energy transfer $\hbar\omega$. Dashed: experimental resolution curve. Experiments carried out at a fixed scattering angle on a time-of-flight spectrometer $\underline{Q}_0$ = scattering vector for $\hbar\omega = 0$. (1 $ps^{-1} \hat{=} 0.66 \cdot 10^{-3}$ eV). (ROWE, RUSH, DE GRAAFF, and FERGUSSON, 1972)

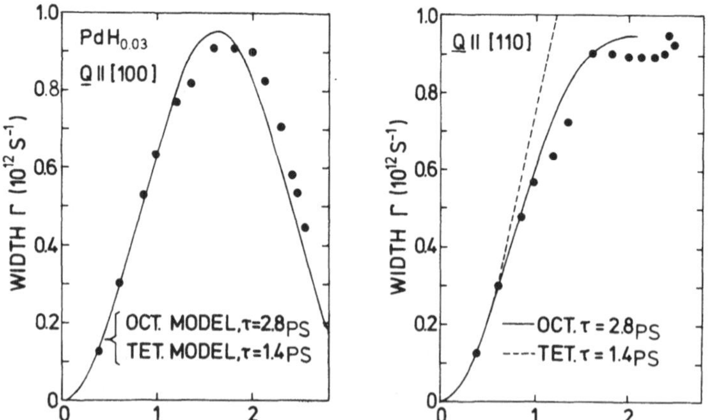


Fig. 5.5 Full width at half maximum $\Gamma$, vs. Q from quasielastic spectra on $\alpha$-$PdH_{0.03}$ (see Fig.5.4). Solid and dashed line from CHUDLEY-ELLIOTT-model with jumps between octahedral and tetrahedral sites, respectively (ROWE, RUSH, DE GRAAF, and FERGUSSON, 1972)

spectrometer with 250 μeV resolution (FWHM), and a very good fit was obtained between the measured spectra and a single Lorentzian as predicted by the Chudley-Elliott theory. From the slope of the width $\Gamma$ vs. $Q^2$ the self-diffusion constant D can be determined. Further experiments of this kind on $\alpha PdH_x$ were reported by CARLILE and ROSS (1974).

Quasielastic scattering on the β-phase of $PdH_x$ (BEG and ROSS, 1970) was compared with magnetic field-gradient spin echo experiments, yielding D, and with nmr spin-lattice relaxation experiments, yielding the correlation time $\tau_c$ for the magnetic fluctuations produced at the diffusing proton (SEYMOUR et al., 1975). For a given model (octahedral jumps) D can be calculated from $\tau_c$. Internal consistency was found between the two magnetic-resonance methods, but disagreement exists with the neutron data, and also with permeation experiments. Recent experiments by NELIN and SKÖLD (1975) on $\beta$-$PdH_x$ show quite convincingly that quasielastic scattering can be described by an octahedral jump model (in disagreement with BEG and ROSS, 1970). In their data evaluation the authors take into account the influence of a finite jump time (see GISSLER and STUMP, 1973); however, the results were not sensitive enough to yield such details of the diffusive step.

### 5.1.3 Hydrogen in the bcc Metals Ta, Nb, and V

In the unit cell of a bcc host lattice there exist $\mu = 3$ non-equivalent interstitial sites of the octahedral type, and 6 sites with tetrahedral geometry (Fig.5.3). Consequently (5.3) has to be replaced by a set of $\mu$ coupled equations, namely

$$\partial P_i(\underline{r}_{\underline{\ell}},t)/\partial t = -(1/\tau)P_i(\underline{r}_{\underline{\ell}},t) + (1/m\tau)\sum_{j,k} P_j(\underline{r}_{\underline{\ell}} + \underline{s}_{ij,k}t) \tag{5.9}$$

where we have introduced sets of jump vectors $\underline{s}_{ij,k}$ which connect a site of type i $(i = 1,..\mu)$ with a neighbouring site of type $j(j = 1,..\mu)$, k labels these jump vectors. The function $P_i(\underline{r}_{\underline{\ell}},t)$ is the probability to find the hydrogen at time t at a site of type i in a cell $\underline{\ell}$. For simplicity, we asumme that the jump probability $1/\tau$ is equal for all equivalent sites. The sum in (5.9) has to be taken over all neighbour sites of the site i in cell $\underline{\ell}$. The initial conditions read here

$$P_i^{(\nu)}(\underline{r}_{\underline{\ell}},0) = \delta_{i\nu}\delta_{\ell\sigma} \quad . \tag{5.10}$$

Consequently, the hydrogen is for $t = 0$ at a site of type $\nu$ within the cell at the origin $\underline{\ell} = 0$. The resulting $P_i^{(\nu)}$ form a set of self-correlation functions $G_{\nu i}(\underline{r}_{\underline{\ell}},t) \equiv P_i^{(\nu)}(\underline{r}_{\underline{\ell}},t)$. They describe the probability of finding the proton at a time $\bar{t}$ at a site i $\bar{\text{in}}$ cell $\underline{\ell}$, if it has started at a site $\nu$ within cell 0 at $t = 0$. The corresponding scattering law consists of a superposition of (at most $\mu$) Lorentzians whose weights and widths are functions of $\underline{Q}$. The mathematical problem of solving (5.9) has been treated by several authors, e.g., ROWE et al. (1971), KEHR (1976).

Instead of using rate equations GISSLER and ROTHER (1970) have developed a random walk method which is very flexible and which may be applied if the single step of diffusion is complicated, e.g., for multiple jumps or for correlations.

We discuss now more recent experiments on *hydrogen in tantalum* ($TaH_{0,02}$) single crystals by ROWE et al. (1974) using a correlation chopper spectrometer (PRICE and SKÖLD, 1970). A fit of the quasielastic spectra with the calculated scattering law was not successful, neither for a tetrahedral nor for an octahedral site model. Also the inclusion of second neighbour jumps or of mixed-site occupancy (de GRAAF et al., 1972) did not improve the results. The best fit of the quasielastic spectra was obtained by a *single* Lorentzian (in contrast to expectation for the bcc lattice), and the resulting half-widths were compared with the half-width as found from the theoretical (multi-Lorentzian) scattering law. Fig.5.6 demonstrates the striking discrepancies; in particular, the width as a function of the $\underline{Q}$-orientation shows a nearly isotropic behaviour, in contrast to theory.

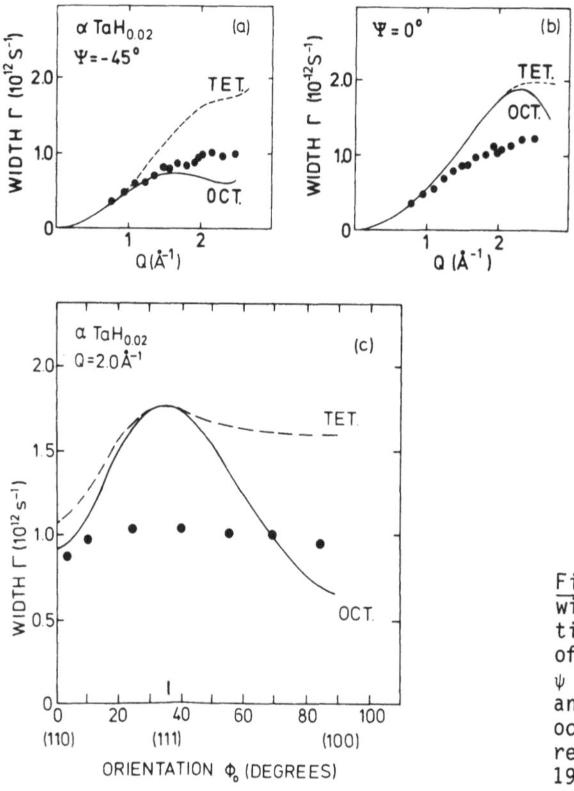


Fig. 5.6 (a,b) Measured quasielastic width $\Gamma$ (Q) for $\alpha$-$TaH_{0.02}$ as a function of scattering vector Q, and (c) of the angle $\Phi_0$ between $\underline{Q}$ and [110]; $\psi$ = angle between $\underline{k}_0$ and [001]. Solid and dashed curves: calculated for octahedral and tetrahedral jump model, respectively (ROWE, RUSH, and FLOTOW, 1974).

On the other hand, for *hydrogen in niobium* single crystals ($NbH_{0,07}$) the expected anisotropy of the quasielastic width was observed (STUMP et al., 1972). For larger Q the results are close to those expected for the tetrahedral site model, but deviations from theory occurred, in particular for $\underline{Q}$ || [100] where the width is smaller than expected.

Experiments on *hydrogen in vanadium* (powder) samples, mainly in the bcc α-phase, were carried out by ROWE et al. (1971), and were extended to extraordinarily large Q-values ( $\leq 4$ Å$^{-1}$) which cannot be obtained with the conventional cold neutron spectrometers (de GRAAF et al., 1972). Experiments on this substance are difficult because of the large incoherent scattering of the host lattice which produces a strong elastic line. For $Q \lesssim 3$ Å$^{-1}$ the results for the quasielastic width show that jumps between near-neighbour tetrahedral sites are the dominating process. For larger Q the width increases more steeply as it can be justified by any simple jump model (actually $\Gamma(\underline{Q})$ should pass over a maximum as Q increases). This behaviour may be caused by the extremely short rest time in the system which can be as small as $2 \cdot 10^{-13}$s. This is comparable with the time needed for a jump between two interstitial sites which is roughly estimated to be

$$\tau_j = s(2M_H/k_BT)^{1/2} \tag{5.11}$$

(s = jump distance). Under these circumstances the Chudley-Elliott model will fail and one expects contributions in the quasielastic spectrum caused by the jump motion itself. These behave approximately like the spectrum of a free gas, where $\Gamma \propto Q\,(k_BT)^{1/2}$, rather than $\Gamma = 2\,Q^2\,D$.

For bcc α-$NbH_x$ a systematic comparison was carried out between measurements of the self-diffusion constant from Γ vs. $Q^2$ at small Q (GISSLER et al., 1970) and from "macroscopic" measurements (G. ALEFELD et al., 1970; VÖLKL, 1972). In these measurements the diffusion constant is determined from the mechanical relaxation of the sample due to hydrogen diffusion, after an external stress has been applied (Gorski effect). For small concentrations, good agreement was found between both methods within their error limits (~ 20%). No agreement was obtained for 33 at. % hydrogen loading and at $T_c$ = 440 K; here the $NbH_x$ system undergoes a second order phase transition which leads from a (α,α') coexistency region below $T_c$ to a homogenous phase above $T_c$. Close to $T_c$ the diffusion constant $D^{(G)}$ from the Gorski effect goes to zero. Obviously, in this method, the decay of a concentration gradient is measured where the chemical potential μ of the interacting protons enters, and the corresponding diffusion coefficient is then (VÖLKL, 1972)

$$D^{(G)} = (\partial\mu/\partial c)\,c\,m_H(c,T) \tag{5.12}$$

where $(\partial\mu/\partial c)$ is the derivative of $\mu$ with respect to the hydrogen concentration c; this quantity approaches zero as $(T - T_c)$. $m_H$ is the mobility of the dissolved protons. On the other hand, the self-diffusion constant from incoherent neutron scattering is related to the motion of an *individual* proton, since coherency between the scattered waves from different protons can be ignored practically.[2] The mutual interaction of the protons is not the driving force so that $\partial\mu/\partial c$ has the value for a dilute system $(k_BT/c)$, and one obtains

$$D = D^{(G)} \frac{(\partial\mu/\partial c)_{c\to 0}}{(\partial\mu/\partial c)_c} \quad . \tag{5.13}$$

For the critical concentration it has been demonstrated that D vs. T, as calculated by (5.13) with $D^{(G)}$ and $(\partial\mu/\partial c)$ from Gorski-effect experiments, agrees very well with D from quasielastic scattering. No critical behaviour of D, and therefore of $m_H$, was observed near $T_c$.

Finally, a geometrical difference between the fcc and the bcc case should be pointed out. From the arguments given in Subsec.5.1.1 it should be noticed that for a bcc lattice the half-width is zero at a Q-value which is the same for jumps between tetrahedral or octahedral sites. This is not so for the fcc lattice: Here the Bravais lattice for tetrahedral and octahedral sites is different. Consequently, the discrimination between both models is more easy than in the bcc case.

### 5.1.4 The Temperature- and Q-Dependence of the Quasielastic Intensity

During its rest time the diffusing proton vibrates around its interstitial sites. There are two components to the vibrations: First of all, modes appear far above the frequencies of the acoustic spectrum of the host lattice, and these are called "localized modes", in spite of the fact that they may have a strongly-Q-dependent frequency due to the mutual hydrogen interaction (ROWE et al., II, 1974; see also Chap.2). In addition to these high frequency modes, the proton follows the vibrations of the host lattice. If we assume that all these oscillations are harmonic, and not correlated with the jump motion, we can take them into account formally by adding an oscillatory term $u^2_{osc}(t)$ in (5.5). Within this approximation

$$F_s(\underline{Q},t) = \exp[-tf(\underline{Q}) - Q^2u^2_{osc}(t)] \quad . \tag{5.14}$$

After Fourier transformation (5.14) leads to the inelastic part of the scattering law, and the resulting Lorentzian in (5.7) has to be multiplied by a Debye-Waller factor

---

[2] Obviously, the measurement of D by incoherent scattering is analogous to the tracer method and to the spin-echo gradient method (see, e.g., ZOGAL and COTTS, 1975). Coherent neutron scattering would yield $D^{(G)}$.

$$I_{qe} = \exp[-\langle u^2 \rangle Q^2] \quad . \tag{5.15}$$

$\langle u^2 \rangle = u^2_{osc}$ $(t \to \infty)$ is the mean square amplitude of the protonic vibrations. For a harmonic Einstein oscillator of frequency $\omega_{loc}/2\pi$ one gets

$$\langle u^2 \rangle = \hbar/2M_H\omega_{loc} + \langle u^2_{host} \rangle \quad \text{for} \quad k_BT \ll \hbar\omega_{loc} \tag{5.16}$$

Here $\langle u^2_{host} \rangle$ is the contribution due to the modes for which the proton follows the vibrations of the host lattice. $\langle u^2_{host} \rangle$ is of the order of $3k_BT/M_{host}\omega_D$ where $\omega_D/2\pi$ is the Debye frequency (KLEY, 1966). For a general discussion of the Debye-Waller factor for localized vibrations we refer to KAUFMAN and LIPKIN (1962).

For most hydrogen metal systems with high diffusion constants investigated so far, experiments have shown that the quasielastic intensity $I_{qe}$ vs. $\underline{Q}$ drops much faster than expected from (5.15) and (5.16) using the experimentally known frequencies. Table 5.1 quotes typical values of $\langle u^2 \rangle$ which were obtained from the Q-dependence of the quasielastic intensity, as described by the formula (5.15). Most of these values are at least 10 times larger than expected theoretically, except in $TaH_x$. More detailed investigations of $I_{qe}(\underline{Q})$ have shown that this function has a steep slope for smaller Q-values (say $\lesssim 2$ Å$^{-1}$), and that the slope decreases for larger Q (GISSLER et al., 1973; WAKABAYASHI et al., 1974). At low temperatures, the Q-dependence approaches the normal Debye-Waller factor behaviour. Fig.5.7 shows intensity measurements on hydrogen in niobium, and Fig.5.8 presents similar results for hydrogen in palladium.

For a quantitative discussion of this subject, the meaning of the quasielastic intensity $I_{qe}$ has to be defined more exactly, in particular at larger Q where the separation of the quasielastic intensity from the phonon "background" is difficult.

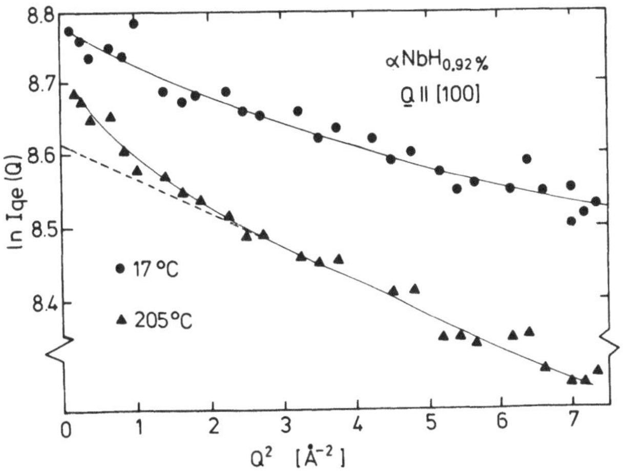


Fig. 5.7 Intensity of the quasielastic line $I_{qe}$ vs. scattering vector $Q$ for $\alpha$-$NbH_{0.0092}$. All intensities were calibrated by a standard scatterer (vanadium foil) (BAUER, SEITZ, HORNER, and SCHMATZ, 1975)

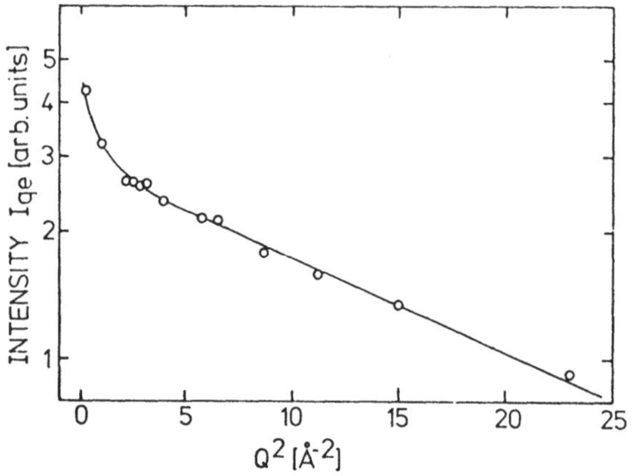


Fig. 5.8 Integrated quasielastic intensity $I_{qe}(Q)$ for hydrogen in $\beta$-$PdH_{0.48}$ An anomalously large slope is observed in the region of smaller Q (NELIN and SKÖLD, 1975)

One possibility is to define $I_{qe}$ as an integration of the measured scattering law $S_s(Q,\omega)$ at constant $\underline{Q}$, over a certain energy window $\Delta E$. This window has to be chosen large enough to comprise most of the quasielastic line, but sufficiently small so that the phonon contributions are small. Consequently, we write

$$I_{qe} = \int_{-\Delta E/2}^{+\Delta E/2} S_s(\underline{Q},\omega)d\omega \quad . \tag{5.17}$$

A more clearly defined procedure is the following: The central part of the measured quasielastic spectrum is fitted with the calculated scattering law for diffusion $S_s^{(D)}$. Integration of this function over energy transfer from $-\infty$ to $+\infty$ defines

$$I'_{qe} = \int_{-\infty}^{+\infty} S_s^{(D)}(\underline{Q},\omega)d\omega \quad . \tag{5.18}$$

Clearly, this definition excludes non-diffusive components of the spectrum with frequencies above $1/\tau$. To calculate (5.18), a knowledge of $S_s^{(D)}(\underline{Q},\omega)$ is needed. This is without problems at small Q where $S_s^{(D)}$ is a single Lorentzian with a width $2Q^2D$ in any case. At larger Q, $S_s^{(D)}$ depends on the details of the diffusive step which introduces uncertainties.

The interpretation of the anomalous behaviour of the quasielastic intensity is still controversial. Nevertheless, certain qualitative conclusions can be drawn. From the time-Fourier transformation of $G_s(\underline{r},t)$ in (1.69) it can be shown that the energy integral in (5.18) is equivalent to an average of $G_s(\underline{r},t)$ over a time interval of the order of the mean time $\tau$ between successive diffusive steps. To formulate this average, the hypothesis is made that the density correlation function is characterized by two time scales: A large one, determined by the mean rest time $\tau$, and a shorter one with $\tau_1$, where $\tau_1 << \tau$, and which may be related to the motion of the proton near and over the saddle point of the barrier for diffusion. Under these circumstances the time average may be written approximately as follows

$$\int \overline{d\underline{r}'\langle n(\underline{r}' - \underline{r},0)n(\underline{r}',t)\rangle} \simeq \int d\underline{r}' \, \bar{n}(\underline{r} - \underline{r}') \, \bar{n}(\underline{r}') \tag{5.19}$$

where $G_s$ has been formulated in terms of the time dependent density $\bar{n}$ $(\underline{r},t)$ (1.46). The bar means an average over a time interval of the order of $\tau$. Fourier integration of (5.19) over $\underline{r}$ leads to (WAKABAYASHI et al., 1974)

$$I_{qe}(\underline{Q}) = | \int d\underline{r} \, \bar{n}(\underline{r}) \exp(i\underline{Q}\underline{r})|^2 \quad . \tag{5.20}$$

Consequently, the quasielastic intensity is proportional to the square of the structure factor for the average density $\bar{n}(\underline{r})$ defined above.[3] If $\tau$ becomes comparable to $\tau_1$ this formulation loses its meaning, and an experimental separation of the quasielastic line from the spectrum due to the other rapid motion is meaningless as well.

The following physical picture can be applied to the formulation (5.20): The proton spends most of its time in the deep potential well at the interstitial site where it is only slightly delocalized by the zero point oscillation. For a certain fraction of time the proton reaches the saddle point of the potential barrier. There it spreads out widely, forming a delocalized distribution which overlaps with the neighbouring interstitial site. From this "state" the proton either returns to the localized ground state of the original site, or it falls into the adjacent site where it is again localized. The delocalized state is supposed to be responsible for the steep slope of $I_{qe}$ vs. Q at small Q. For simplicity it can be assumed that the localized, and the non-localized states, can be described by average distributions $g_0(\underline{r})$ and $g_1(\underline{r})$, with corresponding mean square amplitudes $u_0^2$, $u_1^2$ , and weights $1 - \alpha(T)$, $\alpha(T)$, respectively, so that

$$\bar{n} \, (\underline{r}) = (1 - \alpha) \, g_0(\underline{r}) + \alpha g_1(\underline{r}) \quad . \tag{5.21}$$

According to (5.20) and (5.21) the initial slope of $\ln I_{qe}(\underline{Q})$ is $\alpha(u_0^2 - u_1^2)$. The quasielastic intensity $I_{qe}$ for $NbH_{0.16}$ has been investigated and interpreted in

[3] For small Q this is equivalent with an extrapolation of the exponent in the intermediate scattering law (5.14) in writing $\ln I_{qe} \to Dt-Q^2\langle u^2\rangle$ for $t \to 0$.

terms of such a model (WAKABAYASHI et al., 1974). At 260 °C one obtains $u_0^2 = 0.019$ Å$^2$ and $u_1^2 = 1.0$ Å$^2$, where $u_1$ is of the order of the interstitial distance. Furthermore, one gets $\alpha/(1 - \alpha) = 0.15$ which means that the proton spends about 15% of its time in the delocalized "state".

Another model, based on the concept of localized and transient states, was discussed earlier by GISSLER et al. (1973). In contrast to (5.21), the correlation functions for both states were superimposed, and not the average densities. The resulting scattering law is a sum of two parts with weight factors proportional to $\exp(-Q^2 \langle u_T^2 \rangle)$ and $\exp(-Q^2 \langle u_R^2 \rangle)$ where T and R denote transient and resident state, respectively. No interference terms between the two states appear. Therefore, the model would be valid only if the lifetime of the transient state is comparable to, or longer than, the mean time between successive jumps. An improved version of this model (GISSLER and STUMP, 1973) includes such interferences. However, this model is also unsatisfactory due to the artificial introduction of the Debye-Waller factors for the transient and the resident state (see also FUNKE, 1974).

### 5.1.5 Conclusions on H-Metal Systems; Other Problems of Diffusion in Solids

The present state of quasielastic scattering experiments on hydrogen in metals can be summarized in the following way. For $PdH_x$ the interpretation of the quasielastic width is a clear evidence in favour of a certain jump model, namely octahedral jumps. Nevertheless, certain inconsistencies still exist for the values of D and for the activation energy (Table 5.1). As has been shown, for the bcc system (H in Nb, Ta, and V) there are many inconsistencies with the simple jump model described in Sec. 5.1.2. For $VH_x$ at larger Q the discrepancies may be explained by the fact that the jump time is comparable with the mean rest time.

Part of the discrepancies may be related to the "delocalization" of the proton on its interstitial site which was concluded from the anomalous Debye-Waller factor, and which obviously affects the geometry of the diffusive step (however, this statement should as well hold for $PdH_x$ where such discrepancies have *not* been observed). So far, no general theory exists which takes these effects properly into account. From the experimental point of view it would be important to reach larger scattering vectors, and to study carefully the wings of the quasielastic line where the "rapid" components of the protonic motion may be observable. Unfortunately, large-Q measurements are very difficult due to phonon background and Bragg reflections. An inherent problem in the interpretation of the quasielastic results in the bcc systems is the insufficient knowledge of the interstitial hydrogen lattice; this is analogous to attempting to measure phonon dispersion for crystals whose unit cells are not known! So far, only some indirect information exists on the interstitial lattice, e.g., from diffuse neutron scattering in $NbH_x$ (BAUER and SCHMATZ, 1975) which indicates tetrahedral sites (see also STEWARD, 1975; CARSTANJEN and SIZMANN, 1972).

The delocalization of the proton as discussed before seems to be generally related to fast diffusion. This idea is supported by several observations. For instance, for $VH_x$ a lowering of x from about 60% to 20% leads roughly to a doubling of $\langle u^2 \rangle$ and D, whereas the observed change of the localized-mode frequency is very small (de GRAFF et al., 1972). It should be pointed out that an anomalously steep Debye-Waller factor was also observed for the $Ag^+$ ions in the bcc α-AgI (HOSHINO, 1957), a solid electrolyte with a cationic diffusion constant as high as that of hydrogen in metals at elevated temperatures.

In connection with the observation of fast diffusion it is interesting to notice that, except for $PdH_x$, the activation energy for diffusion in the discussed metals is comparable with, or even smaller than, the observed localized mode energies (Table 5.1)! The observed width of the localized peaks was found to be anomalously large (VERDAN et al., 1968). However, it is not yet clear if this width can be related to the lifetime of the excited state or if it is, at least partially, due to the multiphonon processes (excited at large momentum transfer), due to dispersion, and/or due to disorder in occupancy. This question needs further experiments. It was speculated that hydrogen diffusion might reveal quantum effects (SUSSMAN and WEISSMANN, 1972; STONEHAM, 1975). However, it is not to be expected that these may play a role in the temperature region of the experiments discussed before (KEHR, 1976), but such effects cannot be excluded at lower temperature where the activation energy for hydrogen diffusion in niobium undergoes a large change (SCHAUMANN et al., 1970). This effect was the subject of further investigations by means of quasi-elastic scattering at very high resolution, and was confirmed by these experiments (RICHTER et al., 1977).

Finally, we draw attention to certain selected applications of quasielastic scattering for diffusion problems which seem to be promising, but which are still in a more or less preliminary or developing state. These investigations are related mainly to problems with small self-diffusion constants and, therefore, to the application of backscattering spectrometers.

*1) Hydrogen Diffusion in a Crystal with Impurities.* It is known experimentally that interstitial impurities as N,C, and O in bcc metals act as traps for the dissolved hydrogen. This effect was investigated by studying the incoherent quasielastic spectrum by means of a backscattering spectrometer for $NbH_{0.004}N_{0.007}$ (RICHTER et al., 1976). Such an experiment was feasible due to the very small incoherent scattering cross-section of Nb, and the availability of a high flux reactor. Fig.5.9 a) shows the quasielastic width as a function of Q. For comparison, the quasielastic width for H in pure Nb is shown (extrapolated from experiments of GISSLER, et al., 1970). For small scattering vectors, a Lorentzian-shaped quasielastic line was found whose width is roughly proportional to $Q^2$. From this width the effective self-diffusion coefficient $D_{eff}$ was determined which is considerably smaller than

D for the case of pure niobium. For large Q, the width is far below $Q^2D$, and moreover, the intensity of the observed line $I_1$ is much smaller than the intensity for small Q. This is demonstrated in Fig. 5.9 b). In this figure the intensity is calibrated by an experiment at sufficiently small temperature where the quasielastic line falls entirely within the resolution window (intensity $I_0$).

This behavior is explained naturally in terms of hydrogen trapping on the dissolved nitrogen atoms. A hydrogen atom diffuses partly over the sites of a more or less undisturbed lattice, and partly over sites in the vicinity of the dissolved nitrogen atoms where the binding and activation energies are changed. At small Q, the scattering process averages over *long* diffusive paths, and the full intensity is found in the quasielastic line. As Q increases, the spectral distribution due to *single* diffusive steps is observed. Consequently, the resulting spectrum is a sum of several Lorentzians due to a variety of sites having different rest times. In the quasielastic line, as observed by high resolution, the contributions with the largest rest time dominate. Contributions due to small rest times appear in the wings of the spectrum. They are practically invisible and, obviously, their fraction increases with increasing temperature. From these experiments one estimates that the characteristic times for the trapped hydrogen atoms are several hundred times larger than the mean rest time in the pure niobium. The ratio $I_1/I_0$ is roughly proportional to the fraction of hydrogen atoms which are trapped for times equal to, or larger than, the reciprocal line width ($1/\Gamma \simeq 10^{-9}$s). A simple theoretical interpretation can be given which relates the trapping times, the intensity ratio $I_0/I_1$, and $D_{eff}$ as compared to D for pure niobium.

*2) Hydrogen Diffusion in an Ordered Phase.* In the $\alpha$-phase of $NbH_x$ all tetrahedral sites have an equal probability of occupancy. However, the low temperature $\beta$-phase is characterized by a privileged occupancy of one class of tetrahedral sites. For instance, the upper one of the four sites on the cube face is occupied (Fig.5.3) and these sites form rows along a certain [110] direction in the lattice (which gets a slight orthorhombic distortion). It may be assumed that diffusive steps along such rows are more frequent than jumps between adjacent rows. Following the arguments given in Subsec.5.1.2, one-dimensional diffusion along [110] rows leads to a quasielastic width which disappears at [110] *planes* in reciprocal space (B. ALEFELD, 1975).

*3) Diffusion of Atoms other than Hydrogen.* Part of the scattering of sodium is incoherent. Consequently, between Bragg reflections it is possible to investigate Na diffusion by quasielastic scattering. Experiments of this kind were carried out near the melting temperature on a sodium single crystal at a backscattering spectrometer (AIT-SALEM, 1976). The motion of the Na atoms is induced by vacancy diffusion. Therefore, the Na atoms jump over the regular sites of a bcc lattice. This was confirmed approximately by the observed Q-dependence of the quasielastic

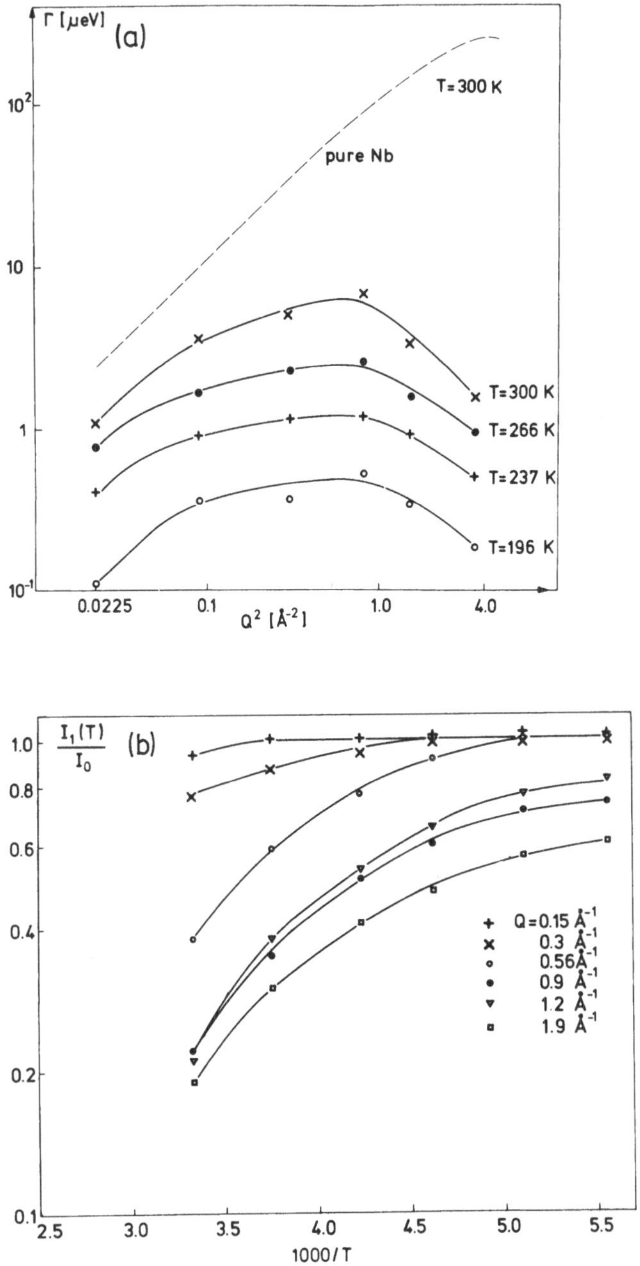


Fig. 5.9a,b Quasielastic scattering on hydrogen diffusing in $NbH_{0.004}N_{0.007}$. (a) Width of the quasielastic spectrum as compared to that of pure Nb. (b) Intensity $I_1$ of the observed quasielastic line for various Q-values, in units of the total intensity $I_0$. At larger Q, this ratio yields the fraction of H atoms trapped at the N impurities for times $\hbar/\Gamma \geq 10^{-9}$s (RICHTER, TÖPLER, and SPRINGER, 1976)

spectra. Near the melting temperature, the spectra are, to a certain extent, affected by divacancy diffusion. This needs a refined theory of the scattering law which does not yet exist.

It should be pointed out that for a *coherent* scatterer, such as potassium, the diffusion of a *vacancy* would be observed rather than diffusion of K-atoms: this can be simply argued by Babinet's theorem. A wide field of future applications will probably be the investigation of diffusion in solid electrolytes. In this case the diffusive step may be similar to that for hydrogen diffusion in metals at elevated temperatures. Such studies have been carried out on bcc α-AgI with respect to the motion of the $Ag^+$ ions. The interpretation of the data is difficult due to the coherent scattering contribution of the silver. Solid electrolytes with Na-ions are better candidates for such investigations (see FUNKE et al., 1974).

## 5.2 Rotations in Molecular Solids

### 5.2.1 Rotations and Hindrance Potential; Scattering Law

In many crystals the molecules and molecular groups have several equivalent orientations. These are separated by potential barriers, and the molecules or groups can rotate between their *different* equilibrium orientations. This leads to orientational disorder of the crystal. Also in cases where *only one* orientation exists, molecular rotations occur, leaving the orientation of the molecule unchanged. For tetrahedral molecules, or ions, the first case corresponds to $90^0$ rotations about a twofold axis, and the second to $120^\circ$ rotations about a threefold axis (Fig.5.10). The second case applies also to the $CH_3$-groups rotating about the bond axis in a potential of threefold symmetry. In many cases the molecular rotations have the character of collective

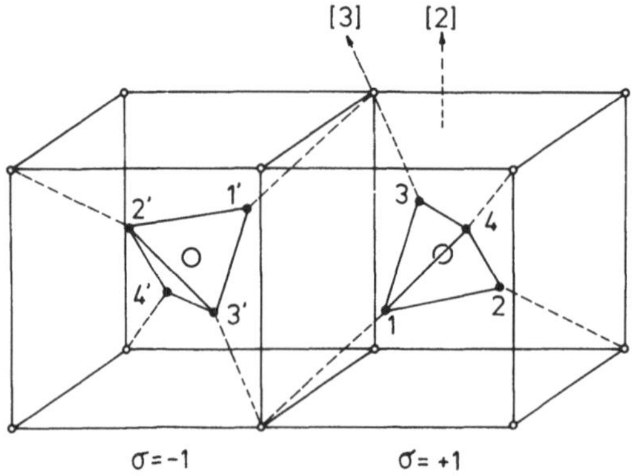


Fig. 5.10 Tetrahedral ions, as $NH_4$, in a cubic lattice. Anions at the cube corners. [2] and [3] are the axes for 90° and 120° rotations. Two equilibrium orientations σ = ± 1 are shown. α = 1,.. 4, and 1'.. 4' are the positions which a proton can occupy for a certain σ

excitations, and are related to the dynamics of phase transitions. Sometimes the rotations are important for the understanding of binding forces and mechanical properties.

This section will be introduced by a classification of the orientational motions with regard to the height of the hindrance potential in relation to the rotational constant $\hbar^2/2J$ (J = moment of inertia). Subsequently, the incoherent scattering spectra for such motions will be discussed, and the physical information which can be obtained from them. For crystals where the rotating molecules are in direct touch with each other, collective motions may appear, for instance coupled rotations, as in solid $H_2$ (MERTENS et al., 1969), or critical fluctuations of the orientations in methane (PRESS et al., 1974, see Subsec.3.22).

The following cases will be considered: 1) Height of the *hindrance potential large* compared to the rotational constant $\hbar^2/2J$ (in typical cases > 1 kcal/mol). Under these conditions the wave functions of the protons in the molecule are essentially localized in the minima of the rotational potential. The corresponding low energy levels of the protonic wave functions are equidistant, as for harmonic librations, and fall below the top of the hindrance potential. In the *high temperature limit*, thermal excitation of the molecule causes rotational jumps between the potential minima, and the average time between orientational jumps follows an Arrhenius law

$$\tau = \tau_0 \exp(E_{act}/k_B T) \tag{5.22}$$

where $E_{act}$ is the difference between the height of the potential and the librational ground state energy. These rotational jumps, as well as transitions between the librational levels, are related to the width of these levels. At sufficiently *low temperatures* the lifetime increases, and a splitting of the librational levels appears which is due to the overlap of the wave functions in adjacent potential minima (*tunneling splitting*). The energies of these levels are obtained as the eigenvalues E of the Schrödinger equation of a rotating molecular group which reads, in the simple case of one-dimensional rotation,

$$-(\hbar^2/2J)d^2\psi/d\phi^2 + [V(\phi) - E]\,\psi = 0 \quad . \tag{5.23}$$

$V(\Phi)$ is the potential which is a periodic function of the angle of rotation, $\Phi$.

2) *Small rotational potential* in comparison to the rotational constant (typical values: below 1 kcal/mol). In the *high temperature* region strong fluctuations of the hindrance potential occur, caused by phonons and/or uncorrelated rotations of neighbouring molecules. In this case, no well-defined equilibrium positions exist. This causes random jumps or, if the steps are small, rotational diffusion. This rotational diffusion, which may occur in certain plastic crystals, is similar to

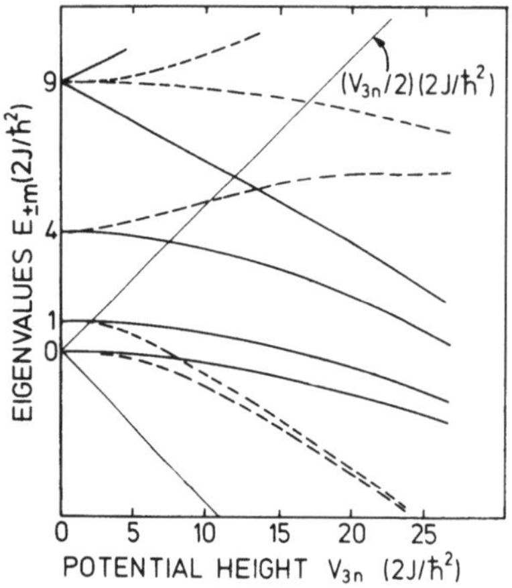


Fig. 5.11 Scheme of energy levels for a linear rotator in a hindrance potential $V = V_{3n} \cos(3n\Phi)$ ($\Phi$ = angle of rotation). $J$ = moment of inertia about the rotational axis. Solid line: $n = 2$. Dashed: $n = 1$. Straight lines: top and bottom of the potential barrier. For increasing $V_{3n}$ the distance between the 3 lowest levels decreases; finally they approach the split ground state of a harmonic librational oscillator. Free rotator means $V_{3n} = 0$ and $E_I = (\hbar^2/2J)\, I^2$ with $I = 0,1,2...$ (ALEFELD, KOLLMAR, DASANNACHARYA, 1975)

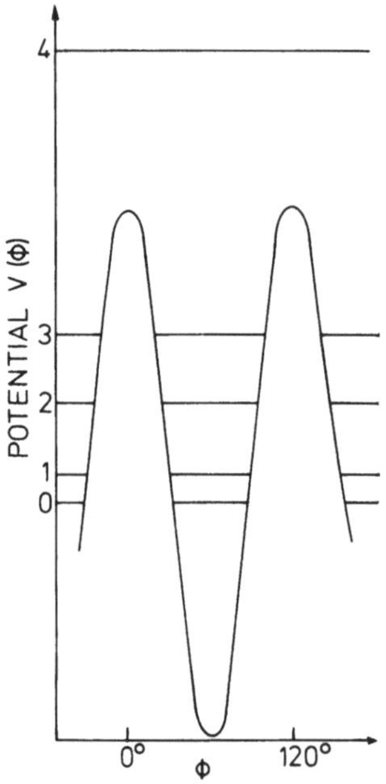


Fig. 5.12 The angular dependence $V(\Phi)$ of the potential corresponding to the spectrum of Fig. 5.16 (schematic)

rotations in a molecular liquid. It may also happen that a molecule alternates between a librational (localized) state, and a more freely rotating transition state.

At *low temperatures* well-defined rotational quantum states appear, and these are shifted in energy respect to the levels of a free rotator. This is shown in Fig. 5.11 for a three- and a sixfold potential

$$V = (V_3/2)\cos 3\Phi \quad \text{or} \quad (V_6/2)\cos 6\Phi \tag{5.24}$$

where (5.23) is a modified Mathieu equation. Regarding the diagram in Fig.5.12, it should be recognized that a level shift of only a few percent occurs with respect to the free-rotator scheme, in spite of the fact that three eigenvalues are already *below* the top of the hindrance barrier. Consequently, such states may be called "*rotational tunneling*", rather than "nearly free rotation". The observation of transitions between such states, as well as transitions due to tunneling splitting, is a very recent and particularly interesting development in neutron spectroscopy.

### 5.2.2 High Temperatures: Rotational Jumps and Rotational Diffusion

For not too low temperatures the molecular rotations can be understood in terms of rotational diffusion, or of rotational jumps with librations during the time between the jumps. Such motions have been investigated by proton magnetic resonance on many substances. Incoherent scattering of neutrons is an alternative and supplementary method. Under certain conditions, neutron scattering yields more information than nmr with regard to the geometry of the rotation: The scattering spectrum can be analysed for various values and orientations of the scattering vector $\underline{Q}$, in particular if single crystals are available. The characteristic times which are accessible experimentally can be extended to frequencies higher than those studied by nmr ($< 10^{-12}$ s). With back-scattering spectrometers, the characteristic times which can be studied are of the order of $10^{-8}$ to $10^{-10}$ s.

In comparing nmr with neutron data we have to recognize that incoherent scattering observes the *transport of a single proton* by the molecular rotation. On the other hand, nmr "sees" essentially the rotation of *vectors connecting pairs of protons* within the molecules. For instance, a $120^\circ$ jump about [111] of a $NH_4$-ion affects all vectors, whereas only three out of four protons are carried around. *Only for a given model*, the characteristic times or rates found from both methods can be related with each other.

Ammonium salts at higher temperatures are typical for systems with ratational jumps between well-defined orientations of their tetrahedral ions. This is a consequence of their high rotational potential barriers: The activation energies are 4.7 kcal/mol for $NH_4Cl$; 4.0 kcal/mol for cubic $NH_4Br$; and 3.9 kcal/mol for rhombic $(NH_4)_2SO_4$ (see WATTON et al., 1972).

For tetrahedral ions $S_s(\underline{Q},\omega)$ will be derived from the self-correlation function of a proton, following the formulation given by MICHEL, II (1973)(see also STOCKMEYER and STILLER, 1968; THIBAUDIER and VOLINO, 1973). For simplicity it will be assumed that the $NH_4$-ion carries out independent orientational motions about the threefold and the fourfold axes (Fig.5.10). In the following, we introduce the quantity $\nu_3$, which is the probability that, per unit time, a rotation occurs about a *certain one* of the threefold axes of the ion; $\nu_4$ has the same meaning with respect to the fourfold axes. The time needed for the reorientational motion is assumed to be negligible. The incoherent scattering law is then (Sec.1)

$$S_s(\underline{Q},\omega) = \int dt\, e^{-i\omega t}\, d\underline{r}d\underline{r}'\, e^{i\underline{Q}(\underline{r}-\underline{r}')} \langle n(\underline{r}',t)n(\underline{r},t)\rangle \qquad (5.25)$$

where <...> is the density autocorrelation function for one of the four equivalent scattering protons. For simplicity, librations will be neglected. Consequently, the proton is localized at the various sites $\underline{R}_{\sigma\alpha}$ which it can occupy. These are defined by the molecular orientation $\sigma = \pm 1$ and the four proton sites $\alpha = 1,2,3,4$, which exist for a given orientation $\sigma$. This leads to

$$n(\underline{r},t) = \sum_{\text{all } \sigma,\alpha} P_{\sigma\alpha}(t)\, \delta(\underline{r} - \underline{R}_{\sigma\alpha}) \quad . \tag{5.26}$$

$P_{\sigma\alpha}(t)$ is the probability of finding a certain proton at a position $R_{\sigma\alpha}$ at time t. This yields,

$$S_s(\underline{Q},\omega) = \int dt\, e^{-i\omega t} \sum_{\sigma\alpha} \sum_{\sigma'\alpha'} e^{i\underline{Q}(\underline{R}_{\sigma\alpha} - \underline{R}_{\sigma'\alpha'})} < P_{\sigma'\alpha'}(t) P_{\sigma\alpha}(0)> \tag{5.27}$$

with

$$<P_{\sigma'\alpha'}(t)\, P_{\sigma\alpha}(0)> = P(\sigma'\alpha';t|\sigma\alpha;0)\; P(\sigma\alpha;0) \tag{5.28}$$

where $P(\sigma'\alpha';t|\sigma\alpha;0)$ is the conditioned probability of finding the proton in a position characterized by $\sigma'\alpha'$ at time t if it was at $\sigma\alpha$ for an earlier time, $t = 0$. $P(\sigma\alpha;0)$ is the *a priori* probability of finding the proton at $\underline{R}_{\sigma\alpha}$ for $t = 0$. Obviously this is 1/8 in the disordered phase. Below $T_c$, $P(\sigma\alpha,0)$ depends on the order parameter. The conditioned probability can be calculated from a system of rate equations with $P(\sigma\alpha;0)$ as initial condition, namely

$$(\partial/\partial t)\, P(\sigma\alpha;t) = -3(\nu_3 + \nu_4)\, P(\sigma\alpha;t) + \nu_3 \sum_{\substack{\alpha' \\ \alpha' \neq \alpha}} P(\sigma\alpha';t) + \nu_4 \sum_{\alpha'} P(-\sigma\alpha';t) \tag{5.29}$$

and one obtains for $\sigma = \sigma'$

$$P(\sigma\alpha';t|\sigma\alpha;0) = (1/8)[1 + (4\delta_{\alpha\alpha'} - 1)\,(e^{-\lambda_1|t|} + e^{-\lambda_2|t|}) + e^{-\lambda_3|t|}] \quad , \tag{5.30}$$

where the three non-zero roots are

$$\lambda_1 = 2\nu_4 + 4\nu_3;\; \lambda_2 = 4(\nu_4 + \nu_3);\; \lambda_3 = 6\nu_4 \quad . \tag{5.31}$$

A similar expression holds for $P(-\sigma\alpha';t|\sigma\alpha;0)$ where the molecular orientation has flipped from $\sigma$ to $-\sigma$. The resulting incoherent scattering law, which we will not quote explicitly, has the general form

$$S_s(\underline{Q},\omega) = f_0(\underline{Q})\ \delta\ (\omega) + \sum_{i=1}^{3} f_i(\underline{Q})\ L\ (\omega,\lambda_i) \qquad (5.32)$$

with $\sum f_i = 1$ and $L(\omega,\lambda) = (\lambda/\pi)/(\omega^2 + \lambda^2)$. The $f_i(\underline{Q})$ can be calculated from the arrangement of the accessible sites $\underline{R}_{\sigma\alpha}$ of the proton. $f_0(\underline{Q})$ can be understood in terms of a diffraction of the neutron wave on the proton which is distributed over the sites $\underline{R}_{\sigma\alpha}$ by the rotational jumps. To observe this quantity experimentally, the coherency time $\hbar/\Delta E$ of the neutron wave must be long compared to the characteristic times $\tau_{rot}$ needed for this motion, i.e., one must have $\hbar/\Delta E > \tau_{rot}$, where $\Delta E$ is the energy resolution of the instrument.[4] For small Q, the structure factor decreases with increasing Q, having an oscillatory behaviour at larger Q. The sum of the elastic and quasieleastic scattering law is unity. More strictly it is equal to the Debye-Waller factor which has been left out for simplicity (on the anisotropy of the molecular Debye-Waller factor see, e.g., WILLIS et al., 1975).

Comprehensive experiments were carried out on various *ammonium salts*, in particular on $NH_4Br$ in its cubic phase (LIVINGSTON et al., 1974). Fig.5.13 shows selected spectra as a function of energy transfer for a certain orientation of the scattering vector with respect to the orientation of the single-crystalline sample. At small Q the resolution-broadened purely elastic line can be easily recognized. The

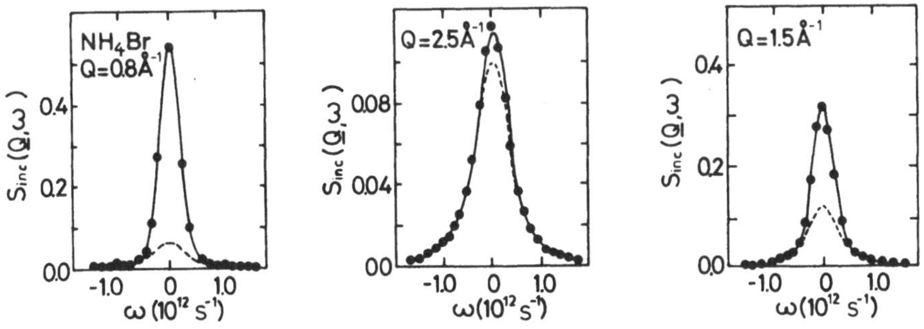


Fig. 5.13 Typical scattering spectra for $NH_4Br$ single crystals at 273 K. Points: experimental values. Solid line: according to theory ($\delta$-function + Lorentzian, (5.32), with $\nu_4$ and $\nu_3$ as fitting parameters). Dashed: Lorentzian components in (5.32), after folding the theory with the instrumental resolution. Incident beam parallel [110]. (LIVINGSTON, ROWE, and RUSH, 1974)

[4] In the above formalism, $f_0$ corresponds to a proton which is equally distributed over the eight corners of a cube. However, for $\nu_4 = 0$ only *one* orientation of the ion exists; because of $L(\omega,\lambda_3) = \delta(\omega)$ the elastic structure factor is then $f_0 + f_3$ (instead of $f_0$) which corresponds to the structure factor of a proton on the four corners of a tetrahedron).

Lorentzian parts of the scattering law appear as small wings. With increasing Q, the intensity of the Lorentzians increases, whereas the strength of the purely elastic line becomes smaller. Without a detailed shape-analysis one could be misled into assuming that the curves were simple Lorentzians whose widths increase with increasing Q! The spectra were fitted with the rotational jump-model derived before, using $\nu_3$, $\nu_4$, and the ordinate scale as disposable parameters. The functions $f_i$ were calculated. As a result it was found that the fourfold rotations are the dominating process. $\nu_3$ is too small to be determined from these experiments. To exclude a model of threefold rotations a calculation with $\nu_4 = 0$ and with $\nu_3$ as a free parameter for the various orientations was carried out; the resulting $\nu_3$ varies strongly for different orientations which is physically unreasonable, and which was not found for the fourfold model.

Similar experiments were carried out on $NH_4Cl$ near the order-disorder phase transition occurring at -30.5 °C (TÖPLER et al., 1976) which is related to the orientation of the ammonium ions (Fig.5.10) and which has been studied by a great number of nmr and other investigations. By scattering on single crystals, suitable orientations and values of the scattering vector were chosen in order to reduce the number of terms in (5.32). This improves the reliability of the data evaluation. For instance, with $\underline{Q} \parallel [111]$ and $Q = 3.8$ Å$^{-1}$, the delta-function at $\omega = 0$ disappears for $T > T_c$, and $S_s$ is dominated by the Lorentzian $L(\omega,\lambda_3)$ whose width is given by the fourfold rotational rate $\nu_4$. On the other hand, with $\underline{Q} \parallel [100]$ and $Q = 1.42$ Å$^{-1}$ one gets essentially $L(\omega,\lambda_1)$, so that $\lambda_1 = 2\nu_4 + 4\nu_3$ is obtained. With a backscattering spectrometer, spectra as well as intensities at energy transfer $\hbar\omega = 0$ were measured to obtain the orientational rates ("fixed window method", see Subsec.5.2.3). In all experimental runs, the intensities were normalized. For this purpose, the total intensity was determined by means of separate experiments at a sufficiently low temperature where the full spectrum (elastic + quasielastic) falls entirely within the resolution window (except phonon contributions which have to be taken into account by additional measurements of the Debye-Waller factor). Consequently, the fitting of the experiments with the theory includes only two disposable parameters, namely $\nu_3$ and $\nu_4$ (the factors $f_i(\underline{Q})$ in (5.32) are determined by the geometry of the ion). This holds for $T > T_c$. Below $T_c$ also the population ratio of privileged and non-privileged orientations enters, as well as the corresponding rates $\nu_4$, $\bar{\nu}_4$, $\nu_3$, and $\bar{\nu}_3$ which makes the evaluation of the spectra extremely complicated. Experiments on $NH_4Cl$ single crystals at 200 °C were reported by SKÖLD and DAHLBORG (1973).

Spin-lattice relaxation rates $1/T_1$ of the proton, as determined by nmr, are related to the reorientational rates $\nu_3$ and $\nu_4$: The quantity $1/T_1$ is proportional to a sum of several Lorentzians $(C/\pi\tau_c)/(\omega_0^2 + \tau_c^{-2})$ where the characteristic times $\tau_c$ depend on $\nu_3$, $\nu_4$; the weight factors are given by the geometry of the rotating ion; $\omega_0$ is the Larmor frequency. Consequently, only one experimental quantity $T_1(T)$ is available to obtain these rates $\nu_3$ and $\nu_4$, and detailed assumptions on their temperature dependence are needed to determine them separately (KODOMA, 1972;

MANDEMA, 1974; MICHEL I, 1973). In the case of $NH_4Cl$, the rates $\nu_3$ and $\nu_4$ for $T > T_c$ from the neutron scattering experiments are several times smaller than the values from nmr investigations. The origin of this discrepancy is not yet understood and needs further discussion.

For *ammonium perchlorate*, nmr and specific heat investigations have suggested a rather low rotational potential. This has been verified by quasielastic scattering experiments on $NH_4ClO_4$ polycrystals (PRASK et al., 1975) where the spectra were interpreted in terms of the model described before. It was found that instantaneous rotations about all threefold axes are the dominating process. As a matter of fact the assumption of four equivalent threefold rotational axes is doubtful since neutron diffraction indicates that one out of four hydrogens is bonded more rigidly than the others. The activation energy from $\nu_3$ vs. T was determined to be only 0.54 kcal/mol, which is unusually small with respect to the other ammonium salts. The low barrier suggested a search for tunneling splittings at low temperatures. Actually, a recent neutron scattering experiment on $NH_4ClO_4$ at 4 K revealed clearly several lines in the region of a few µeV which were attributed to tunneling transitions (ALEFELD and PRAGER, 1976; see also Subsec.5.2.4).

On various *hydrosulfates* investigations of the $SH^-$ rotations were carried out by ROWE et al. (1973) and RUSH et al. (1973). The suggested structure of RbSH can be visualized as a slightly distorted NaCl-lattice (actually, the low temperature phase is trigonal). It was assumed that at each cation site, the axis of the $SH^-$ ion can be randomly aligned in opposite [111] orientations. Consequently, instantaneous jumps of the proton between two sites could occur at higher temperatures. This model leads to a scattering law which can be easily calculated in analogy to (5.25) to (5.27). After performing a polycrystalline average one obtains

$$S_s(Q,\omega) = (1/2)\ \delta(\omega)[1 + j_0(QR)] + \frac{1/\pi\tau}{\omega^2 + (2/\tau)^2}\ [1 - j_0(QR)] \quad . \qquad (5.33)$$

$j_0$ is a spherical Bessel function of order zero. R is the distance between the hydrogen sites (twice the $SH^-$ bond length), and $\tau$ is the average time between successive jumps. The Debye-Waller factor has been ignored. From a fit of the resolution-broadened scattering law with the experimental data the relative intensity contribution of the purely elastic line was extracted. Fig.5.14 shows the godd agreement with the proposed two-site jump model. Obviously, one can exclude reorientational jumps between the eight body diagonals of the (pseudo-cubic) cell. Due to the very simple jump geometry, and the relatively large values of QR in the structure factor, it was possible to draw such conclusions from a *powder* experiment. Similar investigations were performed on NaSH and CsSH.

Many crystals with more or less spherical molecules undergo a solid-solid transition from a normal (or brittle) phase to a phase which is mechanically soft, called "*plastic*". In this phase the molecules rotate very rapidly, nearly as fast as in the

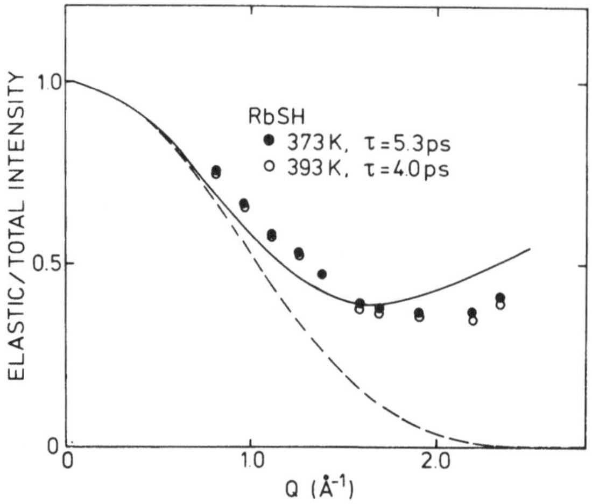


Fig. 5.14 Ratio of purely elastic to total intensity (elastic + quasielastic) for neutron scattering on RbHS. Points: data from experimental spectra. Solid line: two-site jump model. Dashed line: random jumps over 8 equivalent sites (ROWE, LIVINGSTON, RUSH, 1973)

liquid state. For adamantane ($C_{10}H_{16}$) in its plastic phase (above 208 K) a $90^\circ$ jump model has been tentatively applied to interpret the spectra measured on a polycrystal, and the jump rates were evaluated (STOCKMEYER and STILLER, 1968). A decision in favour of a site-jump model could be obtained from more recent single crystal experiments (LECHNER and HEIDEMANN, 1976). Experiments on neopentane $C(CH_3)_4$ indicate that the rotation is, at least partially, an isotropic rotational diffusion (LECHNER, 1972). The rotational barrier quoted in this work seems to be unusually small (see SMITH, 1965). A substance with a rather low hindrance barrier is cyclohexane which was also studied by neutron scattering (LEADBETTER, 1972; EGELSTAFF, 1970). So far, it is not yet clear if molecular rotations in plastic crystals with low barriers are isotropic, or of the site-jump type, a question which could be answered by means of such scattering experiments.

We will not present a detailed discussion of the existing theories for scattering on isotropic rotators, and we restrict ourselves to discuss briefly the so-called *Sears expansion* which formulates the scattering law in terms of rotational correlation functions (SEARS, 1967), and which is very convenient for many applications:

$$S_s^{(rot)}(Q,\omega) = \delta(\omega) j_0^2(QR_0) + \sum_{\ell=1}^{\infty} (2\ell + 1)\, j_\ell^2(QR_0)\, \tilde{F}_\ell(\omega) \quad . \qquad (5.34)$$

The $j_\ell$ are spherical Bessel functions; R is the distance of the proton in the molecule from the center of its rotation. The functions $\tilde{F}$ are the Fourier-transformed orientational autocorrelation functions, namely

$$\tilde{F}_\ell(\omega) = (1/2\pi) \int e^{i\omega t} < P_\ell[\cos\beta(t)]> \quad . \qquad (5.35)$$

Here $\beta(t)$ is the angle through which a vector, fixed at a certain proton in the molecule, rotates within a time interval t. $P_\ell$ is a Legendre polynominal. The $<P_\ell(t)>$ can be calculated explicitly for specific models, e.g., for rotational diffusion, or for isotropic, large-angle jumps (SEARS, 1967). It should be noticed that this expansion holds also for a polycrystal whatever the local symmetry of the molecular rotation is (LASSIER et al., 1973). The term $\ell = 1$ appears also in the calculation of infrared spectra of rotating dipoles; it would be identical for neutron scattering if the molecular dipole vector coincides with the position vector of a scattering proton in the molecule. Under certain conditions the term $\ell = 2$ is related to Raman and nmr spectra (GORDON, 1965 and 1966; LIVINGSTON et al., 1973).

In principle, the neutron spectrum yields *all* correlation functions $F_\ell(t)$. However, it seems that only the first two terms can be properly extracted. Unfortunately, no systematic investigations exist combining neutron and light scattering to obtain more detailed information on the rotations. Progress is to be expected if such experiments can be interpreted in terms of computer simulations of the molecular rotations (e.g., QUENTREC, 1975).

An important quantity in all the experiments is $F_0(\underline{Q})$, the intensity of the purely elastic line. For instantaneous jumps between discrete sites $\underline{R}_i$ we had already

$$F_0(\underline{Q}) = \sum_i \sum_j \exp[i\underline{Q}(\underline{R}_i - \underline{R}_j)] \tag{5.36}$$

and, for isotropic and diffusive rotations of the scattering particle on a sphere

$$F_0(Q) = j_0^2(QR) = \sin^2(QR)/(QR)^2 \quad . \tag{5.37}$$

Of course this result does not depend on the type of motion which means, not on the way in which the asymptotic and isotropic probability distribution is reached. On the other hand, for a free rotation one finds (SEARS, 1967)

$$F_0(Q) = \sum_{\ell=0}^{\infty} j_\ell^2\,(QR) = \mathrm{Si}(2QR)/2QR \tag{5.38}$$

(Si = integral sinus). The difference with respect to the diffusive rotation (5.37) can be understood by the fact that, for a given initial condition, an undisturbed free rotation never reaches an isotropic distribution. Nevertheless, after a sufficiently long time ($\tau_R$) all rotations in condensed matter get randomized. Consequently, one measures the structure factor $j_0^2(QR)$ as soon as the width of the resolution window is smaller than $\hbar/\tau_R$. For a uniaxial rotation on a circle with radius R one gets (DIANOUX and VOLINO, 1975)

$$F_0(Q) = J_0^2(QR \sin\theta) \tag{5.39}$$

where $\theta$ is the angle between $\underline{Q}$ and the axis of rotation ($J_0$ = Bessel function). From the discussion of the structure factor $F_0(\underline{Q})$ one may get the feeling that incoherent scattering is also an efficient method for the determination of the orientational *structure* of the molecules in a crystal. Normally, neutron diffraction will do as well or better, with the exception of plastic crystals where only a few X-ray reflections can be observed, due to the large-amplitude rotations of the molecules. In any case, the main information from incoherent scattering experiments is related to the *dynamics* of the molecular rotations, and the orientational structure should be established by coherent diffraction before investigations on the molecular rotation will be carried out. In this connection it should be emphasized that the incoherent structure factor $F_0(\underline{Q})$ is related to the Fourier transform of the *time average* of a single rotating molecule, whereas the ordinary (coherent) structure factor implies an *ensemble average* as well.

### 5.2.3 Fixed-Window-Method

As a supplementary measurement during the investigation of rotational spectra it is very useful to perform a relatively simple and a less time-consuming experiment, namely the determination of the spectral intensity at $\hbar\omega \simeq 0$. This means an experiment where the energy of the analyzing system coincides with the incident energy. If $W_0(\omega)$ is the resolution function of the spectrometer at $\omega = 0$, the measured intensity is proportional to

$$I(0,\underline{Q}) = \int_{-\infty}^{+\infty} w_0(\omega)\, S_s(\underline{Q},\omega)d\omega \quad . \tag{5.40}$$

For simplicity we assume a rectangular-shaped $w_0(\omega)$ of width $\Delta E$, and a quasielastic spectrum as in (5.33). This leads to

$$I(0,\underline{Q}) = e^{-2W}\{F_0(\underline{Q}) + [1 - F_0(\underline{Q})]\ (2/\pi)\ \mathrm{arctg}(\tau\Delta E/\hbar)\} \quad . \tag{5.41}$$

Qualitatively, the temperature dependence of $I(0,\underline{Q})$ can be generally understood in the following sense. At small temperatures the purely elastic line and the quasielastic spectrum are entirely within the resolution "window" (the phonon spectrum is assumed to fall essentially outside the window). Therefore, the temperature dependence of $I(0,\underline{Q})$ is given by the Debye-Waller factor. As the temperature increases, the quasielastic part of the spectrum broadens and falls more and more beyond the window. Hence $I(0,\underline{Q})$ decreases until the quasielastic component is practically entirely outside, and only the purely elastic line remains within. This yields a step in the intensity of magnitude $F_0(\underline{Q})$. This step occurs in a temperature region where the characteristic time $\tau$ of the rotation agrees approximately with $\hbar/\Delta E$.

To demonstrate this method, Fig.5.15 shows the results for a fixed-window measurement on tetramethyl-ammonium-manganese fluoride $N(CH_3)_4MnCl_3$ (ALEFELD et al., 1976) where the quasielastic spectrum is related to the rotation of the $CH_3$-groups (LASSIER et al., 1973). One has to keep in mind that the step shown in the figure has nothing to do with any phase change related to the dynamical behaviour of the molecules; it is only due to the temperature dependence of the quasielastic spectrum in relation to the width of the resolution window. From such curves, as a function of scattering vector and temperature, the structure factor, the characteristic time, and its activation energy can be calculated, provided that a certain model of the molecular rotation has already been established by a few careful spectrum measurements. It is easy to recognize that the step in the curve of Fig.5.15 bears a certain analogy to the step of the $T_2$-curves in nmr spectroscopy.

The fixed-window method is also applicable for the measurement of the temperature dependence of self-diffusion constants D. For sufficiently small Q the quasielastic spectrum is always a *single* Lorentzian whose intensity is practically independent of temperature ($\exp(-Q^2u^2) \cong 1$), and one gets $I(0,Q) = (2/\pi)\ \mathrm{arctg}\ (\Delta E/\hbar Q^2 D)$. Obviously, for an absolute determination of D, the resolution function must be reliably known.

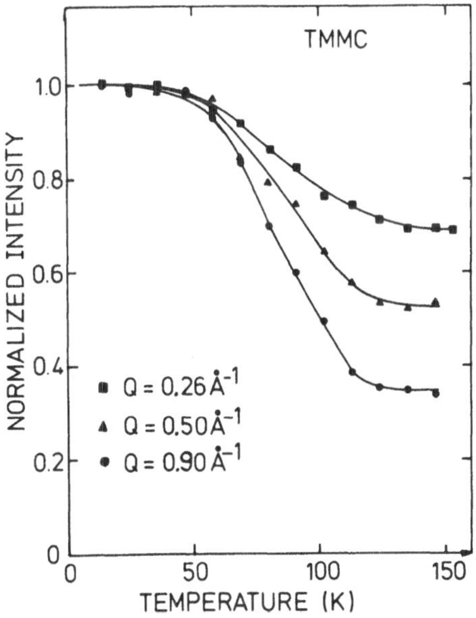


Fig. 5.15 "Fixed window spectroscopy". Intensity I (0,Q) at nominal zero-energy transfer for scattering on TMMC. The step occurs at a temperature where the resolution is comparable with the quasielastic width. The step height gives approximately $F_0(Q)$ (5.41) (ALEFELD et al., 1976)

### 5.2.4 Low Temperature Experiments; Quantum Rotations

Quantum states of molecular groups at low temperature, namely hindered rotations or tunneling transitions, were identified by magnetic resonance methods. One way of investigation is to measure the temperature dependence of the protonic spin-lattice relaxation time $T_1$: Low energy states are responsible for an unsual temperature dependence, as observed for methyl groups at small temperatures (e.g., ZWEERS, 1974;

HAUPT, 1971). Another way of investigation is by means of resonant phenomena in a material which is doped by free radicals with unpaired electrons. By varying an external magnetic field, resonances are observed if the electron Larmor frequency coincides with one of the tunneling transitions, for instance for methyl groups, or for solid $CH_4$ as a host material (CLOUGH and HILL, 1975; GLÄTTLI et al., 1972).

Incoherent neutron scattering on the protons of the rotating groups is the most direct means to observe such transitions, and transition energies above 1 μeV can be observed. First experiments of this kind were carried out on crystalline *4-methyl-pyridine* (ALEFELD et al., 1975)

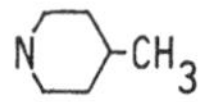


where magnetic resonance experiments have suggested an extremely low rotational potential for the $CH_3$-group. Fig.5.16 shows a typical neutron spectrum measured at 5 K on the energy loss side at Q = 2.5 $Å^{-1}$. In addition to a rather strong elastic line, three weak peaks are observable with energy transfers of 0.52 ± 0.01; 1.41 ± 0.03, and 1.92 ± 0.03 meV. The 0.52 meV peak was also found on the energy gain side where the neutron energy is increased by the scattering process. These peaks were interpreted in terms of the calculated level scheme of Fig.5.11. The measured transition energies can be explained by a sixfold cosine potential with a height of $V_6$ = 15 ± 3 meV or 0.35 kcal/mol. A purely threefold cosine potential can be

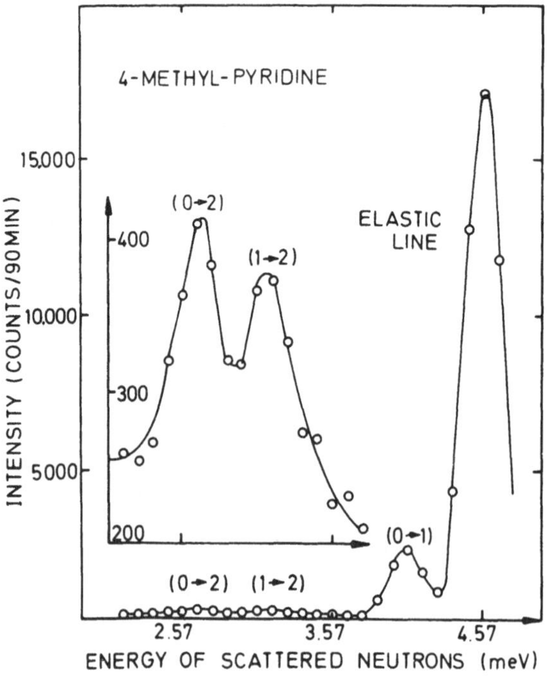


Fig. 5.16 Neutron spectra for transitions between states due to tunneling rotations of the $CH_3$-group in 4-methyl-pyridine. 0,1,2 means ground state, first, and second excited state. The transition energies can be explained in terms of the level scheme Fig. 5.11. Elastic line at 4.57 meV (ALEFELD, KOLLMAR, and DASANNACHARYA, 1975)

excluded. The observed value of $V_6$ is nearly twenty times larger than the "internal" barrier of the free molecule which has also a sixfold symmetry and which is known from microwave experiments. This means that the rotational barrier is essentially due to the crystal field from the surrounding molecules. As a matter of fact, the experiment rules out a purely threefold cosine potential, but it cannot exclude higher Fourier components. From proton magnetic resonance $T_1(T)$-experiments, an energy splitting of 0.60 meV between the ground state and the first excited state of the $CH_3$-rotator was deduced (ZWEERS et al., 1974). The discrepancy with the value quoted before may be attributed to the difficult interpretation of a spin-lattice relaxation time in terms of tunneling states.

Investigations by means of the backscattering spectrometer were performed on the linear molecule *dimethylacethylen*

$$CH_3 - C \equiv C - CH_3$$

in its crystalline phase (ALEFELD and KOLLMAR, 1976). Also for this substance, nmr $T_1$-experiments have suggested a comparatively low activation energy for the rotations ($E_{act} \simeq$ 0.83 kcal/mol). Therefore observable tunneling transitions were expected. Neutron scattering experiments at 4.5 K demonstrated the appearance of well-defined peaks at ± 1.7 μeV (Fig.5.17). They were explained by a tunneling splitting of the librational ground state. No other level was observed when the energy range was extended up to $10^{-4}$ eV. Therefore it was concluded that the potential is threefold which leads to a twofold ground state splitting. With the help of the eigenvalues of the Mathieu-equation this assumption yields a potential height of $V_3$ = 1.04 kcal/mol or 45 meV. The difference between this value and the activation

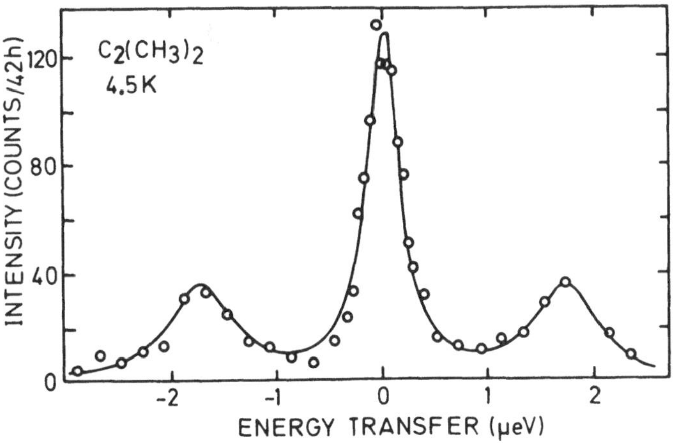


Fig. 5.17 Neutron spectrum on crystalline $C_2(CH_3)_2$. The peaks correspond to transitions between the levels of the librational ground state split by tunneling (ALEFELD and KOLLMAR, 1976)

energy quoted before can be understood by the fact that $E_{act}$ counts from the librational ground state to the top of the hindrance potential. A comparison between $E_{act}$ and the potential height as determined from a *sixfold* potential leads to a significant inconsistency; therefore, the threefold potential is assumed to be approximately correct. More detailed information on the shape of the hindrance potential could be obtained by the inspection of transitions from the ground state to the higher librational states, and from an investigation of the transitions within the multiplets of the higher states. Further information can be evaluated from the Debye-Waller factor of the librations which is essentially related to the curvature $V(\phi)$ near the potential minima.

The tunneling peaks in Fig.5.17 have a width which is larger than the resolution width (as obtained from the central line), and which was observed to increase with temperature. Above 20 K it may be assumed that this is caused by the lifetime of the tunneling states, for instance due to interactions with the phonons. This width should become observable as soon as the transition rate between the librational ground state and the higher librational states becomes comparable with the tunneling frequency. If the temperature is increased further, the tunneling states fade out, and the rotation finally ehanges into a "hopping" or jump-type motion as described earlier (e.g., STONEHAM, 1975).

Obviously, the experiments on 4-methyl-pyridine and on dimethylacethylene represent two limiting cases of non-classical motions of a methyl group at low temperatures: in the first case (left-hand side of the diagram in Fig.5.11) the rotational potential is low, and, consequently, the energy levels are only slightly shifted in relation to those of a one-dimensional free rotator. In the second case the barrier is relatively high, and the lower states of the level scheme are close to that of a harmonic libration. The corresponding librational states are split due to the overlap of the wave functions in the neighbouring potential minima. In the framework of this section, *solid methane* is particularly interesting because *both* these cases of rotations coexist, namely molecules with weak and with strong hindrance potential. At 20.4 K, $CH_4$ undergoes a phase transition from a fcc plastic phase with one randomly orientated molecule per unit cell, to an ordered low temperature phase which is supposed to exist down to zero temperature. In the ordered phase the structure was evaluated on $CD_4$. It has a fcc unit cell with eight molecules (Fig.3.20): Six of them are orientationally *ordered*, similar to the spins in an antiferromagnet. These six molecules form a cage, and the orientational potential acting on the molecule in the center of this cage is very weak: the lowest order interaction (due to electric octupoles) disappears, and only a small cubic crystal field persists which appears also in the disordered pahse. As a consequence, this central molecule is *unoriented*.

This structure is reflected directly by the dynamical behaviour of the molecules (HÜLLER and KROLL, 1975) as observed by incoherent neutron scattering. In the low

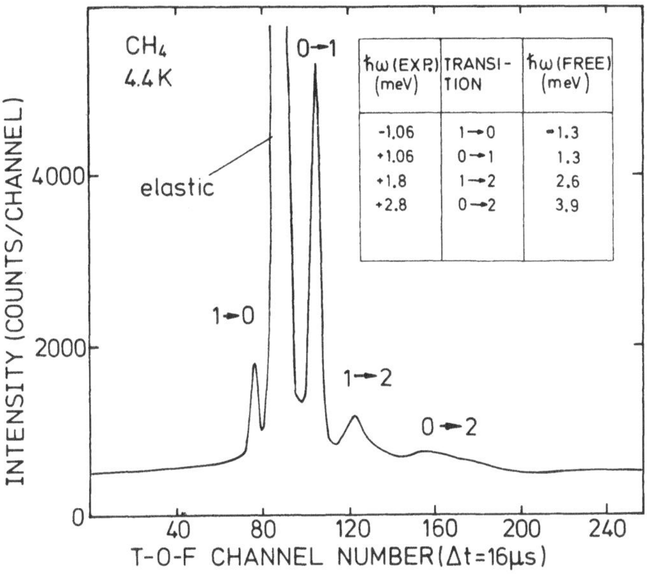


Fig. 5.18 Nearly free rotations in solid $CH_4$ in its low temperature phase. Insert: comparison with the energies of free rotational transitions (KAPULLA and GLÄSER, 1972)

temperature phase, two kinds of spectra are to be expected, and were actually observed. First, nearly free rotations occur for the molecules at the disordered sublattice. These were identified by several authors (e.g., KAPULLA and GLÄSER, 1972). As shown in Fig.5.18, the rotational transitions are slightly shifted with respect to the free rotator, as expected from the influence of the small cubic crystal field (YAMAMOTO, 1968). Secondly, for the molecules on the ordered "antiferromagnetic" sublattice we may expect split librational states, or at least a split librational ground state.

In contrast to the $CH_3$-rotations treated before, in the case of methane the potential can be calculated from first principles. Ignoring the weak cubic crystal field, the potential can be described by an octupole-octupole interaction with the surrounding molecules. The corresponding average orientational potential is then

$$\langle V \rangle = A(T) H_{11}^{(3)} \tag{5.42}$$

where $H_{11}^{(3)}$ is a cubic harmonic and it depends on the Euler angles describing the molecular orientation. Ignoring librations and orientational disorder, the interaction constant $A_3(0)$ can be related to $Q_3$, the octupole moment of the molecule, or to the critical temperature $T_c$ of the phase transition,

$$A(T) = (357/4)(Q_3^2/R^7) = 7k_BT_c \quad . \tag{5.43}$$

R is the nearest neighbour distance. This potential leads to three sublevels of the librational ground state, called A(2), T(1), and E(0), where A is the lowest sublevel (total nuclear spin in brackets). The allowed transitions have energy transfers $E_T - E_A$ and $E_E - E_T$. The transition A - E with $\Delta I = 2$ is forbidden since the neutron has spin 1/2 and can therefore flip one proton only. This selection rule is a peculiarity of neutron scattering which does not exist for light spectroscopy.

For the first time, such tunneling transitions were observed by electron spin-resonance (GLÄTTLI et al., 1972). More recently, neutron scattering experiments (PRESS and KOLLMAR, 1975) on a cold source triple axis spectrometer, with a resolution of 0.04 meV, revealed clearly two peaks on the gain on the loss side of the spectrum, at $\hbar\omega$ = 143 µeV and 73 µeV, with an accuracy of a few µeV. These energies can be attributed to the T-A and to the T-E transitions, respectively. The ratio of the transition frequencies is close to the theoretical value $\omega_{TA}/\omega_{TE} = 2$ for large $A_3$ where the 180° overlap of the protons is small compared to the 120° overlap. Also, the absolute value of the transition frequencies agrees roughly with theory where $A_3$ was calculated either from $T_c$ or from the amplitude of the molecular librations as determined by neutron scattering. There is also agreement with the calculations of KATAOKA et al. (1973). The 143 µeV transition coincides with one of the transitions observed with the resonance method mentioned before. (For a general treatment of the splittings for tetrahedral molecules see NAGAMYA, 1951).

Figure 5.19 shows the higher one of the tunneling transitions and one of the rotational transitions as a function of temperature, together with the corresponding line width. Obviously, the tunneling frequency increases strongly as the phase

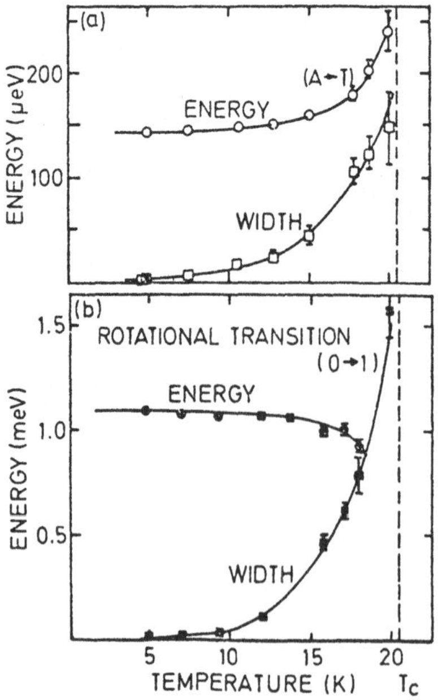


Fig. 5.19 Energy of transitions (a) due to the tunneling splitting, and (b) due to the nearly free rotations of the $CH_4$ molecules in solid methane which can be observed simultaneously in the low temperature phase. Above $T_c$ the orientational structure is disordered. The lower curves show the width of the transition lines (PRESS and KOLLMAR, 1975)

transition temperature is approached: This is due to a reduction of the orientational order parameter which leads to a decrease of the average interaction at the ordered sites, $A_3(T)$. At the same time, the energy of rotational transitions for the non-oriented molecules *decreases*. This can be related to the observed increase of rotational damping as $T_c$ is approached. At $T_c$ both sublattices become equivalent, and, consequently, tunneling and rotational transitions merge together, and all molecules perform the same random, rotational motion. The corresponding neutron spectrum is a Lorentzian-shaped quasielastic line plus a delta function, as observed by KAPULLA and GLÄSER (1972), and methane behaves like the plastic crystals discussed in Subsec.5.2.2.

Finally we mention briefly the weakly hindered three-dimensional rotations in solid hydrogen. Transitions from J = 0 to 1 were observed in the neutron spectra, with an energy change between about 13.7 and 14.7 meV, depending on the ortho-concentration. This spin-flip scattering process is a transition between the para and the ortho state of $H_2$. The rotational hindrance for J = 1 is of the quadrupole-quadrupole type. From the energy difference for the J = 1 state in pure o-$H_2$ and (extrapolated) pure p-$H_2$, the strength of the quadrupole interaction was evaluated (STEIN et al., 1972; SCHOTT, 1970).

As a brief *summary* of this part of the section we make the following statements. In the region of classical rotations the quasielastic spectrum for single crystals yields information on the geometry of the rotations. If this information is established, the corresponding rotational rate(s) can be evaluated, as we believe, with a higher reliability than is generally possible from nmr $T_1$-experiments. At low temperature and in cases where the barrier height is relatively small one obtains a tunneling spectrum instead of the quasielastic line.

The tunneling transitions allow the reconstruction of the energy levels of the tunneling rotations or of the tunneling split librational ground state. Under favourable conditions, the shape of the hindrance potential can be evaluated from this spectrum. In contrast to classical rotations, the calculation of the transition matrix elements or of the scattering cross-section is difficult, in particular because each rotational state is correlated with a certain spin configuration in the molecule. Under these conditions, the separation of the cross-section in a coherent and an incoherent part looses its sense. This is well known for liquid hydrogen, but it has not yet been treated extensively for rotations of more complicated molecules. At the end of this discussion we draw the reader's attention to the interesting evidence that tunneling rotations can manifest themselves also in a "macroscopic" scale: For certain materials a relation was observed between methyl-group tunneling and viscoelastic relaxation (see WILLIAMS et al., 1975).

Finally we treat an interesting aspect of the hydrogen motions in molecular crystals which is only loosely related to the main subject of the sections. It concerns the fluctuation of a proton along *hydrogen bonds* which play an important role in ferroelectrics, as well as in biological molecules.

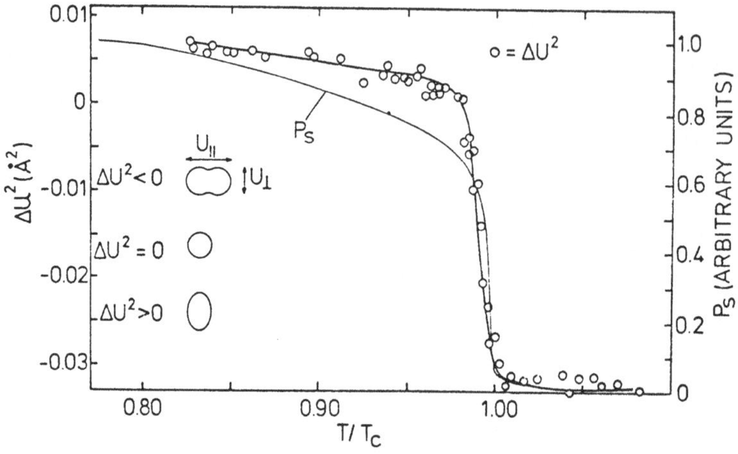


Fig. 5.20 Anisotropy of the incoherent structure factor for protons in $KH_2PO_4$. $\Delta u^2$ gives the deviation of the mean-square amplitude from a spherical distribution of the proton, parallel and perpendicular to the hydrogen bond (along a-axis). Below the ferroelectric phase transition at $T_c$ the protons are localized on one side of the bond: the structure factor is nearly isotropic. For comparison the electric polarization $P_s$ is shown (thin line)(ARSIC-ESKINIA, GRIMM, STILLER, 1972)

This effect has been studied by incoherent neutron scattering in the case of $KH_2PO_4$ where (O-H-O) bonds form connections between adjacent $PO_4$-ions. The elastic line from proton scattering was investigated as a function of the angle between the hydrogen bond axis and the scattering vector $\underset{\sim}{Q}$ (GRIMM et al., 1970). The result is shown in Fig.5.20. In the paraelectric phase, above the transition temperature, a strong anisotropy was observed. This means that the proton fluctuates along the bond and, consequently, the incoherent structure factor corresponds to a strongly elongated density (or probability) distribution. Below $T_c$ the salt is ferroelectric and most of the protons are ordered and localized at the center of the bond. Therefore, the density distribution is more spherical. In principle, the Q-dependence of the elastic line would reveal whether the proton is tunneling between two potential minima across the bond, or if it just performs a strongly anisotropic and anharmonic vibration. Investigations at larger Q could clarify this difference.

## 5.3. Liquids, Liquid Crystals, and Related Problems

### 5.3.1 The Scattering Law for Molecular Liquids

For a liquid with hydrogenous molecules the protons undergo motions which are a superposition of the translational diffusion of the molecule as a whole, and of the various rotations about the molecular axes. In addition, side-group rotations may occur which lead to a further complication of the spectrum. As already mentioned (see Fig.5.2) the separation of these components is difficult; to be more specific it will be feasible only if the widths of the corresponding spectral distributions are sufficiently different. Roughly speaking, a separation may be possible if

$$Q^2D < D_r \quad \text{or} \quad Q^2D < (k_BT/J)^{1/2} \tag{5.44}$$

where D is the self-diffusion constant, and $D_r$ is the rotational diffusion coefficient. The second inequality holds for molecules with a small moment of inertia J, where the width of the rotational spectrum is determined by more or less free rotations. In addition, the experimental resolution width must be smaller than $Q^2D$. As a matter of fact, we can always fulfill the conditions (5.44) by the choice of a sufficiently small Q. However, this makes the observation of the rotational spectrum difficult, since its intensity goes as $1 - [\sin(QR_0)/QR_0]^2 \approx Q^2R_0^2/6$ (5.37). As a consequence, the simultaneous observation and interpretation of the rotational and the translational spectrum is possible only if the liquid has a high viscosity (e.g., glycerol at lower temperature), for small molecules (e.g., liquid $CH_4$), and for liquid crystals. The theories for the scattering law of molecular liquids have been formulated with the simplifying assumption that translational and rotational motions are uncorrelated. Quantitatively, this means that the intermediate scattering law can be factorized as

$$F_s(Q,t) = F^{(rot)}(Q,t)\; F^{(trans)}(Q,t) \quad . \tag{5.45}$$

Consequently, $S_s(Q,\omega)$ is a convolution of the scattering law for a rotating molecule in a solid, $S_s^{(rot)}$ as in (5.34), and a Lorentzian-shaped scattering law for diffusion $S_s^{(D)}(Q,\omega)$ with a half-width

$$\Gamma = 2Q^2D \quad . \tag{5.46}$$

The resulting scattering law has then the form

$$S_s(Q,\omega) = \frac{(\Gamma/\pi)}{\Gamma^2 + \omega^2}\, F_0(Q) + S_s^{(D)} \times S_s'^{(rot)} \tag{5.47}$$

where $S_s'^{(rot)}$ means $S_s^{(rot)}$ without the $\delta(\omega)$-term.

For liquids with strong association, and also in many other cases, the factorization may not be fully justified. More general models have been worked out assuming that the molecule alternates, stepwise, between two kinds of motion. For instance, one may assume that the molecule spends a certain time interval in a rotational or librational state of motion, and, in the next step, spends another time interval by carrying out a purely diffusive motion; then it again returns to the rotational state, and so on. The scattering law for such a Marcovian sequence was formulated first by SINGWI and SJÖLANDER (1960, see also ROSCISZEWSKI, 1974).

In view of the arguments given above we may state that the applicability of incoherent quasielastic scattering for the investigation of liquids is relatively restricted in its usefulness, except for very simple ones (e.g., OLSSON and LARSSON, 1975; for references see LARSSON, 1968 and 1973; SPRINGER, 1972). However, this method may develop into a useful tool for the study of liquid crystals, in particular to investigate the self-diffusion constant of the molecules and its anisotropy, and the rotations of the individual molecules; these rotations have rather short characteristic times ($10^{-10}$- $10^{-12}$ s) which are often difficult to determine by means of other methods. An example of such an experiment will be discussed subsequently.

### 5.3.2 Liquid Crystals, Polymers

Earlier quasielastic scattering experiments on liquid crystals with moderate energy resolution could not separate experimentally the translational and the rotational components of the spectra, and only a fitting procedure with the theoretical scattering law allowed a determination of the parameters of the rotational motion (JANIK et al., 1971 and 1974). From experiments on nematic PAA (p-azoxyanisol) with a backscattering spectrometer at relatively small Q-values an observation of the translational spectrum was possible (TÖLPER et al., 1974). Its half-width was interpreted in terms of the diffusive motion of the molecule, and consistency was obtained with the results of tracer experiments. Moreover, the anisotropy of the diffusive part of the spectrum for an aligned sample was investigated which led to the components of the diffusion tensor. For this purpose, the theory of anisotropic diffusion of ROSCISZEWSI (1972) is used, and it shows that, for an oriented sample, the scattering law is still a Lorentzian, whose width $Q^2D$ has to be replaced by $D^{ij}Q_iQ_j$, where $D^{ij}$ and $Q_i$ are the components of the diffusion tensor and of $\underline{Q}$, respectively. For an unoriented sample with a multidomain structure, it should be emphasized that the scattering law $S_s$ has to be averaged, and not $D^{ij}$. Only for weak anisotropy, (5.46) still holds approximately, where D has to be replaced by the average $D = (1/3)\, D_{\parallel} + (2/3) D_{\perp}$. More recently, scattering experiments to study ments the self-diffusion tensor for a nematic crystal were reported by LEADBETTER et al. (1975).

Both the rotational part of the spectrum, and the incoherent structure factor $F_0(Q)$ of the diffusion-broadened quasielastic line, were studied in the smectic B-phase of TBBA (terephthal-bis buthylaniline) by HERVET et al. (1974). An important improvement was achieved by the use of a partially deuterated compound D-TBBA:

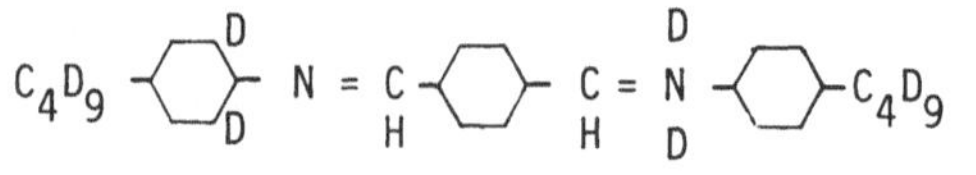 .

For this material, the spectrum will reveal essentially the motion of the aromatic body of the molecule. On the other hand, for normal TBBA the spectrum includes the components from the butyl tail motions. The structure of the smectic B-phase is not yet understood completely. It might either be a three-dimensional crystal with rotational disorder, or a layered structure of two-dimensional crystals gliding on each other. The experiments were carried out on a time-of-flight multichopper at a cold source with a resolution of 33 μeV at 9.5 Å incident wave length. The measured spectra were fitted with the scattering law for a uniaxial cylindrical rotation of the linear molecule about its long axes, assuming jumps of the proton between N orientations equally spaced on a circle of radius $R_0$. The corresponding scattering law was derived earlier for the interpretation of rotations in solid paraffins by BARNES (1973; see also DIANOUX et al., 1975). The incoherent structure factor of the rotation is shown in Fig.5.21a as calculated from a fit of the quasielastic spectrum with the theoretical scattering law. The solid curve corresponds to a radius of gyration of $R_0$ = 2.4 Å (with N = 6 which practically corresponds to the case of continuous rotation). From the width of the quasielastic spectrum, the jump rate was calculated ($1/\tau_r = 5.10^{10}\ s^{-1}$ at 119 °C). It was shown that the assumption of a spherically isotropic rotation does not agree with the experimental data.

Since there is no significant difference between the deuterated and the normal TBBA data, one concludes that the radius of gyration is essentially the same for the butyl-chains and for the molecular body. However, a difference appears with respect to the total intensity which gives the structure factor (or Debye-Waller factor) due to motions *faster* than those described before. In Fig.5.21b the total

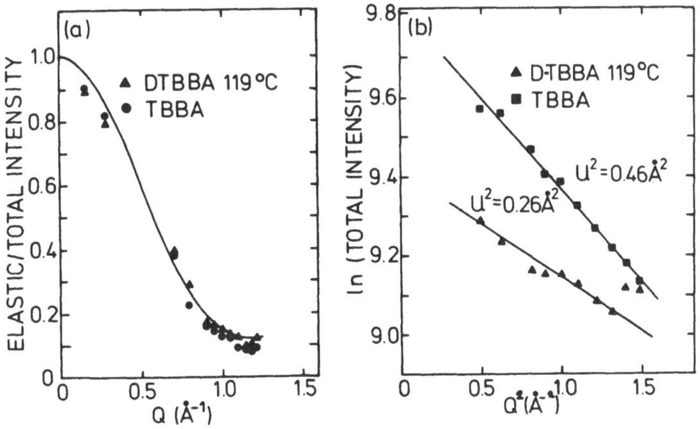


Fig. 5.21a,b Scattering on the smectic B-phase of deuterated and normal TBBA. a) Intensity of the (nearly) elastic line vs. Q. Solid curve: calculated for uniaxial rotation. b) Total intensity as a function of $Q^2$ where $u^2$ = apparent mean-square amplitude of the proton (HERVET, VOLINO, DIANOUX and LECHNER, 1974)

intensity is presented in a lnI vs. $Q^2$-plot where $u^2$ has the meaning of an effective mean square amplitude of the protons for times shorter than $\tau_r$. The difference between the observed $u^2$ for TBBA and for D-TBBA suggests that the amplitude of the rapid tail motion is considerably larger than the motion of the molecular body. The authors conclude that the experiments support the picture of a random *uniaxial* rotation of the whole molecule, together with large amplitude displacements of the butyl-chains out of the plane of this rotation. A more recent and comprehensive study of quasielastic scattering on TBBA was published by VOLINO et al. (1976).

We conclude this section with a few general remarks on the investigation of *non-periodic polymeric motions*, a field of research which has just started, and which is expected to attain increasing importance. Qualitatively, the following kinds of motions have to be considered: First of all, a polymeric chain undergoes a diffusive motion with respect to its center of gravity, provided that the temperature is sufficiently high. This leads to a quasielastic width $Q^2D$ as discussed earlier. In most cases this motion is too slow to be "seen" in the quasielastic spectrum. However, there are other non-periodic motions: If the incoherent scattering law is studied at larger Q (say > 0.5 $Å^{-1}$) we may observe the random motions of an *individual* chain segment, or of a short sequence of repeat units (for instance kink jumps, or "crankshaft" motions). Such motions have been observed in the non-crystalline regions of solid polyethylene by means of the fixed-window-method (Subsec.5.2.3; PETERLIN-NEUMEIER and SPRINGER, 1976). From the incoherent structure factor, and from the width of the quasielastic spectrum, the observed process was interpreted in terms of the well-known $\gamma$-process.

For smaller Q, diffusive motions of a single chain segment over *larger* distances will appear in the spectrum (with the center of gravity practically fixed). Such motions can be treated analytically in their asymptotic limits. For instance, the "internal" diffusion of a segment in a polymeric coil has been treated (DE GENNES, 1967; DUBOIS-VIOLETTE and DE GENNES, 1967): For a proton in a coil which is embedded in a viscous medium, the mean square displacement under the influence of a random force, $< \underline{r}^2(t) >$, is proportional to $t^{1/2}$ (rather than Dt as for ordinary diffusion). It can be shown that the corresponding scattering law is a complicated non-Lorentzian curve whose half-width goes approximately as $\Gamma_{1/2} = 3.\ 10^{-2} DRQ^4$ where D is the center-of-mass diffusion constant. $R^2$ is the mean square radius of the polymeric coil. This relation holds for $R \gg 1/Q \gg b$, where b is the length of one chain segment. Consequently, Q must be of the order of a few $10^{-2}$ $Å^{-1}$, which needs high resolution in $\omega$ *and* in Q! (See ALLEN and HIGGINS, 1973; in this reference, "D" is not to be identified with the "true" diffusion constant, but with some sort of a mobility which depends on the Q-value of the experiment under consideration). Generally speaking there is a wide variety of eigenvalues or characteristic relaxation times describing the random motion of a proton in a polymeric chain, and therefore a great number of components in the composite quasielastic spectrum. For a polymer in solution we expect that these relaxation times are, more or less, continuously spread over a large frequency region, reaching from $Q^2D$ up to $1/\tau$ where $\tau$ is the characteristic time of the motion of a *single* segment. In a bulk polymer, however, we may distinguish a group of high frequency components, to be visible at larger Q, which are due to the motion of individual segments, and a bundle of components with very low frequencies appearing only at small Q. The investigation of these low-Q or long range motions will become one of the interesting applications of high resolution spectroscopy.

*Acknowledgment*

Thanks are due to B. Alefeld for leaving us unpublished results and for many discussions, as well as to H. Stiller for valuable suggestions and discussions.

References

Ait-Salem, M. (1976): Thesis, to be published (internal Jülich report).
Alefeld, B. (1975): pers. communic.
Alefeld, B., Prager, M. (1976): Proc. Congress Ampère (Heidelberg) p. 389
Alefeld, B., Birr, M., Heidemann, A. (1969): Naturwissenschaften 56, 410
Alefeld, B., Kollmar, A., Dasannacharya, B.A. (1976): J. Chem. Phys. 63, 4415
Alefeld, B., Kollmar, A., (1976): Phys. Lett. 57A, 289
Alefeld, B., Prager, M., Springer, T., White, J. (1975): to be published
Alefeld, B., Bohn, G.G., Stump, N. (1975): International Meeting on Hydrogen in Metals, Jülich 1972, Jül-Conf.-6, Vol. I, p. 286
Alefeld, B., Stockmeyer, R. (1973): unpublished
Alefeld, G., Völkl, J., Schaumann, G. (1970): Phys. Stat. Sol. 37, 337
Allen, G., Higgins, J.S. (1973): Rep. Progr. Phys. 36, 1073
Arsic-Eskinja, M., Grimm, H., Stiller, H. (1972): *Neutron Inelastic Scattering Proc. IAEA*, Vienna, p. 825
Barnes, J.D. (1973): J. Chem. Phys. 58, 5193
Bauer, G., Schmatz, W. (1976): Proc. Internat. Conf. on "Effects of Hydrogen on Behavior of Materials" (Proc. AIME).
Bauer, G., Seitz, E., Horner, H., Schmatz, W. (1975): Solid State Commun. 17, 161
Beg, M.M., Ross, D.K. (1970): J. Phys. C: Solid State Phys. 3, 2487
Carlile, C.J., Ross, D.K. (1974): Solid State Commun. 15, 1923
Carstanjen, H.D., Sizmann, R. (1972): Phys. Lett. A40, 93
Chudley, C.T., Elliott, R.J. (1961): Proc. Phys. Soc. 77, 353
Clough, S., Hill, J.R. (1975): J. Phys. C: Solid State Phys. 8, 2274
De Gennes, P.G. (1967): Physics 3, 37, 181
De Graaf, L.A., Rush, J.J., Flotow, H.E., Rowe, J.M. (1972): J. Chem. Phys. 56, 4574
Dianoux, A.J., Volino, F., Hervet, H. (1975): Mol. Phys. 30, 1181
Dubois-Violette, E., De Gennes, P.G. (1967): Physics 3, 181
Egelstaff, P.A. (ed.) (1965): *Thermal Neutron Scattering* (Academic Press, London)
Egelstaff, P.A. (1970): J. Chem. Phys. 53, 2590
Funke, K., Kalus, J., Lechner, R.E. (1974): Solid State Commun. 14, 1021
Gissler, W., Rother, H. (1970): Physica 50, 380
Gissler, W., Stump, N. (1973): Physica 65, 109
Gissler, W., Alefeld, G., Springer, T. (1970): J. Phys. Chem. Solids 31, 2361
Gissler, W., Jay, B., Vinhas, L.A. (1973): Phys. Lett. 43 A, 279
Glättli, H., Sentz, A., Eisenkremer, M. (1972): Phys. Rev. Lett, 28, 871
Gordon, R.G. (1965): J. Chem. Phys. 43, 1307
Gordon, R.G. (1966): J. Chem. Phys. 44, 1830

Grimm, H., Stiller, H., Plesser, Th. (1970): Phys. Stat. Sol. 42, 207
Haupt, J. (1971): Z. Naturforsch. 26a, 1578
Heidemann, A. (1974): Intern. Rept. ILL, Grenoble
Hervet, H., Volino, F., Dianoux, A.J., Lechner, R.E. (1974): J. de Phys. Lettres 35, L-151
Hoshino, S. (1957): J. Phys. Soc. Japan 12, 315
Hossfeld, F., Amadori, R., Scherm, R. (1970): Proc. Panel Conf. on Instrumentation for Neutron Inel. Scatt., Vienna, Int. Atomic Energy Agency
Hüller, A., Kroll, D.M. (1975): J. Chem. Phys. 63, 4495
Janik, J.A., Janik, J.M., Otnes, K., Riste, T. (1971): Mol. Cryst. and Liq. Cryst. 15, 189
Janik, J.A., Janik, J.M., Otnes, K., Rosciszewski, K. (1974): Physica 77, 514
Kapulla, H., Gläser, W. (1972): *Neutron Inelastic Scattering, Proc. IAEA*, Vienna, p. 841
Kataoka, Y., Okeda, K., Yamamoto, T. (1973): Chem. Phys. Lett. 19, 365
Kaufmann, B., Lipkin, H.J. (1962): Ann. Phys. 18, 294
Kehr, K., (1976): Jül-1211, Diffusion von H in Metallen
Kley, W. (1966): Z. Naturforsch. 21a, 1770 (Supplement)
Kodoma, T. (1972): J. Magn. Resonance 7, 137
Larsson, K.E. (1968): *Neutron Inelastic Scattering, Proc. IAEA*, Vienna, Vol. 1, p. 397
Larsson, K.E. (1973): J. Chem. Phys. 59, 4612
Lassier, B., Brot, C., White, J.W. (1973): J. de Phys. 34, 473
Leadbetter, A.J., Litchinski, D., Turnbull, A. (1972): *Neutron Inelastic Scattering, Proc. IAEA*, Vienna, p. 231
Leadbetter, A.J., Temme, F.P., Heidemann, A., Howells, W.S. (1975): Chem. Phys. Lett. 34, 363
Lechner, R.E., Heidemann, A. (1976): personal comm.
Lechner, R.E. (1972): Solid State Commun. 10, 1247
Livingston, R.C., Rothschild, W.G., Rush, J.J. (1973): J. Chem. Phys. 59, 2498
Livingston, R.C., Rowe, J.M., Rush, J.J. (1974): J. Chem. Phys. 60, 4541
Mandema, W., Trappeniers, N.J. (1974): Physica 76, 73 and 123
Mertens, F., Biem, W., Hahn, H. (1969): Z. Physik 220, 1
Mezei, F. (1972): Z. Physik 255, 146
Michel, K.H. (I) (1973): J. Chem. Phys. 58, 142
Michel, K.H. (II) (1973): J. Chem. Phys. 58, 1143
Nagamya, T. (1951): Progr. Theoret. Phys. 6, 702
Nelin, G., Sköld, K. (1975): J. Phys. Chem. Solids 36, 1175
Olsson, L.G., Larsson, K.E. (1975): Physica 80A, 203
Peterlin-Neumaier, T., Springer, T. (1976): J. Polymer. Sci. 14, 1315
Prask, H.J., Trevino, S.F., Rush, J.J. (1975): J. Chem. Phys. 62, 4156
Press, W., Hüller, A., Stiller, H., Stirling, W., Currat, R. (1974): Phys. Rev. Lett. 32, 1354
Press, W., Kollmar, A. (1975): Solid State Commun. 17, 405
Price, D.L., Sköld, K. (1970): Nucl. Instr. Methods 82, 208
Quentrec, B. (1975): Phys. Rev. A12. 282
Richter, D., Alefeld, B., Heidemann, A., Wakabayashi, N. (1977): J. Phys. F, in print
Richter, D., Töpler, J. Springer, T. (1976): J. Phys. F: Metal Phys. 6, L 93
Rosciszewski, K. (1972): Acta Phys. Polon. A (Poland) A41, 549
Rosciszewski, K. (1974): Physica 75, 268
Rowe, J.M., Sköld, K., Flotow, H.E., Rush, J.H. (1971): J. Phys. Chem. Solids 32, 41
Rowe, J.M., Rush, J.J., De Graaf, L.A., Fergusson, G.A. (1972): Phys. Rev. Lett. 29, 1250
Rowe, J.M., Livingston, R.C., Rush, J.J. (1973): J. Chem. Phys. 59, 6652
Rowe, J.M., Rush, J.J., Flotow, H.E. (1974): Phys. Rev. B9, 5039
Rowe, J.M., Rush, J.J., Smith, H.G., Mostoller, M., Flotow, H.E. (1974): Phys. Rev. Lett. 33, 1297
Rush, J.J., Flotow, H.E. (1968): J. Chem. Phys. 48, 3795
Rush, J.J., De Graaf, L.A., Livingston, R.C. (1973): J. Chem. Phys. 58, 3439
Rush, J.J., Livingston, R.C., De Graaf, L.A., Flotow, H.E., Rowe, J.M. (1973): J. Chem. Phys. 59, 6570

Schaumann, G., Völkl, J., Alefeld, G. (1970): Phys. Stat. Sol. 42, 401
Schmatz, W. (1975): pers. communication
Schott, W. (1970): Z. Physik 231, 243
Sears, V.F. (1966): Can J. Phys. 44, 1279, 1299
Sears, V.F. (1967): Can. J. Phys. 45, 237
Seymour, E.F.W., Cotts, R.M. Williams, W.D. (1975): Phys. Rev. Lett. 35, 165
Singwi, K.S., Sjöländer, A. (1960): Phys. Rev. 119, 863
Sköld, K. Nelin, G. (1967): J. Phys. Chem. Solids 28, 2369
Sköld, K. (1968): J. Chem. Phys. 49, 2443
Sköld, K., Dahlborg, U. (1973): Sol. State Commun. 13, 543
Smith, G.W. (1965): J. Chem. Phys. 42, 4229
Springer, T. (1972): Quasielastic Neutron Scattering for the Investigation of Diffusive Motions in Solids and Liquids. In *Springer Tracts in Modern Physics* (Springer, Berlin, Heidelberg, New York) Vol. 64
Stein, H., Stiller, H., Stockmeyer, R. (1972): J. Chem. Phys. 57, 1726
Steward, S.A. (1975): Sol. State Commun. 17, 75
Stockmeyer, R., Stiller, H. (1968): Phys. Stat. Sol. 27, 269
Stoneham, A.M. (1975): Collective Phenomena 2, 99
Stump, N., Gissler, W., Rubin, R. (1972): Phys. Stat. Sol. (b) 54, 295
Sussmann, J.A., Weissman, Y. (1972): Phys. Stat. Sol. (b) 53, 419
Thibaudier, C., Volino, F. (1973): Mol. Phys. 26, 1281; (1975) 30, 1159
Töpler, J., Alefeld, B., Springer, T. (1974): Mol. Cryst. Liq. Cryst. 26, 297
Töpler, J., Richter, D., Springer, T. (1976): to be published
Verdan, G., Rubin, R., Kley, W. (1968): *Neutron Inelastic Scattering, Proc. IAEA*, Vienna, Vol. 1, p. 223.
Volino, F., Dianoux, A.J., Hervet, H. (1976): J. de Phys. (Colloques)
Völkl, J. (1972): Ber. Bunsenges. Physik. Chem. 76, 797
Völkl, J., Alefeld, G. (1975): *Diffusion in Solids: Recent Developments*, ed. by A.S. Novick and J.J. Burton, (Academic Press, New York) p. 231
Wakabayashi, N., Alefeld, B., Kehr, K.W., Springer, T. (1974): Solid State Commun. 15, 503
Watton, A., Sharp, A.R., Petch, H.E., Pintar, M.M. (1972): Phys. Rev. B5, 4281
Williams, J., Shohamy, E., Reich, S., Eisenberg, A. (1975): Phys. Rev. Lett. 35, 951
Willis, B.T.M., Howard, J.A.K. (1975): Acta Cryst. A31, 514
Yamamoto, T. (1968): J. Chem. Phys. 48, 3193
Zogal, O.J., Cotts, R.M. (1975): Phys. Rev. B11, 2443
Zweers, A.E., Brom, H.B., Huiskamp, W.J. (1974): Phys. Lett. 47A, 347

## Recent References with Titles

Kehr, K.W., Richter, D. (1976): Incoherent scattering law for diffusion in crystals with random impurities. Solid State Comm. 20, 477

Eckold, G., Funke, K., Kalus, J., Lechner, R.E. (1976): The diffusive motion of silver ions in α-AgI: Results from quasielastic neutron scattering. J. Phys. Chem. Solids 37, 1097

Schlaak, M., Lassegues, J.C., Heidemann, A., Lechner, R.E. (1977): Reorientations in crystalline $(CH_3)_3NHCl$ studied by quasielastic neutron scattering. Mol. Phys. 33, 111

Leadbetter, A.J., Lechner, R.E. (1977): Neutron Scattering Studies. In *The Plastic Crystalline State*, ed. by J.N. Sherwood (John Wiley and Sons, New York)

Dianoux, A.J., Heidemann, A., Volino, F., Hervet, H. (1976): Self-diffusion and undulation modes in a smectic A liquid crystal. A high-resolution neutron scattering study. Mol. Phys. 32, 1521

Volino, F., Dianoux, A.J., Hervet, H. (1976): Neutron quasielastic scattering study of rotational motions in the smectic C, H and VI phases of TBBA. J. Physique, Coll. C3, suppl. 6, 37, C3-55

Lechner, R.E., Heidemann, A. (1976): Rotational motion in plastic adamantane: The elastic incoherent structure factor. Communications on Physics 1, 213

Prager, M., Alefeld, B. (1976): Rotational tunneling splitting of the $NH_4$ ion in $NH_4ClO_4$ studied by inelastic neutron scattering. J. Chem. Phys. 65, 4927

Prager, M., Alefeld, B., Heidemann, A. (1976): The tunnel splitting of the $NH_4$-vibrational groundstate in $NH_4ClO_4$ measured by inelastic neutron scattering. *Proceedings of the XIXth Congress Ampere*, Heidelberg, p. 389

Müller-Warmuth, W., Schüler, R., Kollmar, A., Prager, M. (1976): Tunneling in methyl substituted pyridines as studied by NMR spinlattice relaxation and inelastic neutron scattering. *Proceedings of the XIXth Congress Ampere*, Heidelberg, p. 345

de Gennes, P.G. (1976): Dynamics of entangeled polymere solutions. Macromolecules 9, 587, 594

# 6. Collective Modes in Classical Monoatomic Liquids

R. D. Mountain

With 8 Figures

The study of collective modes involves the investigation of dynamical processes involving correlated motions of large numbers of particles. As was shown in Chap.1, the coherent scattering of neutrons by a liquid is a measure of the correlated motion of the atoms in the liquid. Thus, the formalism developed in Chap.1 provides a convenient starting point for the discussion of collective effects in liquids. There it was shown for a classical liquid in thermal equilibrium that the scattering law $S(Q,\omega)$ is the power spectrum of the density fluctuations in liquid. The way to be followed in this chapter in the discussion of collective motions in liquids is implicit in this observation. The intermediate scattering function, $F(Q,t)$, is the time correlation function for density fluctuations of wave-vector Q. Techniques for describing time correlation functions have been developed which enable one to emphasize those physical aspects of the dynamics thought to be important. The resulting theoretical description is formally exact. Even though it contains molecular quantities which are difficult to evaluate, the formalism identifies the molecular processes described by these quantities. This is an advantage when interpreting collective mode data.

The study of "collective modes" in liquids by neutron spectroscopy has, until recently, yielded ambiguous results. The is in part due to experimental problems of low resolution, etc., and in part to the absence of a clear concept of how such modes are manifest in neutron spectra. Within the last few years, these difficulties have been resolved and a number of studies of these modes have been published. Our discussion will be restricted to recent measurements. The review by COPLEY and LOVESEY (1975) contains an overview of available measurements.

The first task in the study of collective modes is to provide a useful definition of such modes. Recent developments in the statistical mechanics of dynamical processes provide such a means of defining modes and, quite importantly, also provide a way of relating them to experimentally accessible quantities. These theoretical techniques are commonly known as "generalized hydrodynamics", and form the basis for the discussion of collective modes in liquids in this chapter.

The use of generalized hydrodynamics to describe density fluctuations in the neutron scattering range of wave-vectors makes it possible to have a unified description for both light scattering and neutron scattering studies. Ordinary hydrodynamics provides an accurate description of the density fluctuations in a fluid if 1) there are no other slow processes operating and 2) the wavelength of the fluctuations is large compared with all molecular scales of distance. Condition 1) is violated, for example, when internal degrees of freedom of the molecules couple to the density (MOUNTAIN,1966) and in highly viscous liquids (DEMOULIN, et al., 1974). In such situations the generalization of hydrodynamics appears in the form of a frequency dependent viscosity. These effects are readily observed in the light scattering spectra of molecular liquids. Condition 2) is violated in most neutron scattering studies of liquids since the range of spatial correlation in liquids is on the order of 10 Å or so. In this case the transport coefficients become frequency and wave-vector dependent because there are processes taking place which are not fast compared with the decay of density fluctuations and which reflect the spatial order in the liquid. For noble gas liquids only these effects are of importance. The characterization of the processes which lead to the frequency dependence of the coefficients is an important task for future studies.

The second task in the study of collective modes is to analyze high quality coherent neutron scattering data or molecular dynamics (computer simulation) results within the framework of generalized hydrodynamics. (The exacting task of generating such data is not to be underrated. It falls outside the scope of our discussion). This type of analysis has been applied to neutron scattering results for argon, neon, rubidium and helium and to molecular dynamics simulation results for the Lennard-Jones model fluid and for a model for rubidium. The interpretation of these analyses forms the core of our discussion of collective modes.

The third task in the study of collective modes is the physical interpretation of the experimental results and the development of models which embody such interpretations. This task falls in the category of "unfinished business". The state of this interpretative task provides a measure of our knowledge of collective effects in liquids and is an indicator of scientifically interesting investigations yet to be performed.

## 6.1 Collective Modes

In order to discuss collective modes in liquids, it is necessary to specify the variables which are studied experimentally and then to understand how the dynamical information is contained in the experimental results. The next two subsections are concerned with these two aspects of collective modes.

### 6.1.1 Collective Variables

The general form for the collective variables of interest to liquid studies is

$$X_k = \sum_{j=1}^{N} f(\underline{r}_j, \underline{v}_j) \exp(i\underline{k} \cdot \underline{r}_j) \quad . \tag{6.1}$$

The collective variable X is the spatial Fourier transform of some density $f(\underline{r},\underline{v})$ which is expressed in terms of the positions and velocities of the particles in the liquid. A single particle variable on the other hand depends only on the coordinates of one particle. The time dependence of $X_k$ is implicit in the time dependence of the particle coordinates.

For systems in thermal equilibrium, the objects directly related to experimental quantities are not the collective variables themselves but rather time correlation functions of the variables. Thus we are concerned with quantities of the type

$$C_X(k,t) = \langle X_k(t) \; X_{-k}(0) \rangle \tag{6.2}$$

and the power spectrum of this quantity,

$$C_X(k,\omega) = \int_{-\infty}^{\infty} dt \; e^{-i\omega t} \; C_X(k,t) \quad . \tag{6.3}$$

In (6.2), the angular brackets indicate an equilibrium average of the enclosed quantity. For classical systems, $C_X(k,\omega)$ is the cosine-transform. In the discussion of coherent neutron scattering, we assume that the detailed balance condition has been taken into account so that the classical symmetry in $\omega$ and t exists; cf. (1.9) and (1.47).

Coherent neutron scattering probes density fluctuations. The primary variable of interest is therefore the Fourier component of wave-vector Q of the density fluctuations of the fluid.

The explicit form for the variable is

$$n(Q) = \sum_{j=1}^{N} \exp(i\underline{Q}\underline{r}_j) - \delta_{\underline{Q},\underline{0}} N \quad . \tag{6.4}$$

The experimentally observed quantity is not the density fluctuation itself, but the power spectrum of the density fluctuations,

$$\tilde{S}(Q,\omega) = \int dt \; e^{-i\omega t} \langle n(Q,t) \; n^*(Q,0) \rangle \quad . \tag{6.5}$$

For coherent neutron scattering, the connection between experiment and collective modes is expressed in (6.5) through the time correlation function for density fluctuations of wave-vector Q.

When discussing the dynamics of density fluctuations, it is useful to introduce two other collective variables which are the momentum density $\underset{\sim}{J}(\underset{\sim}{Q})$

$$\underset{\sim}{J}(\underset{\sim}{Q}) = M \sum_{j=1}^{N} \underset{\sim}{v}_j \exp(i\underset{\sim}{Q}\,\underset{\sim}{r}_j) \quad , \tag{6.6}$$

where M is the mass of a particle, and the energy density E(Q),

$$E(Q) = \frac{1}{2} \sum_{j=1}^{N} [Mv_j^2 + \sum_{\ell \neq j} \phi(\underset{\sim}{r}_{j\ell})][\exp(i\underset{\sim}{Q}\underset{\sim}{r}_j) - \delta_{\underset{\sim}{Q},\underset{\sim}{0}}] \quad . \tag{6.7}$$

The delta-function terms have been subtracted off so that the variables are, therefore, fluctuations of wave-vector $\underset{\sim}{Q}$ in the number density, momentum density and energy density of the fluid.

The reason for considering these collective variables is discussed in the next subsection where the dynamics of the fluctuations are introduced. The collective modes of liquids which are accessible to study by neutron scattering and computer simulation are those associated with the time evolution of the time correlation functions of these variables.

There is a particular type of collective variable known as a collective coordinate which merits discussion. A collective coordinate is one which exhibits damped, oscillatory behavior (RAHMAN, 1967). In solids, the collective coordinates are the phonons. Hence in solids the collective modes studied by coherent neutron scattering are directly associated with collective coordinates. This is not the case with liquids. What this means is that the modes studied in liquids are not related simply to the eigenvalues of the Hamiltonian of the system. This in turn means that the dynamical techniques useful in the study of the solid state will not be particularly useful in the study of the liquid state and that other approaches should be employed. One very promising approach is examined in the next subsection.

### 6.1.2 Generalized Hydrodynamics

Density fluctuations can be studied in the long wavelength limit by light scattering. In this limit, linearized hydrodynamics provides an accurate description of the intermediate scattering function. Hydrodynamics provides equations of motion for time correlation functions of the fluctuating quantities, the number density, the momentum density, and the energy density. Hydrodynamics also provides an unambiguous interpretation of the collective modes present in the long wavelength region. Although hydrodynamics is a macroscopic theory, it works well at the intermediate region probed by light scattering because the wavelength of light is large compared to the atomic level details.

The reader may object that the hydrodynamic equations apply to variables, not to correlation functions. The connection to the correlation functions is made in the following way. Imagine a fluid at time $t = 0$ with some fluctuation from equilibrium. The hydrodynamic equations are used to predict the most probable path for the evolution of the system, subject to that initial condition. The time correlation functions are obtained by averaging over all possible initial conditions, subject to the constraint of overall thermal equilibrium. This procedure requires the assumption of thermal equilibrium of the fluid. A clear discussion of this point was made by CALLEN and GREEN (1952).

As the wave-vector of the density fluctuations increases, the accuracy of the hydrodynamic description decreases. This is because the short range space and time correlations, which are important in the region probed by neutron scattering, are not built into the equations of linearized hydrodynamics. The point of view we shall follow in this chapter is that it is useful and desirable to retain the overall structure of the hydrodynamic equations and to modify them so that the short range space and time correlations existing at the atomic level of distances and time intervals are taken into account. The resulting equations represent a generalization of hydrodynamics. In these equations the thermodynamic coefficients appearing in the macroscopic hydrodynamic equations are replaced by wave-vector dependent quantities and the transport coefficients are replaced by wave-vector and frequency dependent quantities. In the low frequency, long wavelength limit, these quantities go smoothly over to the macroscopic coefficients. In this way, a unified description of light scattering and neutron scattering measurements is possible. This also provides a framework for discussing collective modes which is operationally well defined.

There are several ways to go about developing the equations of motion for the collective hydrodynamic variables. MURASÈ (1970), COPLEY and LOVESEY (1975), and SCHOFIELD (1976) have discussed the derivation of such equations in some detail. In this chapter we shall discuss the form of the equations without going into a large amount of analytical detail. We shall appeal to the results of the derivations as needed but will not undertake to reproduce the analysis here.

The linearized hydrodynamic equations for $n(Q)$, $J_\ell(Q)$ and $T(Q)$ are

$$\frac{\partial n(Q)}{\partial t} + iQ\, J_\ell(Q) = 0$$

$$\frac{\partial J_\ell(Q)}{\partial t} = -iQ\frac{p(Q)}{M} - D_\ell Q^2 J_\ell(Q) \qquad (6.8)$$

$$\frac{\partial T(Q)}{\partial t} - B\,\frac{\partial n(Q)}{\partial t} = -\gamma D_T Q^2 T(Q) \quad .$$

The mass of a molecule is M.

Here $J_\ell(Q)$ is the component of $J(Q)$ parallel to Q and $T(Q)$ is the temperature fluctuation variable (SCHOFIELD, 1868),

$$T(Q) = \frac{1}{nC_V}\,[E(Q) - (\partial E/\partial n)\, n(Q)] \quad , \qquad (6.9)$$

which is orthogonal to the number density and the particle current density at $t = 0$. The pressure fluctuation $p(Q)$ is related to $n(Q)$ and $T(Q)$ by the local equilibrium state statement

$$p(Q) = \left(\frac{\partial p}{\partial n}\right)_T n(Q) + \left(\frac{\partial p}{\partial T}\right)_n T(Q) \quad . \qquad (6.10)$$

Other quantities in (6.8) are the longitudinal kinematic viscosity

$$D_\ell = \frac{\frac{4}{3}\eta_s + \eta_v}{M_n} \qquad (6.11)$$

with $\eta_s$ the shear viscosity, and $\eta_v$ the volume viscosity; the thermal diffusivity

$$D_T = \lambda/nC_p \qquad (6.12)$$

with $\lambda$ the thermal conductivity, $C_V$ the specific heat at constant volume and $\gamma = C_p/C_V$ where $C_p$ is the specific heat at constant pressure; and

$$B = \frac{\gamma - 1}{n\beta_T} \qquad (6.13)$$

with $\beta_T$ the thermal expansion coefficient, and n is the density. The quantity $B$ provides the dynamical coupling, via thermal expansion, between the density and temperature fluctuations.

It is a straightforward task, using Laplace transform techniques, to construct the intermediate scattering function from these equations.

Laplace transforms are indicated by

$$\hat{C}(Q,z) = \int_0^\infty dt \exp(-zt)\, C(Q,t) \quad . \tag{6.14}$$

The Laplace transform of the intermediate scattering function is

$$\frac{\hat{F}(Q,z)}{F(Q,0)} = \cfrac{1}{z + \cfrac{\cfrac{v_s^2 Q^2}{\gamma}}{z + D_\ell Q^2 + \cfrac{\cfrac{v_s^2 Q^2}{\gamma}(\gamma - 1)}{z + \gamma D_T Q^2}}} \tag{6.15}$$

where $v_s$ is the velocity of sound. The role of thermal expansion as the coupling coefficient between the thermal diffusion mode and the sound wave modes is readily apparent from this expression since $\gamma - 1$ is proportional to $\beta_T^2$.

The extension of these equations to large (finite) values of wave-vector and frequency leads to two major modifications. The first is that the thermodynamic coefficients, such as $C_v$ and $(\partial p/\partial n)_T$ become wave-vector dependent quantities. This is a reflection of the spatial correlations existing on the molecular level. The second modification involves the transport coefficients which become both wave-vector and frequency dependent. This is a reflection of both the spatial and temporal correlations involved in the dissipation of energy and momentum at the molecular level and that these processes occur on a time scale faster than that associated with the decay of the hydrodynamic variables.

The characterization of these processes is one of the areas for further study opened up by the neutron scattering study of collective modes. Implicit in this generalization is the assumption that n, $\underline{J}$ and T are the only slow variables relevant to the fluctuations in the density.

The consequences of this generalization show up in the intermediate scattering function as

$$\frac{\hat{F}(Q,z)}{F(Q,0)} = \cfrac{1}{z + \cfrac{Q^2 k_B T/MS(Q)}{z + \hat{D}_\ell(Q,z)Q^2 + \cfrac{\frac{Q^2 k_B T\,\Gamma(Q)}{MS(Q)}}{z + \hat{D}_T(Q,z)Q^2}}} \tag{6.16}$$

$\Gamma(Q)$ is the wave-vector dependent generalization of the thermodynamic quantity, $\gamma - 1$. It represents the degree of coupling between the density fluctuation and temperature fluctuation modes as a function of Q. Molecular expressions for $\Gamma(Q)$ involve correlations among four particles (SCHOFIELD, 1968), so it should be considered as a phenomenological quantity given the current state of equilibrium theory. We have neglected terms describing the dynamical coupling between the particle current density and the temperature fluctuations. These terms vanish in the long wavelength limit. Further they are probably not very important for present purposes as they have been ignored, without resulting problems, in the analysis of $\tilde{S}(Q,\omega)$ data. Until the nature of the wave-vector generalization of the thermal expansion coefficient is clarified, there is little point to include such terms.

Equation (6.16) provides the starting point for the generalized hydrodynamics analysis of coherent neutron scattering and molecular dynamics data. The structure factor S(Q) is a known quantity while $\Gamma(Q)$, $\hat{D}_\ell(Q,z)$ and $\hat{D}_T(Q,z)$ are known only in the long wavelength, low frequency limits. Given the limited precision of existing data, only specified forms for these functions have been fit to the data.

The scattering law, $\tilde{S}(Q,\omega)$ is obtained from (6.5) as

$$\tilde{S}(Q,\omega) = \frac{1}{n\pi} \operatorname{Re}\{\hat{F}(Q,i\omega)\} \quad . \tag{6.17}$$

Using this prescription (6.15) yields for small Q the well-known three-peak structure found in light scattering measurements. The central, or Rayleigh, component of the light scattering spectrum comes from the term involving the coefficient $D_T$, while the sound wave, or Brillouin, components, are associated with the entire expression; in particular, the modes for the decay of density fluctuations are determined by the poles of $\hat{F}(Q,z)$ and are known to be, for small Q (see also (1.128))

$$\begin{aligned} z &= -D_T Q^2 \\ z &= \pm i v_s Q - \frac{1}{2} Q^2 \left[D_\ell + D_T(\gamma - 1)\right] \quad . \end{aligned} \tag{6.18}$$

The first pole corresponds to the thermal diffusion, or temperature mode, and the other two correspond to sound wave modes. This tells us that while the continued fraction appearing form of (6.15) is instructive as to the form of the equations of motion for the time correlation functions, it is not a particularly transparent way of representing the dynamical modes of motion if more than one type of mode is present.

To represent the modes, it is more instructive to rewrite (6.15) as the ratio of two polynomials in z:

$$\frac{\hat{F}(Q,z)}{F(Q,0)} = \frac{(z + D_\ell Q^2)(z + \gamma D_T Q^2) + v_s^2 Q^2 (1 - 1/\gamma)}{z(z + D_\ell Q^2)(z + \gamma D_T Q^2) + v_s^2 Q^2 (z + D_T Q^2)} \quad . \tag{6.19}$$

Here the modes are contained in the zeros of the denominator (the poles of $\hat{F}(Q,z)$) and the coupling information is contained in the numerator. Note that if $\gamma = 1$ ($\beta_T = 0$) then the thermal diffusion mode is not present in the dynamics of the density fluctuations. This shows, again, the role of thermal expansion as the coupling mechanism between density fluctuations and the thermal diffusion mode.

We are now in a position to specify how a "mode" is represented in the intermediate scattering function. For linearized hydrodynamics, the modes are determined by the poles of $\hat{F}(Q,z)$. More generally, the modes are specified by the singularities in the complex z-plane of the Laplace transform of the intermediate scattering function. In physical terms, the modes are the terms in the intermediate scattering function which describe the decay of the equilibrium correlations. Thus $\mathcal{F}(Q,t)$ is the sum of a series of terms composed of amplitudes multiplied by functions, such as exponentials or Gaussians, which decay to zero in time.

Writing $\hat{F}(Q,z)$ in the form of (6.16) puts it in the form obtained when the memory function formalism is used with a single slow variable. That is

$$\frac{\hat{F}(Q,z)}{F(Q,0)} = \frac{1}{z + \hat{K}_n(Q,z)} \tag{6.20}$$

defines a "memory function", $\hat{K}_n(Q,z)$, for density fluctuations. While it is possible to write time correlation functions of the hydrodynamic variables in this form and to call the resulting quantities "memory functions", this is not a sensible way to define memory functions as it mixes up the elastic response terms, such as those associated with p(Q) in (6.10), with the memory effects contained in the generalized transport coefficients $\hat{D}_T(Q,z)$ and $\hat{D}_\ell(Q,z)$. This mixing of terms can lead to delta functions and constant terms in the effective memory function and makes it difficult, if not impossible, to exploit the formal definition of the memory functions to focus on the dynamical events responsible for the "memory".

The memory function for the longitudinal current correlation function,

$$C(Q,t) = \langle J_L(Q,t)\, J_L(-Q,0)\rangle / \langle\, |J_L(Q,0)|^2 \rangle$$

is introduced in a slightly different way so that the constant terms are excluded. RAHMAN (1972) discusses the rationale for this in some detail. The memory function $\hat{K}_c(Q,z)$ appears as

$$\hat{C}(Q,z) = \frac{z}{z^2 + z\,\hat{K}_c(Q,z) + Q^2 k_B T/MS(Q)} \quad .$$

Since

$$F(Q,t) = -Q^2 \langle |J(Q)|^2 \rangle \; C(Q,t)$$

(6.16) can also be written as

$$\frac{\hat{F}(Q,z)}{F(Q,0)} = \frac{1}{z + \cfrac{Q^2 k_B T/MS(Q)}{z + \hat{K}_c(Q,z)}} \quad .$$

Generalized hydrodynamics shows that there are in principle two types of dynamical processes (or "modes") operating in the decay of density fluctuations. $\Gamma(Q)$, which is a wave-vector dependent generalization of the thermal expansion coefficient, must be small if the energy diffusion process is to be neglected. Theories with $\Gamma(Q) = 0$ are known as viscoelastic theories and involve only the density modes.

There are two tasks to be accomplished when developing a quantitative generalized hydrodynamics description. One task is to develop representations for $D_\ell(Q,z)$ and $\hat{D}_T(Q,z)$ and to provide a physical basis for these representations. The structure factor, S(Q), plays two roles, both as an initial value and as an "elastic" constant. The other task involves the development of the theory of equilibrium quantities such as S(Q), the Fourier transform of the pair correlation function and of $\Gamma(Q)$, the finite wave-vector generalization of the heat capacity ratio and therefore of the thermal expansion coefficient. Since $\Gamma(Q)$ involves three and four particle correlations this task is of major significance in the theory of the liquid state. RAVECHÉ and MOUNTAIN (1977) discuss this topic in considerable detail.

In the following discussion of neutron scattering experiments and of molecular dynamics calculations, we shall focus on how these tasks have been approached, and on what is known about the underlying dynamics of density fluctuations in monoatomic liquids.

## 6.2. Coherent Scattering

Density fluctuations in insulating and metallic liquids have been studied both by coherent neutron scattering and by computer simulation. In the next few paragraphs, we shall discuss the results of these studies. Then we shall attempt to set forth the status of our knowledge of generalized hydrodynamics as it applies to density fluctuations in these systems. The metals are rather different in their finite wave-vector properties than are the insulators thus providing a number of questions for further study.

### 6.2.1 Insulating Liquids

*Argon.* The careful measurements of SKÖLD et al. (1972) of the coherent and incoherent scattering from liquid argon over the wave-vector range 1.0 - 4.4 $\text{Å}^{-1}$ provide the opportunity to study dynamics in the region where the Q and $\omega$ dependence of generalized transport coefficients is expected to be quite important. The general features of the dynamics are contained in Fig.6.1 where $\tilde{S}(Q,\omega)/S(0)$ contours for the interval 1.0 $\text{Å}^{-1} \le Q < 2.4\ \text{Å}^{-1}$ are shown. There are two regions where narrowing occurs. The first is that around Q = 2 $\text{Å}^{-1}$ associated with the maximum in S(Q). The other narrowing occurs around $Q \simeq 1.2 - 1.6\ \text{Å}^{-1}$ and results from the different Q dependences of S(Q,0) and S(Q) as shown in Fig.6.2. The narrowing at small Q is readily observable in plots of the half-width. The dynamical significance, if any, of this effect is yet to be established.

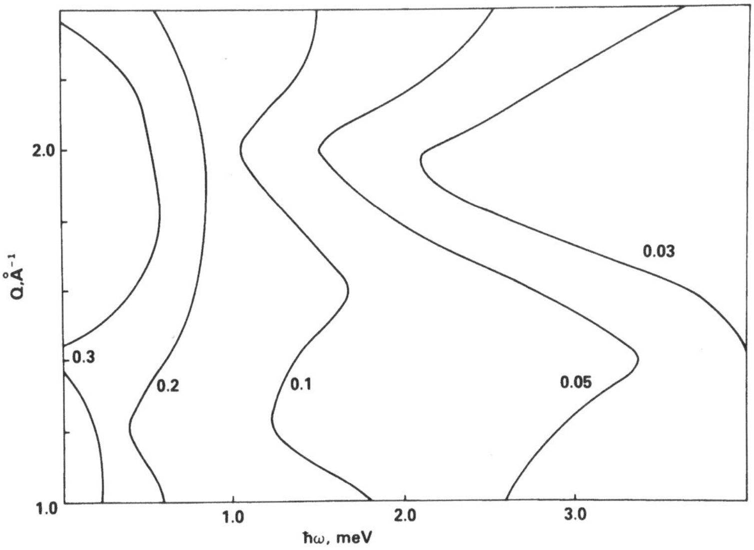


Fig. 6.1 The reduced scattering law $\tilde{S}(Q,\omega)/S(Q)$ for liquid argon as a function of Q and $\omega$. The contours were constructed using the data of SKÖLD et al. (1972) for $S(Q,\omega)$ and the data of YARNELL et al. (1973) for S(Q)

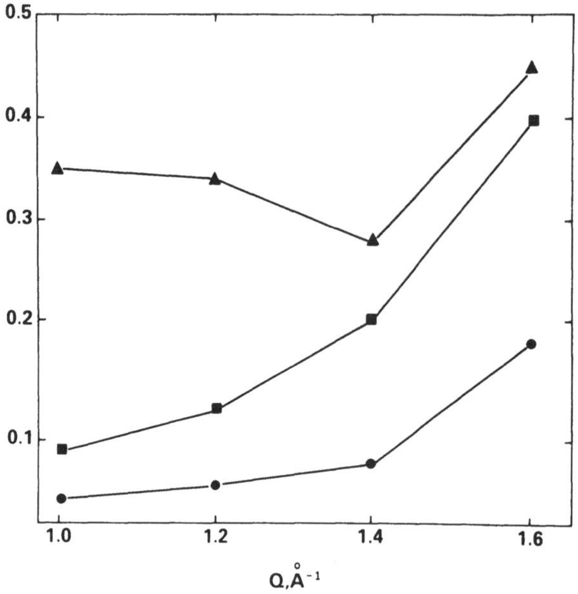


Fig. 6.2 A comparison of S(Q,0) (●) and S(Q) (■) for liquid argon for 1.0 Å$^{-1}$ ≤ Q ≤ 1.6 Å$^{-1}$. The dip in the ratio of these quantities (▲) is indicative of the existence of two narrowings in the scattering law

An examination of the intermediate scattering functions constructed from these data reveals three regions. The first of these lies in the interval 1.0 Å$^{-1}$ ≤ Q ≤ 1.4 Å$^{-1}$ and has two time scales for the decay of density fluctuations. This can be seen in Fig.6.3 where F(Q,t)/F(Q,0) for Q = 1.4 Å$^{-1}$ is displayed as a function of time by the circles. Q = 1.6 Å$^{-1}$ marks the transition to the second region which occupies the interval 1.8 Å$^{-1}$ ≤ Q ≤ 2.2 Å$^{-1}$. There the faster process merges with the slower one, as shown by the triangles in Fig.6.3. This is the Q-interval spanned by the principal maximum in S(Q). The transition to the third region occurs around Q = 2.4 Å$^{-1}$. In the interval 2.6 Å$^{-1}$ ≤ Q ≤ 4.4 Å$^{-1}$, only a fast process is operative, as indicated for Q = 3.8 Å$^{-1}$ by the squares in Fig.6.3. The terms "fast" and "slow" are intended to be descriptive only. The rates of the processes are Q-dependent and the origin of the time scales reflected in F(Q,t) is yet to be assigned with certainty. It is instructive to compare these observations with those of a generalized hydrodynamic analysis.

ROWE and SKÖLD (1972) have compared these data with a number of models for $\tilde{S}(Q,\omega)$. The memory function models they examined can be considered to be generalized hydrodynamics type expressions with $\Gamma(Q) \equiv 0$ and with a model functional form for $D_\ell(Q,t)$. The best results were obtained when $D_\ell(Q,t)$ is assumed to be a Gaussian and the amplitude and decay time were fit to the data. A two relaxation time form proposed by SEARS (1970) gave significantly poorer results, poorer even than a single exponential form fit to the data.

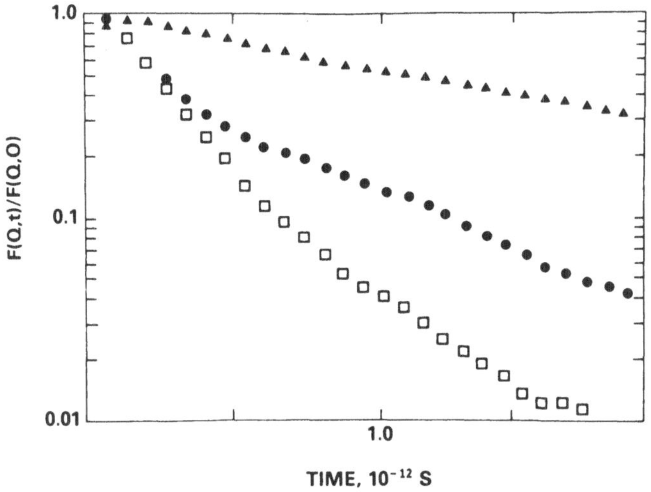


Fig. 6.3 The intermediate scattering function, normalized to unity at t = 0, for argon. The data used to prepare Fig.6.1 were also used here. F(Q,t) is shown for three Q-values: Q = 1.4 $Å^{-1}$ (••), Q = 2.0 $Å^{-1}$ (▲▲) and Q = 3.8 $Å^{-1}$ (□□)

A heuristic interpretation of the decay time can be made by considering the magnitude and Q-dependence of the time. AILAWADI et al. (1971) carried out an analysis for the Lennard-Jones model for liquid argon. They found that a single Gaussian provides a reasonable representation for $D_\ell(Q,t)$. The time constant is on the order of $10^{-13}$ s and is only weakly dependent on Q over the region 0.5 $Å^{-1} \le Q \le 3.5$ $Å^{-1}$. This suggests that the decay time, which is characteristic of the time required to dissipate a flux of momentum in the Q-direction, is determined mainly by close encounters which change the direction of the flow of momentum. Some sort of collective or flow process would be slower and more Q-dependent than this analysis indicates. These heuristic remarks have not yet been incorporated into a model for $D_\ell(Q,t)$.

The density fluctuations for argon in the interval 1.0 $Å^{-1} \le Q \le 4.4$ $Å^{-1}$ might be characterized as exhibiting either strongly overdamped sound modes or modified diffusive modes although neither phrase is accurate. More to the point, the motion is a generalized hydrodynamic one with the generalization of the viscosity time correlation function exhibiting decay dominated by processes with one time scale. The single parameter models do not exhaust the information contained in the data [ROWE, private communication].

Attempts to describe the data without adjustable parameters, i.e., those in which the parameters such as relaxation times are given molecular definitions, have been less successful than have those where the parameters are simply fit to the data. What is not clear from this analysis is whether this is due to the neglect of the energy transport modes through the assertion $\Gamma \equiv 0$ or whether it is due to an inadequate choice of functional form for $D_\ell(Q,t)$. The studies of neon and the Lennard-Jones fluid indicate that the latter suggestion is the more accurate one.

*Neon.* The full range of generalized hydrodynamic phenomena has been observed by neutron scattering in fluid neon (BELL et al., 1973, 1974; BUYERS et al., 1975), albeit not all effects have been observed for a single thermodynamic state.

Brillouin scattering is usually observed by light scattering. BELL et al. (1973) observed Brillouin scattering of neutrons by dense fluid neon in the interval $0.06\ \text{Å}^{-1} \leq Q \leq 0.14\ \text{Å}^{-1}$. Three states with relatively large compressibilities were chosen so that usable counting rates could be realized at small Q's. These data were found to be consistent with (6.15) as predicted by conventional hydrodynamics (BELL et al., 1975) in that the thermodynamic and transport coefficients used in this analysis are consistent with those obtained by macroscopic measurements. This can be seen in Fig.6.4 where the hydrodynamic result, (6.15) has been combined with the instrumental resolution to obtain the solid curve and then found to compare quite satisfactorily with experiment.

A second experiment was performed on liquid neon and covered the interval $0.2\ \text{Å}^{-1} \leq Q \leq 1.50\ \text{Å}^{-1}$ (BELL et al., 1975). These data were analyzed in terms of generalized hydrodynamics using (6.16). Several variations of how departures from conventional hydrodynamics, (6.15), might occur were considered. First $\Gamma$ was assumed independent of Q and second $\Gamma(Q)$ was postulated to be a Lorentzian in Q. No definite conclusion could be drawn as to which is better although the Q-dependent version leads to an improved fit to the data.

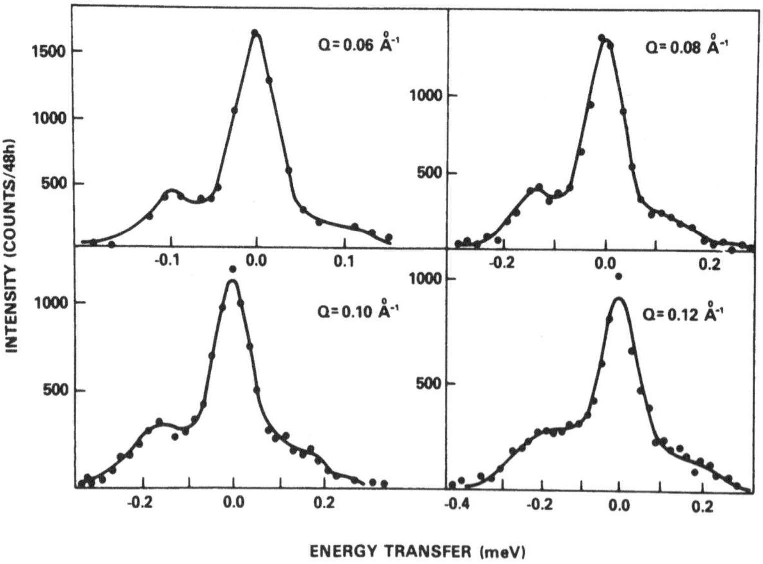


Fig. 6.4 The hydrodynamic theory, (6.15), folded with the instrumental resolution function fit to data obtained in the region $0.06\ \text{Å}^{-1} \leq Q \leq 0.12\ \text{Å}^{-1}$ for neon at $\rho = 0.48\ \text{g/cm}^3$ and $T = 70$ K (BELL et al., 1975). A background subtraction has been made to the data

The final representation of the data assumed specific functional forms for the Q and $\omega$ dependent quantities. These were

$$\begin{aligned} D_T &= D_T^o(1 + Q^2r_o^2)^{-1} \quad , \\ D_\ell &= D_\ell^o\tau_o^{-1}[z + \tau^{-1}(Q)]^{-1} \quad , \\ \tau(Q) &= \tau_o(1 + Q^2R_o^2)^{-1} \quad , \\ \Gamma(Q) &= \alpha_o(1 + Q^2r_\gamma^2)^{-1} \quad , \end{aligned} \qquad (6.21)$$

and

$$D_\ell^o = D_\ell^{oo}[a + bS(0)/S(Q)] \quad .$$

The coefficients a and b were taken from a computer simulation study of the Lennard-Jones fluid (AILAWADI et al., 1971). These representations are strictly empirical; even so they do give a way of assessing the time and distance scales over which generalized hydrodynamic effects are important. (The data were not able to provide any information about the frequency dependence of the thermal diffusivity term $D_T$). Table 6.1 shows these quantities as determined from fitting the data. These are qualitative measures of the processes associated with thermal conduction, viscous dissipation and thermal expansion.

The values of the characteristic distances and times appearing in (6.21) indicate that $D_\ell(Q,t)$ decreases less rapidly with increasing Q than do $\Gamma(Q)$ and $D_T(Q,t)$. These distances are indicative of the minimum spatial extent required for the full development of the correlations involved.

These results for neon can be used to re-examine the earlier discussion of argon. The Lennard-Jones potential parameter $\sigma$ for argon is 3.4 Å and is 2.75 Å for neon, thus Q = 1.5 $\text{Å}^{-1}$ for neon corresponds to Q ≃ 1·0 Å for argon. With this correspondence one sees that the neglect of the heat transport terms ($\Gamma \equiv 0$) in the analysis of the argon data is probably justified. Because the coupling between the energy fluctuations and the density fluctuations becomes small at relatively small values of Q, it is tempting to say that generalized hydrodynamics is of little use for wave-vectors greater than say 1-2 $\text{Å}^{-1}$. Such a view misses the point. Generalized hydrodynamics provides a way of directing attention to the physically important

Table 6.1 Parameters which characterize the generalized hydrodynamics of fluid neon. These coefficients appear in (6.21)

| $r_o$(Å) | $R_o$(Å) | $\tau_o(10^{-13}s)$ | $r_\gamma$(Å) |
|---|---|---|---|
| 2.0 ± 0.5 | 0.66 ± 0.06 | 2.6 ± 0.4 | 3.4 ± 0.5 |

processes involved in the decay of density fluctuations. For $Q > 1\ \text{Å}^{-1}$ in the rare gas liquids, the important physics involves the momentum density current correlations contained in $D_\ell(Q,t)$.

BUYERS et al. (1975) examined the neutron scattering from liquid neon at larger Q values; $0.8\ \text{Å}^{-1} < Q < 12.5\ \text{Å}^{-1}$. These data have not been analyzed in terms of generalized hydrodynamics so the discussion of collective modes is therefore necessarily less definite. For wave-vectors with $Q < 1.6\ \text{Å}^{-1}$ there is an appreciable inelastic wing in the observed neutron spectra. This is cited as evidence of collective motions in the liquid, although no indication as to the nature of the cooperative motion is inferred. In agreement with the studies of argon and neon discussed above, there is no indication of sound wave modes in these rare gas liquids for values of wave-vector greater than $0.3\ \text{Å}^{-1}$.

The results for $Q > 4\ \text{Å}^{-1}$ are indicative of how the collective motions go over to single particle motions as Q increases. This transition is not at all understood for liquids. Such understanding is essential for the development of reliable representations of $\tilde{S}(Q,\omega)$ which span both regions. The Gram-Charlier expansion, an effectively single particle scattering representation of $\tilde{S}(Q,\omega)$, provides a fairly reliable representation of the data for $Q > 6\ \text{Å}^{-1}$. Thus the region between $2\ \text{Å}^{-1}$ and $6\ \text{Å}^{-1}$ is where the collective modes die out.

*Normal* $^4$*He.* Recent neutron scattering measurements of Brillouin structure in liquid helium at 4.2 K and saturated vapor pressure conditions (DASANNACHARYA et al., 1976) indicated that the hydrodynamic description of density fluctuations may also be applicable to liquid helium well above the $\lambda$-transition. For this reason, we include it in our discussion even though liquid helium is not a classical liquid.

In these measurements, a separate Brillouin component could be observed for $Q \leq 0.15\ \text{Å}^{-1}$ but not for $Q = 0.20\ \text{Å}^{-1}$. These data could fit with the hydrodynamic scattering law obtained from (6.15). There is no evidence for dispersion (Q or $\omega$ dependence) in the coefficients although the values of the coefficients differ significantly from their macroscopic values. The results of using (6.15) for $\tilde{S}(Q,\omega)$ are compared with experiment in Fig.6.5. The dashed curve results when macroscopic parameters are used and the solid curve results from the fitted parameters.

The uncertainty in the derived value of $\gamma$ is substantial and the macroscopic value lies with the error bars. The derived value of $D_T$ appears to be somewhat smaller than the macroscopic value. On the other hand the derived sound speed is a bit larger than the thermal value. The derived value of the sound attenuation is slightly smaller than the macroscopic value, a result consistent with the trends in the sound speed.

If these preliminary results should be confirmed, this would imply the existence of processes/effects not included in the hydrodynamic equations of motion.

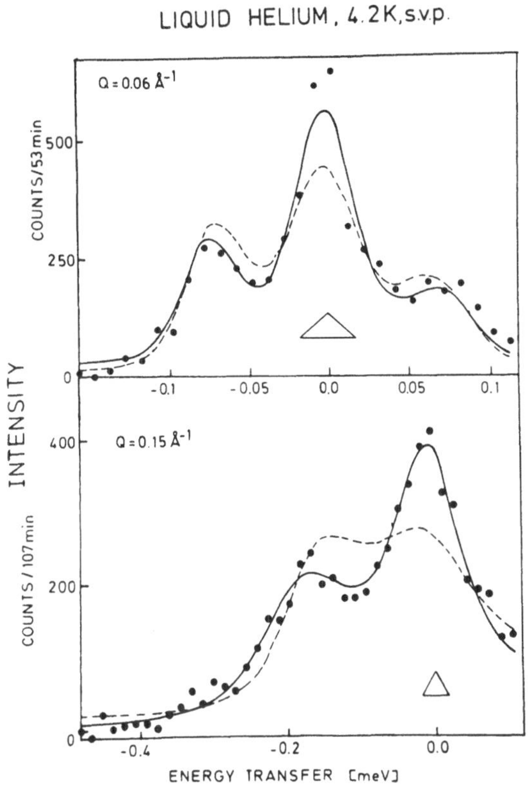


Fig. 6.5 Neutron scattering spectra for $^4$He at 4.2 K, SVP after subtraction of the background. The solid curve is the best fit obtained by adjusting the parameters in (6.15). The broken curve is obtained using macroscopic values (DASANNACHARYA et al., 1976). A background subtraction has been made to the data. The triangles are indicative of the resolution of the measurements

*Lennard-Jones.* The Lennard-Jones model is defined by the two body potential function

$$\phi(r) = 4\varepsilon\left[\left(\frac{\sigma}{r}\right)^{12} - \left(\frac{\sigma}{r}\right)^{6}\right] \quad . \tag{6.22}$$

This system is a testing ground for theories of liquids in that the potential is represented analytically and the system is rather like the rare gas liquids in its thermodynamic and transport properties. Numerical solution of the equations of motion for a collection of Lennard-Jones atoms provides a means of realizing the properties of this system and is commonly referred to as "molecular dynamics".

A study of the collective effects in the Lennard-Jones liquid near the triple point has been performed by LEVESQUE et al. (1973). The results for the density fluctuations are presented in terms of generalized hydrodynamics and is probably the most detailed study of the Q and $\omega$ dependence of the coefficients appearing in (6.16).

The most striking effect is the presence of Brillouin peaks (sound modes in $S(Q,\omega)$ for $Q\sigma \le 0.75$. This corresponds to $Q \le 0.22$ Å$^{-1}$ for argon, $Q \le 0.27$ Å$^{-1}$ for neon and $Q \le 0.29$ Å$^{-1}$ for helium. This is in good agreement with experimental results for argon and neon but not for helium. This further reinforces the suggestion that helium involves effects not contained in the hydrodynamic equations of motion.

Another interesting feature revealed by the molecular dynamics study is the range in Q vector over which the energy mode is strongly coupled to the density fluctuations. This coupling is negligible for $Q\sigma \geq 2$, corresponding to $Q \geq 0.6$ Å$^{-1}$ for argon and $Q \geq 0.72$ Å for neon. Again this is in good agreement with the neutron scattering results. Also it provides further justification for setting $\Gamma(Q) \equiv 0$ in the analysis of argon scattering data performed by ROWE and SKÖLD (1972). The frequency dependence of $D_T(Q,\omega)$ was neglected. The Q dependence obtained from the analysis indicated that $D_T(Q,o)$ is small for $Q\sigma \geq 2$. It would be interesting to pursue this further by the direct construction of the temperature fluctuation time correlation function by molecular dynamics.

The time correlation function $D_\ell(Q,t)$ was parametrized with two relaxation times. The longer time process has a Q dependent amplitude which decreases with increasing Q and is negligible for $Q\sigma \approx 4$. The shorter time has a weak Q dependence but no Q dependence would be ascribed to the longer time due to the limited precision of the computation. The cooperative effect associated with the longer time process has a spatial range on the order of three atomic diameters. Since the time correlation function $D_\ell(Q,t)$ involves the component of the flux of momentum parallel to Q, moving in the Q direction, it is reasonable to infer the existence of strongly correlated atomic motions which result in momentum transfer. The motion is not so highly developed, however, that oscillatory effects are present in the time correlation function. Rather there is a persistence of an initial disturbance for a time long on the molecular scale. This effect is reported to be correlated with the existence of a negative portion in the single particle velocity autocorrelation (LEVESQUE, private communication). It occurs only for states near the fluid-solid phase boundary.

### 6.2.2 Liquid Metals

The liquid alkali metals rubidium and sodium have been examined for collective effects. Sound wave modes are found to persist to much larger Q-values than is the case for the insulating liquids. There are also other collective effects observed in the metals whose character is not yet understood.

The quasi-free electrons in a liquid metal can have large effects on the long wavelength hydrodynamic motion of the fluid. The large values of the thermal conductivity and volume viscosity found for the liquid metals are due to electronic effects. These effects do not appear to directly influence the motions of the ions probed in neutron scattering and molecular dynamics studies. This implies that some important collective modes in the liquid metals are not amenable to study by neutron scattering.

*Rubidium.* The dynamics of liquid rubidium, close to the triple point, has been examined both by neutron scattering (COPLEY and ROWE, 1974a,b) and by molecular dynamics (RAHMAN, 1974a,b). These two sets of studies complement each other and provide a nice example of how these two techniques can be used to advantage. Since the results of both investigations are effectively the same, we shall not distinguish between neutron and molecular dynamics results. This means that the pseudo-potential of PRICE et al. (1970) provides a faithful description of density fluctuations for liquid rubidium. By approximate scaling arguments this conclusion can be reasonably extended to the other liquid alkali metals. While the structure of the density fluctuations is rich in cooperative effects, no generalized hydrodynamic analysis has been performed on them. Even so, a number of qualitative statements on the nature of cooperative motions in the liquid alkali metals are possible.

For $Q < 1\,\mathrm{\AA}^{-1}$, a well defined sound wave mode can be observed. A dispersion curve for this mode is displayed in Fig.1.12 of Chap. 1. Some of the data used to construct this dispersion curve are shown in Fig.1.11 of Chap. 1. For $Q > 1.25\,\mathrm{\AA}^{-1}$, no sound wave mode can be observed. The structure factor for liquid rubidium, $S(Q)$, is shown in Fig.6.4 where the arrow indicates the approximate point where sound waves cease to be a "good" excitation. This occurs for much smaller wavelength than is the case for the rare gas liquids. If the maximum in $S(Q)$ occurs at $Q_m$, the limit for sound wave modes in rubidium is $Q \simeq 2Q_m/3$ while for neon it is $Q \simeq Q_m/10$.

The small Q patterns of $\tilde{S}(Q,\omega)$ are much in appearance like those observed in light scattering. However, linearized hydrodynamics, (6.15), is unable, with reasonable choices for the material parameters, to reproduce even qualitatively the features of $\tilde{S}(Q,\omega)$ found in the small-Q region.

The success of the molecular dynamics calculations in reproducing the neutron scattering results has a very important consequence in that these calculations do not make any explicit reference to the electrons. This means that the direct electronic contribution to the generalized hydrodynamic coefficients is unimportant for $Q \geq 0.2\,\mathrm{\AA}^{-1}$. The implication of this observation is that connection between the macroscopic transport coefficients and the Q and $\omega$ dependent coefficients of liquid rubidium is not well defined by these experiments. The role of the electrons in the dynamics of the fluctuations of the ions in a metal is a topic needing further study.

There are important features which indicate that generalized hydrodynamic effects are significant for $Q \geq 0.2\,\mathrm{\AA}^{-1}$. For the interval $0.2\,\mathrm{\AA}^{-1} < Q < 0.8\,\mathrm{\AA}^{-1}$, the widths of the Rayleigh (central) and Brillouin components are roughly proportional to Q, unlike the $Q^2$ dependence found in the long wavelength limit. Very little dispersion in the sound speed is observed for $Q < 0.5\,\mathrm{\AA}^{-1}$. A "negative dispersion" is found for $0.5\,\mathrm{\AA}^{-1} < Q < 1.2\,\mathrm{\AA}^{-1}$, in keeping with the increase in $S(Q)$ which begins at about $Q \simeq 0.5\,\mathrm{\AA}^{-1}$. Table 6.2 contains a comparison of the experimental values of $\omega_B$, the Brillouin peak angular frequency, and $\omega_0$, the value associated with the "isothermal" sound frequency defined by

$$\omega_0 = [k_BT/MS(Q)]^{1/2}Q \tag{6.23}$$

Table 6.2 Comparison of the Q-dependent Brillouin ($\omega_B$) and "isothermal" ($\omega_0$) frequencies for liquid rubidium

| $Q(A^{-1})$ | 0.30 | 0.45 | 0.60 | 0.75 | 0.90 | 1.00 |
|---|---|---|---|---|---|---|
| $\omega_B(10^{12}s^{-1})$ | $4\frac{1}{2} \pm \frac{1}{4}$ | $6\frac{1}{2} \pm \frac{1}{2}$ | $7\frac{1}{2} \pm \frac{1}{2}$ | $8 \pm \frac{1}{2}$ | $7 \pm \frac{1}{2}$ | $6\frac{1}{2} \pm \frac{1}{2}$ |
| $\omega_0(10^{12}s^{-1})$ | 2.9 | 4.4 | 5.5 | 6.3 | 6.0 | 5.6 |

The specific heat ratio for liquid rubidium is about 1.15. The results in Table 6.2 would suggest that thermal expansion remains more important for larger Q values than it does in the rare gas liquids. This suggestion requires a generalized hydrodynamic analysis of the data before it can be definitely asserted. BARKER and GASKELL (1974) looked briefly at generalized hydrodynamics effects. Their analysis is too sketchy and incomplete to clarify this issue.

The dynamical properties of liquid rubidium of $Q \geq 1.25$ Å$^{-1}$ are also of interest. The reduced structure factor $\tilde{S}(Q,\omega)/S(Q)$ is shown in Fig.6.7. Except for 1.5 Å$^{-1}$ < Q < 1.75 Å$^{-1}$, the region of the principal maximum of S(Q), the scattering data for Q < 2.5 Å$^{-1}$ indicate that F(Q,t) decays with two time scales. This is borne out in the intermediate scattering function shown in Fig.6.8 for Q = 1.375 Å$^{-1}$, Q = 1.625 Å$^{-1}$ and Q = 1.875 Å$^{-1}$. These curves indicate that in addition to a rapid

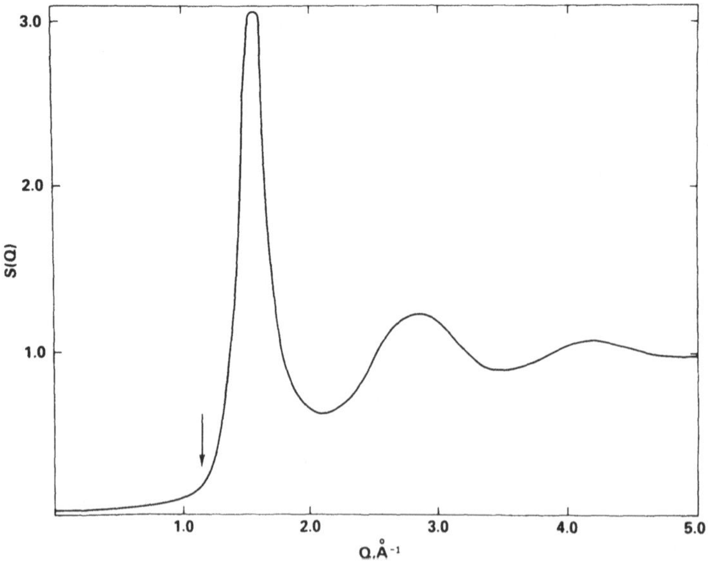


Fig. 6.6 The structure factor of liquid rubidium. This curve was prepared using values supplied by RAHMAN (private communication and 1974a,b). The arrow indicates the limit of sound wave modes

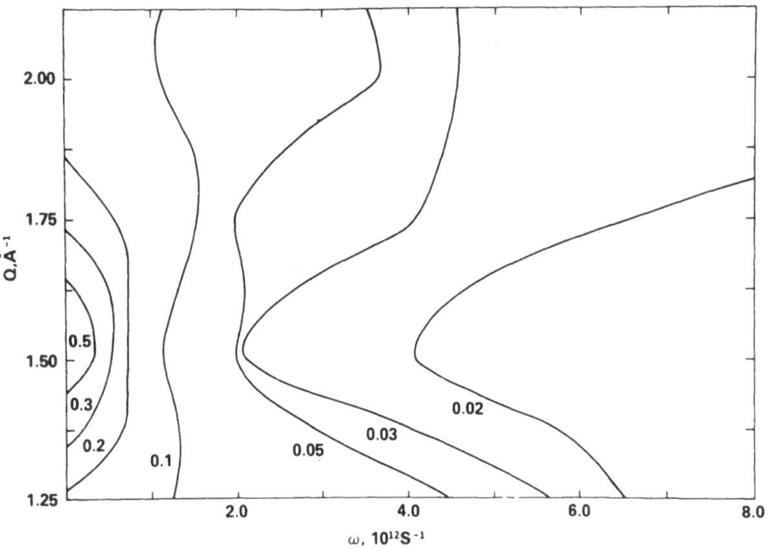


Fig. 6.7 The reduced scattering law $S(Q,\omega)/S(Q)$ for liquid rubidium as a function of Q and $\omega$. The contours were constructed using the parametrized representation of the data of COPLEY and ROWE (1974). The parameters can be found in Table A1 of the review by COPLEY and LOVESEY (1975)

initial decay, there is a slower process working. For $1.5\ \text{Å}^{-1} < Q < 1.75\ \text{Å}^{-1}$, the faster process appears to merge into the slower one. For $Q > 2.5\ \text{Å}^{-1}$, only the fast process seems to be operative. Again, "fast" and "slow" are intended only to be descriptive terms. The reduced scattering law and the intermediate scattering function are shown in Fig.6.7 and 6.8.

The neutron scattering data have been parametrized in terms of a sum of Gaussians by COPLEY and ROWE (1974). The coefficients can be found in Table A1 of the review by COPLEY and LOVESEY (1975). No physical significance exists for these parameters aside from the fact that they provide an accurate way of representing both $S(Q,\omega)$ and $F(Q,t)$ over the ranges appropriate to the experiment.

COPELY and LOVESEY (1975) have examined a viscoelastic characterization of the liquid rubidium results. For $Q \geq 1.25\ \text{Å}^{-1}$, the agreement with the data is reasonable. This model corresponds to a single relaxation time for the memory function, $K_n(Q,t)$, in (6.20). The amplitude and relaxation time were chosen so that the first three frequency moments of $S(Q,\omega)$ are satisfied. From this analysis we conclude that the "fast" and "slow" parts of $F(Q,t)$ are reflections of the Q dependence of the memory function.

*Sodium.* RAHMAN (1972) carried out a molecular dynamics simulation of liquid sodium using the potential of PRICE et al. (1970). Although the reported results are less extensive than those for rubidium, the picture which emerges is similar. This is to be expected given the close relationship of the pseudo-potentials used in the two calculations.

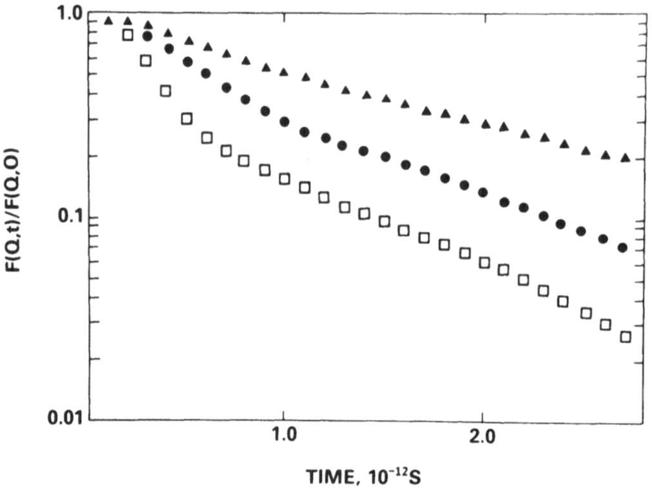


Fig. 6.8 The intermediate scattering function, normalized to unity at t = 0, for liquid rubidium. The parametrization used to construct Fig.6.7 was used here as well. F(Q,t) for three Q-values: Q = 1.375 $Å^{-1}$ (●●), Q = 1.625 $Å^{-1}$ (▲▲) and Q = 1.875 $Å^{-1}$ (□ □)

Fluctuations for two wave-vectors were studied. For Q = 0.23 $Å^{-1}$, a well defined sound wave mode exists. The memory function $K_c(Q,t)$ could be described as the sum of two Gaussians with well separated decay times. This means that generalized hydrodynamic effects are important for relatively small Q values. The sound speed derived from the simulation is within 10% of the macroscopic value.

For Q = 1.38 $Å^{-1}$, a quite different situation obtains. The oscillations in F(Q,t) are almost completely damped out and a single Gaussian for the memory function is more or less adequate.

The changes in the memory function which occur between Q = 0.23 $Å^{-1}$ and Q = 1.38 $Å^{-1}$ are consistent with the suggestion that the coupling of the energy fluctuation mode to the density fluctuation decreases with increasing Q. A more detailed study would be needed to clarify the generalized hydrodynamic behavior of the density fluctuations in this wave-vector region. The fluctuations in the kinetic energy were also examined in this study. It is important to realize that these are not the same as the temperature fluctuations which are dynamically coupled to the density in the generalized hydrodynamic equations. Rather it is a part of the temperature fluctuations and it is not possible to say whether or not it is the dominant piece.

### 6.2.3 Comparison of Systems Studied

The density fluctuations of the rare gas liquids and the alkali metal liquids have been examined in the previous two subsections. Here we shall compare the results for the two systems. The most striking difference occurs at small Q's where the Q intervals over which sound wave modes are observed are quite different, the interval extending to appreciably larger Q's for the metals than for the rare gas liquids. The physical reason for this difference has not been spelled out although it must ultimately result from the relative softness of the two body potential energy function acting between the ions in the metals.

The utility of the generalized hydrodynamics approach in the interpretation of neutron scattering data in the small Q region of the rare gas liquids has been demonstrated in the work of BELL et al. (1975) and of LEVESQUE et al. (1973). The study of sodium by RAHMAN (1972) indicates that this is also the case for the alkali metals although a detailed generalized hydrodynamics analysis has not yet been performed for these systems.

For Q's where sound wave modes are not observed, the two systems are similar in that the memory function with a single time scale provides a not unreasonable characterization of the dynamics. Such a representation does not accurately describe the situation but it reproduces the gross features of the intermediate scattering function, indicating that the coupling to the temperature mode is unimportant.

## 6.3 Transverse Current Correlations

The linearized equation of hydrodynamics provides no coupling between the longitudinal and transverse components of the momentum density. The transverse component, $J_T(Q)$, is defined by

$$\underline{Q}\ \underline{J}_T(Q) = 0 \quad . \tag{6.24}$$

The prediction of hydrodynamics is that the transverse current correlation function is

$$\psi_T(Q,t) = \frac{\langle J_T(Q,t)J_T(-Q,0)\rangle}{\langle |J_T(Q)|^2\rangle} = \exp(-\eta_s Q^2 t/Mn) \quad . \tag{6.25}$$

This correlation function cannot be investigated by neutron scattering because there is no coupling to the density. It can be studied by molecular dynamics and it shows interesting generalized hydrodynamics effects.

RAHMAN (1968) and LEVESQUE et al. (1973) constructed transverse current correlation functions for the Lennard-Jones model. These calculations indicate that for sufficiently small Q values there are no oscillations in $\psi_T(Q,t)$. However, as Q increases, $\psi_T(Q,t)$ begins to show well defined oscillations so that $J_T(Q)$ might be said to be a collective coordinate for large enough Q's. Of course at very large Q's $\psi_T(Q,t)$ must go over to the ideal gas form (RAHMAN, 1967; MOUNTAIN, 1974)

$$\psi_T(Q,t) = \frac{Q^2}{\beta M}\left(1 - \frac{Q^2t^2}{\beta M}\right) \exp\left(-Q^2t^2/2\beta M\right) \quad , \tag{6.26}$$

so the concept of a collective coordinate is not without ambiguity in liquids. What is clear is that the computer results are the true collective effects.

AILAWADI et al. (1971) have examined RAHMAN's (1968) results for $\psi_T(Q,t)$ in terms of a Gaussian memory function. LEVESQUE et al. (1973) on the other hand used a two relaxation time memory function to characterize their results. While both representations of the memory function provided qualitatively correct descriptions of the transverse current correlation functions, neither of them is particularly accurate. Much of the detail associated with the onset of the oscillations in the correlation function is not reproduced. This is apparent when one examines the figures in these papers.

The physical process associated with the memory remains to be described in any detail. This is a general characteristic of the state of affairs in the development of generalized hydrodynamics descriptions.

Although it is reasonable to expect similar effect in the liquid metals, no studies of transverse current correlations have been reported.

## 6.4. Relations Between $S(Q,\omega)$ and $S_s(Q,\omega)$

Single particle motion is probed by incoherent neutron scattering. A commonly asked question is "How can one relate $S(Q,\omega)$ to the incoherent scattering function $S_s(Q,\omega)$?" A substantial number of attempts have been made, beginning with VINEYARD (1958) and the convolution approximation

$$S(Q,\omega) = S(Q)\, S_s(Q,\omega) \quad , \tag{6.27}$$

to do just that. COPLEY and LOVESEY (1975) briefly reviewed some of the more important efforts to express $S(Q,\omega)$ in terms of $S_s(Q,\omega)$.

These studies have shown that it is not easy to represent cooperative motions in liquids in terms of single particle motions. GYORFFY and MARCH (1971) showed that there is a good reason for this difficulty, one which is inherent in the problem.

Following Gyorffy and March, let us introduce the functions

$$I(Q,z) = \int_0^\infty dt e^{-zt}\, \partial F(Q,t)/\partial t$$

and (6.28)

$$I_S(Q,z) = \int_0^\infty dt e^{-zt}\, \partial F_S(Q,t)/\partial t$$

which are the Laplace transforms of the time derivatives of the intermediate scattering functions for coherent and incoherent scattering. Through successive integration by parts these functions can be represented as power series in $z^{-2}$ as

$$I(Q,z) = \sum_{j=1}^{\infty} \alpha_j z^{-2j}$$

and (6.29)

$$I_S(Q,z) = \sum_{j=1}^{\infty} \gamma_j z^{-2j} \quad .$$

The coefficients $\alpha_j$ and $\gamma_j$ are proportional to the 2jth frequency moments of $S(Q,\omega)$ and $S_s(Q,\omega)$. Within the radius of convergence of these series, it is in principle possible to express $I(Q,z)$ as a power series in $I_S(Q,z)$,

$$I(Q,z) = \sum_{n=1}^{\infty} a_n [I_S(Q,z)]^n \quad , \tag{6.30}$$

thereby providing a formal solution to the problem. The coefficients $a_n$ are complicated functionals of I and $I_S$. It is now understandable why the attempts to represent $S(Q,\omega)$ as relatively simple functions of $S_s(Q,\omega)$ have not been particularly successful.

At the heuristic level, there is little reason to expect that the dynamical events probed by coherent scattering are simply related to those events probed by incoherent scattering. The derivation of generalized hydrodynamics shows us that the dynamical processes involved have no simple connection between them. The formalism does not rule out such a connection but it does indicate that such a connection would require special conditions. All in all, the representation of the cooperative motions in terms of single particle motions does not appear to be a fruitful avenue for future studies of collective modes of the liquid state.

## 6.5. Current Issues

From the discussion contained in the previous section of this chapter, it is apparent that the topic of collective modes in liquids is not a closed one. A number of issues have been identified and we shall consider them further here.

The foremost issue is that of developing physical models for the memory functions $D_\ell(Q,t)$ and $D_T(Q,t)$. The emphasis has been on obtaining functional forms for these memory functions subject to the constraints of the frequency moment relations for $S(Q,\omega)$. The moment conditions provide important checks on the memory functions but are of little help in deciding on the form of those functions or in describing the physical situation described by them.

BARKER and GASKELL (1975) developed a scheme for approximately evaluating the memory function for the longitudinal current correlation function. This procedure invokes a substantial number of physical assumptions whose significance has not been evaluated. Even so, the underlying premise that the effects of the strongly repulsive interaction can be separated somehow from the effects of the attractive part of the interaction is an appealing one. The results of their calculation for liquid rubidium are in better agreement for $Q < 1\ \mathring{A}^{-1}$ than any other published calculation.

They find that the "memory function", which includes both $D_\ell(Q,t)$ and $D_T(Q,t)$, has two parts. The piece associated with the strongly repulsive interactions decays rapidly. The remaining parts decay more slowly. This is not a separation into viscosity and thermal diffusion terms so the point of view embodied in the approximations is not strictly a hydrodynamic one. The relative success of this approach suggests that it is worth refining.

A direct calculation of the memory functions for the longitudinal and transverse current memory functions has been made by GÖTZE and LÜCKE (1975). A closed set of equations for the correlation functions was derived by assuming that the dynamics of the memory functions could be expressed in terms of two mode processes. These equations have been solved self-consistently for liquid argon. The resulting power spectra are in qualitative and, for some Q values, quantitative agreement with the neutron scattering and molecular dynamics results. This is encouraging and makes it worth while to explore this work in more detail.

Three approximations go into this theory. The first is the assumption that the dynamics of the memory functions can be expressed in terms of two mode processes. The second assumption is that the two generalized hydrodynamic modes are themselves non-interacting. The third assumption involves the use of a modified superposition approximation for the static three particle correlations which enter into the amplitudes for the two mode processes. It is not possible to independently assess the consequences of these assumptions from the published results. Also, the neglect of the coupling to the energy (temperature) fluctuations limits the applicability of the calculation to the viscoelastic region. For argon this means $Q > 1\ \mathring{A}^{-1}$. Finally, the two mode approximation is not able to make the transition at high Q's to the free gas region. The authors suggest that there are non-trivial difficulties involved in making this transition.

This type of calculation makes it possible to examine in detail the processes involved in the memory functions. The authors make twelve specific comments on the detailed mechanisms brought out in this calculation. We shall not repeat them here and content ourselves with the observation that this calculation is the first one for which explicit insight into the mechanisms described by the memory functions is possible. The complexity of the analysis should not be a deterrent to further effort along these lines.

Kinetic theories of liquids offer another promising way of studying the properties of the memory functions. A variety of formal approaches has been developed (MAZENKO, 1974; JHON and FORSTER, 1975) which should make it possible to examine in detail the mechanisms described by the memory functions. The work of JHON and FORSTER (1975) has yielded good agreement with neutron scattering and molecular dynamics data so we shall restrict our discussion to that theory. This is not intended to imply that other theories cannot lead to good agreement with experiment.

The objects of study in kinetic theory are distribution functions for particle positions and coordinates from which quantities such as $S(Q,\omega)$ can be constructed. The theory of JHON and FORSTER (1975) proceeds along the same lines used to derive the equations of generalized hydrodynamics except that the one-particle phase-space density function for position and momentum provides the variables rather than the particle density, etc. This work differs from earlier studies of this sort in that the energy density is explicitly included in the set of variables. The coupling of the density to the energy density is essential in the small Q region. The resulting equations of motion for the correlation functions of the variables have the same overall form as the generalized hydrodynamic equations, but are richer in detail because the momentum of the particles has not been averaged out.

The memory functions for these kinetic equations are now assumed to vary smoothly from the hydrodynamic to the free particle limits as Q varies form "0" to "∞". This is realized by assuming a simple functional form for the memory function. In this sense, the approach is much like the one used with generalized hydrodynamics. However, the increased detail in the kinetic equations of motion makes it possible to examine the coefficients of generalized hydrodynamics within the context of the model kinetic equations. This has not yet been done although the results should be quite interesting as the computed values of $S(Q,\omega)$ for argon and the Lennard-Jones fluid are in quite good agreement with experiment.

Of the three theories examined in this section, the kinetic theory appears to be the most promising for future advances in the understanding of liquid dynamics. A large amount of work will be needed if these advances are to be realized. This will include both experimental and theoretical studies. A combination of neutron scattering experiments in dense gases (CHEN et al., 1973) and kinetic theory calculations (FURTADO et al., 1975; JHON and FORSTER, 1975) would appear to be a fruitful way to proceed.

Other issues in the study of collective modes involve the static properties $\Gamma(Q)$ and $S(Q)$. Progress in this area will require new developments in the equilibrium theory of the liquid state. RAVECHE and MOUNTAIN (1977) reviewed the status of this topic with emphasis on three particle correlations, the central problem of the equilibrium theory of liquids.

The thermal diffusion mode can be studied indirectly by neutron scattering because Γ(Q) goes to zero as Q increases. A direct investigation of this mode would be possible using molecular dynamics and would complement the studies of collective modes discussed in this chapter.

Generalized hydrodynamics offers a promising way of understanding the complex physical phenomena which are probed by the coherent scattering of neutrons. Once these phenomena are understood for monatomic liquids, this theory can form the basis for analyzing the collective modes which are present in molecular liquids.

## References


Ailawadi, N.K., Rahman, A., Zwanzig, R. (1971): Phys. Rev. A 4, 1616
Barker, M.I., Johnson, M.W., March, N.H., Page, D.I. (1973): in *The properties of Liquid Metals,* Proceedings of the Second Int'l Conference in Tokyo, (Taylor and Francis LTD, London) p. 99
Barker, M.I., Gaskell, T. (1974): J. Phys. C 7, L293
Barker, M.I., Gaskell, T. (1975): J. Phys. C 8, 3715
Bell, H.G., Kollmar, A., Alefeld, B., Springer, T. (1973): Phys. Lett. 45A, 479
Bell, H., Moeller-Wenghoffer, H., Kollmar, A., Stockmeyer, R., Springer, T., Stiller, H. (1975): Phys. Rev. A 11, 316
Buyer, W.J.L., Sears, V.F., Lonngi, P.A., Lonngi, D.A. (1975): Phys. Rev. A 11, 697
Callen, H.B., Greene, R.F. (1952): Phys. Rev. 86, 702
Chen, S.H., Lefevre, Y., Yip, S. (1973): Phys. Rev. A 8, 3163
Copley, J.R.D., Rowe, J.M. (1974a): Phys. Rev. Lett. 32, 49
Copley, J.R.D., Rowe, J.M. (1974b): Phys. Rev. A 9, 1656
Copley, J.R.D., Lovesey, S.W. (1975): Rep. Progr. Phys. 38, 461
Dassanacharya, B.A., Kollmar, A., Springer, T. (1976): Phys. Lett. 55A, 337
Demoulin, C., Montrose, C.J., Ostrowsky, N. (1974): Phys. Rev. A 9, 1740
Furtado, P.M., Mazenko, G.F., Yip, S. (1975): Phys. Rev. A 12, 1653
Götze, W., Lücke, M. (1975): Phys. Rev. A 11, 2173
Gyorffy, L., March, N.H. (1971): Phys. Chem. Liquids 2, 197
Jhon, M.S., Forster, D. (1975): Phys. Rev. A 12, 254
Levesque, D., Verlet, L., Kurkijarvi, J. (1973): Phys. Rev. 7, 1690
Mazenko, G.F. (1974): Phys. Rev. A 9, 360
Mountain, R.D. (1966): J. Res. NBS 70A, 207
Mountain, R.D. (1974): J. Res. NBS 78A, 413
Murase, C. (1970): J. Phys. Soc. Japan 29, 549
Price, D.L., Singwi, K.S., Tosi, M.P. (1970): Phys. Rev. B 2, 2983
Rahman, A. (1967): Phys. Rev. Lett. 19, 240
Rahman, A. (1968): In *Neutron Inelastic Scattering,* Vol. 1, p. 561 (IAEA, Vienna)
Rahman, A. (1972): In *Statistical Mechanics,* ed. by S.A. Rice, K.F. Freed and J.C. Light (University of Chicago Press, Chicago.).
Rahman, A. (1974a): Phys. Rev. Lett. 32, 52
Rahman, A. (1974b): Phys. Rev. A 9, 1667
Raveché, H.J., Mountain, R.D. (1977): *Progress in Liquids,* ed. by C.A. Croxton (John Wiley, London).
Rowe, J.M., Sköld, K. (1972): In *Neutron Inelastic Scattering 1972* (IAEA, Vienna). Note that in an erratum the authors point out an error in their original expression for the Gaussian memory function. The correct expression yields good agreement with experiment.
Schofield, P. (1968): *Physics of Simple Liquids,* ed. by H.N.V. Temperly, J.S. Rowlinson and G.S. Rushbrooke, (North-Holland, Amsterdam) pp. 563-609
Schofield, P. (1976): Specialist Reports-Statistical Mechanics, Vol. II., 1
Sears. V.F. (1970): Can. J. Phys. 48, 616
Sköld, K., Rowe, J.M., Ostrowski, G., Randolph, P.D. (1972): Phys. Rev. A 6, 1107
Vineyard, G.H. (1958): Phys. Rev. 110, 999
Yarnell, J.L., Katz, M.J., Wenzel, R.G., Koenig, S.H. (1973): Phys. Rev. A 7, 2130

Recent References with Titles

Furtado, P.M., Mazenko, G.F., Yip, S. (1976): Effects of correlated collisions on atomic diffusion in a hard-sphere fluid. Phys. Rev. A14, 869

Kahol, P.K., Bansal, R., Pathak, K.N. (1976): Collective excitations in liquid rubidium. Phys. Rev. A14, 408

Mountain, R.D. (1976): Generalized Hydrodynamics. Adv. in Mol. Relaxation Processes 9, 225

Svensson, E.C., Stirling, W.G., Woods, A.D.B., Martel, P. (1976): Low-Momentum-Transfer neutron Scattering from Liquid $^4$He at T=4.2K and P=2.5 MPa. Proceedings of the Conference on Neutron Scattering, Gatlinburg, Tenn., ed. R.M. Moon, p. 1017

Woods, A.D.B., Svensson, E.C., Martel, P. (1976): The Dynamic Structure Factor of Nonsuperfluid Liquid $^4$He. Proceedings of the Conference on Neutron Scattering, Gatlinburg, Tenn. p. 1010

Woods, A.D.B., Svensson, E.C., Martel, P. (1976): The First-to-Zero Sound Transition in Non-Superfluid Liquid $^4$He. Phys. Lett. 57A, 439

# 7. Magnetic Scattering

S. W. Lovesey and J. M. Loveluck

With 15 Figures

The scope of research activity encompassed by the title of this chapter is enormous, and includes, for example, magnons in metals, alloys, and ionic compounds, critical phenomena, crystal field levels, magnon-phonon interactions, magnetic excitons and excitations in spin glasses and amorphous magnets, each of which rightly deserves a chapter to itself. In the space available to us here we attempt only to illustrate the usefulness of neutron scattering in the study of some dynamic effects of current interest, and the areas from which we have chosen examples are prescribed largely by our competence. Moreover, for each area of activity included, we shall in general, concentrate on one, or perhaps two particular cases. We shall consider only one example of a mixed magnetic system; useful reviews of inelastic neutron scattering measurements on mixed magnetic systems have been given by COWLEY and BUYERS (1972), and ELLIOTT et al. (1974). These few words should portend the very restricted nature of our discussion of inelastic magnetic neutron scattering.

Much of the formalism required to describe, and interpret, inelastic scattering measurements is reviewed briefly in the final section of Chap.1. Here we shall consider for the most part specific examples of the formalism. One topic which we cannot illustrate by way of an experiment is the use of polarized neutrons, because, to our knowledge, no such experiments have yet been performed. The problem is one of neutron flux, but the situation is ameliorated by the advent of improved polarisers and analysers, in conjunction with the new generation of high flux reactors, and the technique should become established shortly.

In discussing magnons (or spin waves) it is both convenient and sensible to separate systems which support long range magnetic order from those which do not. We begin in the following section with a discussion of magnons in the insulating ferromagnets EuO and EuS, and then discuss the rutile antiferromagnet $FeF_2$ and the metamagnet $FeCl_2$. This leads to a discussion of magnon-phonon hybridisation created by the ligand crystal field. We then discuss in Sec.7.2 some examples of collective, magnetic excitations in quasi one- and two-dimensional systems, in which long range order is absent.

Magnetic excitations in metals and alloys have been shown to have different features, and different underlying physics, from those in ionic compounds, and our basic understanding of them is still quite rudimentary. The reason for this state of affairs is that, unlike the ionic compounds, a meaningful discussion of the excitations in metals should be based on an itinerant model of the magnetic electrons, and include band structure effects, i.e., it requires an understanding of electron correlations in multiband systems. Several detailed experimental and theoretical investigations have been completed, and in Sec.7.3 we shall pay particular attention to those on ferromagnetic nickel. The experiments have revealed several features which are not fully understood, reflecting the complexity of these systems, and the inadequacy of the theoretical apparatus.

Neutron scattering experiments have been crucial in establishing our present understanding of magnetic rare earth metals and compounds. The effect of the crystal field is particularly important for materials which contain rare earth ions with unquenched orbital angular momentum. In some materials, the magnetic ion has a singlet ground state, and magnetic ordering occurs only at very low temperature. The crystal field energy is therefore dominant in determining the magnetic properties, and the level scheme can be determined by neutron scattering, as discussed in Sec.7.4. In the opposite extreme, the exchange energy dominates the crystal field energy. The effect of the crystal field can be described, in this case, by single site anisotropy energies. We illustrate this situation by reference to terbium, and include also a discussion of magnon-phonon interactions. The third class of rare earth materials which we discuss are those in which there is, effectively, a competition between the exchange and crystal field energies, and we take praseodymium metal as our example.

The investigation of critical phenomena in magnetic systems has a long history, and continues to be an area of great activity. The allure of such problems, because of their fundamental importance in theoretical physics, has spurred experimentalists to several very detailed investigations on both ionic compounds and metals. In Sec.7.6 we illustrate some of the points of progress by reference to experiments on $RbMnF_3$, Ni and Fe.

## 7.1 Magnons in Ionic Compounds

Neutron scattering from magnons, and the theory of magnons in ionic compounds, have been reviewed by several authors. The articles by VAN KRANENDONK and VAN VLECK (1958), WALKER (1963), KEFFER (1967) and IZYOMOV (1963) are particularly relevant to the following discussion.

There are very few ionic ferromagnets, and, of these, the divalent europium chalcogenides EuO and EuS have been studied most thoroughly by neutron scattering, because they are examples of simple Heisenberg ferromagnets. In both these materials, the $Eu^{2+}$ ions form a fcc lattice. If we assume that the magnetic interactions include only nearest ($\underset{\sim}{R}_1$) and next-nearest ($\underset{\sim}{R}_2$) neighbours, then the magnon dispersion relation, (1.167), reduces to

$$\omega_q = 2S\,[12J_1 + 6J_2 - J_1 \sum_1 \cos(\underline{q}\cdot\underline{R}_1) - J_2 \sum_2 \cos(\underline{q}\cdot\underline{R}_2)] \tag{7.1}$$

where $J_1$ and $J_2$ are the two exchange integrals and S is the magnitude of the spin. Experiments by PASSELL et al. (1972) were performed on polycrystalline samples of EuO and EuS, and, consequently, all values of q are observed with equal probability. In view of this, the dispersion relation must be averaged over all directions of $\underline{q}$.

Several groups of experimentalists have attempted to determine values for $J_1$ and $J_2$ from the analysis of data for properties which involve an integral over the whole Brillouin zone, e.g., magnetisation measurements, and have obtained conflicting results. PASSAL et al. (1972) were able to deduce values of the exchange parameters directly, by fitting (7.1)-averaged over all directions of $\underline{q}$ - to the measured dispersion curves. They showed, for example, that the next-nearest exchange integral $J_2$ in EuO is positive whereas the interpretation of specific heat and NMR measurements had led to negative values for $J_2$.

Beginning with the work of DYSON (1956), there have been many calculations of the effect of magnon-magnon interactions on the temperature renormalisation and lifetime of magnons. The line shape, and therefore the lifetime, of the magnons in EuO and EuS has not been studied, but GLINKA et al. (1973) investigated the temperature renormalisation of magnons in EuO. DYSON showed that the interaction could be divided into two parts: a repulsive interaction which arises because of the statistics of spin operators, and called by him the kinematic interaction; and, secondly, an attractive interaction, which arises because it costs less energy for a spin to suffer a deviation if the spins to which it is coupled by the exchange interaction have also undergone deviations, and this he called the dynamic interaction. At temperatures small compared to the critical temperature, the kinematic interaction is small compared to the dynamic interaction. If we neglect the kinematic interaction, and consider effects arising from two magnon processes in the dynamic interaction, then, to first order in the interaction, only the energy of a magnon is affected; i.e., magnon energies are renormalised. We find that (1.167) is replaced by (we set the external field equal to zero)

$$\omega_q = 2S\,[\mathcal{J}(o) - \mathcal{J}(\underline{q})] - \frac{2}{N}\sum_{\underline{q}'} [\mathcal{J}(o) + \mathcal{J}(\underline{q}-\underline{q}') - \mathcal{J}(\underline{q}) - \mathcal{J}(\underline{q}')]\,n_{\underline{q}'} \tag{7.2}$$

where

$$n_{\underline{q}} = [\exp(\beta\omega_{\underline{q}}) - 1]^{-1} \quad , \quad \beta = 1/k_BT \quad . \tag{7.3}$$

Equations (7.2) and (7.3) together give a transcendental equation for the renormalised magnon dispersion, $\omega_q \equiv \omega_q(T)$. For nearest neighbour exchange interactions and small q it can be shown that

$$\omega_q \simeq D(T)q^2 \tag{7.4}$$

where D(T) is a temperature dependent stiffness constant, which decreases from its zero temperature value as the temperature is increased through a term proportional to $T^{5/2}$.

The measurements by GLINKA et al. (1973) on EuO were analysed by them in terms of the foregoing expressions, using nearest, and next nearest exchange constants with values 0.052 and 0.013 meV. No solution of the transcendental equation was found for temperatures exceeding 63 K, whereas the observed Curie temperature is 69.1 K. Good agreement between the theory and observed magnon energies is found at all $\underline{q}$, as can be seen from Fig.7.1.

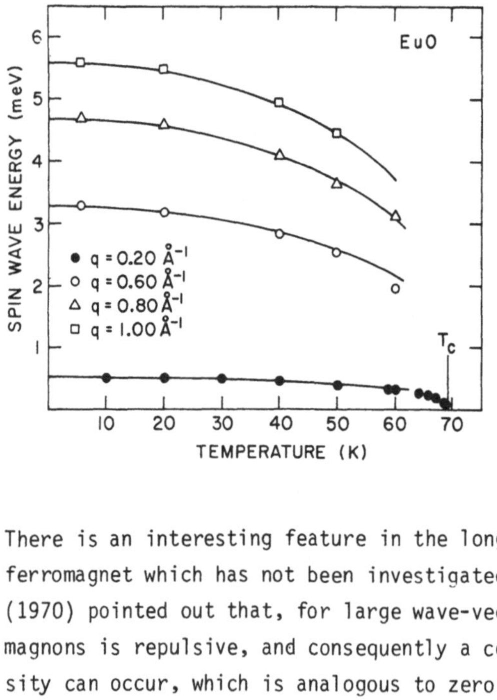


Fig. 7.1 Measured magnon energies in EuO for various wave-vectors, normalized to their values at 5 K, are shown as a function of temperature. The solid curve is calculated from (7.2) and (7.3). After GLINKA et al. (1973)

There is an interesting feature in the longitudinal susceptibility of a Heisenberg ferromagnet which has not been investigated experimentally. NATOLI and RANNINGER (1970) pointed out that, for large wave-vectors, the dynamical interaction between magnons is repulsive, and consequently a collective oscillation of the magnon density can occur, which is analogous to zero sound in Fermi liquids, and to plasmons in an electron gas. The authors report a calculation for a simple cubic ferromagnet which shows the existence of such a collective oscillation near the zone boundary, at an energy slightly larger than the maximum magnon energy. The longitudinal and transverse components of the magnon susceptibility are easily separated because they appear in the cross-section multiplied by the orientation factors $(1 - \tilde{Q}_z^2)$ and and $(1 + \tilde{Q}_z^2)$, respectively. Consequently, when the scattering vector is parallel to the axis of quantisation (the z-axis) only the transverse susceptibility contributes to the scattering, whereas when the scattering vector is perpendicular to the axis of quantisation the two susceptibilities contribute with equal weight.

Many ionic antiferromagnets have been studied by neutron scattering. Here we shall concentrate on two ferrous compounds, namely $FeF_2$ and $FeCl_2$.

The compound $FeF_2$ has a rutile structure, in which the $Fe^{2+}$ ions form a body-centred lattice, and each magnetic ion is surrounded by six fluorine atoms. In the ordered state, which occurs for $T < 79$ K, the ions align along the c-axis, with the corner and body-centre ions pointing in opposite directions. The basic difference between $FeF_2$ and the much-studied antiferromagnet $MnF_2$ (cf. KEFFER, 1967) is the single electron outside the half-filled shell in the ferrous ion. The motion of this electron is influenced strongly by the ligand crystal field, with the result that there is a large single-ion anisotropy, and hybridisation between the magnons and phonons.

Some measurements of the magnon dispersion at 4.2 K by RAINFORD et al. (1972) are shown in Fig.7.2. The exchange interactions are dominated by those between nearest neighbours on opposite sublattices, i.e., between the ions at the corner and body-centre positions. Denoting this exchange integral by J, the measured magnon dispersion is represented, to a good approximation, by (cf. (1.173))

$$\omega_{\underline{q}} = 1\partial JS[(1 + h_A)^2 - \gamma_{\underline{q}}^2]^{1/2} \tag{7.5}$$

where

$$\gamma_{\underline{q}} = \cos(\tfrac{1}{2}aq_x)\,\cos(\tfrac{1}{2}aq_y)\,\cos(\tfrac{1}{2}cq_z) \tag{7.6}$$

and the reduced single site anisotropy constant

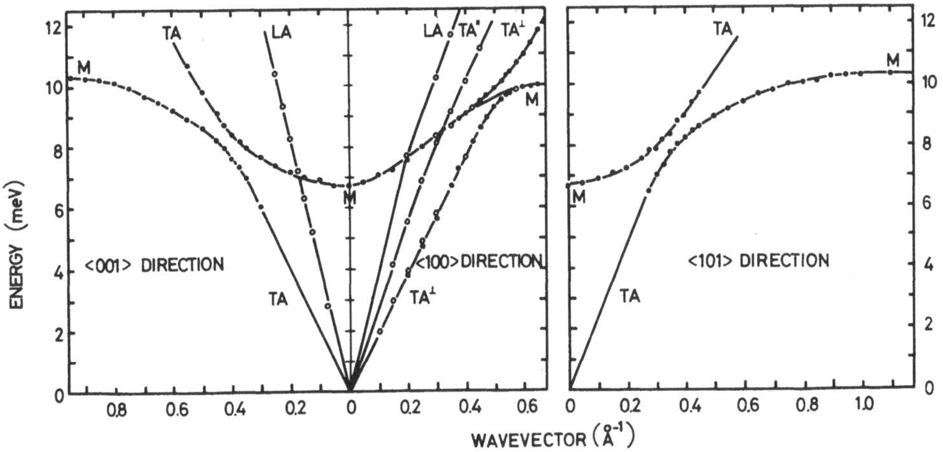


Fig. 7.2 Measured magnon and phonon energies in $FeF_2$, at 4.2 K, are shown as a function of wave-vector for propagation in the (001), (100) and (101) directions. After RAINFORD et al. (1972)

$$h_A = (2S - 1)A/16JS$$
$$\simeq 0.34 \quad . \qquad (7.7)$$

The single-site anisotropy originates mainly from the effect of the ligand crystal field whereas, in $MnF_2$, the effect of the crystal field on the $Mn^{2+}$ is minimal, and the energy gap in the dispersion relation at q = 0 originates from dipolar interactions.

RAINFORD et al. (1972) studied carefully the region of the nominal intersection of the magnons and acoustic phonons. They observed a significant hybridisation between the transverse acoustic phonons and magnons, as shown in Fig.7.2. In the high symmetry (001) direction, they observed two clearly resolved peaks, as shown in

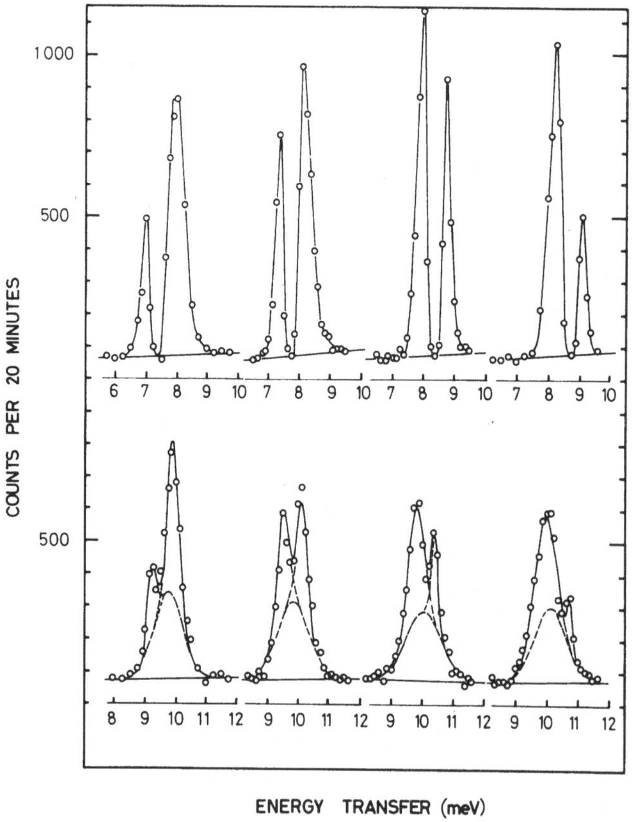


Fig. 7.3 Intensity of neutrons scattered magnetically from $FeF_2$, at 4.2 K, shown as a function of energy transfer for various wave-vectors in the (001) and (100) directions. The curves in the upper part of the diagram are for q in the (001) direction, with q = 0.35, 0.375, 0.40 and 0.45 $Å^{-1}$, and in the lower part of the diagram q is in the (100) direction, with q = 0.5, 0.525, 0.55 and 0.575 $Å^{-1}$. Solid lines are line shapes fitted to the sum of three Gaussian curves (shown as dashed lines) whose areas are constrained to be equal to the cross-section calculated by LOVESEY (1972). After RAINFORD et al. (1972)

Fig. 7.3. For $\underline{q}$ along the (100) direction there was evidence of a third peak predicted by LOVESEY (1972), although the three peaks could not be resolved clearly.

The physical mechanism responsible for the hybridisation was discussed first by VAN VLECK (1940) in his analysis of paramagnetic relaxation. The vibrations of the ligand field modulate the orbital motion of the single electron outside the half-filled shell in $Fe^{2+}$, and this, in turn, modifies the spin dynamics via the spin-orbit coupling. An analysis of the coupling shows that, at low temperatures, there is a term linear in both the spin operators, and the phonon operators, which results in a hybridisation of the magnons and phonons. The solid curves in Fig.7.2 are the results of a calculation by LOVESEY (1972), which uses only parameters determined by neutron and resonance experiments.

Group theoretic arguments can be used to generate selection rules for the character of the excitations which can undergo hybridisation. These selection rules tell us only *which* excitations can hybridise for a given wave-vector; the specific form of the interaction may not actually generate all the hybridisations which are allowed. The selection rules for $FeF_2$ and $FeCl_2$ for special wave-vectors have been calculated by MONTGOMERY and CRACKNELL (1973) and CRACKNELL (1974).

Prior to the experiment on $FeF_2$, described above, and on $UO_2$ by COWLEY and DOLLING (1968), most theoretical discussions of magnon-phonon interactions concerned the interaction which arises because of the modulation in the exchange integral due to lattice vibrations. The exchange integral is expanded in a Taylor series in the displacements $\underline{u}_\ell$ about the equilibrium lattice positions, $\underline{\ell}$, and, to first order in the displacements, we have

$$J(\underline{\ell} + \underline{u}_\ell, \underline{\ell}' + \underline{u}_{\ell'}) \simeq J(\underline{\ell} - \underline{\ell}') + (\underline{u}_\ell - \underline{u}_{\ell'}) \cdot \underline{\nabla} J(\underline{\ell} - \underline{\ell}') + \dots \quad . \tag{7.8}$$

We see that this interaction results in additional terms in the Hamiltonian which are linear in the phonon operators and quadratic in the spin operators. Therefore it does not lead to a hybridisation effect, but only a broadening of the magnon line shape.

The second antiferromagnet we shall discuss belongs to a special class of magnets called metamagnets. $FeCl_2$ is composed of hexagonal layers of ferromagnetically aligned ferrous ions with a strong single site anisotropy confining the moments to the hexagonal axis. A very weak antiferromagnetic coupling exists between the hexagonal layers, so the overall magnetic structure is antiferromagnetic, with a Néel temperature of 23.5 K. If a field is applied parallel to the c-axis, then at a field of some 10 kG the weak antiferromagnetic inter-layer coupling is overcome and the system undergoes a metamagnetic transition to a structure in which all the ions are aligned along the c-axis.[1] In contrast to this, a conventional antiferromagnet

[1] The phase diagram of $FeCl_2$ has been measured by neutron scattering by BIRGENEAU et al. (1974). A detailed theoretical discussion is given by KINCAID and COHEN (1975).

placed in an external field displays a spin-flop phase, as described, for example, by KEFFER (1967).

BIRGENEAU et al. (1972b) showed that magnons with wave-vectors parallel to the c-axis show no observable dispersion, whereas excitations in the hexagonal layers show a pronounced dispersion which is described well by a simple magnon theory. A calculation by LOVESEY (1974) showed, on the basis of an analysis of thermal conductivity measurements by LAURENCE and PETITGRAND (1973), that the splitting of the magnon and phonon dispersions at the nominal intersection was quite small, and it was pointed out that an external field would increase the separation between the hybridised modes, because a field parallel to the c-axis lifts the double degeneracy of the antiferromagnetic magnons (cf. (1.172)). Increasing the separation of the hybridised modes also enhances the possibility of resolving the three peaked structure predicted for certain directions of $\underset{\sim}{q}$ in $FeF_2$ and $FeCl_2$. In Fig.7.4a we show some data collected by ZIEBECK and HOUMANN (1975, private communication) from $FeCl_2$ in a magnetic field of 8.9 kG, and a wave-vector in the x-direction equal to 20% of the zone boundary value. In the absence of a field, the central mode is simply an unperturbed magnon, with a weight half of that which would occur in the absence of

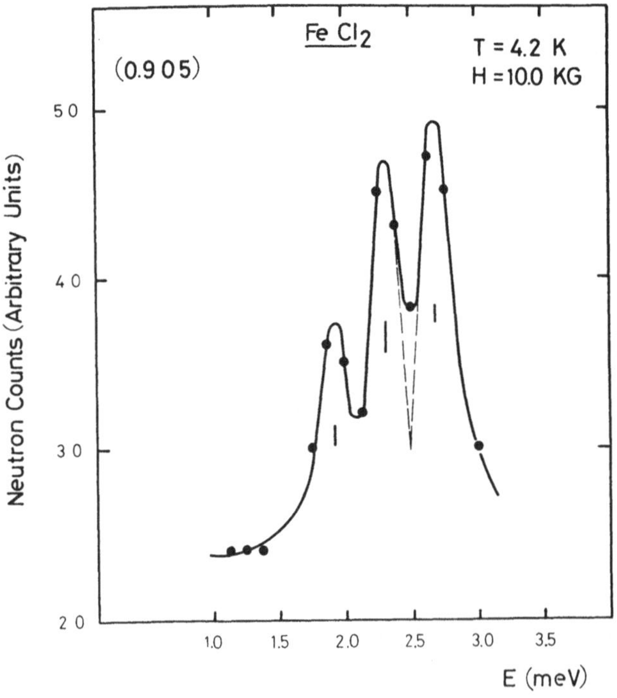


Fig. 7.4a Intensity of neutrons scattered magnetically from $FeCl_2$, at 4.2 K, with a component of the magnetic field = 8.9 kG parallel to the c-axis, and a wave-vector in the x-direction equal to 20% of the zone boundary value

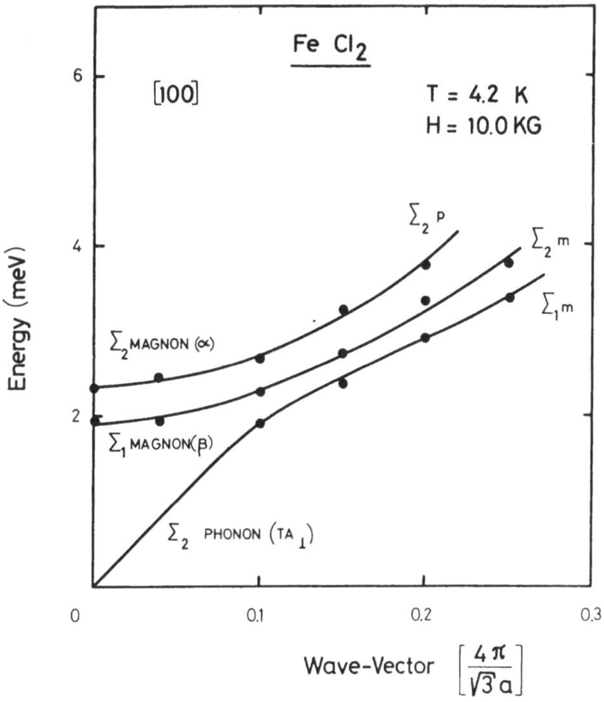


Fig. 7.4b Dispersion of the modes for propagation in the x-direction; the zone boundary is at 0.5 with the reduced wave-vector used in the diagram. After ZIEBECK and HOUMANN (private communication)

hybridisation, the weight being independent of wave-vector in this case; the additional weight is shared between the two hybridised modes. However, in the presence of a field, even the weight for the central mode changes with wave-vector. The dispersion of the magnetic excitation for q along the x-direction is summarised in Fig.7.4b.

## 7.2 Quasi One- and Two-Dimensional Systems

Theoretical physicists have a long-standing interest in low-dimensional models, stemming from the possibility of exact calculations for many-body systems, and also their use as a test-bed for approximate theories. However, in the study of magnetic properties, the recent burgeoning of theoretical activity has been stimulated by a flow of experimental information on a variety of real materials exhibiting quasi one-dimensional (1D) and two-dimensional (2D) magnetic behaviour. The experimental investigations have revealed striking magnetic properties, and the theoretical interpretation of these results, without the necessity of uncontrolled approximations, such as RPA, which often attend calculations in 3D, has attracted considerable interest.

There now exists abundant evidence, some of which will be presented below, that many materials exhibit a 1D or 2D magnetic structure, at least in some restricted, but readily accessible, temperature range. A thorough survey, and guide to the literature, is furnished by the review article of DE JONGH and MIEDEMA (1974). The manifestation of 1D (or 2D) magnetic properties reflects a strong intrachain (intraplane) exchange coupling of magnetic atoms, with a relatively weak interchain (interplane) coupling. Clearly, this situation can arise if there is a relatively large physical separation, by groups of non-magnetic atoms, of linear chains (planes) of magnetic atoms, with relatively short superexchange paths between the magnetic atoms within the chains (planes). The residual interchain (interplane) interaction (due, for example, to dipolar forces) leads to 3D magnetic behaviour at a sufficiently low temperature, $T_c$, the value of $T_c$ being a measure of the strength of the interchain (interplane) coupling relative to the intrachain (intraplane) coupling.

There are striking differences, in both the thermodynamic and dynamic behaviour, between 3D and lower dimensional systems. Effects which may be marginal, or subtle, in 3D are often exaggerated, or even qualitatively different, in 1D or 2D. These differences can be traced to two basic characteristics of 1D and (most) 2D systems: the absence of long-range-order (LRO) for any finite temperature, together with an enhancement of short-range-order (SRO). Theoretical proofs of the absence of LRO above T = 0, for 1D systems with finite-range forces, have been given by LANDAU and LIFSHITZ (1958) and by MERMIN and WAGNER (1966). In 2D the implications of the MERMIN-WAGNER theorem are less transparent, and will be discussed in Sec.7.2.2. The enhanced SRO in low-dimensional systems is evinced by neutron scattering measurements, which we shall discuss below, exhibiting the persistence of magnon-like excitations in the absence of LRO. This feature is not restricted to low-dimensional systems, and such excitations have been observed, for example, in the paramagnetic phase of 3D $RbMnF_3$, as discussion in Chap.1. However, the temperature range over which short-range-order effects are important is considerably extended in 1D and 2D magnetic systems, and it is instructive to investigate this feature.

Following HONE and RICHARDS (1974). we note that, for a magnetic system in which z spins S are coupled to a given spin by the exchange interaction J, SRO begins to develop when the characteristic interaction energy, typified by the Weiss temperature $\theta = 2JzS(S + 1)/3$, becomes comparable to thermal energies $k_BT$. In 3D, one finds that $\theta/T_c \simeq 1$, indicating that SRO rapidly induces LRO, whereas, in lower dimensions, this process is inhibited by long wavelength thermal fluctuations, which depress the transition temperature and, in particular in 1D $\theta/T_c \to \infty$.

In the following subsections, we confine ourselves to a limited number of examples of 1D and 2D magnetic structures, selected to illustrate the dominant features which have been observed in neutron scattering experiments, and the theoretical techniques which have been employed in their interpretation. In restricting our discussion in this way, we divorce ourselves from a wealth of further experimental information on low-dimensional systems, together with attempts at theoretical interpretation, including thermodynamic properties, and NMR, EPR and light-scattering measurements. For surveys and guides to the literature which include these areas, we refer the reader to the review articles by DE JONGH and MIEDEMA (1974), HONE and RICHARDS (1974) and STEINER et al. (1976). Finally, it should be mentioned that we restrict our attention to systems which are well described by localised model Hamiltonians, and exclude, for example, various organic charge-transfer crystals, which are best described by an itinerant model.

### 7.2.1 Magnetic Linear-Chain Systems

Experimental and theoretical interest in 1D magnetic structures has concentrated on Heisenberg systems, rather than Ising chains. It appears that materials which approximate 1D Ising chains are rather difficult to realise, but there are other reasons for this bias of interest, including the fact that, for a given ratio of intrachain to interchain interaction, Heisenberg chains remain 1D at much lower temperatures than a corresponding Ising chain; furthermore, the dynamic behaviour of the Ising chain is somewhat prosiac.[2] Consequently, we will focus our attention on systems described by the nearest neighbour Heisenberg Hamiltonian

$$H = -\frac{1}{2} J \sum_{\ell,\delta = \pm a} \hat{\underline{S}}_\ell \cdot \hat{\underline{S}}_{\ell+\delta} + B \sum_\ell (\hat{S}^z_\ell)^2 \tag{7.9}$$

in which a is the lattice constant, and the second term allows for the possibility of single-site anisotropy.

The first example which we shall consider is $(CH_3)_4NMnCl_3$(TMMC), for which a wealth of experimental data has accumulated, indicating that at temperatures above $T_c = 0.84$ K, the 3D ordering temperature, it behaves as a 1D antiferromagnet ($J < 0$ in (7.9)), with $S = 5/2$ and $B \simeq 0$. The persistence of 1D magnetic behaviour down to this extremely low temperature, which has been established both by static susceptibility measurements (DINGLE et al., 1969; WALKER et al., 1972) and by neutron diffraction studies (BIRGENEAU et al., 1972a), renders this system particularly interesting as a model 1D structure. Results of quasielastic scattering

[2]The magnetic properties of $CoCl_2.2H_2O$ can be interpreted in terms of weakly interacting Ising-like chains with small Heisenberg and transverse anisotropies. This model successfully describes the magnetic field dependence of the multi-magnon bound state spectrum, observed by optical absorption (TORRENCE and TINKHAM, 1969), and the dispersion of the one-magnon mode, measured by inelastic neutron scattering (KJEMS et al., 1975).

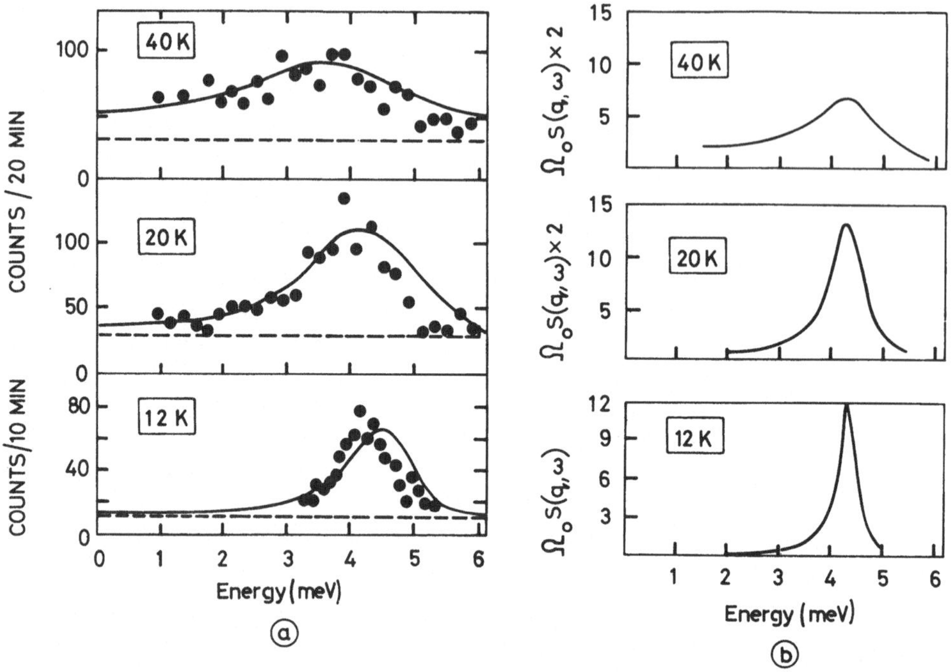


Fig. 7.5a The scattered intensity from TMMC for $aq = 0.75\pi$, measured by HUTCHINGS et al. (1972), is shown for several temperatures (solid circles); the solid curve corresponds to the computer simulation results of WINDSOR and LOCKE-WHEATON (1976).

Fig. 7.5b For comparison, we show calculated results of LOVESEY and MESERVE (1972) for $\Omega_0 S(q,\omega)$ for the same wave-vector and temperatures as in (a); here $\Omega_0 = 2J\,[S(S+1)/3]^{1/2}$. Note the change of scale for T = 12 K which allows for the change in experimental counting time

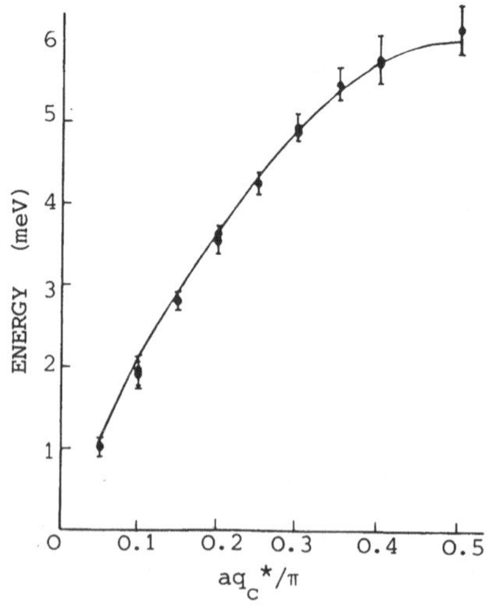


Fig. 7.6 Experimental results (solid circles, with error bars) for the dispersion relation of the magnon-like mode in TMMC at 4.4 K, as measured by HUTCHINGS et al. (1972), are compared with the calculated results of LOVESEY and MESERVE (solid line)

experiments provide direct evidence for the 1D nature of TMMC. In the quasielastic approximation, the neutron scattering cross-section is proportional to $<\hat{\underline{S}}_q \cdot \hat{\underline{S}}_{-q}>$ (cf. Chap.1), and, for a 1D magnetic structure, this relation implies that the cross-section is constant in planes perpendicular to the direction of the chains. This 1D analogue of Bragg peaks[3] was strikingly demonstrated by the results of BIRGENEAU et al. (1971) on deuterated TMMC. Furthermore, the large value of S = 5/2 makes feasible a quantitative comparison between the experimental measurements of $<\hat{\underline{S}}_q \cdot \hat{\underline{S}}_{-q}>$ and the exact theoretical calculation of FISHER (1964) for a classical, isotropic Heisenberg linear magnet with nearest-neighbour interactions. HUTCHINGS et al. (1972) concluded that the experimental data could be accurately fitted by FISHER'S result, at all temperatures between 1.1 K and 40 K. HUTCHINGS et al. also investigated the inelastic neutron scattering cross-section for deuterated TMMC, and it was these measurements which provided the first direct, and striking, evidence for the existence of well-defined magnon-like modes in a magnetically 1D system. Fig.7.5 shows the magnetic scattering intensity measured at a wave-vector $aq = 0.75\pi$, for several temperatures. At low temperatures a well-defined peak, corresponding to a collective excitation, is quite apparent, while the excitation is seen to broaden as the temperature is raised, but with a minimal shift in the peak position. For low temperatures, the dispersion relation of the collective mode was found to follow a sine curve typical of linear antiferromagnetic magnons (see Fig.7.6). The experimentally measured dispersion relation was fitted to the theoretical expression obtained using renormalised spin-wave theory (see, e.g., OGUCHI, 1960), and the value of the nearest-neighbour exchange interaction, obtained in this way, was found to be in good agreement with that obtained from the susceptibility measurements.

The theoretical interpretation of the spin dynamics of magnetic systems is formulated in terms of the relaxation shape function $F_\alpha(q,\omega)$ ($\alpha$ = x,y or z), which, as indicated in Chap.1, is intimately related to the inelastic magnetic neutron scattering cross-section. The experimental evidence outlined above points clearly to the existence of a sharp collective mode in TMMC at low temperatures. Consequently, at low temperatures, we can represent $F(q,\omega)$ (the subscript $\alpha$ being redundant for the isotropic system), to a first approximation, by a pair of delta functions at $\pm\omega_q$,

[3] An illuminating discussion of the magnetic Bragg scattering expected from 1D and 2D structures, which give rise to Bragg planes and lines (or "ridges"), is given by SKALYO et al. (1970).

corresponding to energy loss and gain. The collective mode frequency $\omega_q$ may be determined from the result

$$\omega_q^2 = \langle\omega^2\rangle = \int_{-\infty}^{\infty} d\omega\ \omega^2 F(q,\omega) = \frac{4NJ}{\chi(q)} (1 - \cos aq) \langle\hat{S}_0^z\ \hat{S}_1^z\rangle \tag{7.10}$$

(cf. Chap.1, (1.186)). For the case of TMMC, with S = 5/2, the static properties should be adequately described by a *classical* linear chain model, in which case exact analytic expressions for $\langle S_0^z\ S_1^z\rangle$ and $\chi(q)$ are available from the work of FISHER (1964), and indeed, as $T \to 0$ this yields

$$\omega_q = 2J[S(S + 1)]^{1/2} |\sin aq| \tag{7.11}$$

in agreement with the measured dispersion relation of HUTCHINGS et al. at 1.9 K. For a system without LRO, the inclusion of damping of the collective mode is of paramount importance, and may be expected to modify significantly the dynamics, particularly at small wave-vectors, even at moderate temperatures. Intuitively, one might expect that, for low temperatures, the width of the mode is governed by the fluctuations about $\langle\omega^2\rangle$, and that the magnitude of such fluctations would be determined by the fourth moment $\langle\omega^4\rangle$ of $F(q,\omega)$. A more complete description of the dynamics at finite temperatures, including the shift, due to the increased line width, in the collective mode frequency away from the result (7.10), involves a study of the equation of motion for the spin variable, $S_q^z$ say, in terms of a fluctuating local field. The formulation due to MORI (1965a,b), of the equation of motion of a dynamical variable, lends itself naturally to a description of this nature. We are unable, in this limited space, to do full justice to the Mori formalism, but some insight may be obtained by considering its relation to a Langevin equation. In terms of the analogous problem of the motion of a particle in a fluid, the Langevin equation separates the net force on a particle into a secular term, corresponding to the viscous drag on the particle, and a fluctuating force, the time-averaged correlation of which is related to the damping constant. The Mori equation of motion for a dynamical variable, or, indeed, a *set* of dynamical variables, retains this separation of the net force into a secular and fluctuating part, but has the additional feature that the driving term is expressed as a convolution of the dynamical variable with a time-dependent kernel, which describes the non-Markovian nature of the temporal evolution of the dynamical variable, and is appropriately referred to as a memory function. The resulting description in terms of a "generalised" Langevin equation, can be broadened to describe a *set* of dynamical variables, and a judicious choice of this set of variables is crucial, especially in deriving approximations to the memory function, the objective being to achieve a close approximation to a Markovian equation of motion. Clearly, in the interpretation of neutron scattering experiments, $\hat{S}^z_q$ must be included, since we wish to calculate the relaxation function, (1.185). In principle, the spin current, which is proportional to $\dot{\hat{S}}^z_q$ (which is a conserved quantity), should also be included, but, in the absence of a magnetic field, dynamical effects involving $\dot{\hat{S}}^z_q$ can be expressed in terms of $\hat{S}^z_q$.

The ensuing theoretical development involves the pursuit of an approximation to the memory function, and two basic approaches have been followed. The mode-mode coupling theory (see, e.g., POMEAU and RESIBOIS, 1975) represents an attempt to derive a microscopic approximation for the memory function, but is restricted to small wave-vectors, and hence excludes effects due to SRO, which are essential in a description of collective modes in systems without LRO. McCLEAN and BLUME (1973) confronted this difficulty by the introduction of a short-range order parameter, and were able to obtain at least qualitative agreement with the experimental results of HUTCHINGS et al.

The second approach follows the development of MORI (1965b), who showed that the memory function itself satisfied an equation of motion, similar to that for the relaxation function, but involving a higher order memory function. In fact, MORI showed that a hierarchy of coupled equations, involving higher order memory functions, could be constructed, and that this set of equations was equivalent to a continued fraction expansion for the relaxation function. For the case we are interested in, the coefficients in this continued fraction expansion are related to the even moments $\langle\omega^{2n}\rangle$ of the relaxation shape function $F(q,\omega)$, and these, in turn, can be expressed in terms of static spin correlation functions, which can be evaluated exactly for the classical, isotropic Heisenberg chain. In this way, the continued fraction approach can, potentially, take advantage of exact information on the static properties, in contrast, for example, to conventional equation of motion calculations, in which the aim is often the more ambitious one of a self-consistent theory.

To obtain a closed form for the relaxation function, one seeks a physically reasonable termination of the continued fraction expansion. For the interpretation of neutron scattering experiments, an approximation which describes correctly the short time behaviour is appropriate, and this is ensured by demanding that the low order moments of the relaxation shape function be given correctly. Examination of the coefficients in the continued fraction expansion reveals that, beyond the second level, they are only weakly q-dependent, and this suggests that a termination involving an approximation to the second order memory function should be adequate. Two alternative terminations at this level have been proposed: that due to TOMITA and MASHIYAMA (1972) involves a Gaussian approximation to the second-order memory function, the strength and width of the Gaussian being determined by the second, fourth and sixth moments, and hence requiring two- four- and six-spin correlation functions; the "three-pole" approximation of LOVESEY and MESERVE (1972, 1973) suggests an exponential decay, with a relaxation time determined self-consistently in terms of

$\langle\omega^2\rangle$ and $\langle\omega^4\rangle$. The form of the relaxation time determined in this way has been shown by WINDSOR and LOCKE-WHEATON (1976) to give optimum agreement with computer simulation results. The termination of TOMITA and MASHIYAMA has the advantage that it satisfies the first four moment relations exactly (whereas that of LOVESEY and MESERVE satisfies only the first three) but results in a rather complicated expression for the relaxation function, and this precludes a simple analysis of its salient features. Some predictions of the two theories are summarised in Table 7.1.

Table 7.1 Predictions by various theories of critical wave-vector, diffusion constant, and line widths for one-dimensional magnets

| | Ferromagnetic coupling $D(\infty)/Ja^2[S(S+1)]^{1/2}$ | $q_c(T)$ | $\Gamma_q(T)$ | $\Delta(T)$ | Antiferromagnetic coupling $q_c^*(T)$ | $\Gamma_q(T)$ | $\Delta_q(T)$ |
|---|---|---|---|---|---|---|---|
| McLEAN and BLUME (1973) | 1.38 [a] | $\kappa$ | $q^2$ | $T$ | $\kappa$ | $T$ | $T$ |
| LOVESEY and MESERVE (1972, 1973) | 0.72 | $T^{1/2} > \kappa$ | $q^2T$ | $T^{3/2}$ | $T^{1/2} > \kappa$ | $T^{3/2}$ | $T^{3/2}$ |
| TOMITA and MASHIYAMA (1972) | 0.96 | $T^{1/2}$ | $q^2$ | $T$ | $T^{1/2}$ | $T$ | $T$ |

[a] Computer simulation experiments yield $D(\infty)/Ja^2[S(S+1)]^{1/2} = 1.33 \pm 0.10$ (LURIE et al., 1974); 1.2 (WINDSOR and LOCKE-WHEATON, 1976).

$\Gamma_q$ is the width of the central peak, $q < q_c$

$\Delta_q$ is the width of the collective mode, $q > q_c$

$q_c$ is the dynamic inverse correlation length, $\kappa$ the static inverse correlation length; at low T, $a\kappa \approx k_BT/[JS(S+1)]$. for a classical one-dimensional magnet

$D(\infty)$ is the diffusion constant at infinite temperature

Both approximations result in qualitative, and, in some respects, semi-quantitative agreement with the experimental results. The more transparent expression for $F(q,\omega)$ of LOVESEY and MESERVE can be shown to yield the following behaviour:

a) as $T \to 0$, the expression for $F(q,\omega)$ reduces to a pair of delta functions at $\pm\omega_q$, with $\omega_q$ given by (7.10), in agreement with the qualitative argument presented earlier;

b) as the temperature is raised the line width of the collective mode increases, while a peak develops about $\omega = 0$, which, for small q, has a Lorentzian form; for any finite temperature there is a critical wave-vector $q_c^*$ (for an antiferromagnet, $q^*$ is measured from the magnetic zone boundary), and it is only for $q^* > q_c^*$ that a collective mode peak appears.[4]

[4] Within the three-pole approximation, a simple relation, involving the second and fourth moments, determines $q_c^*$ as a function of temperature. LOVESEY and MESERVE refer to $q_c^*$ as the inverse dynamic correlation length, to distinguish it from the inverse static correlation length $\kappa$.

Figure 7.6 demonstrates the excellent agreement obtained for the dispersion relation of the magnon-like mode in TMMC at 4.4 K, while Fig.7.5 illustrates the agreement obtained for the line width and intensity as a function of temperature.

As shown in Table 7.1, there are discrepancies between the theoretical predictions for the variation with temperature of the width of the collective mode. Recent inelastic neutron-scattering measurements on TMMC, by HUTCHINGS and WINDSOR (1977), concentrated on the temperature dependence of the energy and line width of the collective mode. Their results indicate an approximately linear dependence on temperature of the line width in the range 4.3 - 20 K. Comparison with results of the theory of LOVESEY and MESERVE shows that their measurements fall outside the region where the $T^{3/2}$ temperature variation is predicted, and that the theory does approach a linear variation at higher temperatures; however, the magnitude of the theoretically predicted broadening is about half the measured values. The magnitude of the temperature renormalisation of the collective mode energy predicted by LOVESEY (1974) was also found to be systematically smaller than the measured values. The temperature dependence of both the energy and width of the collective mode predicted by McLEAN and BLUME (1973) was found to be in reasonably good agreement with the experimental measurements. The measurements were also found to be at variance with computer simulation results, by WINDSOR and LOCKE-WHEATON (1976), for the line width (at 10 K and 17 K), the computed line widths being significantly smaller than those measured (especially at 17 K), and the renormalisation of the energy at 10 K, the value from the computer simulation study being much larger than the measured value; these discrepancies suggest a possible inadequacy of the theoretical model for TMMC. For the case of antiferromagnetic coupling, the three pole approximation predicts a Lorentzian form about $\omega = 0$ for small values of $q^*$, with a width which goes to zero as $T^{3/2}$ as $T \to 0$. This temperature dependence appears to be at variance with NMR measurements[5] by HONE et al. (1974), which suggest that, for $q^* = 0$, the width of the central peak varies with temperature as $T^2$ in the range $4\text{ K} < T < 12\text{ K}$. However, the limiting low temperature behaviour predicted by the three-pole approximation may not be attained until rather lower temperatures, and, in fact, RICHARDS (1972) found that the prediction of LOVESEY and MESERVE is consistent with their NMR measurements on TMMC in the temperature range $1.2\text{ K} < T < 4\text{ K}$.

At high temperatures, it appears that the spin dynamics is adequately described in terms of a spin diffusion picture. The various approximations are discussed in the review by STEINER et al. (1976), and we will not dwell on this subject, other than to point out that the various approximations yield a rather wide variation in the value of the diffusion constant D (see Table 7.1), but with the more sophisticated theories giving good agreement with the computer simulation results of LURIE et al. (1974) and WINDSOR and LOCKE-WHEATON (1976) for a classical chain. The value

[5]The region close to $q^* = 0$ is not accessible to neutron scattering measurements, due to the presence of nuclear incoherent scattering.

of D obtained from computer simulation experiments is in rather poor agreement with that obtained from NMR measurements on TMMC, but it should be mentioned that there are rather large uncertainties associated with the NMR result.

Finally we mention the work of KRETZEN et al. (1974), which represents a significant theoretical advance, in that the authors attempted a first principles calculation of the time-dependent spin pair correlation function from the classical equations of motion. The technique employed was pioneered by VILLAIN (1974), who considered an easy plane system (see below), which is a somewhat simpler case, since the order parameter is two dimensional, in contrast to the 3D order parameter of the isotropic system; as for the calculation of VILLAIN, the results are restricted to low temperature. The results obtained for the self-correlation function $\langle \underline{S}_0(t) \cdot \underline{S}_0(0) \rangle$ were found to be in qualitative accord with those obtained by computer simulation experiments by LURIE et al. (1974), even at a rather high temperature, although quantitative agreement was lacking. The dynamical structure factor $S(q,\omega)$ could only be calculated in the continuum limit, and this yielded two Lorentzian peaks at $\pm\omega_q$, with $\omega_q$ in agreement with (7.10), the widths of the Lorentzians being proportional to temperature.

We turn next to one of the few *ferromagnetic* linear chain compounds, $CsNiF_3$, which has spin $S = 1$. This system has the additional interesting feature of a large single-site anisotropy, which has a pronounced effect on the character of the collective excitations. The three-dimensional ordering temperature of 2.61 K (STEINER et al., 1971), implies somewhat stronger interchain coupling than in TMMC, but is still sufficiently low that collective excitations could be observed over a considerable temperature range (STEINER et al., 1975). In terms of the parameters J and B, introduced in (7.9) to describe the exchange interaction and single-site anisotropy, STEINER et al. found that, at low temperatures, the dispersion relation for the collective mode could be accurately fitted to the following form (with q along the c-axis)

$$\omega_q = 2S\,J[(1 - \cos aq)(1 - \cos aq + B/J)]^{1/2} \tag{7.12}$$

with $B/J \simeq 0.2$.

One immediate consequence of the single-site anisotropy is that, for B positive, the xy-plane forms an easy plane for the spins. Consequently, the dynamics for the spin components parallel and perpendicular to the easy plane are markedly different. This was indeed found to be the case by STEINER et al., whose inelastic neutron scattering measurements on $CsNiF_3$ revealed the existence of two types of collective mode, one associated with fluctuations in the in-plane correlation function $\langle \hat{S}^x_q(t)\ \hat{S}^x_{-q}(0) \rangle$, the other with out-of-plane correlation function $\langle \hat{S}^z_q(t)\ \hat{S}^z_{-q}(0) \rangle$. The in-plane collective mode was found to give rise to a broad peak, with a lineshape similar to that found in TMMC, the implication being that the single-site anisotropy has a minimal effect on the in-plane fluctuations. In contrast, the lifetime of the out-of-plane fluctuations is increased substantially, with the result that the width of the collective mode peak in the neutron cross-section was smaller than the available energy resolution in the experiments of STEINER et al. Moreover, the intensity of the out-of-plane collective mode was found to decrease dramatically with temperature, changing from an intense peak at 5 K, which dominates the in-plane mode, to an unmeasurably small intensity at 12 K, indicating that the line width of the out-of-plane mode increases rapidly with temperature. A physical picture for the decrease in line width of the out-of-plane mode, for $B > 0$, is that the anisotropy energy acts as a potential barrier against fluctuations in $\langle \hat{S}^z_q(t)\ \hat{S}^z_{-q}(0)\rangle$, whereas it has minimal effect on the fluctuations of the in-plane correlation functions, since the motion of the spin components in the plane is not hindered by this potential barrier.

A recent theoretical paper by LOVELUCK and LOVESEY (1975) sought to determine whether the three-pole approximation to the Mori continued fraction approach, which, as we have seen, gave a satisfactory description of the spin dynamics in TMMC, could account for the rather striking features of the out-of-plane mode.

The theoretical formalism of LOVELUCK and LOVESEY is similar to that employed by LOVESEY and MESERVE, but with some important differences, the most basic of which is that the presence of the single-site anisotropy necessitates a description in terms of two relaxation functions, $F_z(q,\omega)$ describing the out-of-plane spin dynamics, and $F_x(q,\omega)$ describing the dynamics in the xy-plane. The corresponding moments will be denoted by $\langle\omega^n\rangle_z$ and $\langle\omega^n\rangle_x$, respectively. A second complication is that the formalism due to FISHER (1964), which yields analytic expressions for static spin-correlation-functions for an *isotropic* classical Heisenberg chain, cannot be applied in the presence of a single-site anisotropy. However, it was shown by LOVELUCK et al. (1975) that the required correlation functions could be evaluated, without approximation, using a transfer-matrix formalism.

LOVELUCK and LOVESEY concentrated on a description of the out-of-plane collective mode, since it was here that the observed behaviour was qualitatively different from that found in TMMC, and we now summarise the results obtained in this work.

At low temperatures, the dispersion of the in-plane and out-of-plane modes is given by the square roots of $\langle\omega^2\rangle_x$ and $\langle\omega^2\rangle_z$, respectively, and, for moderate values of the anisotropy constant, it was found that these quantities were almost identical, and in agreement with (7.12). As the temperature is increased, the collective mode peak is broadened by fluctuations, in agreement with the experimental results. It was found that the line width of the out-of-plane mode, for the value of B/J appropriate to $CsNiF_3$, was greatly reduced from its value in a corresponding *isotropic* ferromagnetic chain. In fact, the calculated line width was found to be less than the instrumental resolution for temperatures up to 12 K, in accord with the experimental findings of STEINER et al. The temperature renormalisation of the collective mode

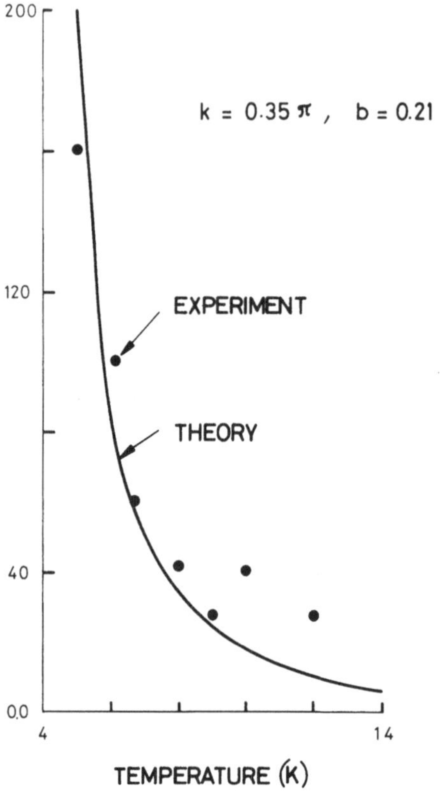


Fig. 7.7 The value of $S(q,\omega)$ at the position of the collective mode peak for $q = 0.35\pi$ is plotted as a function of temperature; parameters appropriate to $CsNiF_3$ were used, and the corresponding experimental measurements of STEINER et al. (1975) are indicated by solid circles. The vertical scale is in arbitrary units

frequency was found to be minimal, for the wave-vector studied, which is also in agreement with experiment. Furthermore, it was found that, for a given temperature, the critical wave-vector $q_c$, above which a collective mode is predicted, was markedly decreased from the value in a corresponding isotropic ferromagnetic chain, indicating that the single-site anisotropy significantly enhances the onset of the collective mode. Finally, in Fig.7.7 we show the calculated intensity of the collective mode as a function of temperature, together with experimental results. Agreement between theory and experiment is seen to be excellent, especially when one allows for the fact the experimental points for higher temperatures have considerable uncertainty associated with them, since the out-of-plane peak is not clearly distinguishable from the in-plane peak at these temperatures.

Additional insight into the interpretation of the results of LOVELUCK and LOVESEY has been provided by ALLROTH and MIKESKA (1975) in terms of an analogue of the Debye-Waller factor, the temperature dependence of which reflects the shift of the scattering intensity from the one-magnon excitation at low temperature to the multi-magnon background as the temperature is increased.

It was shown by STEINER et al. that their experimental findings could also be interpreted in terms of the theory of VILLAIN (1974), who introduces a "semi-polar" representation of the spin operators, together with a harmonic approximation, which makes possible, under certain conditions which appear to be satisfied in the case of $CsNiF_3$ at low temperatures, the definition of a collective mode for easy-plane systems. The dispersion relation, at low temperatures, obtained by VILLAIN is consistent with that of LOVELUCK and LOVESEY, but the line width of the out-of-plane mode is not evaluated explicitly; however, the line-width of the in-plane mode was evaluated by STEINER et al., using VILLAINS's theory, and found to be in good agreement with their experimental measurements.

The compound $CuCl_2 \cdot 2N(C_5D_5)$, dichlorobis pyridine copper II (CPC), is of interest, since it is an excellent approximation to an idealised 1D isotropic Heisenberg antiferromagnet with S = 1/2. Evidence for the validity of this model for CPC, together with the results of neutron scattering experiments, was reported by ENDOH et al. (1974).

For an S = 1/2 system, quantum effects are expected to be important, and one cannot rely on the theories of classical chains which we have discussed above. There is, however, an exact result for this system, due to DES CLOIZEAUX and PEARSON (1962), which states that the lowest excited states of the system obey the dispersion relation

$$\omega_q = \pi\frac{J}{2}|\sin aq| \tag{7.13}$$

and thus predicts a collective excitation, the dispersion relation being similar to that given above for a *classical* antiferromagnetic chain, apart from a different numerical constant. ENDOH et al. did indeed observe well-defined collective excitations at low temperatures, which follow very precisely the dispersion relation, (7.13). Apart from the exact calculation on finite S = 1/2 chains by RICHARDS and CARBONI (1972), there is at present no published calculation, to the best of our knowledge, of $F(q,\omega)$ for the S = 1/2 system. ENDOH et al. resorted to a comparison with the experimental data on TMMC, suitably scaled to allow for the differences in spin and exchange interaction. Rather surprisingly, quite close agreement was observed, indicating that quantum effects do not play a crucial role in determining the spin dynamics.

The presence of a single-site anisotropy has been shown above to give rise to some significant effects, and leads naturally to the consideration of other types of anisotropy, in particular, that induced by a magnetic field. While, as yet, no experiments have been reported on linear chain systems in a magnetic field, such studies are certainly feasible with the advent of large superconducting magnets, capable of inducing significant anisotropy. The theoretical consequences of an external field, applied, for example, in the z-direction, are also of considerable interest, since the presence of this type of anisotropy instigates a coupling between the spin and energy densitites. Within the framework of the Mori formalism, a spin-energy coupling of this nature necessitates an enlargement of the set of

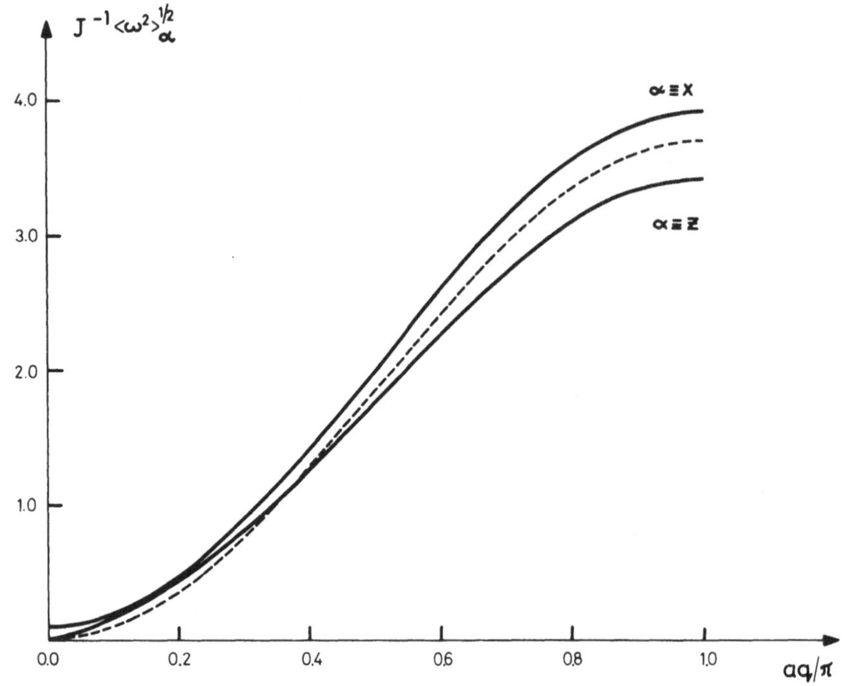


Fig. 7.8 Calculated results for $J^{-1}\langle\omega^2\rangle_\alpha^{1/2}$ ($\alpha$ = x,y) are plotted (solid lines) for a ferromagnetic linear chain in the presence of an applied field; the results correspond to a reduced temperature $T^* \equiv k_BT/J = 0.1$ and a magnetic field H given by $g\mu_BH/J = 0.1$. The corresponding result in the absence of a field is indicated by a broken line

dynamical variables considered, from the single variable $\hat{S}^z_q$ to the set of three variables $\hat{S}^z_q$, $\hat{S}^z_q$ and the energy density. While the Mori formalism embraces this situation[6] it is not clear that an approximation with the simplicity of, for example, the three-pole approximation is adequate, and this topic has not yet been fully explored. However, the second moments, $\langle\omega^2\rangle_x$ and $\langle\omega^2\rangle_z$, of the relaxation shape functions $F_x(q,\omega)$ and $F_z(q,\omega)$ can be calculated readily, and, as we have seen above, these yield the collective-mode dispersion relations at low temperature if the effects of energy fluctuations are neglected. Such calculations indicate that even a moderate field can lead to pronounced effects, and, in particular, gives rise to a pronounced flattening of the dispersion of the out-of-plane mode, with a much less pronounced effect on the in-plane mode, as shown in Fig.7.8. The figure also demonstrates the appearance of a gap at q = 0 for the transverse mode, but it is possible that damping of the mode will render this gap unobservable.

[6]In fact, the analogous situation in liquids, where the corresponding variables are the particle density, current density and thermal energy density, has been studied with the framework of the Mori formalism, and is reviewed by COPLEY and LOVESEY (1975)

## 7.2.2 Layered Magnetic Systems

For 2D magnetic systems, the bias of theoretical and experimental studies is noticeably different from those concerning 1D structures. A short summary of some of the salient features of 2D magnetic systems, in particular those aspects which distinguish them from 1D structures, may assist in understanding this shift of emphasis.

We recall that a principal driving force behind the theoretical effort devoted to 1D systems has been the possibility of exact calculations for static quantities, which serve as a firm basis for dynamical theories. For 2D magnetic systems, apart from the 2D Ising model[7] for which (as in 1D) the dynamics is rather unexciting, this feature is absent. However, spin-wave theories can still be applied, and there have been several satisfactory studies; for example, the magnetic excitations in the 2D ferromagnet $FeCl_2$ have been studied, in the context of renormalised spin-wave theory, by LOVESEY (1974b), as discussed in Sec.7.1.

The occurrence of a phase transition to an ordered state, at finite T, for the 2D Ising model (ONSAGER, 1944) exposes a further important contrast with 1D models, for which LRO is not possible except at $T = 0$. The implication of the MERMIN-WAGNER theorem, for 2D systems, is that, provided that a fairly weak condition on the range of the exchange interactions is satisfied, a spontaneous magnetisation is precluded for the 2D isotropic Heisenberg and XY models. However, the situation is complicated by the fact that this result is valid only in the absence of a preferred direction, arising from the presence of anisotropy. In real systems, a small amount of either intralayer anisotropy or interlayer coupling is always present, both of which may be represented by a local field, and allowing the possibility of a phase transition and LRO. Moreover, there is nothing in the theorem of MERMIN and WAGNER which prohibits, for example, a divergence in the susceptibility *without* LRO. In fact, STANLEY and KAPLAN (1966) have analysed the high temperature series expansion for the uniform susceptibility of a 2D Heisenberg model with $S > 1/2$, and find evidence for a divergence in this quantity at a finite temperature, $T_{SK}$.

While the investigation of dynamic properties in 2D has not been neglected, the theoretical considerations outlined above have prompted a considerable experimental effort towards clarifying our understanding of magnetic order in 2D structures; these investigations will be illustrated by the example of $K_2NiF_4$, which has received detailed study.

---

[7]A close approximation to a 2D Ising model is realised by the system $CoCs_3Br_5$. However, as far as we are aware, there have been no neutron scattering studies on this system, although, for example, the magnetic specific heat has been measured, as a function of temperature, and found to follow very closely the logarithmic singularity predicted by ONSAGER's exact solution for a quadratic Ising lattice (WIELENGA et al., 1967, MESS et al., 1967).

The study of *disordered* 2D magnetic systems has also been instigated, and we consider this more recent development in the latter part of this section. Excellent examples of well-defined, fully miscible 2D mixed magnetic systems are now available (e.g., $Rb_2Mn_xNi_{1-x}F_4$), which offer (to quote ALS-NIELSON et al., 1975) "an opportunity of studying the static and dynamic response of a random system which has both the simplicity of a simple quadratic lattice structure and known interactions which are predominantly nearest neighbour and isotropic".

An extensive study of $K_2NiF_4$ by elastic and inelastic neutron scattering techniques has been reported by BIRGENEAU et al. (1970a, 1970b, 1971), and by SKALYO et al. (1969), and there are also complementary but less detailed studies of the isostructural compounds $Rb_2MnF_4$, $Rb_2FeF_4$ and $K_2MnF_4$ by BIRGENEAU et al. (1970a, 1973). The conclusions drawn from these investigations can be summarised as follows. Above $T_N$ = 97.23 K, $K_2NiF_4$ behaves as a 2D antiferromagnet without LRO; at $T = T_N$, it undergoes a phase transition to a state with 3D LRO, but with the rather curious feature that the phase transition can be characterised as a *2D* continuous magnetic phase transition. The evidence for the 2D character above $T_N$ was twofold: first, in the paramagnetic phase, Bragg *ridges*, rather than peaks, were observed; secondly, a magnon-like mode was observed above $T_N$ with negligible dispersion in directions perpendicular to the layers in reciprocal space (see Fig.7.9), and indicated that the ratio of interlayer to intralayer exchange coupling was less than $3 \times 10^{-3}$. The latter measurements also provided an estimate of the anisotropy energy (from the q = 0 energy gap), yielding an estimate of ∿0.2% of the exchange coupling; this

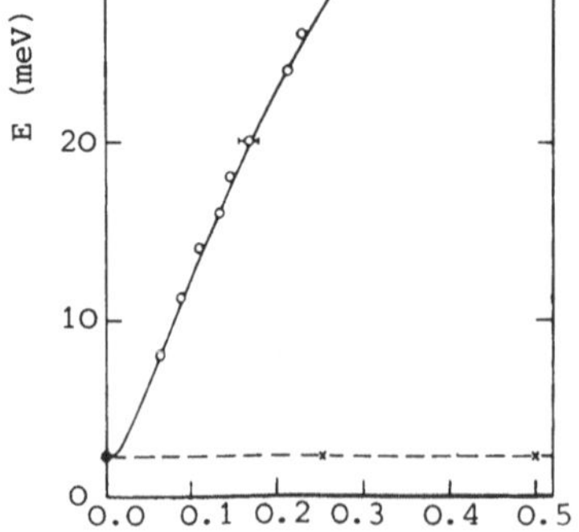

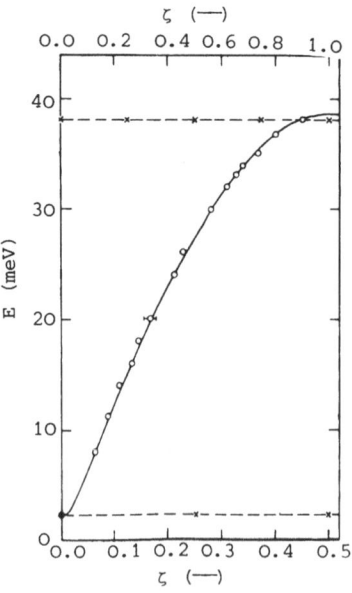

Fig. 7.9 The dispersion relation for the magnon-like mode in the (010) zone for $K_2NiF_4$ at 5 K, as measured by SKALYO et al. (1969). The solid line is for (ζ,0,0), while the dashed lines are for (0,0,ζ) at 2.4 meV and (0.45,0,ζ) at 38.1 meV

value is almost an order of magnitude larger than would arise from dipolar forces, indicating the presence of single-ion anisotropy. As for the linear chain systems, strong SRO above $T_N$ was indicated, since the magnon-like mode was found to persist well above $T_N$, and remained well defined, and with little temperature renormalisation, at 105 K, although by 146 K the mode was no longer observable.

In an attempt to elucidate the nature of the phase transition, BIRGENEAU et al. (1970b, 1971) made detailed studies of the Bragg peaks for $T < T_N$, and of the wave-vector dependent susceptibility for $T > T_N$. For $T < T_N$, they found that the sublattice magnetisation followed a simple power law

$$M(T)/M(0) = B(1-T/T_N)^{\beta} \tag{7.14}$$

with $\beta = 0.138 \pm 0.004$, which is closer to the 2D Ising value of 0.125 than to 3D values. For $T > T_N$, the critical scattering indicated that the critical fluctuations were entirely 2D, and only the Ising component $\chi_z$ was found to diverge. One puzzling feature, which has not yet been adequately explained, is that the exponents obtained for the correlation length and susceptibility divergences were found to be close to the mean-field values, $\nu = 0.5$, $\gamma = 1.0$, which are markedly different from the Ising values, $\nu = 1.0$, $\gamma = 1.75$.[8]

It can be argued (see, e.g., DE JONGH and MIEDEMA, 1974) that the Ising-like behaviour arises from anisotropy effects, which give rise to an "easy axis". DE JONGH and MIEDEMA analysed the results for $K_2NiF_4$, together with data for other layered antiferromagnets; by considering the transition temperature $T_N$ as a function of anisotropy $\alpha$ (the ratio of the effective fields associated with the exchange and anisotropy), they showed, by extrapolation to $\alpha = 0$, that the results are consistent with the estimate for the Stanley-Kaplan transition temperature, $T_{SK}$. However, this extrapolation is not completely convincing, and the analysis neglects, for example, the possibility of terms in the Hamiltonian which are cubic in the spin operators, which might be important in the interpretation of experimental results corresponding to very small values of their anisotropy parameter, which lumps together all anisotropy effects in an effective field. Moreover, their conclusions are at variance with recent theoretical work (see, for example, BREZIN and ZINN JUSTIN, 1976) which implies that there is no Stanley-Kaplan transition for the isotropic Heisenberg model, although, in contrast, it has been shown by ZITTARTZ (1976) that such a phase transition does take place for the XY model.

It is expected that interlayer coupling must eventually lead to 3D ordering at sufficiently low temperatures, and that cross-over effects should be observed between the 2D Ising-like transition and the 3D ordering temperature. Such effects have been

---

[8]It has been suggested by ALS-NIELSEN et al. (1975) that this discrepancy arises from an incorrect subtraction of the transverse scattering, and that more recent experiments do indicate Ising-like exponents.

observed in 2D ferromagnetic systems, but not in antiferromagnetic systems, and these results are analyzed in a recent paper by DE JONGH and STANLEY (1976).

Apart from the application of spin-wave theory, theoretical studies of the dynamic behaviour in planar systems are conspicuously lacking. The theory of VILLAIN (1974) can be applied to easy-plane systems, but no direct confrontation between experiment and theory has yet occurred. The relatively small renormalisation of the magnon-like mode above $T_N$ has been discussed, for example, by DE JONGH and MIEDEMA (1974), who argue that the magnitude of renormalisation effects depends on the ratio of thermal to exchange energy, and that, for example, this quantity is about eight times as large for (3D) $MnF_2$ as for $K_2NiF_4$ at their respective transition temperatures, basically because of the lowering of $T_N$ in the 2D material.

The motivation for studying 2D random magnetic systems was discussed in the introduction to this section, and will not be elaborated on here, except to add that, as for 1D systems, the low dimensionality favours computer simulation experiments on arrays of much larger linear dimensions than those possible for 3D models; indeed, such studies have been performed for a 2D random antiferromagnet and will be discussed below.

An extensive study of the binary mixed antiferromagnet $Rb_2Mn_{0.5}Ni_{0.5}F_4$, using elastic and inelastic neutron scattering techniques, has been undertaken at Brookhaven National Laboratory, and the results are reported in two papers by BIRGENEAU et al. (1975) and ALS-NIELSEN et al. (1975). The pure materials $Rb_2MnF_4$ (S = 5/2) and $Rb_2NiF_4$ (S = 1). are 2D antiferromagnets, very similar to $K_2NiF_4$. The elastic scattering experiments on the mixed crystal indicated that the $K_2NiF_4$ structure was retained above 67 K. Information from both Bragg and diffuse scattering demonstrated that the Mn and Ni atoms are randomly distributed in the mixed crystal, with no noticeable clustering effects. The pure crystals can be characterised by nearest neighbour intraplanar exchange coupling, and reliable estimates are available for the Mn-Mn and Ni-Ni exchange constants from investigations on the pure crystals. Since the lattice constants of $Rb_2MnF_4$ and $RbNiF_4$ differ by only ∿5%, the Mn-Mn and Ni-Ni exchange constants may be assumed to be unchanged in the mixed system, leaving only the Mn-Ni exchange constant undetermined; the latter was assumed to be given by the geometric mean of the Mn-Mn and Ni-Ni exchange constants. With this assumption, good agreement was obtained between the measured collective mode dispersion, and that calculated using a "mean-crystal" model, in which the system is treated as a four sublattice antiferromagnet.

The magnetic structure measurements of ALS-NIELSEN et al. exhibited similar features to those observed in $K_2NiF_4$: above ∿67 K no 3D magnetic scattering was observed; however, strong 2D magnetic scattering occurred, and persisted up to 120 K. Critical scattering indicated that both the static susceptibility $\chi_x(0)$ and inverse correlation length, $\kappa$, exhibited power law behaviour with 2D Ising-like exponents. Below ∿67 K the persistence of 2D critical scattering indicated that the 2D structure is retained, but with the simultaneous appearance of Bragg peaks indicating 3D LRO, analogous to the behaviour observed in $K_2NiF_4$. However, in contrast to the observations in $K_2NiF_4$, the phase transition at $T_N \sim 64$ K was found to be considerably rounded; ALS-NIELSEN et al. showed that their data for the sublattice magnetisation were consistent with a power law behaviour, with $\beta \simeq 0.2$, but with a smeared transition temperature. A smeared phase transition is expected in a disordered system, although, apart from rather unrealistic model calculations, explicit theoretical analysis is lacking in this area.

The spin dynamics of the system were also investigated, and measurements revealed the presence of *two* collective excitations. Both modes were found to have the characteristics of well-defined 2D magnon-like excitations: negligible dependence on the coordinate normal to the magnetic planes and pronounced dispersion, large compared to the width of the mode, in the plane (see Fig.7.10). As mentioned above, good agreement was obtained between the measured dispersion and a mean-crystal model calculation, with no adjustable parameters.

It was pointed out by ALS-NIELSEN et al. that, although the simple mean-crystal model gave a reasonable description of the dispersion relations, the measured profiles and relative intensities of the two modes could not be accounted for by this

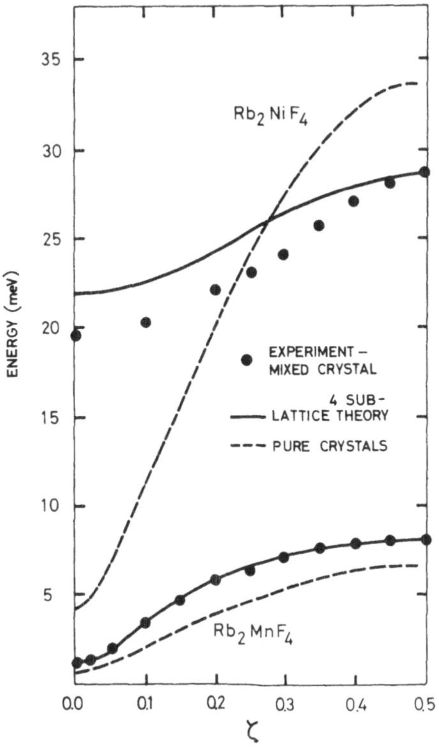


Fig. 7.10 Dispersion relations in the $(\zeta,0,0)$ direction for $Rb_2MnF_4$, $Rb_2NiF_4$ and $Rb_2Mn_{0.5}Ni_{0.5}F_4$ at 7 K. The dispersion curves for the pure crystals (broken lines) are calculated spin-wave dispersion relations assuming nearest neighbour interactions. The solid lines are the results of the four sublattice calculation, referred to in the text.(After ALS-NIELSEN et al., 1975)

model, nor by Ising cluster models. They suggested that, in view of the fact that the system is one for which the interactions are simple and well characterised by nearest neighbour intraplanar exchange coupling, it provides an excellent opportunity for a confrontation between experiment and more sophisticated theories, in particular the coherent potential approximation (CPA), which has been very successful in other areas. Indeed, this possibility has not escaped the notice of theoreticians, and there are calculations for 2D random antiferromagnets by SCHLITING and KREY (1975) and by COOMBS and COWLEY (1975), which are based on CPA, but include, in some approximation, off-diagonal randomness and cluster effects. Both theories correctly predicted the existence of two well-defined magnon-like modes for 2D random antiferromagnetic systems. SCHLICHTING and KREY considered the system $K_2Ni_{0.5}Mn_{0.5}F_4$, which has exchange constants very similar to those of the mixed Rb system. The qualitative features of their results are in agreement with experiment, but no quantitative comparisons were made; however, calculated densities of states were shown to be in good agreement with those obtained in the computer simulation experiments of HUBER (1974). The system $RbMn_{0.5}Ni_{0.5}F_4$ was considered explicitly in the work of COOMBS and COWLEY (1975), who obtained satisfactory agreement with the low temperature experimental measurements. The results were found to be somewhat dependent on the different approximations chosen for treating the off-diagonal terms, and it seems that further investigation of the criteria for choosing a satisfactory approximation would be useful. No quantitative comparison was made between theory and the experimental results for the widths, line shapes and relative intensities of the two modes, which would present a more stringent test. In addition, the temperature dependence of the excitations has not yet been considered theoretically. COOMBS and COWLEY also presented results of calculations for the system $K_2Mn_{0.75}Zn_{0.25}F_4$, and predicted, in contrast to theories which treat all magnetic ions equivalently, a many peaked structure in the inelastic neutron scattering cross-section. This prediction clearly invites an experimental investigation.

Finally, as mentioned above, the possibility of computer simulation experiments for 2D disordered systems, which yield more accurate results than are possible in 3D, has been exploited by HUBER (1974), ALBEN and THORPE (1975) and by KIRKPATRICK and HARRIS (1975). The work by HUBER yielded only the density of states of magnetic excitations, while that of ALBEN and THORPE, which involved numerical integration of the equations of motion for Monte Carlo samples, gave $S(\underline{q},\omega)$ directly. The latter authors were able to obtain excellent agreement with experimentally measured dispersion relations. Agreement with the experimental intensity, from inelastic scans, was found to be not quite so precise. In particular, a difference of some 30% was found between the experimental and computer simulation values for the relative weights in the two magnon-like modes; the authors conjecture that the discrepancy may be due to the use of a linear spin-wave approximation in the derivation of the equations of motion of the system.

More comprehensive computer simulation studies on Monte Carlo samples have been performed by KIRKPATRICK and HARRIS (1975) for the system $Rb_2Mn_xNi_{1-x}F_4$, for arbitrary values of x. The extent of the agreement between experimental data and these calculations, for the case x = 0.5, was similar to that obtained by ALBEN and THORPE, the largest discrepancies again occurring for the rather sensitive test of the relative weights in the two modes. These authors also showed that the four-sublattice "mean-crystal" model could be generalised to arbitrary x, and that this approximation gave satisfactory agreement with the Monte Carlo studies. The calculations suggest further experimental investigation of the concentration dependence of various features of the spectrum, for example, the concentration dependence of the energy gap between the two modes, which was calculated both by Monte Carlo simulations and by a low frequency continuum approximation, similar to the hydrodynamic theory of HALPERIN and HOHENBERG (1969).

In conclusion, it may be said that a number of fruitful experimental and theoretical studies of low-dimensional systems have been made, and that many instances of qualitative, and in some cases quantitative, agreement between experiment and theory have occurred. However, the enquiries are by no means exhausted, and are possibly entering an even more interesting phase, with the possibility of more detailed confrontation between theory and experiment, and also the study of more complex situations, such as the behaviour of low-dimensional systems in the presence of magnetic fields and/or magnetic disorder.

## 7.3 Magnons in Metals

Itinerant theories of magnetism in metals have been reviewed by HERRING (1967). A principal aim of the theories is to resolve the apparent mismatch between the localised nature of magnetic metals, as seen, for example, in the compact distribution of magnetisation about lattice sites observed in the 3D metals, and the itinerant and band structure effects inherent in interpretations of transport coefficients and Fermi surface effects. The key to understanding these various properties lies in a correct description of electron correlation in multiband metals. While very detailed calculations have been completed by COOKE (1973) and COOKE and DAVIS (1972), for example, some very interesting features remain unexplained, as we shall see.

The neutron interaction with electrons is represented by the sum of two terms. The first arises from the spin of the electron, and the second from its translational motion. The effect of the latter can be neglected in most cases of interest, as shown by[9] LOVESEY and WINDSOR (1971), and consequently the calculation of the

[9]Some interesting aspects of the orbital interaction for systems in a magnetic field have been explored by ELLIOTT and KLEPPMAN (1975).

cross-section is reduced to the determination of the response of band electrons to a fluctuation in their spin density. Magnetic excitations correspond, in this picture, to the formation of a bound state of two excitations, each of which is a mode composed of an electron and a hole, the electron and hole having oppositely aligned spin orientations. These modes may be thought of as single particle spin-flip excitations which are frequently referred to as Stoner excitations. The formation of a bound state of two single-particle spin-flip excitations occurs, of course, as a direct consequence of the electron-electron interaction in the metal. Consequently, an important quantity in a discussion is the density of states of the spin-flip excitations. Fig.7.11 shows the density of states for ferromagnetic iron, for a wave-vector approximately one-third of the distance across the zone, in the x-direction. We see that the very pronounced structure in the Stoner density of states occurs at energies much larger than those observed in neutron scattering - the region of interest in neutron scattering corresponds to a very small portion of the diagram in the left-hand corner.

Calculations of the magnetic susceptibility for interacting electrons by COOKE and DAVIS (1972), using band structures for iron and nickel, show an excitation with an almost isotropic dispersion curve of the form

$$\omega_q = Dq^2(1 - \beta q^2) \tag{7.15}$$

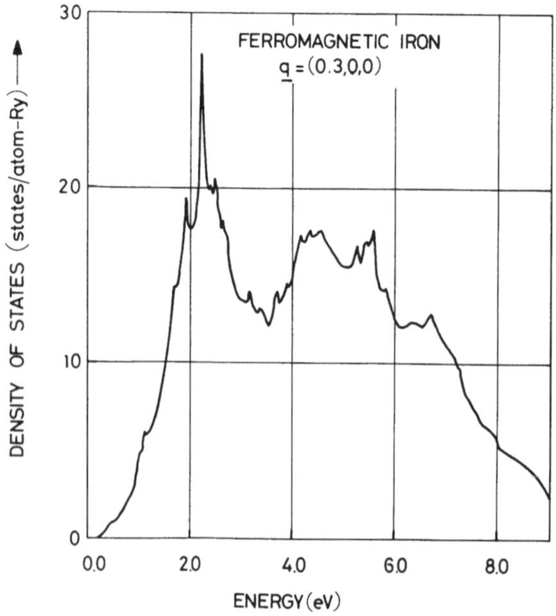


Fig. 7.11 Density of Stoner excitations in ferromagnetic iron for a wave-vector in the (100) direction equal to 30% of the zone boundary wave-vector. After LYNN (1974), thesis, Georgia Institute of Technology

for small wave-vectors. The calculated results are consistent with data obtained from neutron scattering experiments by COLLINS et al. (1969) and MINKIEWICZ et al. (1969), who deduced the values D = 280 meV Å$^2$ and $\beta$ = 0.96 Å$^2$ for Fe, and D = 400 meV Å$^2$ and $\beta$ = -10.5 Å$^2$ for Ni. The measurements were made for energy transfers up to 70 meV, and this energy corresponds to a wave-vector of approximately 40% of the zone boundary wave-vector for Fe, and 27% for Ni.

Surveys of the inelastic scattering from Ni, for all wave-vectors and moderate energy transfers, were reported by LOWDE and WINDSOR (1970) for a wide range of temperatures. This detailed study showed substantial scattering away from the excitation, which appears in surveys as a ridge in the intensity plotted against energy and wave-vector transfer. The ridge was found to persist, more or less without any change, as the temperature was raised through the Curie temperature. The work of LOWDE and WINDSOR (1970) therefore raised interesting questions, some of which have been pursued by MOOK and NICKLOW (1973), MOOK et al. (1973) and LYNN (1975).

These authors measured the intensity of the magnetic excitation in Fe and Ni as a function of energy transfer $\omega$ and temperature. They found, for both materials, that the intensity as a function of energy transfer decreased quite rapidly for $\omega \sim$ 90-100 meV. The fall-off in intensity occurs at slightly different energies for different directions in the Brillouin zone. Fig.7.12 shows the excitation intensity versus energy transfer in Ni, for the (111) direction, at a series of temperatures. We see from this figure that the energy at which the fall-off occurs is essentially independent of temperature. A second interesting feature of the temperature dependence is the existence of well-defined, short wavelength excitations above the Curie temperature. As an example, we show in Fig.7.13 the excitation observed at an energy transfer of 12.4 meV in Ni at several temperatures; the wave-vector is plotted in reduced units and it should be multiplied by 3.08 Å$^{-1}$ to obtain conventional units.

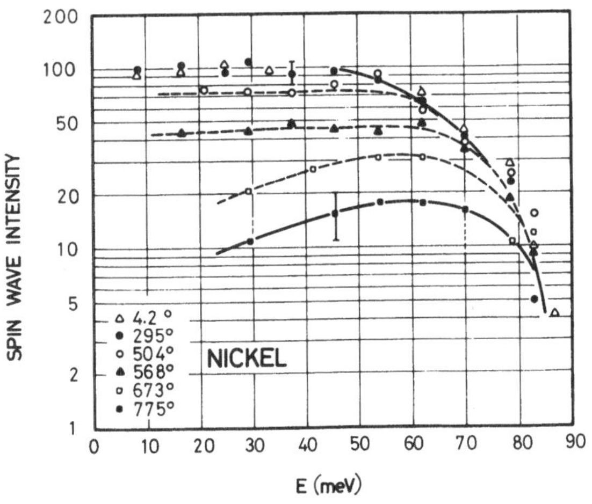


Fig. 7.12 Spin wave intensity for nickel as a function of energy transfer, at six different temperatures, for wave-vectors in the (111) direction. Reference as in Fig.7.11; see also, MOOK et al. (1973)

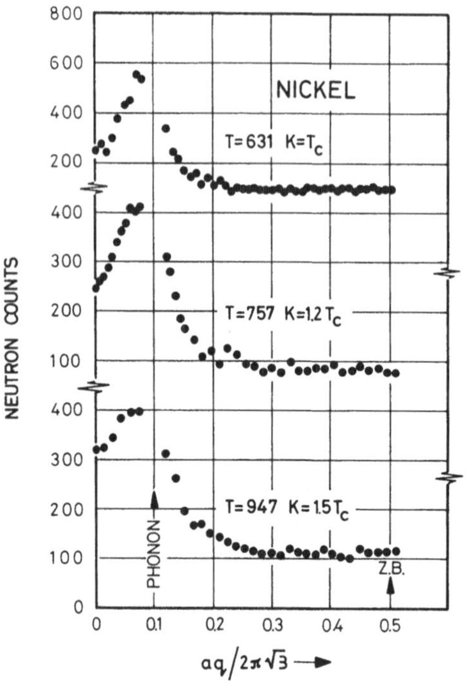


Fig. 7.13 Intensity of neutrons scattered magnetically from nickel as a function of reduced wave-vector in the (111) direction. After MOOK et al. (1974)

The fall-off in intensity with increasing energy transfer has been correlated with *fine* structure in the density of states for the Stoner excitations. COOKE (private communication) finds that the calculated intensity is quite sensitive to changes in the atomic potential used to generate the band structure, whereas the stiffness constant D is relatively insensitive to the details of the potential. The calculations have not been carried through as a function of temperature because of the difficulty of treating the interactions between the magnetic excitations. The existence of a short-wavelength magnetic excitation above the Curie temperature can be attributed to the onset of highly developed, short-range magnetic correlations. The work of HUBBARD (1971), for example, which is discussed in Chap.1, shows the existence of structure in the paramagnetic relaxation function for a Heisenberg ferromagnet. The temperature independence of the value of the energy at which the intensity fall-off occurs (cf. Fig.7.12) remains a mystery on which, thus far, little light has been shed. The pertinent value of the energy can be correlated with fine detail in the Stoner density of states, as mentioned already, but this changes appreciably with temperature. It seems likely therefore, that short-range magnetic correlations play an important role in the temperature dependence of magnetic excitations in metals, but the point has not been elucidated.

## 7.4 Crystal Field Levels in Rare Earth Compounds

The magnetic moment on the atoms in rare earth metals arises from unpaired, 4f electrons. The wave function of these electrons is spatially strongly localised, and consequently the direct overlap of the magnetic wave functions on different sites is negligible, e.g., the mean radius of the 4f shell wave function of Tb is believed to be 0.4 Å, which is small compared to typical lattice spacings. It is usually a good approximation to treat the total spin and orbital angular momentum J as a good quantum number, and the effect of the crystal field is to partially, or totally, remove the 2J + 1 degeneracy of the ground state. Moreover, some rare earth ions have a singlet ground state, and therefore order magnetically only if the exchange interaction exceeds a critical value, as discussed in the following section. If the ions do not order, even at very low temperatures, it can be concluded that the exchange interactions are weak compared to crystal field interactions. This means, amongst other things, that the cross-section for inelastic scattering is given, to a good approximation, by that of an assembly of isolated magnetic ions subject only to a crystal field. The scattering in this case is present at all values of the scattering vector $|\underline{Q}|$, as in paramagnetic scattering.

For compounds which satisfy the conditions discussed in the preceding paragraph, inelastic neutron scattering affords a means of determining the crystal field levels which, in metals, is not always possible using light scattering. TUBERFIELD et al. (1970) successfully studied some simple, singlet ground state compounds and, since then, many compounds have been examined, by several groups of workers.

In describing the type of measurement which is involved, we shall consider the simple case of praseodymium in a cubic field, which is appropriate to the compounds studied by TUBERFIELD et al. The wave functions of the rare earth ion can be labelled by the irreducible representations of the cubic group, denoted here by $\Gamma_n$; a wave function is written $|\Gamma_n \nu\rangle$, where the label $\nu$ distinguishes the degenerate wave functions. Calculation of the cross-section reduces, essentially, to the calculation of matrix elements of the operator $\hat{\underline{D}}$, defined by (1.153), between the various states. Because the orbital angular momentum of the rare earth ions is not quenched by the crystal field, this is, at first sight, a formidable undertaking. However, just as in optical spectroscopy, there are strong and weak transitions. The strong transitions are those which occur for zero wave-vector, that is to say, those allowed in the dipole approximation (see, for example, BALCAR and LOVESEY, 1970, and BALCAR et al., 1970). In this instance, the interaction operator, (1.153), can be replaced by

$$\frac{1}{2}\, g\, F(Q)\hat{\underline{J}} \tag{7.16}$$

where g is the Landé splitting factor, and F(Q) is the appropriate atomic form factor (see, for example, MARSHALL and LOVESEY, 1971). It follows from (7.16) that we need to determine the matrix elements of $\hat{J}_\alpha$ between the states $|\Gamma_n\rangle$ and $|\Gamma_{n'}\rangle$, say. This is a well-known problem in optical spectroscopy, and it is readily answered with a little group theory (BIRGENEAU, 1972). For $Pr^{3+}$ in a cubic crystal field there are four transitions allowed within the dipole approximation.

TUBERFIELD et al. studied seven praseodymium chalcogenides and pnictides and analysed the observed crystal field transitions in terms of a crystal field model. All of the crystal field levels could be quantitatively accounted for in terms of a nearest-neighbour point-charge model with an effective charge of -2. This result was most surprising, in view of the well-known inadequacy of the point-charge model for most insulating compounds.

Following the work by TUBERFIELD et al. (1970), a veritable industry has developed concerned with the investigation of crystal-field levels, which we do not attempt to review.

## 7.5 Magnons and Excitons in Rare Earth Metals

Reviews of the magnetic properties of the rare earths can be found in the book edited by ELLIOTT (1972). A very simple picture suffices to explain most of the magnetic properties of rare earth metals and alloys. In this picture, the magnetic moments are treated as well localised, and, in general, the total spin and orbital angular momentum, J, is treated as a good quantum number. The array of localised magnetic moments is then considered to be immersed in a sea of conduction electrons, to which each atom contributes its three outer electrons. It is really only cerium and ytterbium which have properties which cannot be explained adequately within this simple picture of the magnetic and conduction electrons.

The elucidation of the physics behind the indirect exchange interaction between rare earth atoms is associated with the names of RUDERMANN, KITTEL, KASUYA and YOSHIDA, and it is often referred to simply as the RKKY interaction. In essence, the magnetic moment on a rare earth atom perturbs the conduction electrons, and the perturbation is sensed by other magnetic atoms. In view of this, we anticipate that the exchange interaction is long ranged, and reflects the band properties of the electrons, insofar as they affect the magnetic susceptibility of the conduction electrons, and, from these properties of the exchange, we would expect that complicated magnetic orderings will be observed, which is indeed the case. The role of the exchange interaction in determining the magnetic structure was pointed out in an elegant paper by YOSHIMORI (1959). Reviews of observed orderings have been given by NAGAMIYA (1967) and KOEHLER (1972). A second important feature of the rare earths is that, except for $Gd^{3+}$ and $Eu^{2+}$, there is a large orbital component to the magnetic moment of the ions, and consequently anisotropy effects are important. We shall see also that dynamic magneto-elastic interactions can be important in the proper interpretation of the measured magnon dispersion curves.

We shall take Tb as our main example; various aspects of the dynamic properties studied at the Research Establishment Risø are collated, together with references to earlier work, in papers by JENSEN et al. (1975), JENSEN and HOUMANN (1975) and HOUMANN et al. (1975). Terbium, like most other rare earth metals, forms a hexagonal close-packed structure. There are, therefore, two atoms per unit cell. Below 216 K terbium is ferromagnetic; in the temperature range 216-225 K it forms an antiferromagnetic spiral structure, and above 225 K it is paramagnetic. The 4f electrons, with S = L = 3 and J = 6, do not form a spherical distribution, but are expanded in a plane normal to the total angular momentum. The large single ion anisotropy, which results from the unquenched orbital angular momentum, can be described by a set of terms of the form (the z-axis is taken to coincide with the crystallographic c-axis)

$$\sum_m \{B(\hat{S}^z_m)^2 - \frac{1}{2} G [(\hat{S}^x_m + i\hat{S}^y_m)^6 + (\hat{S}^x_m - i\hat{S}^y_m)^6]\} \quad . \tag{7.17}$$

The first term accounts for the tendency of the moments to lie in the hexagonal basal plane, while the second terms reflect the existence of a preferred direction within the basal plane. With B and G positive, the anisotropy energy is a minimum when the moments are aligned in the basal plane, and parallel to the x-axis. Consequently, the spin raising and lowering operators are expressed as

$$\hat{S}^{\pm} = \hat{S}^y \pm i\hat{S}^z$$

The magnon dispersion, and inelastic neutron cross-section are derived by MARSHALL and LOVESEY (1971), for example. The dispersion curves measured in Tb at 4.2 K, for wave-vectors parallel to the symmetry lines in the Brillouin zone, are shown in Fig.7.14. The single-site anisotropy terms lead to an energy gap in the dispersion of the acoustic magnon at zero wave-vector. Note that the maximum energy of the optical magnon branch is small compared to the maximum magnon energies observed in 3d transition metals.

We mentioned at the outset of this section that dynamic magnetic-elastic effects are important in some rare earth metals, and Fig.7.14 contains evidence of magnon-phonon hybridisation. In the a-direction there is a strong hybridisation of phonons and acoustic magnons around k = 0.6 $Å^{-1}$, for example. The phonon involved is a longitudinal acoustic phonon, whereas in $FeF_2$ and $FeCl_2$, discussed in preceding sections, only transverse phonons are observed to interact with magnons, for wave-vectors along symmetry directions. Hybridisation is seen also in the b-direction, where a hybridisation of the acoustic magnon and optical phonon is observed. The measurements shown in Fig.7.14 have been interpreted by JENSEN (1971) in terms of a model in which the magnon-phonon interaction is derived from an expression for the single site anisotropy, which is akin to the expression (7.17). The author investigates

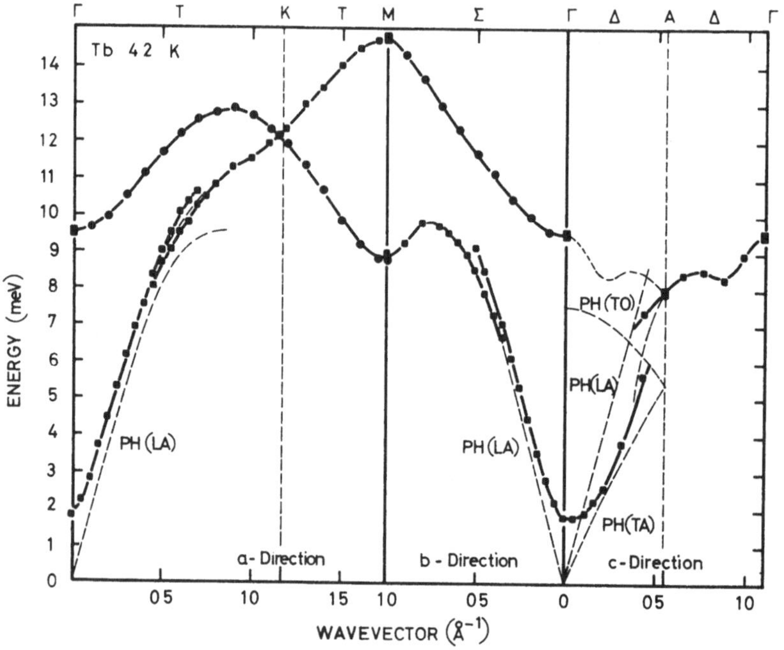


Fig. 7.14 Magnon dispersion relations for Tb at 4.2 K along symmetry lines in the Brillouin zone. The effects of magnon-phonon hybridisation are also shown. After MACKINTOSH and MØLLER (1972)

also the modification of the velocity of acoustic sound waves in terbium, which result from the magneto-elastic interaction.

As a final topic in our discussion of magnons in terbium, we consider the reported work on the investigation of the anisotropic nature of the exchange interaction. It was recognised very early on in the study of rare earth metals with unquenched orbital angular momentum that the exchange interaction was likely to be anisotropic, but, given that a quantitative estimate of the isotropic exchange interaction is difficult, it is understandable that workers in the field did not attempt an estimate of the anisotropic terms, for which the orbital component of the 4f electron's motion must be considered explicitly. Experimental evidence for anisotropic components of the exchange interaction in Tb was given by MØLLER et al. (1972). The experiment measured the magnon dispersion as a function of magnetic field applied along the easy direction in the basal plane. The authors point out that the anisotropic, unlike the isotropic, exchange interaction contributes to an energy gap at zero wave-vector. This complicates the comparison between anisotropy constants measured by macroscopic techniques and by neutron scattering measurements, except, of course, in gadolinium, where the exchange interaction is expected to be isotropic, to a very good approximation.

In Sec.7.4 we considered briefly the measurement of the crystal field levels in metals by neutron scattering. We shall now consider some of the intriguing effects, pointed out by TRAMMELL (1960, 1963) which can occur in systems where the exchange interaction competes effectively with the crystal field.

The basic concepts are illustrated by the simple example of an isolated ion which has two, singlet crystal field levels, separated by an energy $\Delta$. Such an ion does not possess a magnetic moment. However, the application of a magnetic field admixes the two singlet states, and thereby induces a Van Vleck susceptibility. An exchange field has an analogous effect, and, for a sufficiently large exchange interaction, the system will order ferromagnetically. The magnetic ordering is seen, for example, as a peak in the magnetic specific heat superposed on a broad maximum or Schottky anomaly, associated with the crystal field splitting.

We shall refer to the magnetic excitations in these systems as magnetic excitons, and they can exist both in the paramagnetic, and magnetically ordered régimes. By analogy with the soft mode behaviour, observed in structural phase transitions, we might expect there to be a softening of magnetic excitations, in induced moment systems, at the transition between the paramagnetic and ordered régimes.

The double hexagonal phase of praseodymium metal is an example of a singlet ground state system which might order because of magnetic exchange interactions. Indeed, features observed in several experiments have been interpreted as evidence for an antiferromagnetic ordering at very low temperatures. However, recent neutron scattering measurements on a single crystal of Pr by HOUMANN et al. (1975) give evidence that the exchange interaction does not exceed the critical value (at $T = 0$) necessary to drive the system to an ordered state. In fact, they estimate that the exchange is approximately 90% of the critical value. Fig.7.15 shows the temperature dependence of the magnetic excitons for three values of the wave-vector. The exciton in the (100) direction, for $|\underline{q}| = 0.25$ Å$^{-1}$, is seen to decrease rapidly with temperature and reaches a minimum at about 7 K. The authors point out that this wave-vector corresponds very closely to the modulation vector which describes the magnetic ordering in dilute Pr-Nd alloys, and, consequently, they suggest that this excitation be considered as the incipient magnetic soft mode in Pr. The authors also report significant, field dependent, hybridisation between the magnetic excitations and acoustic phonons, but we do not pause to discuss these results here.

Magnetic excitations in the singlet ground state system $Pr_3Tl$ have been investigated experimentally by HOLDEN and BUYERS (1974) and BUYERS et al. (1975). The concept of a soft mode for singlet ground state systems is discussed carefully by BUYERS (1974).

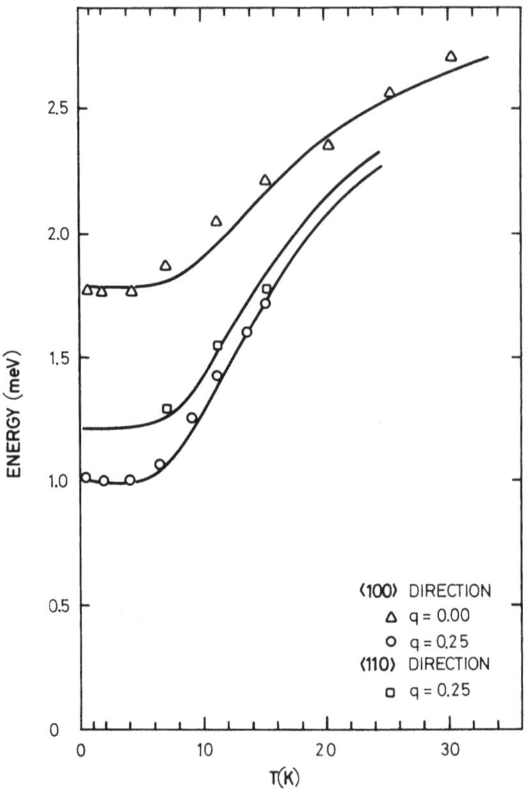


Fig. 7.15 Temperature dependence of selected magnetic excitations in Pr, together with the results of a theory described by HOUMANN et al. (1975)

## 7.6 Critical Magnetic Scattering

No discussion of inelastic magnetic neutron scattering can reasonably omit mention of studies of critical magnetic effects. On the other hand, the field has become very specialized. The recent theoretical techniques devised by K.G. Wilson at Cornell University, which have given a completely fresh approach to the theory of critical phenomena, stem from field-theoretic developments which are quite sophisticated. An overview of the developments in magnetic critical phenomena is given by FISHER (1974). Prior to Wilson's work, HALPERIN and HOHENBERG (1969) developed a so-called dynamic scaling theory, which has received substantial experimental support.

One of the detailed experimental investigations of dynamic critical phenomena, which support the ideas of HALPERIN and HOHENBERG, is the study of critical scattering from the cubic antiferromagnet $RbMnF_3$ by TUCCIARONE et al. (1971). Interpretation of the results is facilitated by writing the cross-section in terms of the relaxation function, introduced in Chap.1. The cross-section can be written as ($\alpha$ = x, y or z)

$$\frac{d^2\sigma}{d\Omega dE'} = \left(\frac{\gamma e^2}{m_e c^2}\right)^2 \frac{k'}{k} \left[\frac{1}{2} gF(Q)\right]^2 \sum_\alpha (1 - \tilde{Q}_\alpha^2) S^\alpha(\underline{Q},\omega) \tag{7.18}$$

where the scattering function $S^\alpha(Q,\omega)$ is given in terms of the relaxation shape function $F_\alpha(Q,\omega)$, (1.185),

$$S^\alpha(\underline{Q},\omega) = \chi_\alpha(\underline{Q}) \left\{\frac{\omega}{1 - \exp(-\beta\omega)}\right\} F_\alpha(\underline{Q},\omega) \quad .$$

Here, $\chi_\alpha(Q)$ is the isothermal magnetic susceptibility, (1.180). The critical exponents $\eta$ and $\nu$, entering this expression, were determined by quasielastic critical scattering to be 0.055 ± 0.010, and 0.707 ± 0.011, respectively, whereas the molecular field results for these two indices are 0, and 0.5.

In the vicinity of the critical temperature, it is convenient to discuss separately the regions for which[10] $q^* \ll \kappa$, where phenomena occur over distances large compared to the correlation length $1/\kappa$, and $q^* \gg \kappa$. The latter region is that in which phenomena occur which are associated with distances, or wavelengths, which are small compared to the correlation length $1/\kappa$, but large compared to all other relevant lengths, such as the lattice parameters. For the regions where $q^* \ll \kappa$ hydrodynamic theory can be used to calculate the relaxation function, c.f. ENZ (1974) and references therein. HALPERIN and HOHENBERG (1969) made certain assumptions about the analytic form of the relaxation function, which enabled them to calculate its form in the critical region, $q^* \gg \kappa$, from a knowledge of its behaviour in the hydrodynamic region.

The experiment by TUCCIARONE et al. showed that, for $T = T_N$ the critical temperature, the relaxation shape function $F(q,\omega)$ for $q^* = 0.05$ Å$^{-1}$ is a single peaked function centered at $\omega = 0$. On increasing $q^*$ to a maximum value of 0.25 Å$^{-1}$, the relaxation function develops distinct side-peaks. A detailed analysis of the data showed it to be in accord with the predictions of dynamic scaling theory. In the hydrodynamic region, and for $T > T_N$, the data is consistent with a single peaked function centered at $\omega = 0$. The width of the line, for $q^* = 0$, should vary with temperature as $\kappa^{1.5}$, according to the theory, and TUCCIARONE et al. found that their data were consistent with a variation $\kappa^{1.46}$. Measurements taken in the ordered magnetic phase, $T < T_N$, show the existence of a central peak at $\omega = 0$, in addition to the magnon scattering. The interested reader is urged to compare the results for the isotropic antiferromagnet $RbMnF_3$, discussed here, with measurements on the uniaxial antiferromagnet $MnF_2$ studied by SCHULHOF et al. (1970).

---

[10] We denote a wave-vector measured relative to a superlattice Bragg position by an asterisk.

There have been several experiments on ferromagnetic materials. Neutron scattering experiments on iron and nickel are reviewed by MINKIEWICZ (1971). The original experiments by COLLINS et al. (1969) and MINKIEWICZ et al. (1969) were in general accord with the dynamic-scaling theory, except for the temperature dependence of the spin diffusion constant in the hydrodynamic region ($T > T_c$), and the absence of a central component in the longitudinal susceptibility for $T < T_c$. The first difficulty was resolved after a careful analysis of the experimental conditions necessary for the measurement of dynamic effects in the hydrodynamic region; the conclusion was that the experiments were not, in fact, in a region which satisfies $q \ll \kappa$. This story illustrates the subtlety of making precise experiments of critical phenomena. The absence of a central peak in the longitudinal susceptibility, in the ordered phase, seems to be a characteristic of ferromagnets, since PASSEL et al. (1972) reached a similar conclusion in their study of EuO.

## 7.7 Hyperfine Field Measurements

The measurement and interpretation of hyperfine field distributions in insulating and metallic magnets has provided valuable information on the microscopic origin of magnetism. It is therefore quite appropriate for a discussion of such measurements to be included in a discussion of magnetism, but it might seem surprising, at first, to include such a topic in a text devoted to inelastic neutron scattering, for the measurement of hyperfine field distributions is, traditionally, a province of resonance technique, such as NMR. The reason for this is that, although neutron scattering affords a probe for nuclear states, nuclear Zeeman energies are ~μeV, which is very small on the scale of thermal neutron energies. This situation has been changed with the advent of spectrometers which are capable of measuring energy changes of a fraction of an μeV with good accuracy, and using such spectrometers it has proved possible to measure hyperfine field distributions in several compounds by measuring the cross-section for inelastic incoherent scattering. We shall not dwell here on the interpretation of hyperfine field distributions; good reviews are given by JACCARINO (1967), and contributors to a book edited by FREEMAN and FRANKEL (1967). Rather, we discuss the basis of the neutron measurements, and comment on the merits and shortcomings of the method as a technique for hyperfine field measurements.

It was demonstrated in Chap.1 that in incoherent nuclear scattering the neutrons change their spin state, even though the cross-section as a whole is independent of the polarization state of the incident nuclei if the nuclei are randomly oriented. In the simplest case, when there is a single isotope ($I \neq 0$) responsible for scattering, 2/3 rds of the neutrons change their spin state, or in other words undergo a spin-flip transition. Given that the scattering processes can be treated as purely s-wave scattering, to a good approximation, the conservation of angular momentum requires that the nuclei change their magnetic quantum number $M$ to $M \pm 1$. If the nuclei are subject to an electromagnetic field which removes the degeneracy of their magnetic ground state, then the spin-flip scattering is accompanied by a change in the nuclear energy of one unit, $\Delta_M$. The partial differential cross-section for this spin-flip process is readily obtained from the general formulae of Sec.1.4, and the result, given first by HEIDEMANN (1970), is

$$\frac{d^2\sigma}{d\Omega dE'} = \frac{k'}{k} \cdot \frac{1}{12\pi} \cdot \sigma_i \exp[-2W(Q)]\delta(\omega - \Delta_M) \quad . \qquad (7.19)$$

A nuclear elastic process always accompanies the one described by (7.19), and the measured spectrum corresponds to the sum of the two processes, convoluted with the resolution function of the instrument, together with other sources of diffuse quasielastic scattering, e.g., the magnetic diffuse scattering which arises from thermal fluctuations in the magnetic moments. Because the process of interest described by the cross-section (7.19) changes the state of polarization of the neutrons it could, in principle, be identified with the use of polarization analysis.

The spin-flip technique described here can only be used to study the hyperfine fields associated with nuclei which have a large spin incoherent cross-section. This was one of the reasons why Heidemann choase a vanadium compound for the first experiment to demonstrate the technique. On the other hand, the interpretation of the neutron data can be easier than for resonance scattering if, for example, there are signals arising from Bloch walls and domains. Broad hyperfine field distributions, which occur in amorphous magnets and mixed, high concentration systems, for example, can be measured easily, in contrast to NMR. However, the neutron scattering technique does not afford the same high energy resolution of NMR. Some of the nuclei which can be investigated by the neutron technique are listed in the following table.

| Nucleus | Spin I | $\sigma_i$ [barn] | $\Delta^a$ [μeV] |
|---|---|---|---|
| $H^1$ | 1/2 | 80 | 1.77 |
| $Li^7$ | 3/2 | 0.6 | 0.69 |
| $Na^{23}$ | 3/2 | 1.4 | 0.47 |
| $Sc^{45}$ | 7/2 | 4.4 | 0.43 |
| $V^{51}$ | 7/2 | 5.1 | 0.46 |
| $Co^{59}$ | 7/2 | 6.4 | 0.42 |
| $Zr^{91}$ | 5/2 | 0.7 | 0.17 |
| $La^{139}$ | 7/2 | 1.7 | 0.25 |

a) Calculated for hyperfine field = 100 kG

*Acknowledgements*

We are grateful to a number of colleagues for helpful discussions on various topics discussed in this chapter. In particular, we thank Dr. J. VILLAIN for his critical reading of various sections of the script, and Dr. A. HEIDEMANN for discussions on the backscattering spectrometer, and the measurement of hyperfine field distributions.

References


Alben, R., Thorpe, M.F. (1975): J. Phys. C8, L275
Allroth, E., Mikeska, H.J. (1975): J. Phys. C8, L523
Als-Nielsen, J., Birgeneau, R.J., Guggenheim, H.J., Shirane, G. (1975): Phys. Rev. B12, 4963
Balcar, E., Lovesey, S.W. (1970): Phys. Lett. 31A, 67
Balcar, E., Lovesey, S.W., Wedgwood, F.A. (1970): J. Phys. C3, 1292
Birgeneau, R.J. (1972): J. Phys. Chem. Solids 33, 59
Birgeneau, R.J., Dingle, R., Hutchings, M.T., Shirane, G., Holt, S.L. (1971): Phys. Rev. Lett. 26, 718
Birgeneau, R.J., Guggenheim, H.J., Shirane, G. (1970a): Phys. Rev. B1, 2211; (1973) Phys. Rev. B8, 304
Birgeneau, R.J., Shirane, G., Blume, M., Koehler, W.C. (1974): Phys. Rev. Lett. 33, 1098
Birgenau, R.J., Shirane, G., Kitchens, T.A. (1972a): *Proc. 13th Low Temperature Conf.*, (Boulder, Colorado) Vol. 2, p. 371
Birgeneau, R.J., Skalyo, J.Jr., Shirane, G. (1970b): J. Appl. Phys. 41, 1303; (1971) Phys. Rev. B3, 1736
Birgeneau, R.J., Walker, L.R., Guggenheim, H.J., Als-Nielsen, J., Shirane, G. (1975): J. Phys. C8, L328.
Birgeneau, R.J., Yelon, W.B., Cohen, E., Makovsky, J. (1972b): Phys. Rev. B5, 2607
Brezin, E., Zinn-Justin, J. (1976): Phys. Rev. Lett. 36, 691.
Buyers, W.J.L. (1974): AIP Conference Proceedings 24, 27
Buyers, W.J.L., Holden, T.M., Perreault, A. (1975): Phys. Rev. B11, 266
Coombs, G.J., Cowley, R.A. (1975): J. Phys. C8, 1889
Collins, M.F., Minkiewicz, V.J., Nathans, R., Passell, L., Shirane, G. (1969): Phys. Rev. 179, 417
Cooke, J.F. (1973): Phys. Rev. B7, 1108
Cooke, J.F., Davis, H.L. (1972): AIP Conf. Proc. 10, 1218
Copley, J.R.D., Lovesey, S.W. (1975): Rep. Progr. Phys. 38, 461
Cowley, R.A., Buyers, W.J.L. (1972): Rev. Mod. Phys. 44, 406
Cowley, R.A., Dolling, G. (1968): Phys. Rev. 167, 464
Cracknell, A.P. (1974): J. Phys. C7, 4323
De Jongh, L.J., Miedema, A.R. (1974): Adv. Phys. 23, 1-260
De Jongh, L.J., Stanley, H.E. (1976): Phys. Rev. Lett. 36, 817
Des Cloizeaux, J., Pearson, J.J. (1962): Phys. Rev. 128, 2131
Dingle, R., Lines, M.E., Holt, S.L. (1969): Phys. Rev. 187, 643
Dyson. F.J. (1956): Phys. Rev. 102, 1217, 1230
Elliott, R.J. (1972): *Magnetic Properties of Rare Earth Metals* (Plenum Press, London, New York)
Elliott, R.J., Kleppmann, W.G. (1975): J. Phys. C8, 2737
Elliott, R.J., Krumhansl, J.A., Leath, P.L. (1974): Rev. Mod. Phys. 46, 465
Endoh, Y., Shirane, G., Birgeneau, R.J., Richards, P.M., Holt, S.L. (1974): Phys. Rev. Lett. 32, 170
Enz, C.P. (1974): Rev. Mod. Phys. 46, 705
Fisher, M.E. (1964): Am. J. Phys. 32, 343
Fisher, M.E. (1974): Rev. Mod. Phys. 46, 597
Freeman, A.J., Frankel, R.B. (1967): *Hyperfine Interactions* (Academic Press, New York)
Glinka, C.J., Minkiewicz, V.J., Passell, L., Shafer, M.W. (1973): AIP Conference Proc. 18, 1060
Halperin, B.I., Hohenberg, P.C. (1969): Phys. Rev. 177, 952
Heidemann, A. (1970): Z. Physik 238, 208
Herring, C. (1967): *Magnetism*, Vol. 4, ed. by G.T. Rado and H. Suhl (Academic Press, New York)
Holden, T.M., Buyers, W.J.L. (1974): Phys. Rev. B9, 3797
Hone, D.W., Scherer, C., Borsa, F. (1974): Phys. Rev. B9, 965
Hone, D.W., Richards, P.M. (1974): Ann. Rev. Mat. Sci. 4, 337
Houmann, J.G., Chapellier, M., Mackintosh, A.R., Bak, P., McMasters, O.D., Gschneidner, K.A. (1975): Phys. Rev. Lett. 34, 587
Houmann, J.G., Jensen, J., Touberg, P. (1975): Phys. Rev. B12, 332

Hubbard, J. (1971): J. Phys. C4, 53
Huber, D.L. (1974): Solid State Commun. 14, 1153
Hutchings, M.T., Shirane, G., Birgeneau, R.J., Holt, S.L. (1972): Phys. Rev. B5, 1999
Hutchings, M.T., Windsor, C.G. (1977): J. Phys. C10, 313
Izyomov, Y.A. (1963): Uspekhi 16, 359
Jaccarino, V. (1967): Proceedings International School of Physics, Enrico Fermi (Academic Press, New York)
Jensen, J. (1971): Intern. J. Magnetism 1, 271
Jensen, J., Houmann, J.G. (1975): Phys. Rev. B12, 320
Jensen, J., Houmann, J.G. Møller, H.B. (1975): Phys. Rev. B12, 303
Keffer, F. (1967): Spin Waves. In *Handbuch der Physik*, Vol. XVIII/2, ed by H.P.J. Wijn (Springer, Berlin, Heidelberg, New York)
Kincaid, J.M., Cohen, E.G.D. (1975): Phys. Lett. C22, 58
Kirkpatrick, S., Harris, A.B. (1975): Phys. Rev. B12, 4980
Kjems, J.K., Als-Nielsen, J., Fogedby, H. (1975): Phys. Rev. B12, 5190
Koehler, W.C. (1972): *Magnetic Properties of Rare Earth Metals*, ed. by R.J. Elliott (Plenum Press, London and New York)
Kretzen, H.H., Mikeska, H.J., Patzak, E. (1974): Z. Physik. 271, 269
Landau, L.D., Lifshitz, E.M. (1958): *Statistical Physics* (Pergamon Press, London) p. 482
Laurence, G., Petitgrand, D. (1973): Phys. Rev. B8, 2130
Loveluck, J.M., Lovesey, S.W. (1975): J. Phys. C8, 3857
Loveluck, J.M., Lovesey, S.W., Aubry, S. (1975): J. Phys. C8, 3841
Lovesey, S.W. (1972): J. Phys. C5, 2769
Lovesey, S.W. (1974): J. Phys. C7, 2008
Lovesey, S.W. (1974b): J. Phys. C7, 2049
Lovesey, S.W., Meserve, R.A. (1972): Phys. Rev. Lett. 28, 614
Lovesey, S.W., Meserve, R.A. (1973): J. Phys. C6, 79
Lovesey, S.W., Windsor, C.G. (1971): Phys. Rev. B4, 3048
Lowde, R.D., Windsor, C.G. (1970): Adv. Phys. 19, 813
Lurie, N.A., Huber, D.L., Blume, M. (1974): Phys. Rev. B9, 2171
Lynn, J.W. (1975): Phys. Rev. B11, 2624
Mackintosh, A.R., Møller, H.B. (1972): *Magnetic Properties of Rare Earth Metals*, ed. by R.J. Elliott (Plenum Press, London, New York)
Marshall, W., Lovesey, S.W. (1971): *Theory of Thermal Neutron Scattering* (O.U.P., Oxford)
McLean, F.B., Blume, M. (1973): Phys. Rev. B7, 1149
Mermin, N.D., Wagner, H. (1966): Phys. Rev. Lett. 17, 1133
Mess, K.W., Lagendijk, E., Curtis, D.A., Huiskamp, W.J. (1967): Physica 34, 126
Minkiewicz, V.J. (1971): Intern. J. Magnetism 1, 149
Minkiewicz, V.J., Collins, M.F., Nathans, R., Shirane, G. (1969): Phys. Rev. 182, 624
Møller, H.B., Houmann, J.G., Jensen, J., Mackintosh, A.R. (1972): *Neutron Inelastic Scattering* (IAEA, Grenoble) pp. 603
Montgomery, H., Cracknell, A.P. (1973): J. Phys. C6, 3156
Mook, H.A., Lynn, J.W., Nicklow, R.M. (1973): Phys. Rev. Lett. 30, 556
Mook, H.A., Lynn, J.W., Nicklow, R.M. (1974): AIP Conference Proceedings 18, 781
Mook, H.A. Nicklow, R.M.(1973): Phys. Rev. B7, 336
Mori, H. (1965a): Prog. Theoret. Phys. 33, 423
Mori, H. (1965b): Prog. Theoret. Phys. 34, 399
Nagamiya, T. (1967): *Solid State Physics*, Vol. 20, ed. by F.Seitz, D. Turnbull and H. Ehrenreich (Academic Press, New York, London)
Natoli, C.R., Ranninger, J. (1970): Phys. Lett. 32A, 11
Oguchi, T. (1960): Phys. Rev. 117, 117
Onsager, L. (1944): Phys. Rev. 65, 117
Passell, L., Als-Nielsen, J., Dietrich, O.W. (1972): *Inelastic Neutron Scattering*, (IAEA, Grenoble)
Passell, L., Dietrich, O.W., Als-Nielsen, J. (1972): AIP Conference Proc.5, 1251
Pomeau, Y., Resibois, P. (1975): Pnys. Rept. 19, 64
Rainford, B.D., Houmann, J.G., Guggenheim, H.J. (1972): *Neutron Inelastic Scattering*, (IAEA, Grenoble)

Richards, P.M. (1972): Phys. Rev. Lett. 28, 1646
Richards, P.M., Carboni, F. (1972): Phys. Rev. B5, 2014
Schlichting, H.J., Krey, U. (1975): Z. Physik B20, 375
Schulhof, M.P., Heller, P., Nathans, R., Linz, A. (1970): Phys. Rev. B1, 2304
Skalyo, Jr. G., Shirane, G., Birgeneau, R.J., Guggenheim, H.J. (1969): Phys. Rev. Lett. 23, 1394
Skalyo, Jr. J., Shirane, G., Friedberg, S.A., Kobayashi, H. (1970): Phys. Rev. B2, 1310
Stanley, H.E., Kaplan, T.A. (1966): Phys. Rev. Lett. 17, 913
Steiner, M., Dorner, B., Villain, J. (1975): J. Phys. C8, 165
Steiner, M., Kruger, W., Babel, D. (1971): Solid State Commun. 9, 1603
Steiner, M., Villain, J., Windsor, C.G., (1976): Adv. Phys. 25, 87
Tomita, H., Mashiyama, H. (1972): Prog. Theoret. Phys. 48, 1133
Torrance, J.B., Tinkham, M. (1969): Phys. Rev. 187, 587 and 595
Trammell, G.T. (1960): J. Appl. Phys. 3625; (1963) Phys. Rev. 131, 932
Tuberfield, K.C., Passell, L., Birgeneau, R.J., Bucher, E. (1970): Phys. Rev. Lett. 25, 752
Tucciarone, A., Lau, H.Y., Corliss, L.M., Delapalme, A., Hastings, J.M. (1971): Phys. Rev. B4, 3206
Van Kranendonk, J., Van Vleck, J.H. (1958): Rev. Mod. Phys. 30, 1
Van Vleck, J.H. (1940): Phys. Rev. 57, 426
Villain, J. (1974): J. de Phys. 35, 27
Walker, L.R. (1963): *Magnetism* Vol. 1, ed. by G.T. Rado and H. Suhl (Academic Press New Jork and London)
Walker, L.R., Dietz, R.E. Andres, K., Darack, S. (1972): Solid State Commun. 11, 593
Wielinga, R.F., Blöte, H.W.J., Roest, J.A., Huiskamp, W.J. (1967): Physica 34, 223
Windsor, C.G., Locke-Wheaton, J. (1976): J. Phys. C9, 2749
Yoshimori, K. (1959): J. Phys. Soc. Japan 14, 807
Zittartz, J. (1976): Z. Physik B23, 55

## Recent References with Titles

Cooke, J.F., Lynn, J.W., Davis, H.L. (1976): Numerical investigation of spin waves in ferromagnetic iron. Solid State Comm. 20, 799

Dietrich, O.W., Henderson, A.J., Meyer, H. (1975): Spin-Wave Analysis of Specific Heat and Magnetization of EuO and EuS. Phys. Rev. B12, 2844

Dietrich, O.W., Meyer, G., Cowley, R.A., Shirane, G. (1975): Line shape of the magnetic excitations in substitutionally disordered antiferromagnets. Phys. Rev. Lett. 35, 1735

Ma, S.K. (1976): Modern Theory of Critical Phenomena. (W. Benjamin, London)

Dietrich, O.W., Als-Nielsen, J., Passell, L. (1976): Neutron scattering from Heisenberg ferromagnets EuO and EuS III spin dynamics of EuO. Phys. Rev. B14, 4932

Shirane, G., Birgeneau, R.J. (1977): Neutron scattering studies of low dimensional magnetic systems. Physica 86-88 B&C (Part II), 639 (Proc. of Int. Conference on Magnetism, Amsterdam, 1976)

Cowley, R.A., Shirane, G., Birgeneau, R.J., Guggenheim, H.J. (1977): Magnetic correlations in $Rb_2Mn_{0.5}Mg_{0.5}F_4$. Physica 86-88 B&C (Part II), 727 (Proc. of Int. Conference on Magnetism, Amsterdam, 1976)

Steiner, M., Kjems, J.K. (1977): Spin waves in $CsNiF_3$ with an applied magnetic field. To be published

Lovesey, S.W., Loveluck, J.M. (1976): Wavelength-dependent fluctuations in classical paramagnets in a magnetic field: II dynamical properties. J. Phys. C : 9, 3659

# Subject Index

## Springer Tracts in Modern Physics

**Ergebnisse der exakten Naturwissenschaften**

Editor: G. Höhler
Associate Editor: E. A. Niekisch

Volume 64
**Quasielastic Neutron Scattering for the Investigation of Diffusive Motions in Solids and Liquids**
T. Springer

36 fig.II, 100 pages. 1972.
ISBN 3-540-05808-7

*Contents:* Scattering Theory - Methodical and Experimental Aspects- Monoatomic Liquids with Continuous Diffusion - Jump Diffusion in Liquids - Diffusion of Hydrogen in Metals - Rotational Diffusion in Molecular Solids - Molecular Liquids - Polymers and other Complicated Systems - Effects of Coherent Scattering - Quasielastic Scattering and other Methods

Volume 80
**Neutron Physics**

40 figs., 11 tables. VII, 135 pages
1977
ISBN 3-540-08022-8

*Contents:* *L. Koester,* Neutron Scattering Lengths and Fundamental Neutron Interactions. - *A. Steyerl,* Very Low Energy Neutrons.

Springer-Verlag
Berlin Heidelberg New York

## Topics in Applied Physics

Founded by H. K. V. Lotsch

Volume 8
**Light Scattering in Solids**
Editor: M. Cardona

111 figs. 3 tables. XIII, 339 pages.
1975
ISBN 3-540-07354-X

*Contents:* *M. Cardona,* Introduction.- *A. Pinczuk, E. Burstein,* Fundamentals of Inelastic Light Scattering in Semiconductors and Insulators. - *R.M. Martin, L.M. Falicov,* Resonant Raman Scattering. - *M.V. Klein,* Electronic Raman Scattering. - *M.H. Brodsky,* Raman Scattering in Amorphous Semiconductors. - *A.S. Pine,* Brillouin Scattering in Semiconductors. - *Y.-R. Shen,* Stimulated Raman Scattering.

Volume 15
**Radiationless Processes in Molecules and Condensed Phases**
Editor: F. K. Fong

67 figs. XIII, 360 pages. 1976.
ISBN 3-540-07830-4

*Contents:* Introduction - Energy Dependence of Electronic Relaxation Processes in Polyatomic Molecules - Vibrational Relaxation of Molecules in Condensed Media - Up-Conversion and Excited State Energy Transfer in Rare-Earth Doped Materials - Exciton Percolation in Molecular Alloys and Aggregates

A monthly journal

# Applied Physics

| | |
|---|---|
| Board of Editors | **S. Amelinckx,** Mol. · **V. P. Chebotayev,** Novosibirsk<br>**R. Gomer,** Chicago, Ill. · **H. Ibach,** Jülich<br>**V. S. Letokhov,** Moskau · **H. K. V. Lotsch,** Heidelberg<br>**H. J. Queisser,** Stuttgart · **F. P. Schäfer,** Göttingen<br>**A. Seeger,** Stuttgart · **K. Shimoda,** Tokyo<br>**T. Tamir,** Brooklyn, N.Y. · **W. T. Welford,** London<br>**H. P. J. Wijn,** Eindhoven |
| Coverage | application-oriented experimental and theoretical physics:<br>*Solid-State Physics* · *Quantum Electronics*<br>*Surface Physics* · *Laser Spectroscopy*<br>*Chemisorption* · *Photophysical Chemistry*<br>*Microwave Acoustics* · *Optical Physics*<br>*Electrophysics* · *Integrated Optics* |
| Special Features | **rapid** publication (3–4 months)<br>**no** page charge for **concise** reports<br>prepublication of titles and abstracts<br>**microfiche** edition available as well |
| Languages | Mostly English |
| Articles | original reports, and short communications<br>review and/or tutorial papers |
| Manuscripts | to Springer-Verlag (Attn. H. Lotsch), P.O. Box 105 280<br>D-69 Heidelberg 1, F.R. Germany |

Place North-American orders with:
Springer-Verlag New York Inc., 175 Fifth Avenue, New York. N.Y. 10010, USA

Springer-Verlag
Berlin Heidelberg New York